AF896356

Bioremediation of Environmental Pollutants

Deep Chandra Suyal • Ravindra Soni
Editors

Bioremediation of Environmental Pollutants

Emerging Trends and Strategies

Springer

Editors
Deep Chandra Suyal
Department of Microbiology
Akal College of Basic Sciences,
Eternal University
Sirmour, India

Ravindra Soni
Department of Agricultural Microbiology
Indira Gandhi Krishi Vishwa Vidyalaya
Raipur, India

ISBN 978-3-030-86168-1 ISBN 978-3-030-86169-8 (eBook)
https://doi.org/10.1007/978-3-030-86169-8

© The Editor(s) (if applicable) and The Author(s), under exclusive license to Springer Nature Switzerland AG 2022
This work is subject to copyright. All rights are solely and exclusively licensed by the Publisher, whether the whole or part of the material is concerned, specifically the rights of translation, reprinting, reuse of illustrations, recitation, broadcasting, reproduction on microfilms or in any other physical way, and transmission or information storage and retrieval, electronic adaptation, computer software, or by similar or dissimilar methodology now known or hereafter developed.
The use of general descriptive names, registered names, trademarks, service marks, etc. in this publication does not imply, even in the absence of a specific statement, that such names are exempt from the relevant protective laws and regulations and therefore free for general use.
The publisher, the authors, and the editors are safe to assume that the advice and information in this book are believed to be true and accurate at the date of publication. Neither the publisher nor the authors or the editors give a warranty, expressed or implied, with respect to the material contained herein or for any errors or omissions that may have been made. The publisher remains neutral with regard to jurisdictional claims in published maps and institutional affiliations.

This Springer imprint is published by the registered company Springer Nature Switzerland AG.
The registered company address is: Gewerbestrasse 11, 6330 Cham, Switzerland

Preface

Presently, the world is facing a crisis of waste collection and management. The existing technologies are incapable of disposing of the huge waste that is constantly being generated. Moreover, several of them are damaging the environment too, knowingly or unknowingly. Therefore, there is an urgent need for eco-friendly approaches to solving these issues effectively. In this regard, bioremediation has shown potential to some extent; though it has a long way to go.

This book documents the latest innovations, and technological advancements in the field of bioremediation, especially for its monitoring and assessment. It comprises 17 chapters, starting with basic concepts of waste management, challenges, and associated opportunities. It is followed by the role of microorganisms and microbial fuel cells to remediate the pollutants. Further, the critical bioremediation process parameters and their optimization strategies are also being discussed. The monitoring of the bioremediation process is an important aspect. Microbial biosensors have shown their potential for it and, therefore, are included in this book. Further, the importance of fungi is well known for biodegradation and waste management; therefore, recent advancements in this field are being covered. The role of genetically modified microorganisms in bioremediation, their genetic and protein networking, and *in silico* approaches are also covered. Few chapters discuss the advancements of existing waste management strategies, viz. landfilling approaches, aerobic granular technology, and immobilized enzyme reactors. Further, the role of chelating compounds, biochar, and bio-inoculants in the bioremediation strategies has been covered. Also, the spectroscopy technique and its exploration for environmental sustainability have been discussed. The advanced molecular technologies are included at the end that can be explored for bioremediation studies in the future. Conclusively, the book covers the latest technologies and innovations which are being explored for environmental clean-up and sustainability.

Sirmour, India — Deep Chandra Suyal
Raipur, India — Ravindra Soni

Contents

Chapter 1
Waste Management: Challenges and Opportunities

Neha Badola and Jaspal Singh Chauhan

1.1 Introduction

'Waste' includes all those substance or materials which is discarded by a holder. The generation of waste is always corresponding to the different activities that take place under urbanization, economic development, and population growth (Kaza et al. 2018; Bhatt et al. 2020). Changing life style in large cities has enhanced the demand of products that release a large amount of waste. For example, with the advent of mobile phones, a large amount of electronic waste has been invaded our earth. Not only the cities but also the rural areas are encroached by the huge mountains of garbage. Solid waste is one of the crucial issues that the world is facing today. Useless and undesirable waste arising by various human and animal activities is termed as solid waste. Alam and Ahmade (2013) stated the characteristics of solid waste as Corrosive, Reactive, Ignitable, and Toxic, yet these characteristics differ from waste to waste. Solid waste has the potential to cause negative impacts on human and animal health and can also significantly affect economic development of a Nation. Proper disposal of discarded waste is a critical challenge in many countries throughout the globe. Unscientific disposal and treatment of solid waste has already covered much of our land and oceans. Presently, the availability of space required to discharge solid waste is an urgent matter of concern. The problem of solid waste has started since many decades. During the earlier time, the large amount of generated solid waste was either organic or consist of metals, glass, and other particles. Most of such waste was biodegraded and mixed into the soil or reused and recycled. But with increasing population, the products like plastic which is non-biodegradable was introduced as a substitute for many things and hence today from micro to mega-size products, the application of plastic is almost seen. Plastic waste is mostly

N. Badola · J. S. Chauhan (✉)
Aquatic Ecology Lab, Department of Himalayan Aquatic Biodiversity, Hemvati Nandan Bahuguna Garhwal University (A Central Univesirty), Srinagar-Garhwal, Uttarakhand, India

© The Author(s), under exclusive license to Springer Nature Switzerland AG 2022
D. C. Suyal, R. Soni (eds.), *Bioremediation of Environmental Pollutants*,
https://doi.org/10.1007/978-3-030-86169-8_1

non-biodegradable in nature and thereby manipulating environment making the problem worse. Also, after industrialization and the rise of the electronic industry, the electronic waste with hazardous potential has enhanced.

As a solution to solid waste many developed countries have opted scientific solid waste management, treatment and strategies. These include recycling, reuse, recovery landfills, incineration, and biological treatment. These management methods also have negative impacts and also require funds, trained personnel, and availability of land. Solid waste can be a serious threat to both living and non-living being, if proper management strategies are not opted. However, it can act as a vital resource and save natural resources, if used with proper scientific technique. Scientists from the world are working hard to find a solution for solid waste and in this search, they have assessed that waste can be used positively to generate income. This chapter here discuss about solid waste handling, management and challenges associated with it and also explores the hidden opportunities in the waste.

1.2 Classification of the Solid Waste

Solid waste can be categorised into various categories according to their origin, content and hazardous potential nature (Fig. 1.1). Based on origin, important categories are Domestic or Residential, Commercial, Municipal, Institutional, Industrial, Agriculture, and so on. Based on the contents of solid waste, it is categorized as Natural Biodegradable matter, Non-biodegradable matter, Partially biodegradable matter, Toxic organic substance, Recalcitrant matter, Metals, Metalloids, etc. (Hamer 2003). While based on hazardous potential, Batteries, E-waste, Pesticides, Herbicides, Paints and Radioactive waste. Hazardous waste has mostly dangerous content and therefore can cause serious or life-threatening impacts on the living and non-living being. So, precautions are required during the process of its collection, handling, management, and disposal.

Solid waste contains different types of constituents like paper, wood, glass, metal, food, etc. The contribution of different kinds of constituents of waste in the total waste generation is different. As per the report published by world bank in 2018, the major contributor of solid waste is food and green waste (44%) which comes under biodegradable waste, followed by paper and cardboard waste (17%) (Kaza et al. 2018). Rest of the miscellaneous waste contributes 14% of the total waste. While plastic waste contributes 12% and is the fourth largest contributor. Plastic waste is non-biodegradable and not easy to manage and therefore now it has been considered as a critical issue as crucial as climate change. Glass; metals; wood; rubber and leather contribute to 5%, 4%, 2% and 2% respectively of total waste (Fig. 1.2). This chart clearly indicates that a large part of the waste is biodegradable but cannot be utilized because of improper segregation. The same is applied to recyclable waste that includes content of metal, glass, plastic, wood, rubber, and leather.

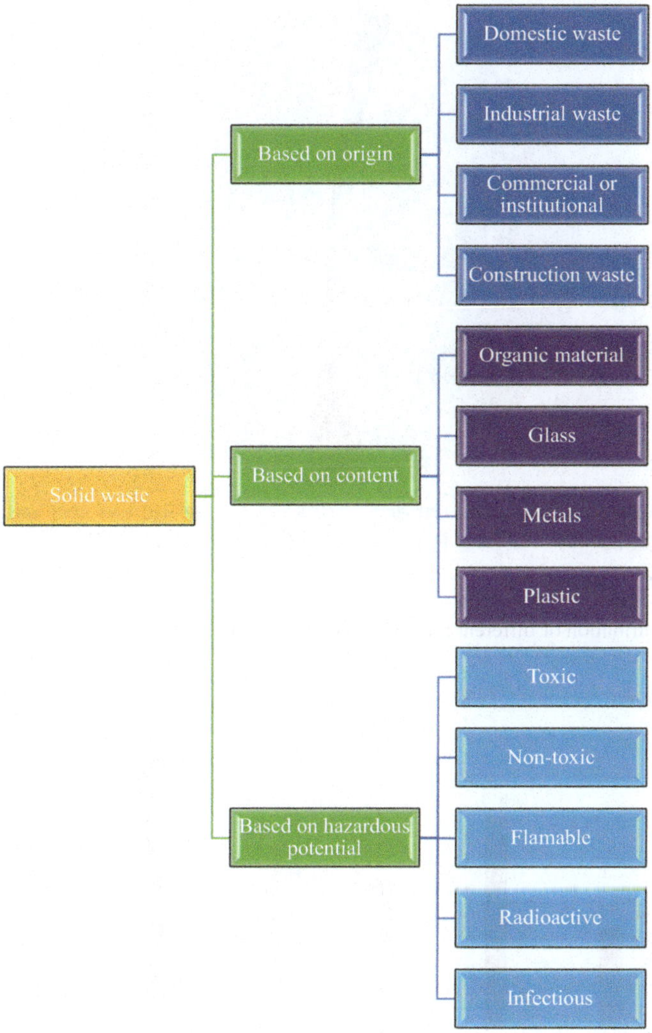

Fig. 1.1 Categorization of the solid waste

1.3 Solid Waste Status and Problems

Through-out the world the range of waste contributed per person per day is recorded to be 0.11–4.54 kg (averages 0.74 kg). High income countries in the world with only 16% of the total population are recorded to generate about 683 million tonnes (34%) of the global waste. East Asia and the pacific generate the highest amount of solid waste as per statistics of world bank (Fig. 1.3). A forecast estimate says that by 2050, the generation of waste will rise by 70%, if strict necessary actions are not imposed. In low-income countries over 90% of waste is often disposed in the unregulated

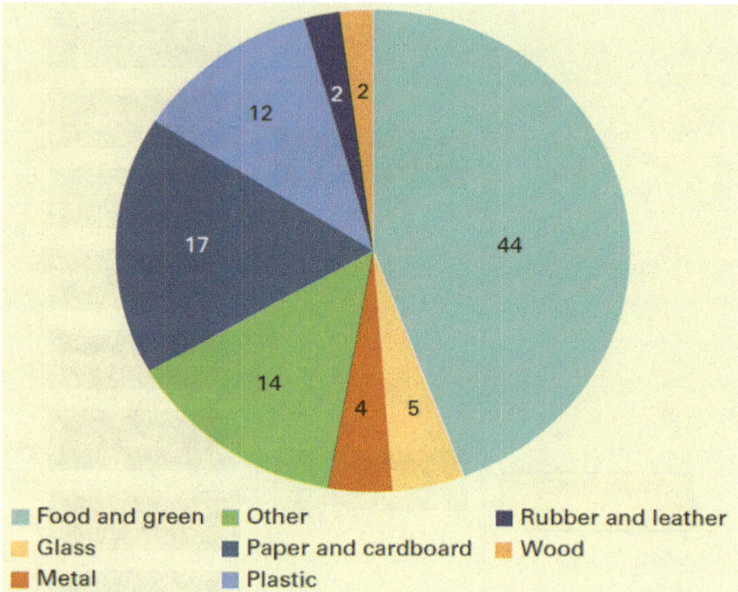

Fig. 1.2 Contribution of different type of waste (Kaza et al. 2018)

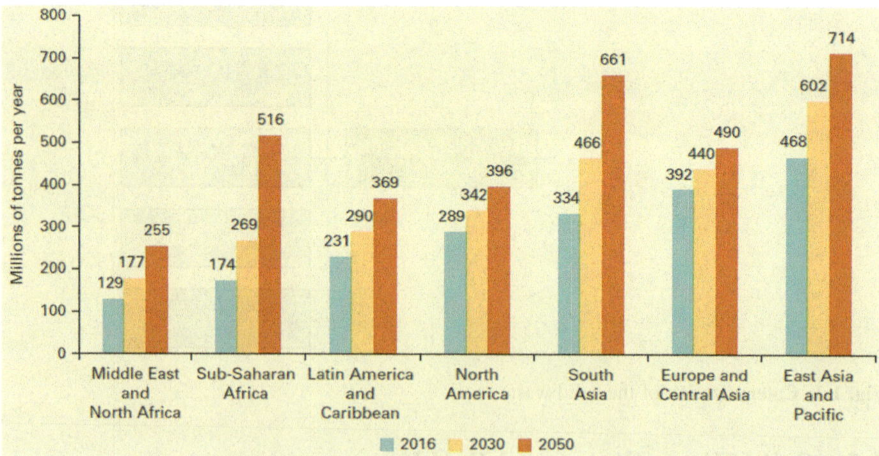

Fig. 1.3 Projected waste generation, by different regions (millions of tonnes/year) (Kaza et al. 2018)

dumps or openly burned. Improper disposal of solid wastes in the environment leads to water, air and soil pollution. Further, waste acts as breeding place for many disease-causing vectors and hence threats living being. Solid waste disposed without proper strategies leaches down and impair drinking groundwater. Solid waste like plastics and rubber when burnt in open environment create fumes that effect air quality in the atmosphere. Potential green-house gaseous like carbon dioxide and methane which are formed due to bio-degradation and burning of waste contribute significantly to the climate change. According to an estimate in 2016, solid waste disposal and treatment released about 1.6 billion tonnes of CO_2 equivalent green-house gases (GHG) that was about 5% of the total global emissions. This happened because of open dumping and disposal of solid waste in improper landfills without any precautionary measures. Approximately about 50% of GHG emissions are contributed by only food waste. The emission of GHG (in term of equivalent CO_2) from Solid waste is projected to increase about 2.38 billion tonnes per year by 2050 if proper improvements are not done in this sector (Kaza et al. 2018).

1.4 Solid Waste Management

Waste management refers to all those activities and actions that are involved in proper handling of waste materials (Wan et al. 2019). Waste management is the dire necessity to live in a safe and healthy environment. It has been observed that many developing countries are lacking proper management of solid waste and as a consequence they are facing the risk of environmental, biological and economic losses (Sharholy et al. 2008). Few common reasons for improper waste management are low waste collection rate, desegregated waste, random dumping and open burning of waste without any criteria, lack of proper dustbins at required places, etc. Waste management practices helps in controlling the negative impact of waste on the human health, organisms, and the environment. Several processes like monitoring, collection, transport, processing, recycling, and disposal are conducted to make waste management efficient. Figure 1.4 shows percentage contribution of waste treatment method opted globally. Data clearly shows that majorly waste is processed for open dumping which is really a menace for future.

Efficient solid waste management requires proper planning, implementation, assessment and strategies for each categories of solid waste (Abdel-Shafy and Mansour 2018). Good waste management strategies provide clean environment and reduce disease spread. The process associated to tackle solid waste starts from the point of generation to final disposal, consist of mainly five steps (Fig. 1.5). Each step requires skilled labour and personnel to deal the waste.

1.4.1 Collection

The term "collection" refers gathering or picking up all type of solid wastes from the different sources. The collection of solid waste usually is time taking and a complex process. Mostly door to door waste collection method is used to collect waste from a common community dustbin to prohibit random littering. Wastes that is collected during the cleaning of streets in an area are also stored in the community bins

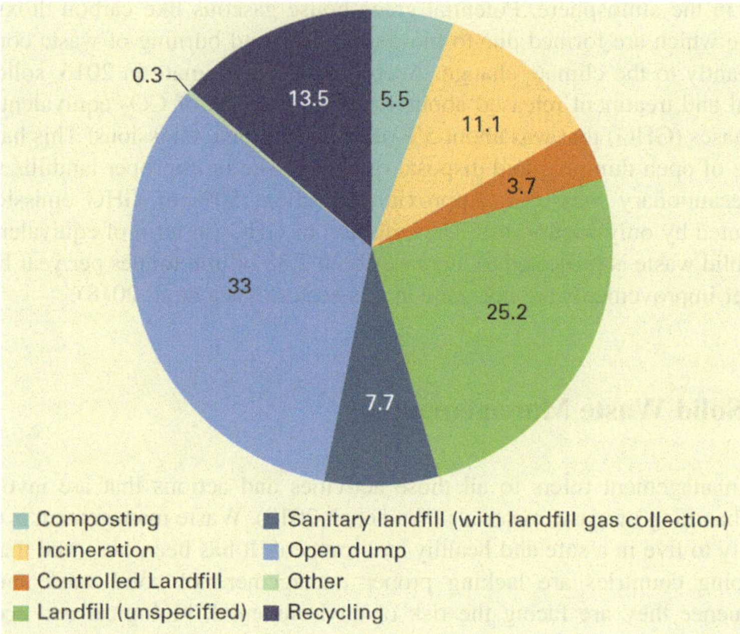

Fig. 1.4 Global treatment and disposal of waste (%) (Kaza et al. 2018)

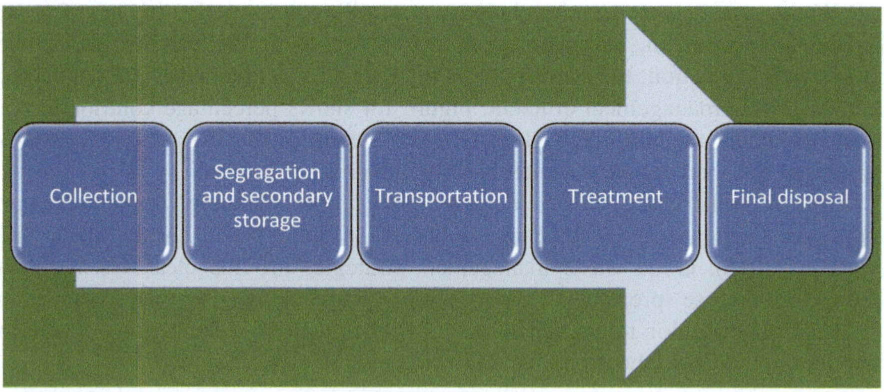

Fig. 1.5 Methods of waste management

(Kumar et al. 2009). Collection of waste needs skilled manpower and a perfect transportation system. Waste collection is usually carried out with the help of handcarts, tricycles, etc. (Annepu 2012). Mostly the responsibility of door-to-door collection of solid wastes is given to local government bodies in developing countries.

1.4.2 Segregation and Secondary Storage

The solid waste mostly consists of different type of waste hence, it is important to separate them. Segregation includes separation of solid waste majorly into biodegradable, non-biodegradable and recyclable. Each type of solid waste needs a specific strategy for the management and disposal. Source separation is considered as a significant component of the Integrated Waste Management system. Segregation takes place at the source only because there it is easy to separate the waste due to its less quantity which after collection will pile up and became impossible task. Special care should be taken so that segregated waste should not be spilled and mixed, mostly during loading, transport and storage process.

1.4.3 Transportation

It is an vital stage which require a good capital and operating cost and also affect the primary collection stage and processing (Parekh et al. 2015). Various transportation methods like hand carts, animal carts, truck, etc. are used for the process of collection of the solid waste. In the developing countries hand carts and animal carts are commonly used for the transportation of waste. These methods are time taking and require much effort but are prevalent in developing or poor countries. In developed countries trains and ships are also used to carry and dispose of waste outside the community to a selected landfill or marine dumping sites. During the transportation of solid waste from the source, the waste should be properly covered to avoid the risk of exposure and spillage in the environment. Further, there should be a separate compartment for driver so that proper care of his health can be given preference.

1.4.4 Treatment

Solid waste management includes the different stages of activities like segregation (separation) reduction, modification, recycling, treatment and disposal (Hamer 2003). The nature of waste i.e. biodegradable non-biodegradable, combustible, etc. are important features which are considered while processing it for any type of treatment (Fig. 1.6).

Earlier solid waste was mostly managed in a very irresponsible ways like open dumping at any random landfill sites, marine dumping or open burning, etc. Today, many advance methods are available for the waste treatment before its disposal such as pyrolysis, gasification, recycling, etc. Some common methods of solid waste management, traditional to advance are explained below:

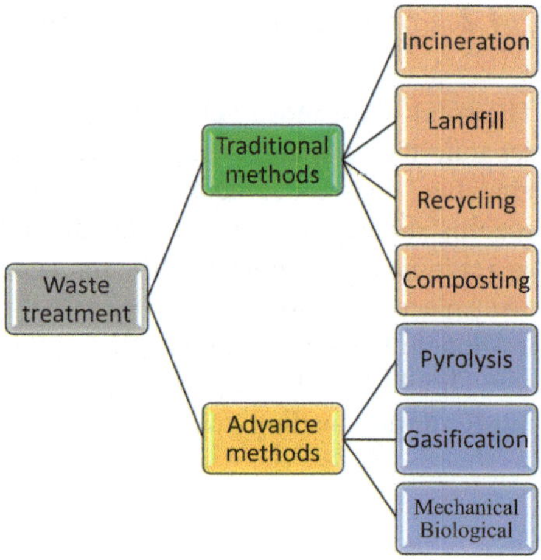

Fig. 1.6 Waste treatment methods

1.4.4.1 Landfill

Conventionally, every type of waste was considered for landfill treatment and site was selected near to waste production region only. No defined criteria were set to assess the suitability of site for landfilling. The area away from the population was directly used for landfilling. The basic operational criteria for landfill were just disposal at a minimal cost. Traditionally landfilling was conducted without considering the problems that may arise in future due to waste. As a result, many problems associated with traditional landfill methods were observed like, release of noxious gas and vapour, bad odour, dust, leachate production and rodent infestation (Hamer 2003). Now, scientists have updated traditional landfill method at a minimum harm to the environment and living beings. Most of the new engineered landfills are designed to extract the produced biogas by undersealing, capping and installing suitable wellheads for the collection of gas and leachate. From landfills mainly methane (CH_4) and carbon dioxide (CO_2) gases are released due to anaerobic degradation of waste. These gases have significant contribution in the greenhouse effect. Methane produced from landfill is about 13% of global CH_4 emission and is about 818 million metric tons per annum in terms of CO_2 equivalent (Rachel et al. 2007). Sanitary landfills and bioreactor landfill methods are commonly used methods that provide good result for waste disposal. These landfill methods are designed to reduce the risk associated with waste after landfilling. Bioreactor landfills uses superior microbiological techniques to gear up waste decomposition at high rate.

1.4.4.2 Incineration

Incineration is considered a significant method for solid waste treatment. It is the combustion of waste at a high temperature in a modern designed furnace under controlled operating conditions. This method suggests proper removal of inert and non-combustible materials like glass, chinaware, stones, marbles, metals, and soils, before burning solid waste. The process of incineration is capable of reducing the original volume of waste up to 80–90% (Alam and Ahmade 2013). The demerits of incineration are the release of carcinogenic and toxic compounds, fly ash, and gases released due to partial combustion. However, modern methods are trying to compensate for these demerits by doing suitable modifications (Hamer 2003).

1.4.4.3 Pyrolysis

Pyrolysis is defined as the process in which solid waste is treated to a high temperature in the absence of oxygen. In pyrolysis, wastes are decomposed into multiple chemical compounds. Pyrolysis products always result in as solid, liquid, and non-condensable gases (Hamer 2003). Efforts are done to recover few chemicals during the pyrolysis of waste by the destructive distillation method. It is an endothermic process (absorb heat) and hence differs from the conventional incineration method which is exothermic in nature. Waste rich in organic materials like paper and cardboard yields combustible gases during pyrolysis that can be further used as fuel.

1.4.4.4 Gasification

In the Gasification process, partial oxidation of solid waste is conducted in the presence of less amount of oxygen so that fuel gases like carbon monoxide and hydrogen may be generated. The gasification of waste is conducted in a gasifier with air as the oxidant, without actually burning it and the end products released are low-energy gases typically containing (by volume) 20% CO, 15% H_2, 10% CO_2 and 2% CH_4 (Hamer 2003). Gasification is a promising and efficient technique to use solid waste in a better way. Almost all kinds of waste are transformed into gases like H_2, CO, CO_2, and CH_4 through the process of gasification. These gas products can be used for the generation of heat, energy, oil (Hameed et al. 2021).

1.4.4.5 Recycling

Recycling is defined as reprocessing and recovery of waste materials in the form of a new product. The waste generated and then collected can be processed for recycling, where the waste material is chemically converted to a new product (Annepu 2012). Common recyclable materials are plastics, iron and steel scrap, aluminum cans, paper, glass bottles, and wood. It helps in the reduction of solid waste deposited in landfills. It also diminishes the pollution issues consequential from waste disposal. Recycling helps in the protection of nature and natural resources and also supports the bio-diversity to maintain equilibrium and sustainability throughout the global eco-system (Ayodele et al. 2018; Robaina et al. 2020). Earlier recycling has been proposed as a major tool to combat the problem of solid waste in different nations. But after the implementation of this method, few limitations were also observed (Ebreo et al. 1999). As per a fact, a 1% increase in the recycling of municipal solid waste not only contributes to the economic growth of a national but also helps to reduce carbon emissions by 0.317% (0.157%) and 0.209% (0.087%) in the long run and short-run respectively (Razzaq et al. 2021).

1.4.4.6 Pulverization

Pulverization is the processing in which the size and volume of solid waste are reduced by the crushing and grinding technique. In this method, waste is powdered in grinding machines to reduces its volume and physical characters. Pulverization increases the surface area of waste and hence makes it more suitable for bacterial action during composting. It also facilitates the handling of waste for moisture content and aeration. It has been observed that the powdered waste is usually odourless and also less attractive to insects.

1.4.4.7 Mechanical Biological Treatment

The mechanical-biological (MB) treatment process of waste is an advanced form of waste treatment. This method has been popular in European countries due to its good results. In MB treatment solid waste is treated by a different mechanical and biological process that helps in the reduction and stabilization of waste and also restricts emissions of different gases from the final disposal. It has the potential to reduce waste content up to 40–60% that is to be disposed at landfills (Kaartinen 2004). Typically, the mechanical treatment includes separation, shredding, and crushing of the solid waste into fractions. The MB treatment plants implement both anaerobic digestion and aerobic process for the treatment of different types of solid waste. Further, the waste then undergoes treatment like composting, anaerobic digestion, combustion, recycling, etc.

1.4.5 Final Disposal

The waste which cannot be reused, recycled, recovered and prevented is finally disposed (Wan et al. 2019). Disposal is marked as the end process of the waste management system and the final fate for all those solid wastes, declared as 'useless' waste. Disposal of waste is done at an isolated place away from living beings so that infectious element can be kept away. All the efforts are done to guarantee the safety of human and environment (Guangyu 2009). The cheapest and widely implemented method of waste disposal is the landfills among all waste management techniques.

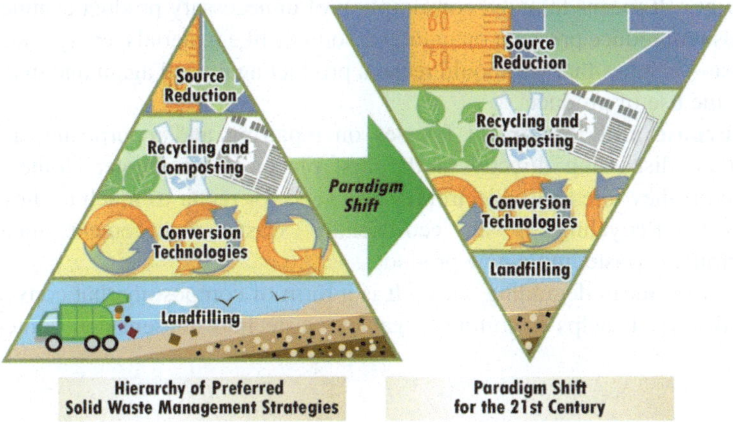

Fig. 1.7 Hierarchy of solid waste management existing *versus* need of time (Hauck 2014)

The demand of time is to overcome problems of solid waste with the new management methods and strategies. Hierarchy of solid waste management strategies clearly shows that conventionally most of the solid waste ends up in landfill site and a very less was used for recycling or recovery and other beneficial works. But in this modern world, we have the urgent need to invert it so that at source maximum reduction could be achieved and at the end less waste remains to be managed. After that the waste generates must go for recycling and composting. Efficient technologies also warranted for management and treatment of waste so that a minimum may end up in landfill sites (Fig. 1.7).

1.5 Circular Economy and 9 R's Principle

The idea of a circular economy is seen as a logical stand against the linear economy. In the circular economy, the original quality of the waste content of a discarded product remains stable and hence it is used again for preparing the same set of products. In this way, very less natural resources will be required for the preparation of new product, and discarded products will not be a waste (Potting et al. 2017). Waste management plan includes 7R's principles, which are significant in the waste reduction. The 7R's of waste management is Rethink, Refuse, Reduce, Reuse, Repurpose, Recycle, and Rot.

Rethink—It is very important to think many times before we make a choice for a particular item, so that the safety of this planet may be ensured. Rethinking makes us to realize that natural resources are limited and need to use wisely and sustainably.

Refuse—It means not to buy single use products that are non-recyclable. If every individual considers it their responsibilities to refuse pollution creating products, much of the problem will be solved.

Reduce—It means to avoid consumption of unnecessary product as much as we can. This will reduce pressure on natural resources like materials, energy, and water.

Reuse—It states that we should reuse a product again and again and discourages single-time use of products.

Repurpose and Repair—It focuses on repairing and repurposing a product rather than discarding directly in the first place. For example, clothes can be restitched if they tore apart, electronics also send for repair as much as they can.

Recycle—Recycling is an eco-friendly waste management method of transforming a waste into a new product.

Rot—Rot, mean degrading waste. It is a form of composting that convert waste into fertilizers. It helps in turning organics waste into nutrient-rich content good for soil.

1.6 Solid Waste as a Resource: Seeking Opportunities

In this modern era, scientists are continuously exploring methods that can convert waste into a valuable resource. A few researches have been already implemented to get valuable products from the solid waste such as energy generation, reuse in construction activities, biofertilizers and biogas from organic or biodegradable waste, and so many other possibilities (Fig. 1.8). If managed properly, the high volume of waste generated could act as resource and hence will not be a problem for future. Various methods are discussed below which proves solid waste as opportunities.

1.6.1 Solid Waste as a Fuel

Conversion of the waste into fuel is the best innovations for the waste management industry. If this is harnessed properly it will surely be a boon for upcoming generations. This will be reducing a large amount of waste to enter the process of landfilling. Solid waste as fuel has become a popular trend in many countries. Process of pyrolysis, liquefaction and gasification of solid waste can produce energy-dense bio-oil and H_2-rich syngas. On the other hand, digestion of municipal solid waste under anaerobic conditions produce biogas and active sludge (Nanda and Berruti 2020). Figure 1.9 is illustration of the process which converts waste into energy and bio-products. Organic matter and plastics are also explored by scientist for generation of fuel. The researchers from Institute of Organic Chemistry, Shanghai and University of California, Irvine have designed a recycling method that has potential to dissolves bonds of polyethylene plastic and form petroleum and other fuel products (Jia et al. 2016). Similarly, in 2018, researchers from Swansea University assessed a method to transform plastic waste into hydrogen fuel. They discovered that by adding a light-absorbing photo-catalyst in plastic waste that

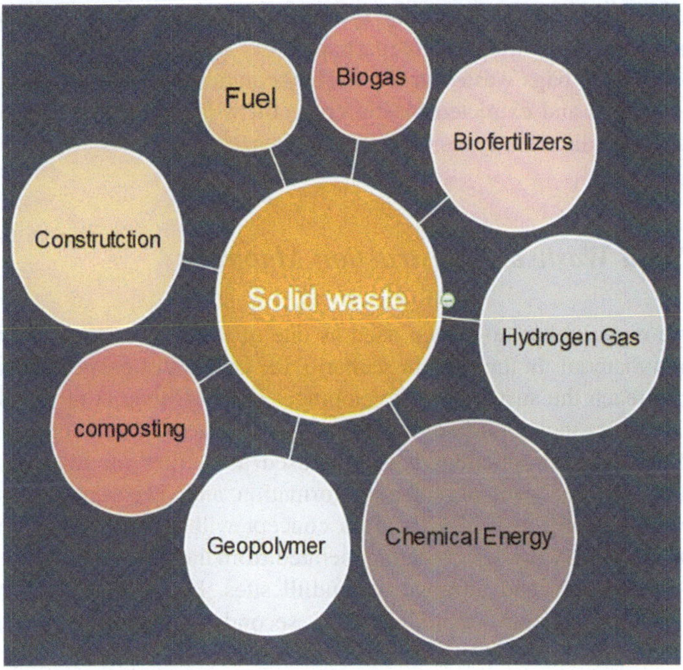

Fig. 1.8 Different opportunities generated by solid waste

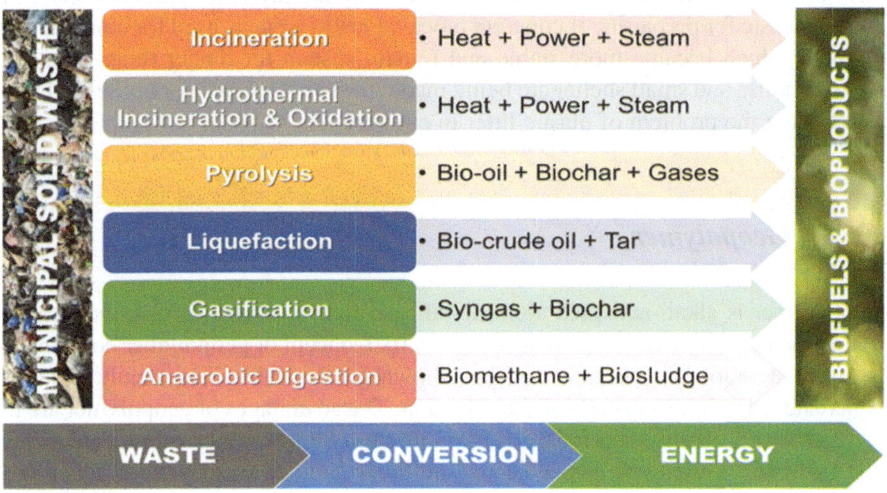

Fig. 1.9 Energy from products of different waste management methods (Nanda and Berruti 2020)

absorbs sunlight and convert it into chemical energy forming fuel. The combination of plastic waste and the photo-catalyst in an alkaline solution when exposed to sunlight, break the plastic releasing the bubbles of hydrogen gas (Uekert et al. 2018).

Researchers, Sharma and Rajagopalan from Sustainable Technology Center, Illinois in collaboration with the US Department of Agriculture, have successfully transformed plastic bags waste into fuel. They conducted pyrolysis of high-density polyethylene bags and extracted plastic crude oil which after distillation produces gasoline and two different types of diesels (Yates 2014).

1.6.2 Solid Waste as Construction Materials

Solid waste material has also been used as one of the constituents or substitute of construction material. In the present scenario, the concept of green construction has enhanced to reach the sustainable development model (Meng et al. 2018). Bottom ash generated after incineration of municipal solid waste has been considered as raw material for cement production. It is also used as a filler for road construction, construction material such as concrete formation and, glass and glass-ceramics (Siddique 2010; Tang et al. 2020). This concept will help to reduce the load of landfill sites as mostly the ash content generated from the incineration unit is usually considered as useless and dumped to landfill sites. Huang et al., suggested that replacing primary (virgin) materials with secondary (recycled) material helps dropping pressures on landfill sites. In the study, waste glass, steel slag, tyres and plastics were explored by researchers as substitute for pavements asphalt construction (Yates 2014). Meng et al. (2018), focussed on wastes like recycled concrete, soda lime glass, crushed brick, crumb rubber, cathode ray tube, glass and ceramic and tile waste for preparing of concrete blocks. Plastic waste is used for construction of roads which became more stable and remain sustain for longer time. Even the roofs of house and small shelter are being made from plastic bricks or bottles. It helps in reducing the problem of plastic litter in environment.

1.6.3 Geopolymer

Geopolymer is alkali-activated alumino-silicate material considered as the alternative to the ordinary cement (Nawaz et al. 2020). Concept of geopolymer composites is good alternative in the field and has shown good results. It is mainly used for concreate manufacturing and soil stabilization. The advantages of geopolymer are its strength and durability and further it has minimum impacts on environment (Provis 2014). In the present era, geopolymers are considered as sustainable practice and hence its demand in the construction industry is rising (Nawaz et al. 2020). In past few years, there has been extensive research on solid wastes to be utilized as fibers, aggregates, precursors, binder, mortar, bricks, paving blocks and concrete, etc. to prepare the geopolymers. Geopolymers prepared from solid wastes have shown potential in the field of concrete, adsorbents, fireproof materials, catalysts, impermeable materials, and energy storage materials, etc. Significantly, geopolymers are capable to restrict the heavy metals in the solid wastes and hence are considered as green material. The Geo-polymerization technique has been found to be very helpful in the safe disposal of non-ferrous metals (Aluminium, Copper, Lead, Nickle and Zinc) which are mostly dumped at landfill. The preparation of this green material has few challenges like technological, administrational and economical. To make geopolymers commercial at large scale, combined effort from the government, enterprises, and the public is warranted (Ren et al. 2020).

1.6.4 Biogas

Biogas is a fuel gas obtained by anaerobic degradation of organic biomass (manure, municipal solid waste, and sewage sludge). The major constituents of biogas are CH_4 and CO_2 (Asgari et al. 2011). Mostly biodegradable and wet solid waste present in the landfill and dumping areas release gases like methane and carbon dioxide. And if these gases can be extracted properly, they can be used as the source of energy. This concept not only reduce the problem of waste management but also serve to tackle the issue of pollution, fossil fuels and global warming. It turns solid waste into resources if done in an efficient way. The digestion of waste for biogas generation can be natural or under controlled conditions. Many countries like India, China, Taiwan, Brazil, Singapore, etc. are using biogas in the different field like heating purposes and electricity generation. In addition, the residues generated during biogas production may be used as low-grade fertilizers. In Sweden many city-buses are fuelled by biogas, and special gas stations are designed for distribution of biogas (Igoni et al. 2008).

1.6.5 Compositing

Composting is defined as the process in which oxidation of organic matter is done in the presence of air. The oxidized product is known as compost and have high nutrient value. The organic matter content in the solid waste remains mostly high, so it is used for composting. The compost improves physical, chemical and biological properties of soil thereby increasing crop yield. Composts improves the nutrient, porosity, water holding capacity, aggregation, electrical conductivity, cation exchange capacity of the soil (Shiralipour et al. 1992; Goel et al. 2020). To do large composting, power-driven composting units have been installed in many big cities of India (Bhide and Shekdar 1998). As in Bengaluru, Vadodara, Mumbai, Delhi, and Kanpur, mechanical composting units of 150 to 300 tons/day capacities are installed for efficient results (Sharholy et al. 2007).

1.6.6 Plasma Arc Recycling

Plasma arc technology is an advance waste treatment technology that require high electrical energy and high temperature. The plasma arc in a waste plant heats the waste to temperatures in range between 1000 and 15,000 °C. At this temperature the organic matter is broken down into simpler atoms/molecules and inorganic components are melted into a glassy slag. In presence of oxygen, the organic matter will be converted into CO_2 and water and anaerobically will produce syngas (mixture of CO and H_2). The syngas has high calorific values and so can be extracted away and used to produced energy. The remaining by-products of the process is mostly used in different construction work in civil engineering (George and Karthi 2010).

1.7 Challenges

Solid waste and its proper management are a strenuous challenge of this modern world. The reason being the generation of a high quantity of waste because of continuous population increase, urbanization, and the booming economy (Tang et al. 2020). Solid waste management is a more challenging task in developing countries in comparison to developed countries. The main reason for it is lack of effective waste treatment techniques, lack of financial input for waste disposal, lack of proper knowledge about waste, lack of awareness, and legal framework (Guerrero et al. 2013). As per the world bank estimate, the world will generate about 2.59 billion tonnes of waste annually by 2030, which will be expected to be 3.40 billion tonnes by 2050 (Kaza et al. 2018). For the best management of waste, we need have to tackle the issues of population and urbanization and then safe disposal of large quantities of wastes (Ravichandran and Venkatesan 2021). Some important challenges in the process of solid waste management are given in Fig. 1.10. There are many problems associated with mismanagement of solid waste like greenhouse gas emission, disease, leachate and soil pollution, and decreased recreational value of the area. Proper management of solid waste is also a daunting task and rise so many problems which needed to be addressed. Detailed discussion on such challenges is given below.

1.7.1 Population

Population is a major factor in the generation huge piles of waste on the earth. It has been observed that developing countries with large population faces many types of difficulties while managing generated solid waste. The metro cities with high population and their enhanced standard of living have given rise to various

Fig. 1.10 Challenges in solid waste management

categories of wastes (Rana et al. 2015). While countries with low population are able to follow the zero waste strategies to easily manage their waste.

1.7.2 Consumerism

Consumerism is a social and economic status that encourages the acquisition of different goods and services in increasing amounts by the population. Consumerism is also one of the key factors affecting waste generation. Because of the increasing influence of media and the demands of modern world, waste is increasing. People are buying and consuming goods in large amount. Tempting advertisements are motivating people to purchase more and more and as a result, more things end up as a waste.

1.7.3 Plastic Production

As compared to plastic production its waste management accounts for very low and therefore most of waste end up mismanaged (Goel et al. 2008). Increased and mismanaged plastic waste is alarming situation for the world. Lebreton and Andrady in their study recorded that globally 60 and 99 million metric tonnes (Mt) of mismanaged plastic waste were produced in year 2015. And in future the load will accelerate majorly in the African and Asian continents and it can only be reduced if proper management steps are taken at proper time (Lebreton and Andrady 2019; Debbarma et al. 2017).

1.7.4 Growing Waste Management Cost

Financial factor is an important part for the solid waste management. For best management of solid waste operation cost is quite high. As per a report high-income country, operating costs for integrated waste management exceed $100 per tonne while it is only about $35 per tonne in Lower-income countries (Kaza et al. 2018). In few countries, local governments subsidies upto 50% of investment incurred in the waste management to the private sector. Mostly municipalities of the different countries have failed to process solid waste for the proper management due to financial factors. A high budget for the service (Sharholy et al. 2007), the limited resources and the unawareness of the users to pay for the management service has hampered waste management services (Sujauddin et al. 2008).

1.7.5 Poor Waste Management Facilities

It has been reported that collection, transfer and transport practices of the waste management system are seriously affected by factors like improper bin collection systems, poor route planning, lack of information about collection schedule (Hazra and Goel 2009), deficient infrastructure (Moghadam et al. 2009), poor roads and availability of number of vehicles for collection (Henry et al. 2006). Sharholy et al. (2008) suggested the involvement of informal sectors and micro-enterprises in the management organization for improving the efficacy of the waste management system. Other technical factors influencing the better functioning of system are lack of technical skills among personnel within municipalities and government authorities (Hazra and Goel 2009), insufficient technologies and reliable data (Mrayyan and Hamdi 2006).

1.7.6 Lack of Legal and Basic Framework

Lack of legal framework makes solid waste management a loose task in many countries. Without any legal frameworks, no one can be subjected responsible for open waste disposal. Also, weak legislation is unable to force proper waste segregation and operation. This subject risk not only the environment but also the people around the vicinity of waste. The waste categories under toxic and hazardous categories are really a menace, if not disposed of properly. Most developing countries are lacking in proper policies to restrict waste in the environment. Globally, the poor management of municipal solid waste is also attributed to a lack of proper infrastructure, appropriate governance, and environmental policing (Razzaq et al. 2021). The regulation about waste management have been formulated by about two-thirds of countries, yet their enforcement varies drastically. Regulatory measures include identifying and categorizing different waste and forcing compulsory segregation, transport, storage, treatment, and disposal practices. International law includes various agreements related to international transport and disposal of hazardous waste.

1.7.7 Impact of Pandemic

COVID-19 pandemic greatly impacted the waste management strategies in the world and made the process more complex. The waste generated during a pandemic is a mostly medical type and harms the environment and living beings. This type of waste increased abruptly during pandemic need a very cautious approach to manage and discard. The pandemic has also increased the single-use plastic problem, which will surely have some critical impacts in the long term. Lockdown has also reduced the waste collection and recycling process make the situation worse.

1.8 Status of Waste Management in India

India is also one of the largest producers of the waste in the world. The high rate of increase in population, unplanned rapid urban growth, rapid industrialization, change of lifestyle to make life comfortable are the common reasons for generations of unmanageable amount of waste in India. And as a consequence, in many of the metro cities huge piles of solid waste are deteriorating the environmental conditions (Rana et al. 2015). The characteristics and category of solid waste generated in the Indian cities almost vary based on population size and the geographical location (Ray and Rahaman 2016). Different form waste need to tackled by separate rules to be strictly implemented at ground.

As per the annual report of India 1,52,076 TPD solid waste generated in India (Table 1.1). The management system of solid wastes in India covers the full cycle from waste generation, collection, resource recovery and recycling, transportation to processing or disposal of waste. In few urban cities, the solid waste management is done in Public-Private Partnership that has open doors for doing business with waste (CPCB Annual Report 2018–19). The informal sectors play a major role in waste management by ensuring more recycling and less landfills treatment of waste in India.

Efforts are also done in the field of technological advancement for better processing, treatment and final disposal of the solid waste. Energy-from-waste is the most focused part of the waste management strategies because it helps to convert

Table 1.1 Status of the Solid Waste Management in all States/Union Territories during year 2018–2019 (CPCB Annual Report, 2018–2019)

Solid Waste Generation	1,52,076 TPD
Solid waste collection	1,49,748 TPD (98.5%)
MSW treated	55,759 TPD
Landfilled waste	50,161 TPD
Source segregation initiated	24 Nos
Total landfill sites operational	22 Nos
Waste processing facilities set up	2028 Nos
Waste processing facilities operational	160 Nos
Identified landfill sites	1161 Nos
Operational landfill sites	37 Nos

the waste into renewable form of energy and also reduces the huge volume of the solid waste from land disposal. But proper functioning and maintenance of many wastes to energy plants in India are constraints to achieve the target of sustainability. Awareness, segregation and characterization of waste, financial support, proper implementation of rules, efficient technological solution, research for waste-to-energy are few constraints for effective solid waste management in India.

In India the solid waste management was initiated in early1960s. During that time government was providing loans for installing composting plant to convert municipal solid waste into compost. Municipal Solid Waste (Management and Handling) Rules were formulated in 2000 were applicable for collection, segregation, storage, transportation, treatment and disposal of municipal solid waste. Till dates the important laws related to solid waste management and handling are framed by Govt. of India are given below

- Environmental (Protection) Act (1986)
- Biomedical Waste (Management and Handling) Rules (1989)
- Biomedical Waste (Management and Handling) Rules (1998)
- Recycled Plastic manufactured & usage Rules (1999)
- Solid Waste Management in Class 1 cities in India-Guided by Supreme Court of India (1999)
- Municipal Solid Waste (Management and Handling) Rules (2000)
- Batteries (Management and Handling) Rule (2001)
- Plastic Waste Rules, (2011)
- E-waste Rules, (2011)
- Solid Waste Management Rules (SWM), (2016)

Though, there are laws and regulation in India still there is still a proper infrastructure lacking especially in rural and backward areas. Most of the waste end up in landfill areas causing so much trouble and affecting environment and human health.

1.9 Conclusion

Waste is a big threat to the world and thus its proper management is urgently warranted by the world. An efficient solid-waste management system can solve the problem of solid waste and can also turn it into an important resource. Hence, the development of sustainable and integrated solid waste management strategies is the need of the hour. More emphasis should be given to reuse, recycle, and other opportunities that can convert waste into valuable products. These techniques should be specific for each type of waste. The important step in solid waste management is to create awareness among people to ensure that they may not generate garbage on the roads and hence do not create open dumpsites. Further, their support in waste segregation and collection will reduce the load on municipal workers. Solid-waste management is a multidimensional issue that requires integrated efforts of people, politicians, stakeholders, and administers. Improving solid waste management in all countries requires efforts to raise public awareness, increase funding, improve research, and create infrastructure. The role of the 7R principle seeks to be very important in controlling waste. The responsibility of the government to design a proper legal framework and its strict implementation is also seen as a significant tool for waste reduction.

References

Abdel-Shafy HI, Mansour MS (2018) Solid waste issue: sources, composition, disposal, recycling, and valorization. Egypt J Pet 27:1275–1290

Alam P, Ahmade K (2013) Impact of solid waste on health and the environment. Int J Sustain Dev Green Econ 2:165–168

Annepu RK (2012) Sustainable solid waste management in India. Master's thesis, Columbia University, New York

Asgari J, Safavi MK, Mortazaeinezahad F (2011) Landfill biogas production process. Int Conf Food Eng Biotechnol 6:208–221

Ayodele T, Alao M, Ogunjuyigbe A (2018) Recyclable resources from municipal solid waste: assessment of its energy, economic and environmental benefits in Nigeria. Resour Conserv Recycl 134:165–173

Bhatt P, Rene ER, Kumar AJ, Kumar AJ, Zhang W, Chen S (2020) Binding interaction of allethrin with esterase: bioremediation potential and mechanism. Bioresour Technol 315:13845

Bhide A, Shekdar A (1998) Solid waste management in Indian urban centers. Int Solid Waste Assoc Times 1:26–28

CPCB Annual Report (2018–19) Central Pollution Control Board, https://cpcb.nic.in/. Accessed 21 May 2021

Debbarma P, Raghuwanshi S, Singh J, Suyal DC, Zaidi MGH, Goel R (2017) Comparative in situ biodegradation studies of polyhydroxybutyrate film composites. 3Biotech 7(178):1–9. https://doi.org/10.1007/s13205-017-0789-3

Ebreo A, Hershey J, Vining J (1999) Reducing solid waste: linking recycling to environmentally responsible consumerism. Environ Behav 31:107–135

George EH, Karthi L (2010) Review on plasma arc technology in waste management. Fourth Cusat national conference on recent advances in civil engineering. Cochin, 2010

Goel R, Zaidi MGH, Soni R, Kusumlata, Shouche YS (2008) Implication of *Arthrobacter* and *Enterobacter* species for polycarbonate degradation. Int Biodeterior Biodegrad 61(2):167–172
Goel R, Debbarma P, Kumar P, Suyal DC, Kumar S, Mahapatra BS (2020) Assessment of soil chemical quality, soil microbial population and plant growth parameters under organic and conventional rice–wheat cropping system. Agric Res 10:193–204
Guangyu Y (2009) Disposal of solid wastes. Point sources of pollution: Local effects and their control, vol II, p 296
Guerrero LA, Maas G, Hogland W (2013) Solid waste management challenges for cities in developing countries. Waste Manag 33:220–232
Hameed Z et al (2021) Gasification of municipal solid waste blends with biomass for energy production and resources recovery: current status, hybrid technologies and innovative prospects. Renew Sust Energ Rev 136:110375
Hamer G (2003) Solid waste treatment and disposal: effects on public health and environmental safety. Biotechnol Adv 22:71–79
Hauck PPE (2014) Evolution of integrated solid waste management systems enhanced with municipal utilities and green energy production. In: Waste Advantage Magazine, Palm Beach Gardens, Fl
Hazra T, Goel S (2009) Solid waste management in Kolkata, India: practices and challenges. Waste Manag 29:470–478
Henry RK, Yongsheng Z, JunD (2006) Municipal solid waste management challenges in developing countries–Kenyan case study. Waste Manag 26:92–100
Igoni AH, Ayotamuno M, Eze C, Ogaji S, Probert S (2008) Designs of anaerobic digesters for producing biogas from municipal solid-waste. Appl Energy 85:430–438
Jia X, Qin C, Friedberger T, Guan Z, Huang Z (2016) Efficient and selective degradation of polyethylenes into liquid fuels and waxes under mild conditions. Sci Adv 2:e1501591
Kaartinen T (2004) Sustainable disposal of residual fractions of MSW to future landfills. Technical, University of Helsinki, Espoo
Kaza S, Lisa CY, Perinaz BT, Woerden FV (2018) What a waste 2.0: a global snapshot of solid waste management to 2050. Urban Development. World Bank, Washington, DC
Kumar S et al (2009) Assessment of the status of municipal solid waste management in metro cities, state capitals, class I cities, and class II towns in India: an insight. Waste Manag 29:883–895
Lebreton L, Andrady A (2019) Future scenarios of global plastic waste generation and disposal. Palgrave Commun 5:1–11
Meng Y, Ling TC, Mo KH (2018) Recycling of wastes for value-added applications in concrete blocks: an overview. Resour Conserv Recycl 138:298–312
Moghadam MA, Mokhtarani N, Mokhtarani B (2009) Municipal solid waste management in Rasht City, Iran. Waste Manag 29:485–489
Mrayyan B, Hamdi MR (2006) Management approaches to integrated solid waste in industrialized zones in Jordan: a case of Zarqa City. Waste Manag 26:195–205
Nanda S, Berruti FA (2020) A technical review of bioenergy and resource recovery from municipal solid waste. J Hazard Mater 403:123970
Nawaz M, Heitor A, Sivakumar M (2020) Geopolymers in construction-recent developments. Constr Build Mater 260:120472
Parekh H, Yadav K, Yadav S, Shah N (2015) Identification and assigning weight of indicator influencing performance of municipal solid waste management using AHP. KSCE J Civ Eng 19:36–45
Potting J, Hekkert M, Worrell E, Hanemaaijer A (2017) Circular economy: measuring innovation in the product chain. Policy report, PBL Netherlands Environmental Assessment Agency, The Hague, 2544
Provis JL (2014) Geopolymers and other alkali activated materials: why, how, and what? Mater Struct 47:11–25

Rachel G, Damodaran N, Panesar B, Leatherwood C, Asnani PU (2007) Methane to markets and landfill gas energy in India. In: Singh H et al (eds) Proceeding of the international conference on sustainable solid waste management. Springer, Cham, pp 519–525

Rana R, Ganguly R, Gupta AK (2015) An assessment of solid waste management system in Chandigarh City, India. EJGE 20(6):147–1572

Ravichandran C, Venkatesan G (2021) Towards sustainable solid waste management-challenges and opportunities. In: Rahman ROA, Chaudhary MH (eds.) Handbook of advanced approaches towards pollution prevention and control. Elsevier, London pp. 67–103

Ray M, Rahaman M (2016) An overview of legal framework for waste management system in India with special allusion to SWM rules, 2016. Int J Interdis Multidis Stud 4:13–19

Razzaq A, Sharif A, Najmi A, Tseng ML, Lim MK (2021) Dynamic and causality interrelationships from municipal solid waste recycling to economic growth, carbon emissions and energy efficiency using a novel bootstrapping autoregressive distributed lag. Resour Conserv Recycl 166:105372

Ren B, Zhao Y, Bai H, Kang S, Zhang T, Song S (2020) Eco-friendly geopolymer prepared from solid wastes: a critical review. Chemosphere:128900

Robaina M, Murillo K, Rocha E, Villar J (2020) Circular economy in plastic waste-efficiency analysis of European countries. Sci Total Environ 730:139038

Sharholy M, Ahmad K, Vaishya RC, Gupta RD (2007) Municipal solid waste characteristics and management in Allahabad, India. Waste Manag 27:490–496

Sharholy M, Ahmad K, Mahmood G, Trivedi R (2008) Municipal solid waste management in Indian cities–a review. Waste Manag 28:459–467

Shiralipour A, McConnell DB, Smith WH (1992) Physical and chemical properties of soils as affected by municipal solid waste compost application. Biomass Bioenergy 3:261–266

Siddique R (2010) Use of municipal solid waste ash in concrete. Resour Conserv Recycl 55:83–91

Sujauddin M, Huda S, Hoque AR (2008) Household solid waste characteristics and management in Chittagong, Bangladesh. Waste Manag 28:1688–1695

Tang Z, Li W, Tam VW, Xue C (2020) Advanced progress in recycling municipal and construction solid wastes for manufacturing sustainable construction materials. Resour Conserv Recycl: X 6:100036

Uekert T, Kuehnel MF, Wakerley DW, Reisner E (2018) Plastic waste as a feedstock for solar-driven H 2 generation. Energy Environ Sci 11:2853–2857

Wan C, Shen GQ, Choi S (2019) Waste management strategies for sustainable development. In: Leal Filho W (ed) Encyclopedia of sustainability in higher education. Springer, Cham, pp 1–9

Yates D (2014) Plastic shopping bags make a fine diesel fuel, researchers report. Illinois News Bureau. https://www.sciencedaily.com/releases/2014/02/140212132853.htm

Chapter 2
Microbes Assisted Bioremediation: A Green Technology to Remediate Pollutants

Yamini Tak, Manpreet Kaur, Jyotsana Tilgam, Harjeet Kaur, Rajendra Kumar, and Chirag Gautam

2.1 Introduction

The leakage of chemicals from industrialization and related practices has caused numerous issues in the environment as a result of multiple human activities. As a result of industrialization and urbanization, a vast volume of sewage, as well as radioactive waste, pumped into the atmosphere, polluting the land, water and air (Ferronato and Torretta 2019). Contaminated soil, as well as poisonous land and surface water, is a significant concern all over the world, and the effects of these harmful substances on the environment are long-lasting. Hydrocarbon deposition in soil has a negative impact on the local environment, and bioaccumulation in plants and living things can results in mutations (Ali et al. 2019). Conventional method of remediation may have risk of handling of contaminants and pollutants. Bioremediation method is more efficient, less harmful and cost-effective way of removing contaminated materials from the environment. Bioremediation is the process of converting hazardous wastes into another form which is not hazardous to living

Y. Tak (✉)
Biochemistry, Agriculture University, Kota, Rajasthan, India

M. Kaur
Biochemistry, Punjab Agricultural University, Ludhiana, Punjab, India

J. Tilgam
Agril. Biotechnology, ICAR-NBAIM, Maunath Bhanjan, Uttar Pradesh, India

H. Kaur
Department of Agronomy, Punjab Agricultural University, Ludhiana, Punjab, India

R. Kumar
MBDDS College, Siswali, Baran, Rajasthan, India

C. Gautam
Plant Pathology, Agriculture University, Kota, Rajasthan, India

© The Author(s), under exclusive license to Springer Nature Switzerland AG 2022
D. C. Suyal, R. Soni (eds.), *Bioremediation of Environmental Pollutants*,
https://doi.org/10.1007/978-3-030-86169-8_2

being and can be reused (Azubuike et al. 2016; Bhatt et al. 2020). Microbes can be found in their natural habitat in ecosystem, and they can be a low-cost alternative to conventional methods. Due to optimistic genetic material microorganism such as fungi, bacteria are eco-friendly in nature and they have capacity to decode environmental combination. By oxidizing, degrading, or immobilizing compounds such as hydrocarbons, petroleum pollutants, heavy metal, oil, dye, pesticides and sewage waste microbes may convert them into non-hazardous compounds (Hyde et al. 2019; Bhatt et al. 2021a, b). Bioremediation is primarily dependent on enzymatic process of microbes, which are influenced by environmental factors such as pH of soil, temperature and oxygen concentration. Microbes can function in both in situ and ex situ conditions; on the other hand severe environmental contaminants can be scrubbed by these organisms (Ojuederie and Babalola 2017). Microbes that have been genetically modified, either directly or implicitly, can increases the bioactivity of remediation and they have a greater potential to degrade the hazardous compounds. Bioremediation is low cost and green technology that can be used to manage vast areas of soil and ground water (Sales da Silva et al. 2020). Synthetic biology can help to mine genes from electronic database and to understand the interrelationship between the catabolic and metabolic reactions in order to synthetically explore the microbial community. The function of microorganisms in remediation, their processes, methods and factor influencing microbial assisted bioremediation are the key topics of this chapter.

2.2 Bioremediation and Its Methods

Bioremediation is an outstanding green technology that eases to elimination and degradation of toxic contaminants to non-toxic or less hazardous chemicals or substances under controlled conditions. It's primarily a microbe-mediated, sluggish, and natural operation. This is positive for both bacteria and the ecosystem because chemical toxins or radioactive chemicals are degraded, providing microbes with nutrients for their metabolism (Hakeem et al. 2020). Bioremediation can occurs aerobically or an-aerobically in which complete or partial removal of hazardous compounds into gases, water and other inorganic compounds which are safe for environment and living beings. Bioremediation may be done in situ or ex situ, depending on the form of pollutants and their concentrations (Sales da Silva et al. 2020). We will effectively restrict pollutant concentrations by choosing the right bioremediation technique.

2.2.1 Biostimulation

Biostimulation is the mechanism of increasing the activity of endogenous microbes by injecting particular electron acceptors, donors, or nutrients into the soil or ground water. In this process boost the intrinsic or natural habitat microbial community and degradation of contaminants by accumulation of nutrients or other restricting factors (Kanissery and Sims 2011). By providing the fertilizers, growth enhancing supplements, mineral, oxygen and temperature metabolic growth of indigenous microbes like soil bacteria, fungi is stimulated. Nutrients such as nitrogen, phosphorus and potassium are main building block of living organisms and they requires for survival of microbes to gain their cellular biomass and energy (Jacoby et al. 2017). The biostimulation system is primarily used to remove petroleum contaminants from soil; however, it is also dependent on the availability of oxygen, pH concentration, and soil temperature (Tyagi et al. 2010). The oxidation of chlorobenzene and 1,2-dichlorobenzene in the soil is improved by activating native microbes with organic supplements and nutrients (Kurt and Spain 2013). The main thing to be considered that is delivery of supplements in a way that permits the supplements to be easily available to subsurface microorganisms.

2.2.2 *Bioaugmentation*

The introduction of endogenous or exogenous microorganisms capable of degrading contaminants by raising their enzymatic concentration at the gene level is known as bioaugmentation. This method is used mainly for oil pollutants, by improving the degradation capacity of contaminated pollutant by the specific strain or consortia or any genetically modified organism (Nzila et al. 2016). Due to their specific DNA modification and varied metabolic profile, genetically engineered microbes degrade toxins more quickly and efficiently than native microbes. Microbes are collected from the infected site, separated, genetically engineered, and reintroduced to the same location (Das and Chandran 2011). Bioaugmentation with bacteria consortium will degrade 2,4 dichlorophenol on a laboratory scale. Rate of cellulose hydrolysis and H_2 productions from carboxymethyl cellulose fermentation can enhanced when bioaugmented with cocultures Clostridium *acetobutylicum* X_9 and *Ethanoigenens harbinense* B_{49} (Ren et al. 2008). By boosting soil mineralization, *A. xylosoxidans* 2BC8 and *S. maltophilia* JR62 bioaugment fluorine and pyrene contaminated soil (Villaverde et al. 2019).

2.2.3 *Bioattenutaion*

Bioattenutaion is a natural mechanism in which microorganisms use various biological mechanisms to degrade environmental toxins and chemicals. Microbes use biochemical pathways to minimise the mass, length, and toxicity of pollutants in the atmosphere. Diffusion, depression, transformation, solubalization, advection, sorption, volatilization, ion exchange, and other chemical reactions may also be used to degrade toxicants. Microbes in soil and ground water can totally dissolve chemicals, converting them to harmless gases and water (Azubuike et al. 2016). As an organic carbon management master plan in aquifer storage and recovery systems, bioattenuation with indigenous soil microbes may be an optimising option. Denitrifying bacteria such as *Noviherbaspirillum denitrificans* and iron reducing bacteria such as *Geobacter* spp. act as organic degraders in the ASR systems (Nguyen et al. 2021).

2.2.4 Biosparging

Biosparging is the process of injecting air under pressure to provide oxygen and nutrients to a certain zone in order to boost native microbial growth and destroy organic constituents. It is mostly used in conjunction with the bioventing process, which involves releasing air into the saturated zone in order to improve aerobic degradation and mineralization (Parween et al. 2017). It is mostly used to degrade dissolved petroleum constituents in ground water as well as the capillary friendge. Dissolved oxygen, nitrate, redox potential, cultivable heterotrophs, and sulphate are increased by the biosparging process, while dissolved ferrous iron, sulphide, cultivable anaerobes, and methanogens are decreased. It accelerates aquifer oxygenation to promote aerobic biodegradation by pumping air below the region of pollution at near well spacing (Kao et al. 2008).

2.2.5 Bioventing

It's an in situ bioremediation process in which bacteria and consortia improved hydrocarbon biodegradation in soil by increasing oxygen flow in the unsaturated environment. It mainly degrades the soil adsorb fuel residual or pollutants. Bioventing is a technique that uses low air flow rates to provide the oxygen needed for biodegradation while minimising the volatilization and release of pollutants into the atmosphere. The bioventing approach is most useful for removing petroleum hydrocarbons, toxic pesticides, non-chlorinated solvents, and organic pollutants from soils. The bioventing method can be hampered by low temperatures and low soil moisture (Azubuike et al. 2016). Combination of bioventing with biosparging can accelerate the remediation of hydrocarbon and TCE vapors in the vadose zone from the ground surface at Plattsburgh Air Force Base where soils are heavily contaminated with JP-4 and solvents (Ahmadnezhad et al. 2021).

2.3 Mechanism of Microbial Bioremediation

Bioremediation's main goal is to stimulate microorganisms with nutrients, air, or oxygen to help them kill pollutants. Bacteria, fungi, algae, and yeast are microorganisms that survive in their natural environments and remove metals, dyes, oils, chemicals, and petroleum hydrocarbons from the atmosphere (Das and Chandran 2011; Giri et al. 2017a, b). Microbes require an energy source from the environment so they can be a prime agent for bioremediation. There is numerous type of contaminates are present in polluted site according to chemical nature of contaminants and polluted site microbes degrade or detoxify the contaminants (Juwarkar et al. 2010). Various microbes secrete a variety of enzymes that can destroy organic harmful compounds in the environment. Some contaminants are detoxified using the oxidoreductase enzyme, which converts toxic compounds into harmless items. Oxidoreductase enzymes are mostly used to remove phenolic materials and toxic metals from the environment (Karigar and Rao 2011). Filamentous fungi easily reach to the soil pollutants and with the helpful of mycelium they secrete laccase and peroxidase enzyme that will detoxify the contaminants (Khatoon et al. 2021). Some soil bacteria contain oxygenase enzyme which degrade aromatic pollutants from ecosystem. To remediate organophosphates and oil spill some bacteria secrete hydrolytic enzymes such as lipase, amylase and proteases hydrolyse the bond of toxicants and detoxify them (Xu et al. 2018). White rot fungi like *Trametes versicolor*, *Peniophora gigantea*, *Pycnosporus coccineus*, and *Phanerochaete chrysosporium* degrade pesticides like chlorpyrifos, linuron, metribuzin, and simazine, as well as trifluralin, to produce water tension, extracellular enzymes, and breakdown mixture of pesticides (Gouma et al. 2014). Some bacteria uptake and reflux the pollutants, assimilate, biosorption to membrane, immobilize, form complex, precipitate and release agents. Heavy metals like zinc, cadmium, and copper can be remedied by the purple nonsulfur bacteria *Rhodobacter sphaeroides* and *Rhodobium marinum* (Li et al. 2016). Engineered *Pseudomonas putida* designed with mechanism such as altered metabolism, properties of membrane, pumps or some genetic manipulation that degrade toluene (Samin et al. 2014).

Heavy metals are integrated into soil bacteria's cellular walls or metal binding proteins by functional groups such as hydroxyl, carbonyl, or phosphate groups, resulting in metabolic processes being altered to impede metal uptake and poisonous metals being transformed into harmless compounds by various enzymatic reactions (Kapahi and Sachdeva 2019). Yeast uses an ion exchange process to bioremediate the hazardous water. Microalgae including *Spirulina* sp., *Chlorella*, *Cladophora*, and *Phaeodactylum tricornutum* may detoxify heavy metals from acid mine drainage by acting as "hyper-accumulators" and "hyper-adsorbents" (Bwapwa et al. 2017). Microalgae supply oxygen, release enzyme and provide nutrients to aerobic bacteria for its metabolism and these bacterium strains degrade oil and industrial waste. Conjugated consortia of microalgae-bacteria degrade caffeine and ibuprofen faster as compared to single bacteria consortia (Sutherland and Ralph 2019).

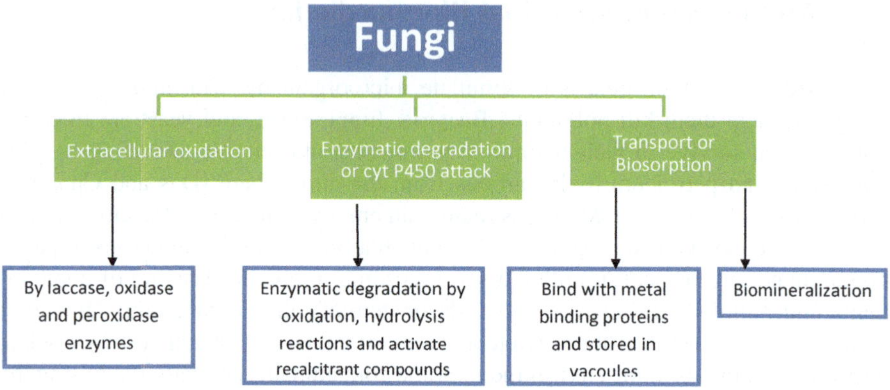

Fig. 2.1 Fungal bioremediation mechanism to remediate pollutants

To detoxify environmental toxins, fungi have their own complex enzymatic pathway that includes monooxygenases, glutathione transferase, and the cytochrom P450 complex. Fungi's P450 complexes in the cytosol and mitochondria can serve as a steriospecific catalyst for non-activated hydrocarbons. White-rot fungus *P. chrysosporium* contain CYP63A2 P450 monooxygenase oxidizes aliphatic carbon of oil and PAHs (Lah et al. 2011). Pesticides and herbicides can be degraded by fungi using dehydrogenation, hydroxylation, and esterification reactions. *Penicillium digitatum*, *Penicillium chrysogenum* and *Fusarium solani* degrade polychlorinated biphenyls and Rhizopus oryzae detoxify pentachlorophenol vai dechlorination and methylation reactions (Aranciaga et al. 2012). *L. squarrosulus* activate lignocellulosic modifying enzymes such as laccase, peroxidases and hydrogen peroxide producing enzymes and degrade industrial waste (Chukwuma et al. 2020).

Yeast, such as *S. cerevisiae*, reduces environmental effluents by reducing metal binding potential, allowing radioactive metals to be immobilized in biomass, extracted, recycled, and processed into environmentally safe compounds. The negatively charged cell wall of yeast binds to heavy metals through electrostatic interactions, while exoploysaccarides increase the cell wall's bisorption ability (Machado et al. 2009). Under laboratory or field conditions *Candida digboiensis* bioaugmented with oily sludge contaminated soil showed more capacity to humiliate petroleum hydrocarbon (Sood et al. 2009). The mannan–protein layer in the outer cell wall and the glucan–chitin layer in the inner cell wall of *Kluveromyces marxianus*, *Candida* spp., and *Saccharomyces cerevisiae* play a greater role in heavy metal detoxification (Fig. 2.1). These yeast cells have specific enzyme and metal binding capacity remove 73–90% of Cu at suitable pH concentrations (Garcia-Rubio et al. 2020).

2.4 Applications of Microbial Bioremediation

Bioremediation is a biological process that uses microbes or enzymes to alter, convert, detoxify, or vitiate harmful compounds into a harmless state (Ndeddy Aka and Babalola 2016; Okoduwa et al. 2017). Various microbes assisted mechanisms or pathway have been identified and described in the remediation of environmental contaminants.

2.4.1 Remediation of Heavy Metals by Microbes

Heavy metals become a serious ecological threat all over the world and one of the utmost environmental challenges because of their non-biodegradability and bioaccumulation (Igiri et al. 2018; Yuan et al. 2021). An advance of cost-effective, innovative, proficient and eco-friendly methodology for heavy metals (Cr, Hg, Cd, and Pb) remediation for the revitalization of the environment is need of hour. In this view, latest developments in microbes-dependent heavy metal ought to impelled bioremediation as a forthcoming alternate to orthodox procedures. In recent years, researchers have demonstrated diverse bioremediation procedures; but in arrears of variant pollutants in nature, not any solitary technique of bioremediation aids as a theriac to refurbish polluted environments. But due to their immense potential, microbes are biochemically discovered equally an impending agent to resist heavy metals and therefore, microbial remediation has turn out to be a highlight of new exploration in bioremediation technology. As native microbes existing in contaminated environs has the ability to face all the bioremediation and biodegradation associated defies as long as surroundings are appropriate for their metabolic and development processes (Azubuike et al. 2016). The role of various microbes in remediation of heavy metals has been reported in Table 2.1.

2.4.1.1 Mechanism of Heavy Metals Detoxification by Microbes

Enzymes, metal oxidation, synthesis of exopolysaccharides and metal chelators including metallothioneins, methylation, extrusion, biotransformation, precipitation, volatilization, surface complex action, electrostatic interaction demethylation, redox process intracellular and extracellular sequestration are some of the mechanisms used by microorganisms to cope with metal toxicity and detoxification (Wu et al. 2010; Yang et al. 2015; Shuaib et al. 2021). The anionic structures found on the cell surface of microorganisms (hydroxyl, phosphoryl, carboxyl, alcohol, amine, thioether, thiol, ester, sulfonate, and sulfydryl groups) confer negative charge, allowing them to bind to metal cations (Gavrilescu 2004).

Table 2.1 Role of microbes in heavy metals remediation

S. No.	Metal	Microbe	References
1.	Cr	*Bacillus metylotrophicus*	Tang et al. (2021)
		Leiotrametesflavida	
		Klebsiella sp.	Hossan et al. (2020)
		Bacillus cereeus	Nayak et al. (2018)
		Sporosarcinasaromensis	Ran et al. (2016)
		Bacillus subtilis	Kim et al. (2015)
		Acinetobacter sp.	Bhattacharya et al. (2014)
		Staphylococcus	Kumar et al. (2011)
		Pseudomonas aeruginosa	Benazir et al. (2010)
		P. aeruginosa and B. subtilis	
		S. cerevisiae and P. aeruginosa	Benazir et al. (2010)
		Saccharomyces cerevisiae	
		Aspergillus sp.	Congeevaram et al. (2007)
		Spirogyra sp.	Mane and Bhosle (2012)
		Spirulina sp.	
2.	Pb	*Methylobacteriumorganophilum*	Bharagava and Mishra (2018)
		Cellulosimicrobium sp.	
		Streptomyces sp.	Kumar et al. (2011)
		Staphylococcus sp.	
		Nostoc sp.	Kumaran et al. (2011)
		Bacillus firmus	Salehizadeh and Shojaosadati (2003)
3.	Cu	*Spirogyra* sp.	Mane and Bhosle (2012)
		Desulfovibriodesulfuricans	Congeevaram et al. (2007)
4.	Co	*Vibriofluvialis*	Jafari et al. (2015)
5.	Zn	*Bacillus firmus*	Salehizadeh and Shojaosadati (2003)
		Pseudomonas sp.	Kumaran et al. (2011)
6.	Hg	*Klebsiellapneumoniae*	Jafari et al. (2015)

(a) Biosorption

The word "biosorption" refers to a group of approaches in which microbial biomass, either alive or dead, collects heavy metals from the environment. Microbes have sorption properties as a result of the presence of active chemical groups with a negative charge on their outer surface, indicating that they are capable of active metal linking (Ubando et al. 2021). Microbial cells derived heavy metals uptake through biosorption process can be categorized into surface adsorption mechanism which is metabolism-independent, occurs on exterior surface of cells by gathering metals and binding them with extracellular polymers, and metabolism-dependent mechanism, which depends on metal infiltration to the middle parts of the cells and comprises sequestration, redox reaction, and species-transformation methods (Vijayaraghavan and Yun 2008).

(b) Bioaccumulation

Bioaccumulation is a toxicokinetic process which occurs when the rate of contaminant absorption is greater than the degree of losing and it disturbs the living organisms' sensitivity to chemicals (Mishra and Malik 2013; Diaconu et al. 2020). Bioaccumulation organisms possess the advanced tolerance level to several contaminants and reveal loftier ability of biotransformation that is converting the toxic compounds into innocuous form which empowers them to ease the contaminant toxicity despite the fact of keeping it confined.

(c) Extracellular Sequestration

The metal ions accretion by components of cells either in the periplasm or their complexation as unsolvable composites is known as extracellular sequestration (Han et al. 2020). Cu-resistant strain of *Pseudomonas syringae* produced periplasmic and outer membrane proteins which are Cu-inducible and bind microbial colonies and Cu ions (Cha and Cooksey 1991). The precipitation of metals is a type of sequestration occurring extracellularly. Many microbes have been reported which convert toxic compounds into harmless or less toxic metals such as *Geobacter* spp. and *Desulfuromonas* spp. which are iron and sulfur reducing bacteria, respectively. Under anaerobic conditions, *Klebsiella planticola* converts thiosulfate into hydrogen sulfide and precipitated Cd ions as insoluble sulfides (Sharma et al. 2000). Similarly *P. aeruginosa* and *Vibrio harveyi* precipitated Cd and soluble divalent lead, respectively (Wang et al. 2002; Mire et al. 2004).

(d) Intracellular Sequesteration

The metal ions complexation resulting from their contact with apparent ligands trailed by deliberate transportation into the cytoplasm of cells with the help of several compounds is known as intracellular sequestration (Chen et al. 2020). It has been reported in literature that microbial cells are capable of accumulating metals intracellularly such as fungi's cell wall exterior act as a ligand for metal ion labeling due to the presence of specific negatively charged functional groups on their surface resulting in the inorganic metals obliteration (Jha et al. 2011). Cd-resistant *P. putida* strain intracellularly sequestered Cu, Cd, Zn zinc ions by aid of low molecular weight and cysteine-rich proteins (Higham et al. 1986) and *Rhizobium leguminosarum* sequestered Cd ions with the help of glutathione (Lima et al. 2006).

(e) Biotransformation

Reduction, oxidation, methylation and de-methylation are the biochemical processes involved in the microbial derived heavy metals transformations with the help of their enzymatic system and these reactions can occur in extracellular environments, vacuoles and on cell surface as well (Mohd et al. 2019). It is of great importance if it involves reduction of high-value and significant metal ions such as tannery sewers isolated Gram positive bacteria converts Cr (VI) which is highly toxic to less toxic Cr (III).

(f) Methylation

Microbial methylation plays a major role in remediation of heavy metals. Toxicity of metals gets increased by methylation due to enhancement of lipophilicity ultimately increasing the permeability across cell membrane (Ma et al. 2019). Most of the methylated compounds are volatile in nature such as methyl-Hg which is produced by methylation of Hg (II) by certain bacterial species such as *Pseudomonas*, *Bacillus*, *Clostridium* and *Escherichia*; dimethyl selenide and arsines formed by methylation of Se and As, respectively (Ramasamy et al. 2006).

(g) Bioleaching of Metals
Bioleaching is the process which involves metal compounds transformation into certainly soluble substances especially sulphides easing the task of metal ions elimination. Bioleaching is primarily employed in biohydrometallurgy and in oxide and sulphide minerals. The ability of microbes to perform bioleaching is due to the formation of several organic compounds and release of complexion molecules (Tripathi et al. 2019).

2.4.2 *Treatment of Solid, Liquid and Gaseous Waste*

Based on its composition, environmental waste can be classified as solid waste, liquid waste, or gaseous waste (Ngoc and Schnitzer 2009). Plastic bottles, medical waste, styrofoam containers, and other waste created by urban, agricultural, and commercial areas are examples of solid waste (Singh et al. 2017). Unwanted liquid waste is often dumped into the water ecosystem, and is referred to as liquid waste. Liquid waste includes chemicals, oils, and liquid effluents created by clothing, pulp/paper, tanneries, and other industries (Hudson-Edwards et al. 2011). Similarly, gaseous waste refers to excessive waste that is emitted into the air in the form of toxic gases.

2.4.2.1 Microbial Degradation of Solid Waste

The oldest and most often used process for solid waste disposal is biological composting (or composting), also known as nature's means of recycling (Mani et al. 2020). It is a biological process which involves the aerobic decomposition of complex organic waste from various sources such as plant/garden waste, food scrap, paper, coffee grounds etc., into humus-like substance by the action of various microbes (bacteria, fungi etc.) (Diaz et al. 2020). The resulting substance, known as manure or compost, is exceptionally high in carbon and nitrogen, making it an ideal organic fertilizer for plants (Suyal et al. 2021). Composting can be further classified into aerobic composting and anaerobic composting (Kalemelawa et al. 2012). Aerobic composting includes the decomposition of waste in the presence of air, which allows the degrading microbes to quickly decompose the waste with no smell (Mani et al. 2020). However, anaerobic composting is referred to the biodegradation of solid waste in the absence of air resulting in the production of methane (CH_4), ammonia (NH_4^+) and carbon dioxide (CO_2). When opposed to aerobic

Table 2.2 List of major microbial (bacterial and fungal) communities involved in different phases of compositing (Srivastava et al. 2016)

Composting phases	Major microbial communities during different composting phases
Mesophilic phase (Early)	**Bacteria**: *Pseudomonas, Bacillus, Flavobacterium, Clostridium, Serratia, Enterobacter* and *Klebsiella.* **Fungi**: *Alternaria, Cladosporium, Mucor, Aspergillus, Humicola* and *Penicillium*
Thermophilic phase	**Bacteria**: *Bacillus* and *Thermus* **Fungi**: *Aspergillus, Mucor, Chaetomium, Humicola, Absidia, Sporotrichum, Thermoascus* and Yeast **Actinomycetes**: *Streptomyces, Thermoactinomyces* and *Thermomonospora*
Mesophilic phase (Late) (maturation phase)	**Bacteria**: B*acillus, Flavobacterium, Pseudomonas* and *Cellulomonas* **Fungi**: *Alternaria, Aspergillus, Bipolaris* and *Fusarium* **Actinomycetes**: *Streptomyces* and *Thermopolyspora*

composting, anaerobic composting takes longer to decompose the waste. The compositing process is started by a mesophilic microorganism that quickly breaks down the waste's soluble components. As soon as the temperature rises above 40 °C, the thermophilic microbes replace the mesophilic microbes and the rapid breakdown of proteins, fats and complex carbohydrates accelerate (Lin et al. 2018). As the process of compositing reaches its final stage, the temperature steadily decreases, and mesophilic microorganisms once again take over for the final phase of "curing" or maturation of the remaining organic matter (Matsinhe 2011). The Table 2.2 lists various microorganisms associated with various stages of the composting process.

Composting is the most natural way for managing solid waste, as well as improving the texture and moisture-holding capability of the soil (Banerjee et al. 2019). The addition of compost to the soil helps conserve nutrients and balances the soil's acidic/alkaline composition, reducing the need for artificial fertilizers (Kumar 2011; Singh 2013).

2.4.2.2 Microbial Degradation of Liquid Waste

Based on the degree of pollution reduction, wastewater treatment can be classified as preliminary treatment, major, secondary, or tertiary treatment (Sonune and Ghate 2004).

A. **Preliminary treatment**: It is the first step of wastewater treatment (following recycling and influent pumping), and it entails the removal of garbage and other hazardous materials in order to prevent clogging or jamming of treatment plant machinery. Several procedures are used in the preliminary treatment to aid in the removal of specific types of substance from the wastewater (Singh 2013). Screening, shredding, grit elimination, pre-aeration, and chemical incorporation are among the procedures (Mani et al. 2020).

   (a) **Screening**: This stage necessitates the use of a perforated bar screen to help capture large debris from the wastewater flow, such as rags, containers, bricks, and trees (Asthana et al. 2017).
   (b) **Shredding**: It utilizes a minimizing or cutting device to cut the solid into sizes that can enter the treatment plant machinery (Mani et al. 2020).
   (c) **Grit removal**: This process is carried out in a grit chamber or by aeration, velocity, the centrifugal force of sludge. In this step, the solid waste is separated from the sludge and wastewater (Quansah et al. 2018). Heavy inorganic solids such as gravel, clay, sand, metal filings, seeds and similar material are removed (Singh 2013).
   (d) **Pre-aeration**: This move involves aerating the effluents under septic conditions in order to attain and sustain an aerobic state. It also means that the biological oxygen requirement (BOD) is reduced (Mani et al. 2020).
   (e) **Chemical treatment**: The effluent stream is treated with chemicals like peroxide, chlorine, acids, mineral salts, bases and enzymes to reduce the BOD, odour, greases etc. (Al-Ghouti et al. 2019).

B. **Primary treatment**: The main aim of primary treatment is to eliminate and minimize suspended and floatable solids, as well as organic materials, by physical isolation (Singh 2013). This procedure can be done in rectangular or circular clearing tanks, and each round of primary treatment can reduce BOD by 25%–30%, settleable solids by 95%, and overall suspended solids by 50%–60% (Mani et al. 2020).

C. **Secondary treatment (Biological treatment)**: After primary care, the effluent is further passed to secondary treatment, with the primary goal of removing biodegradable organic matter. The secondary treatment employs aerobic microorganisms, mostly bacteria, to metabolise the organic matter in the wastewater and produce inorganic end products (CO_2, NH_3, and H_2O) (Samer 2015). A continuous supply of oxygen to the microorganism is a vital requirement for effective wastewater treatment. The secondary treatment can be further divided into two categories, named fixed-film systems and suspended growth systems (Machineni 2019). Microbial biomass or slime added to a media is used in a fixed film device (which may be stone, redwood or any synthetic material). As soon as the slime enters into contact with the wastewater, the microorganism destroys and oxidises the biological components of the effluent. The suspended growth mechanism, on the other hand, makes use of biological growth that has been combined with wastewater (Mani et al. 2020). An aeration bath, secondary basin, settling basin, and clarifier are the essential components of biological sewage treatment in activated sludge. Continuous mechanical stirring provides aeration to the waste treatment (Samstag et al. 2016). The microorganism concentration is maintained by reintroducing a small portion of sludge from the previous run. After primary settling, the incoming effluent is continually fed into an aeration tank with the desired concentration of microorganisms. To allow the bacteria to metabolise the organic compounds into simpler compounds, the sewage effluent is allowed to sit in an aeration tank for several hours. Returning the sludge to the aeration tank will encourage the growth of bacterial microbes, allowing fresh waste to be treated. The partly filtered water is sent to a separate sedimentation tank for bacteria elimination. Before being released into collecting water, the effluent is normally treated with chlorine in the sedimentation tank to remove pathogenic/harmful bacteria (Singh 2013).

D. **Tertiary treatment**: The tertiary treatment is the last step in the sewage treatment process, and it aims to improve the condition of wastewater before it is released into the atmosphere (Naidoo and Olaniran 2014). This treatment further reduces the BOD, eliminates inorganic compounds (nitrates and phosphates) as well as colour and other pollutants not satisfactorily removed by secondary treatment processes. Carbon adsorption, coagulation and sedimentation, ion exchange, membrane filtration etc. can be employed in tertiary treatment (Gunatilake 2015).

2.4.3 Microbial Degradation of Hydrocarbons

Petroleum-based renewable energy has been the major source of energy for both humans and industries (Hassan and Kalam 2013). However, the release of these hydrocarbons into the environment either accidentally or due to human activities often causes extensive damage by contaminating the water and soil. Excessive accumulation of hydrocarbon toxins in plants and animals will result in serious mutations or even death (Ball and Truskewycz 2013). Despite the fact that many methods have been used in the past to handle hydrocarbon emissions, successful decomposition of hydrocarbons from soil and water is still considered difficult due to their complicated nature (Gargouri et al. 2014). In nature, the hydrocarbon compounds are primarily degraded by bacteria (*Pseudomonas*, *Marinobacter*, *Micrococcus* etc.), yeast (*Candida*) and fungi (*Aspergillus*, *Penicillium*, *Fusarium* etc.) (Goel et al. 2008; Xu et al. 2018). Earlier researchers stated that biodegradation of complex hydrocarbon often requires the collaboration of more microbial species with broad enzymatic capacities. Hydrocarbon-degrading bacterial and fungal species and the type of hydrocarbon components they degrade are listed in Table 2.3.

The mechanism of degradation of hydrocarbons by microorganisms essentially involved specific intracellular enzymes and can be subdivided into five main steps (Joutey et al. 2013). First, attachment of the hydrocarbon substrate to microbial cell; Second, emulsification of petroleum hydrocarbon pollutants by surfactants secreted by microbes; Third, the adsorption of the emulsified hydrocarbon on the cell

Table 2.3 Petroleum hydrocarbon-degrading microbes and their desired degradation substrates

	Petroleum hydrocarbon components	Microbial species	References
Bacterial species	Aliphatic n-alkanes (C6-C40)	*Dietzia* sp.	Wang et al. (2011)
	Aliphatic n-alkanes (C15–C36)	*Geobacillus thermodenitrifican*	Abbasian et al. (2015)
	Aliphatic Cyclohexane	*Rhodococcus* sp.	Lee and Cho (2008)
	Aliphatic Branched and Normal Alkanes	*Gordoniasihwensis*	Brown et al. (2016)
	Aromatics Mono-/ Polyaromatics	*Achromobacter xylosoxidans*	Ma et al. (2015)
	Aromatics Mono-/ Polyaromatics	*Aeribacillus pallidus*	Mnif et al. (2014)
	Aromatic Monoaromatics	*Mycobacterium cosmeticum*	Zhang et al. (2013)
	Aromatic Monoaromatics	*Pseudomonas aeruginosa*	Mukherjee et al. (2010)
	Aromatic Polyaromatics	*Bacillus licheniformis* *Bacillus mojavensis*	Eskandari et al. (2017)
	Aromatic Polyaromatics	*Sphingomonas, Sphingobium and Novosphingobium*	Ghosal et al. (2016)
	Asphaltenes	*Citrobacter* sp., *Enterobacter* sp., *Staphylococcus* sp., *Lysinibacillus* sp., *Bacillus* sp., *Pseudomonas* sp.	Jahromi et al. (2014)
Fungal species	Naphthalene	*Trichoderma harzianum*	Mollea et al. (2005)
	n-hexadecane	*Aspergillus niger*	Volke-Sepúlveda et al. (2003)
	Benzo{a}pyrene	*Aspergillus ochraceus*	Passarini et al. (2011)
	Crude oil	*Penicillium sp. RMA1 and RMA2*	Al-Hawash et al. (2018a)
	Different PHs	*Aspergillus sp. RFC-1*	Al-Hawash et al. (2018b)

membrane surface of micro-organism followed by the entry of emulsified product into the cell membrane via active/passive transport or endocytosis (Das and Chandran 2011). Lastly, the petroleum hydrocarbon entering the cell undergoes an enzymatic reaction with the corresponding enzyme to accomplish the purpose of pollutant degradation. The quick and complete degradation of most of the petroleum hydrocarbon requires aerobic conditions (Das and Chandran 2011). The mechanism

Fig. 2.2 Diagrammatic representation of aerobic degradation of hydrocarbons by micro-organism (Das and Chandran 2011)

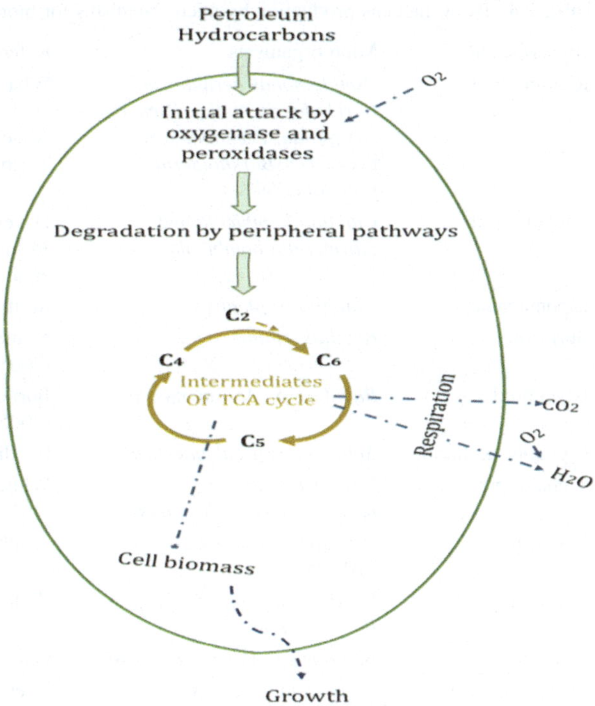

and the main principle of aerobic degradation of hydrocarbons inside the cell is described in Fig. 2.2.

The initial step of intracellular attack of organic pollutants is an oxidative process, with oxygenases and peroxidases catalysing the activation and incorporation of oxygen respectively (Das and Chandran 2011). When molecular oxygen is introduced to a hydrocarbon, primary alcohol is formed, which is then oxidised to aldehyde and fatty acid (Varjani 2017). The peripheral degradation pathways convert organic contaminants into intermediates of central intermediary metabolism, such as the tricarboxylic acid cycle, in a step-by-step process. Cell biomass is made up of key precursor metabolites such as acetyl-CoA, succinate, and pyruvate. The sugars needed for various biosynthesis and growth processes are produced by gluconeogenesis (Noor et al. 2010). Biosurfactants developed by microorganisms have recently shown promise in the degradation of petroleum-based products. Biosurfactants are amphiphilic extracellular metabolites that can lower interfacial stress, break up oil particles, and turn them to harmless substances (Patel et al. 2019). Biosurfactants are classified chemically as lipopeptides (polymyxin, gramicidin, surfactin, serrawettin, subtilisin, viscosin), glycolipids (rhamnolipid, trehalolipid, sophorolipid), phospholipids, surface proteins (vesicles, fimbriae), fatty acids, and polymers (emulsan, biodispersan, liposan) (Desai and Banat 1997; Gudiña et al. 2016). However, among the broad range of biosurfactants, only three have been well-defined rhamnolipid, sophorolipid, and surfactin (Table 2.4).

Table 2.4 Biosurfactants production by microorganisms for bioremediation

Biosurfactants	Microorganisms	References
Rhamnolipids	*Pseudomonas aeruginosa, Burkholderia mallei, Pseudomonas putida, Pseudomonas fluorescensBurkholderia pseudomallei*	Whang et al. (2008), Mahmound et al. (2008), Kumar et al. (2008), Abdel-Mawgoud et al. (2010), Toribio et al. (2010)
Sophorolipid	*Candida bombicola* and *Starmerella bombicola*	Daverey and Pakshirajan (2009), Joshi-Navare et al. (2013), Elshafie et al. (2015)
Lipomannan	*Candida tropicalis*	Muthusamy et al. (2008)
Surfactin	*Bacillus subtilis*	Youssef et al. (2007), Whang et al. (2008)
Glycolipid	*Bacillus* sp., *Aeromonas* sp.	Ilori et al. (2005), Tabatabaee et al. (2005)
Lipopolysaccharides	*Acinetobacter calcoaceticus*	Fondi et al. (2016)
Trehalolipid	*Mycobacterium* sp., *Rhodococcus erythropolis*	White et al. (2013), Kügler et al. (2015)
Viscosin	*Pseudomonasfluorescens* SBW25	Alsohim et al. (2014)
Iturin (A-E)	*Bacillus amyloliquefaciens, Bacillus subtilis*	Zeriouh et al. (2011)
Fengycin	*Bacillus* sp., *Paenibacillus* sp.	Guo et al. (2014)
Serrawettin W1	*Serratia marcescens*	Li et al. (2005)

2.4.4 Microbial Bioremediation of Dyes

Synthetic dyes are utilized in various industries including textiles, food, paper, tanning, cosmetics, plastics, and biomedical industries. Treated or untreated industrial discharge containing recalcitrant dyes are contaminating water resources because of their color and by the production of poisonous and cancerous compounds (Kandelbauer and Guebitz 2005). Physicochemical methods of dye remediation are costly, chemical, and energy-consuming. Alternatively, microbial-based remediation efficient, cost-effective, eco-friendly approach. Azo dyes exhibit strong adverse effects on live organisms, among other classes of dyes (Ihsanullah et al. 2020). Microbes are able to degrade azo dyes into their aromatic amines and convert these aromatic amines into simpler organic compounds too (Ajaz et al. 2020). Diverse strains of microorganisms can uptake or secrete enzymes that eliminate or detoxify hazardous contaminants from the polluted location. Yeasts, fungi, bacteria, and algae are capable of neutralizing industrial effluent (Table 2.5). Dye decolorization and elimination Bioremediation in reactor performed by application of (a) mixed culture, (b) isolated microbes, (c) purified enzymes to decolorize the effluents (Kandelbauer and Guebitz 2005).

Table 2.5 Examples of microorganisms and enzymes involving dye remediation

Example	Micro-organism	Enzymes
Fungi	*Schizophyllum commune* IBL-062, *Ceriporiopsis subvermispora, Cyathus stercoreus, Cladosporium cladosporioides, Pleurotus oxysporum, P. ostreatus, Phellinus pini, Aspergillus allahabadii, Curvularia clavata* NZ27, *Aspergillus niger, Aspergillus sulphureus, Rhizopus nigricans, Pleurotus florida, Trametes hirsute, Bjerkanderaadusta, Trametes versicolor* and *P. eryngii* F032	Lignin peroxidase, manganese peroxidase, and laccase and other lignolytic enzymes
Algae	*Chlorella vulgaris, Coelastrella* sp., *Haematococcus* sp., *Dermatocarpon vellereceum, Spirogyra, Nostoclinckia* and *Oscillatoria rubescens*	Oxygenase, reductase
Bacteria	*Pseudomonas putida, Pseudomonas rettgeri strain* HSL1, *Pseudomonas specie* SUK1, *Bacillus subtilis, P. putida, P. extremorientalis* BU118, *Brevibacillus laterosporus* MTCC 2298, *Galactomyces geotrichum* MTCC 1360*Pseudomonas* spp., *Tsukamurella* sp. J8025, *B. subtilis* SPR42, *P. fluorescens, Georgenia* sp. CC-NMPT-T3, *Bacillus cereus, Bacillus megaterium, Micrococcus luteus, Aeromonas hydrophila, Lysinibacillus* sp. KMK-Aand*Bacillus pumilus*	Azoreductase, hydroxylase, oxygenase, laccase
Yeast	*Candida oleophila, Candida tropicalis Pseudozyma rugulosa, Candida krusei,* and *Debaryomyces polymorphus*	*Manganese-dependent peroxidase (MnP), oxygenase*

2.4.5 Microbial Bioremediation of Agricultural Chemicals

Agricultural chemicals are used to boost agricultural productivity by reducing disease, pathogen, and weed losses. Pesticides, herbicides, fungicides, and fertilizers are among the many agrochemicals used in modern agriculture, which are now becoming the reason of environment pollution (Liu et al. 2019). Field contamination with Agrochemicals is becoming worldwide problem to be solved imperatively. The use of microorganisms in agrochemical waste bioremediation is gaining popularity. Microorganism mediated bioremediation of agrochemical pollution has now received great interest. Various microbes isolated and screened from different sources have exhibited their efficacy to neutralize agrochemicals residues into non-hazardous compound in the environment (Table 2.6).

Table 2.6 Microbes having remediation potential for agrochemicals

Microorganisms	Agrochemicals	References
Acenetobactor, Pseudomonas, Enterobacter, Photobacterium, Moraxella, Yersinia	Chlorpyrifos and methyl parathion	Ravi et al. (2015)
Pseudomonas pseudomallei	Glyphosate	Peñaloza-Vazquez et al. (1995)
Pseudomonas putida, Acinetobacter, Arthrobacter	Ridomil MZ 68 MG, Fitoraz WP 76, Decis 2.5 EC	Mónica et al. (2016)
Bacillus pumilis, Pseudomonas frederiksbergensis, Acinetobacter radioresistens, Serratia liquefaciens, Serratia marcescens	Endosulfan, malation, Chlorpyrifos, Diazinon	Hussaini et al. (2013)
Enterobacter, *Synechocystis, Brucella, Stenotrophomonas*	Chlorpyrifos	Niti et al. (2013)
Proteus, Vibrio, Yersinia, Serratia	Tetrachlorvinphos	Niti et al. (2013)
Bacillus, Pseudomonas, Arthrobacter, Aspergillus, Mycobacterium, Micrococcus, Streptomyces	Oxyfluorfen	Mohamed et al. (2011)
Pseudomonas, Micrococcus	Cypermethrin	Niti et al. (2013)

2.5 Genetic Engineering of Microbial Cultures for Environmental Pollution Regulation

Many microbial strains have improved using genetic engineering to mitigate environmental contamination effectively (Strong et al. 2000; Singh et al. 2011; Yang et al. 2012). Genetic manipulation of microbial cultures can magnify their catalytic potentialities by optimizing enzymes and pathways using the following strategies. (1) Identification and cloning of extremely effective pollutant degrading microbial genes, (2) overexpression of enzymes with novel catalytic potentialities in microbes, (3) expression of different degrading genes in a particular receiver to create recombinant microbe, and (4) protoplast fusion for incorporation of benefits of two different pollution degrading parents (Table 2.7). In addition, through omics approaches, mining of genes and pathways involving in degradation has offered more opportunities to create genetically improved pollutant degrading microbes (Debbarma et al. 2017; Liu et al. 2019; Bilal and Iqbal 2020).

2.6 Conclusion and Future Prospective

Bioremediation is a technique for scrubbing the pollutants from environment by indigenous or exogenous microbial community. Microbe assisted bioremediation is simple, less costly and more friendly to remediate pollutants. There is need to

Table 2.7 Genetic engineering of microbes for environmental pollution regulation

Genetically engineered strains	Gene/enzyme	Pollutants	References
Escherichia coli (E. coli)	*atzA*/atrazine chlorohydrolase	Atrazine	Strong et al. (2000)
E. coli	*bphA2cA1c*/salicylate oxygenase	PAHs	Cho et al. (2005)
Pseudomonasputida U ΔfadBA-phaZ	*phaZ*/PHPhA depolymerase	PAHs	Sandoval et al. (2005)
P. putida KT2440-DOP	*Tpd*/Triazophos hydrolase	Organophosphorus pesticides and aromatic hydrocarbons	
Sphingomonas sp. BHC-A-mpd	*Mpd*/methyl parathion hydrolase gene	Hexachlorocyclohexane (HCH) and methyl parathion	Lu et al. (2008)
Pseudomonasputida PaW340/pDH5	Dehalogenase	4-chlorobenzoic acid	Massa et al. (2009)
Streptomycescoelicolor M145-AH	*alkB*/alkane hydroxylase	n-alkanes	Gallo et al. (2012)

develop or construct genetically modified microbes to degrade pollutant is of main interest, this is alternative green method to degrade pollutants. It gives long elastic effect on marine and earth ecosystem; therefore, it is effective mitigation method of environmental toxicants. As expanding our knowledge about metabolic profiling of microbes and enzyme structure and degradation pathway, in future we can design new genetically modified microbes. Bioremediation would be supported by a better understanding of microbes' genetic makeup and metabolic reactions. Complete electronic database should be there which will be helpful to detoxify pollutants from particular site. Researcher's scientist can outlook to explore new genetically modifies superior microbial strains or species which have more potential to degrade pollutants in natural habitat. Microbes recognize and degrade pollutant by enzymatic reactions such as oxidation, reduction, isomerization, hydrolysis and elimination at a specific temperature and pH. Hence, this can be conclude that microbial assisted biodegradation of toxic and hazardous pollutant is environmental friendly, safe and can restore physicochemical activities of soil and ground water. This can transparently show that further research of bioremediation methods and microbes can helpful to control terrestrial and marine waste. Proper environmental conditions such as nutrients, oxygen, pH, temperature can enhance the degradation potential of microorganisms. To make absolute improvements and develop novel technologies of bioremediation there is need to do significant work on enzyme characterization and improvement of microbial consortia.

References

Abbasian F, Lockington R, Mallavarapu M, Naidu R (2015) A comprehensive review of aliphatic hydrocarbon biodegradation by bacteria. Appl Biochem Biotechnol 176:670–699

Abdel-Mawgoud AM, Lépine F, Déziel E (2010) Rhamnolipids: diversity of structures, microbial origins and roles. Appl Microbiol Biotechnol 86(5):1323–1336

Ahmadnezhad Z, Vaezihir A, Schüth C, Zarrini G (2021) Combination of zeolite barrier and bio sparging techniques to enhance efficiency of organic hydrocarbon remediation in a model of shallow groundwater. Chemosphere 273:128555. https://doi.org/10.1016/j.chemosphere.2020.128555

Ajaz M, Shakeel S, Rehman A (2020) Microbial use for azo dye degradation—a strategy for dye bioremediation. Int Microbiol 23(2):149–159

Al-Ghouti MA, Al-Kaabi MA, Ashfaq MY, Dana DA (2019) Produced water characteristics, treatment and reuse: a review. J Water Process Eng 28:222–239

Al-Hawash AB, Alkooranee JT, Abbood HA, Zhang J, Sun J, Zhang X, Ma F (2018a) Isolation and characterization of two crude oil-degrading fungi strains from Rumaila oil field. Iraq Biotechnol Rep 17:104–109

Al-Hawash AB, Zhang X, Ma F (2018b) Removal and biodegradation of different petroleum hydrocarbons using the filamentous fungus *Aspergillus* sp. RFC-1. Microbiology open e00619

Ali H, Khan E, Ilahi I (2019) Environmental chemistry and ecotoxicology of hazardous heavy metals: environmental persistence, toxicity, and bioaccumulation. J Chem:1–14. https://doi.org/10.1155/2019/6730305

Alsohim AS, Taylor TB, Barrett GA, Gallie J, Zhang XX, Altamirano-Junqueira AE, Johnson LJ, Rainey PB, Jackson RW (2014) The biosurfactant viscosin produced by *Pseudomonas fluorescens* SBW 25 aids spreading motility and plant growth promotion. Environ Microbiol 16(7):2267–2281

Aranciaga N, Durruty I, González J, Wolski E (2012) Aerobic biotransformation of 2, 4, 6–trichlorophenol by *Penicillium chrysogenum* in aqueous batch culture: degradation and residual phytotoxicity. Water SA 38(5). https://doi.org/10.4314/wsa.v38i5.5

Asthana M, Kumar A, Sharma B S (2017) Wastewater treatment. In: Principles and applications of environmental biotechnology for a sustainable future. Springer, Singapore, pp 173–232

Azubuike CC, Chikere CB, Okpokwasili GC (2016) Bioremediation techniques–classification based on site of application: principles, advantages, limitations and prospects. World J Microbiol Biotechnol 32(11). https://doi.org/10.1007/s11274-016-2137-x

Ball A, Truskewycz A (2013) Polyaromatic hydrocarbon exposure: an ecological impact ambiguity. Environ Sci Pollut Res 20(7):4311–4326

Banerjee P, Hazra A, Ghosh P, Ganguly A, Murmu NC, Chatterjee PK (2019) Solid waste management in India: a brief review. In: Ghosh S (ed) Waste management and resource efficiency. Springer, Singapore

Benazir JF, Suganthi R, Rajvel D, Pooja MP, Mathithumilan B (2010) Bioremediation of chromium in tannery effluent by microbial consortia. Afr J Biotechnol 9(21):3140–3143

Bharagava RN, Mishra S (2018) Hexavalent chromium reduction potential of *Cellulosimicrobium* sp. isolated from common effluent treatment plant of tannery industries. Ecotoxicol Environ Safe 147:102–109

Bhatt P, Rene ER, Kumar AJ, Kumar AJ, Zhang W, Chen S (2020) Binding interaction of allethrin with esterase: bioremediation potential and mechanism. Bioresour Technol 315:13845

Bhatt P, Bhatt K, Sharma A, Zhang W, Mishra S, Chen S (2021a) Biotechnological basis of microbial consortia for the removal of pesticides from the environment. Crit Rev Biotechnol 41(3):317–338

Bhatt P, Joshi T, Bhatt K, Zhang W, Huang Y, Chen S (2021b) Binding interaction of glyphosate oxidoreductase and C-P lyase: molecular docking and molecular dynamics simulation studies. J Hazard Mater 5:409:124927

Bhattacharya A, Gupta A, Kaur A, Malik D (2014) Efficacy of Acinetobacter sp. B9 for simultaneous removal of phenol and hexavalent chromium from co-contaminated system. App Microbiol Biotechnol 98(23):9829–9841

Bilal M, Iqbal HM (2020) Microbial bioremediation as a robust process to mitigate pollutants of environmental concern. Case Stud Chem Environ Eng 2:100011

Brown LM, Gunasekera TS, Striebich RC, Ruiz ON (2016) Draft genome sequence of Gordonia sihwensis strain 9, a branched alkane-degrading bacterium. Genome Announc 4:e00622–16

Bwapwa JK, Jaiyeola AT, Chetty R (2017) Bioremediation of acid mine drainage using algae strains: a review. South Afr J Chem Eng 24:62–70. https://doi.org/10.1016/j.sajce.2017.06.005

Cha JS, Cooksey DA (1991) Copper resistance in *Pseudomonas syringae* mediated by periplasmic and outer membrane proteins. Proc Natl Acad Sci USA 88(20):8915–8919

Chen X, Zhao Y, Zhang C, Zhang D, Yao C, Meng Q, Zhao R, Wei Z (2020) Speciation, toxicity mechanism and remediation ways of heavy metals during composting: a novel theoretical microbial remediation method is proposed. J Environl Mngmt 272:111109

Cho O, Choi KY, Zylstra GJ, Kim YS, Kim SK, Lee JH, Sohn HY, Kwon GS, Kim YM, Kim E (2005) Catabolic role of a three-component salicylate oxygenase from *Sphingomonas yanoikuyae* B1 in polycyclic aromatic hydrocarbon degradation. Biochem Biophys Res Commun 327(3):656–662

Chukwuma OB, Rafatullah M, Tajarudin HA, Ismail N (2020) Lignocellulolytic enzymes in biotechnological and industrial processes: a review. Sustainability 12(18):7282. https://doi.org/10.3390/su12187282

Congeevaram S, Dhanarani S, Park J, Dexilin M, Thamaraiselvi K (2007) Biosorption of chromium and nickel by heavy metal resistant fungal and bacterial isolates. J Hazard Mat 146 (1–2):270–277

Das N, Chandran P (2011) Microbial degradation of petroleum hydrocarbon contaminants: an overview. Biotechnol Res Int 1–13. https://doi.org/10.4061/2011/941810

Daverey A, Pakshirajan K (2009) Production of sophorolipids by the yeast *Candida bombicola* using simple and low-cost fermentative media. Food Res Int 42(4):499–504

Debbarma P, Raghuwanshi S, Singh J, Suyal DC, Zaidi MGH, Goel R (2017) Comparative in situ biodegradation studies of polyhydroxybutyrate film composites. 3Biotech 7(178):1–9. https://doi.org/10.1007/s13205-017-0789-3

Desai JD, Banat IM (1997) Microbial production of surfactants and their commercial potential. Microbiol Mol Biol Rev 61(1):47–64

Diaconu M, Pavel LV, Hlihor RM, Rosca M, Fertu DI, Lenz M, Corvini PX, Gavrilescu M (2020) Characterization of heavy metal toxicity in some plants and microorganisms-a preliminary approach for environmental bioremediation. New Biotechnol 56:130–139

Diaz LF, Golueke CG, Savage GM, Eggerth LL (2020) Composting and recycling municipal solid waste. CRC Press

Elshafie AE, Joshi SJ, Al-Wahaibi YM, Al-Bemani AS, Al-Bahry SN, Al-Maqbali D, Banat IM (2015) Sophorolipids production by *Candida bombicola* ATCC 22214 and its potential application in microbial enhanced oil recovery. Front Microbiol 6:1324

Eskandari S, Hoodaji M, Tahmourespour A, Abdollahi A, Mohammadian-Baghi T, Eslamian S (2017) Bioremediation of polycyclic aromatic hydrocarbons by *Bacillus licheniformis* ATHE9 and *Bacillus mojavensis* ATHE13 as newly strains isolated from oil-contaminated soil. J Geogr Environ Earth Sci Int 11:1–11

Ferronato N, Torretta V (2019) Waste mismanagement in developing countries: a review of global issues. Int J Environ Res Public Health 16(6):1060. https://doi.org/10.3390/ijerph16061060

Fondi M, Maida I, Perrin E, Orlandini V, La Torre L, Bosi E, Negroni A, Zanaroli G, Fava F, Decorosi F, Fani R (2016) Genomic and phenotypic characterization of the species *Acinetobacter venetianus*. Sci Rep 6(1):1–12

Gallo G, Piccolo LL, Renzone G, La Rosa R, Scaloni A, Quatrini P, Puglia AM (2012) Differential proteomic analysis of an engineered Streptomyces coelicolor strain reveals metabolic pathways supporting growth on n-hexadecane. Appl Microbiol Biotechnol 94(5):1289–1301

Garcia-Rubio R, de Oliveira HC, Rivera J, Trevijano-Contador N (2020) The fungal cell wall: candida, cryptococcus, and *aspergillus* species. Front Microbiol 10. https://doi.org/10.3389/fmicb.2019.02993

Gargouri B, Karray F, Mhiri N, Aloui F, Sayadi S (2014) Bioremediation of petroleum hydrocarbons-contaminated soil by bacterial consortium isolated from an industrial wastewater treatment plant. J Chem Technol Biotechnol 89(7):978–987

Gavrilescu M (2004) Removal of heavy metals from the environment by biosorption. Eng Life Sci 4 (3):219–232

Ghosal D, Ghosh S, Dutta TK, Ahn Y (2016) Current state of knowledge in microbial degradation of polycyclic aromatic hydrocarbons (PAHs): a review. Front Microbiol 7:1369

Giri K, Rai JPN, Pandey S, Mishra G, Kumar R, Suyal DC (2017a) Performance evaluation of isoproturon-degrading indigenous bacterial isolates in soil microcosm. Chem Ecol 33 (9):817–825. https://doi.org/10.1080/02757540.2017.1393535

Giri K, Suyal DC, Mishra G, Pandey S, Kumar R, Meena DK, Rai JPN (2017b) Biodegradation of isoproturon by *Bacillus pumilus* K1 isolated from foothill agroecosystem of North West Himalaya. Proc Natl Acad Sci India Sect B-Biol Sci 87(3):839–848. https://doi.org/10.1007/s40011-015-0667-x

Goel R, Zaidi MGH, Soni R, Kusumlata, Shouche YS (2008) Implication of A*rthrobacter* and *Enterobacter* species for polycarbonate degradation. Int Biodeterior Biodegrad 61(2):167–172

Gouma S, Fragoeiro S, Bastos AC, Magan N (2014) Bacterial and fungal bioremediation strategies. In: Microbial Biodegradation and Bioremediation, pp 301–323. https://doi.org/10.1016/b978-0-12-800021-2.00013-3

Gudiña EJ, Teixeira JA, Rodrigues LR (2016) Biosurfactants produced by marine microorganisms with therapeutic applications. Mar Drugs 14(2):38

Gunatilake SK (2015) Methods of removing heavy metals from industrial wastewater. Methods 1 (1):14

Guo Q, Dong W, Li S, Lu X, Wang P, Zhang X, Wang Y, Ma P (2014) Fengycin produced by *Bacillus subtilis* NCD-2 plays a major role in biocontrol of cotton seedling damping-off disease. Microbiol Res 169(7–8):533–540

Hakeem KR, Bhat RA, Qadri H (eds) (2020) Bioremediation biotechnology. https://doi.org/10.1007/978-3-030-35691-0

Han H, Cai H, Wang X, Hu X, Chen Z, Yao L (2020) Heavy metal-immobilizing bacteria increase the biomass and reduce the Cd and Pb uptake by pakchoi (*Brassica chinensis* L.) in heavy metal-contaminated soil. Ecotoxicol Environ Safe 195:110375

Hassan MH, Kalam MA (2013) An overview of biofuel as a renewable energy source: development and challenges. Procedia Eng 56:39–53

Higham DP, Sadler PJ, Scawen MD (1986) Cadmium-binding proteins in pseudomonas putida: Pseudothioneins. Environ Health Perspect 65:5–11

Hossan S, Hossain S, Islam MR, Kabir MH, Ali S, Islam MS, Imran KM et al (2020) Bioremediation of hexavalent chromium by chromium resistant bacteria reduces phytotoxicity. Int J Environ Res Pub Health 17(17):6013

Hudson-Edwards KA, Jamieson HE, Lottermoser BG (2011) Mine wastes: past, present, future. Elements 7(6):375–380

Hussaini SZ, Shaker M, Iqbal MA (2013) Isolation of bacterial for degradation of selected pesticides. Adv Biores 4(3):82–85

Hyde KD, Xu J, Rapior S, Jeewon R, Lumyong S, Niego AGT et al (2019) The amazing potential of fungi: 50 ways we can exploit fungi industrially. Fungal Divers. https://doi.org/10.1007/s13225-019-00430-9

Igiri BE, Okoduwa S I, Idoko GO, Akabuogu EP, Adeyi AO, Ejiogu IK (2018) Toxicity and bioremediation of heavy metals contaminated ecosystem from tannery wastewater: a review. J Toxicol 1–16

Ihsanullah I, Jamal A, Ilyas M, Zubair M, Khan G, Atieh MA (2020) Bioremediation of dyes: current status and prospects. J Water Process Eng 38:101680

Ilori MO, Amobi CJ, Odocha AC (2005) Factors affecting biosurfactant production by oil degrading Aeromonas spp. isolated from a tropical environment. Chemosphere 61(7):985–992

Jacoby R, Peukert M, Succurro A, Koprivova A, Kopriva S (2017) The role of soil microorganisms in plant mineral nutrition—current knowledge and future directions. Front Plant Sci 8. https://doi.org/10.3389/fpls.2017.01617

Jafari SA, Cheraghi S, Mirbakhsh M, Mirza R, Maryamabadi A (2015) Employing response surface methodology for optimization of mercury bioremediation by *Vibrio parahaemolyticus* PG02 in coastal sediments of Bushehr, Iran. CLEAN–Soil, Air, Water 43(1):118–126

Jahromi H, Fazaelipoor MH, Ayatollahi S, Niazi A (2014) Asphaltenes biodegradation under shaking and static conditions. Fuel 117:230–235

Jha S, Dikshit S, Pandy G (2011) Comparative study of agitation rate and stationary phase for the removal of Cu^{2+} by *A. lentulus*. Int J Pharm Biol Sci 2:208–211

Joshi-Navare K, Khanvilkar P, Prabhune A (2013) Jatropha oil derived sophorolipids: production and characterization as laundry detergent additive. Biochem Res Int

Joutey NT, Bahafid W, Sayel H, El Ghachtouli N (2013) Biodegradation: involved microorganisms and genetically engineered microorganisms. Biodegradation-Life Sci 289–320

Juwarkar AA, Singh SK, Mudhoo AA (2010) Comprehensive overview of elements in bioremediation. Rev Environ Sci Biotechnol 9:215–288

Kalemelawa F, Nishihara E, Endo T, Ahmad Z, Yeasmin R, Tenywa MM, Yamamoto S (2012) An evaluation of aerobic and anaerobic composting of banana peels treated with different inoculums for soil nutrient replenishment. Bioresour Technol 126:375–382

Kandelbauer A, Guebitz GM (2005) Bioremediation for the decolorization of textile dyes—a review. Environ Chem:269–288

Kanissery RG, Sims GK (2011) Biostimulation for the enhanced degradation of herbicides in soil. App Environ Soil Sci 1–10. https://doi.org/10.1155/2011/843450

Kao CM, Chen CY, Chen SC, Chien HY, Chen YL (2008) Application of in situ biosparging to remediate a petroleum-hydrocarbon spill site: field and microbial evaluation. Chemosphere 70 (8):1492–1499. https://doi.org/10.1016/j.chemosphere.2007.08.029

Kapahi M, Sachdeva S (2019) Bioremediation options for heavy metal pollution. J Health Poll 9 (24):191203. https://doi.org/10.5696/2156-9614-9.24.191203

Karigar CS, Rao SS (2011) Role of microbial enzymes in the bioremediation of pollutants: a review. Enzyme Res:1–11. https://doi.org/10.4061/2011/805187

Khatoon H, Rai JPN, Jillani A (2021) Role of fungi in bioremediation of contaminated soil. In: Fungi bio-prospects in sustainable agriculture, environment and nano-technology, pp 121–156

Kim IH, Choi JH, JooJO KYK, Choi JW, Oh BK (2015) Development of a microbe-zeolite carrier for the effective elimination of heavy metals from seawater. J Microbiol Biotechnol 25 (9):1542–1546

Kügler JH, Roes-Hill L, Syldatk C, Hausmann R (2015) Surfactants tailored by the class actinobacteria. Front Microbiol 6:212

Kumar S (2011) Composting of municipal solid waste. Crit Rev Biotechnol 31(2):112–136

Kumar M, León V, Materano ADS, Ilzins OA, Luis L (2008) Biosurfactant production and hydrocarbon-degradation by halotolerant and thermotolerant *Pseudomonas* sp. World J Microbiol Biotechnol 24(7):1047–1057

Kumar R, Bhatia D, Singh R, Rani S, Bishnoi NR (2011) Sorption of heavy metals from electroplating effluent using immobilized biomass *Trichoderma viride* in a continuous packed-bed column. Int Biodeterior Biodegradation 65(8):1133–1139

Kumaran NS, Sundaramanicam A, Bragadeeswaran S (2011) Adsorption studies on heavy metals by isolated cyano bacterial strain (*Nostoc* sp.) from Uppanar estuarine water, southeast coast of India. J App Sci Res 7(11):1609–1615

Kurt Z, Spain JC (2013) Biodegradation of chlorobenzene, 1,2-dichlorobenzene, and 1,4-dichlorobenzene in the vadose zone. Environ Sci Technol 47(13):6846–6854. https://doi.org/10.1021/es3049465

Lah L, Podobnik B, Novak M, Korošec B, Berne S, Vogelsang M, Kraševec N, Zupanec N, Stojan J, Bohlmann J, Komel R (2011) The versatility of the fungal cytochrome P450 monooxygenase system is instrumental in xenobiotic detoxification. Mole Microbiol 81 (5):1374–1389. https://doi.org/10.1111/j.1365-2958.2011.07772.x

Lee EH, Cho KS (2008) Characterization of cyclohexane and hexanedegradation by Rhodococcus sp. EC1. Chemosphere 71:1738–1744

Li H, Tanikawa T, Sato Y, Nakagawa Y, Matsuyama T (2005) Serratia marcescens gene required for surfactant serrawettin W1 production encodes putative aminolipid synthetase belonging to nonribosomal peptide synthetase family. Microbiol Immunol 49(4):303–310

Li X, Peng W, Jia Y, Lu L, Fan W (2016) Removal of cadmium and zinc from contaminated wastewater using Rhodobacter sphaeroides. Water Sci Technol 75(11):2489–2498. https://doi.org/10.2166/wst.2016.608

Lima AIG, Corticeiro SC, de Almeida EM (2006) Glutathione-mediated cadmium sequestration in *Rhizobium leguminosarum*. Enzyme Microb Technol 39(4):763–769

Lin L, Xu F, Ge X, Li Y (2018) Improving the sustainability of organic waste management practices in the food-energy-water nexus: a comparative review of anaerobic digestion and composting. Renew Sustain Energ Rev 89:151–167

Liu L, Bilal M, Duan X, Iqbal HM (2019) Mitigation of environmental pollution by genetically engineered bacteria—current challenges and future perspectives. Sci Total Environ 667:444–454

Lu P, Hong YF, Hong Q, Jiang X, Li SP (2008) Construction of a stable genetically engineered microorganism for degrading HCH & methyl parathion and its characteristics. Huan Jing Ke Xue 29(7):1973–1976 (Chinese)

Ma YL, Lu W, Wan LL, Luo N (2015) Elucidation of fluoranthene degradative characteristics in a newly isolated Achromobacter xylosoxidans DN002. Appl Biochem Biotechnol 175:1294–1305

Ma M, Du H, Wang D (2019) Mercury methylation by anaerobic microorganisms: a review. Crit Rev Environ Sci Technol 49(20):1893–1936

Machado MD, Janssens S, Soares HMVM, Soares EV (2009) Removal of heavy metals using a brewer's yeast strain of *Saccharomyces cerevisiae*: advantages of using dead biomass. J App Microbiol 106(6):1792–1804. https://doi.org/10.1111/j.1365-2672.2009.04170.x

Machineni L (2019) Review on biological wastewater treatment and resources recovery: attached and suspended growth systems. Water Sci Technol 80(11):2013–2026

Mahmound A, Aziza Y, Abdeltif A, Rachida M (2008) Biosurfactant production by Bacillus strain injected in the petroleum reservoirs. J Ind Microbiol Biotechnol 35(2):1303–1306

Mane PC, Bhosle AB (2012) Bioremoval of some metals by living algae *Spirogyra* sp. and *Spirullina* sp. from aqueous solution. Int J Environ Res 6(2):571–576

Mani S, Chowdhary P, Zainith S (2020) Microbes mediated approaches for environmental waste management. Microorg Sustain Environ Health 17

Massa V, Infantino A, Radice F, Orlandi V, Tavecchio F, Giudici R, Conti F, Urbini G, Barbieri P (2009) Efficiency of natural and engineered bacterial strains in the degradation of 4-chlorobenzoic acid in soil slurry. Int Biodeterior Biodegradation 63(1):112–115

Matsinhe CDC (2011) Studies on the influence of organic waste biodegradability in the composting process (Doctoral dissertation)

Mire CE, Tourjee JA, O'Brien WF, Ramanujachary KV, Hecht GB (2004) Lead precipitation by *Vibrio harveyi*: evidence for novel quorum-sensing interactions. App Environ Microbiol 70 (2):855–864

Mishra A, Malik A (2013) Recent advances in microbial metal bioaccumulation. Crit Rev Environ Sci Technol 43(11):1162–1222

Mnif S, Sayadi S, Chamkha M (2014) Biodegradative potential and characterization of a novel aromatic-degrading bacterium isolated from ageothermal oil field under saline and thermophilic conditions. Int Biodeterior Biodegrad 86:258–264

Mohamed AT, El-Hussein AA, El-Siddig MA, Osman AG (2011) Degradation of oxyfluorfen herbicide by soil microorganisms biodegradation of herbicides. Biotechnol 10(3):274–279
Mohd S, Kushwaha AS, Shukla J, Mandrah K, Shankar J, Arjaria N, Saxena PN et al (2019) Fungal mediated biotransformation reduces toxicity of arsenic to soil dwelling microorganism and plant. Ecotoxicol Environ Safe 176:108–118
Mollea C, Bosco F, Ruggeri B (2005) Fungal biodegradation of naphthalene: microcosms studies. Chemosphere 60(5):636–643
Mónica P, Darwin RO, Manjunatha B, Zúñiga JJ, Diego R, Bryan RB, Maddela NR (2016) Evaluation of various pesticides-degrading pure bacterial cultures isolated from pesticide-contaminated soils in Ecuador. Afr J Biotechnol 15(40):2224–2233
Mukherjee S, Bardolui NK, Karim S, Patnaik VV, Nandy RK, Bag PK (2010) Isolation and characterization of a monoaromatic hydrocarbon degrading bacterium, *Pseudomonas aeruginosa* from crude oil. J Environ Sci Health 45:1048–1053
Muthusamy K, Gopalakrishnan S, Ravi TK, Sivachidambaram P (2008) Biosurfactants: properties, commercial production and application. Curr Sci:736–747
Naidoo S, Olaniran AO (2014) Treated wastewater effluent as a source of microbial pollution of surface water resources. Int J Env Res Public Health 11(1):249–270
Nayak AK, Panda SS, Basu A, Dhal NK (2018) Enhancement of toxic Cr (VI), Fe, and other heavy metals phytoremediation by the synergistic combination of native *Bacillus cereus* strain and *Vetiveria zizanioides* L. Int J Phytoremediation 20(7):682–691
Ndeddy Aka RJ, Babalola OO (2016) Effect of bacterial inoculation of strains of *Pseudomonas aeruginosa*, *Alcaligenes feacalis* and *Bacillus subtilis* on germination, growth and heavy metal (Cd, Cr, and Ni) uptake of *brassica juncea*. Int J Phytoremediation 18(2):200–209
Ngoc UN, Schnitzer H (2009) Sustainable solutions for solid waste management in southeast Asian countries. Waste Manag 29(6):1982–1995
Nguyen HT, Kim Y, Choi JW, Jeong S, Cho K (2021) Soil microbial communities-mediated bioattenuation in simulated aquifer storage and recovery (ASR) condition: long-term study. Environ Res 197:111069. https://doi.org/10.1016/j.envres.2021.111069
Niti C, Sunita S, Kamlesh K, Rakesh K (2013) Bioremediation: an emerging technology for remediation of pesticides. Res J Chem Environ 17(4)
Noor E, Eden E, Milo R, Alon U (2010) Central carbon metabolism as a minimal biochemical walk between precursors for biomass and energy. Mol Cell 39(5):809–820
Nzila A, Razzak S, Zhu J (2016) Bioaugmentation: an emerging strategy of industrial wastewater treatment for reuse and discharge. Int J Environ Res Public Health 13(9):846. https://doi.org/10.3390/ijerph13090846
Ojuederie O, Babalola O (2017) Microbial and plant-assisted bioremediation of heavy metal polluted environments: a review. Int J Environ Res Public Health 14(12):1504. https://doi.org/10.3390/ijerph14121504
Okoduwa SIR, Igiri B, Udeh CB, Edenta C, Gauje B (2017) Tannery effluent treatment by yeast species isolates from watermelon. Toxics 5(1):6
Parween T, Bhandari P, Sharma R, Jan S, Siddiqui ZH, Patanjali PK (2017) Bioremediation: a sustainable tool to prevent pesticide pollution. In: Modern age environmental problems and their remediation, pp 215–227. https://doi.org/10.1007/978-3-319-64501-8_12
Passarini MR, Rodrigues MV, da Silva M, Sette LD (2011) Marine-derived filamentous fungi and their potential application for polycyclic aromatic hydrocarbon bioremediation. Mar Pollut Bull 62(2):364–370
Patel S, Homaei A, Patil S, Daverey A (2019) Microbial biosurfactants for oil spill remediation: pitfalls and potentials. Appl Microbiol Biotechnol 103(1):27–37
Peñaloza-Vazquez A, Mena GL, Herrera-Estrella L, Bailey AM (1995) Cloning and sequencing of the genes involved in glyphosate utilization by *Pseudomonas pseudomallei*. Appl Environ Microbiol 61(2):538–543
Quansah A, Ntaryamira T, Rwemera J (2018) Sludge wastewater management by conventional treatment process: case study-Bujumbura municipal sewage. Int J Sci 7:52–65

Ramasamy K, Kamaludeen S, Parwin B (2006) Bioremediation of metals microbial processes and techniques. In: Singh SN, Tripathi RD (eds) Environmental bioremediation technologies. Springer, New York, pp 173–187

Ran ZHAO, Bi WANG, Cai QT, Li XX, Min LIU, Dong HU, Chun FAN (2016) Bioremediation of hexavalent chromium pollution by *Sporosarcinasaromensis* M52 isolated from offshore sediments in Xiamen, China. Biomed Environ Sci 29(2):127–136

Ravi RK, Bhawana P, Fulekar MH (2015) Bioremediation of persistent pesticides in rice field soil environment using surface soil treatment reactor. Int J Curr Microbiol Appl Sci 4(2):359–369

Ren N, Wang A, Gao L, Xin L, Lee DJ, Su A (2008) Bioaugmented hydrogen production from carboxymethyl cellulose and partially delignified corn stalks using isolated cultures. Int J Hydrogen Energy 33(19):5250–5255. https://doi.org/10.1016/j.ijhydene.2008.05.020

Salehizadeh H, Shojaosadati SA (2003) Removal of metal ions from aqueous solution by polysaccharide produced from *Bacillus firmus*. Water Res 37(17):4231–4235

Sales da Silva IG, Gomes de Almeida FC, Padilha da Rocha e Silva NM, Casazza AA, Converti A, Asfora Sarubbo L (2020) Soil bioremediation: overview of technologies and trends. Energies 13 (18):4664. https://doi.org/10.3390/en13184664

Samer M (2015) Biological and chemical wastewater treatment processes. In: Waste water treatment engineering, pp 150

Samin G, Pavlova M, Arif MI, Postema CP, Damborsky J, Janssen DB (2014) A *Pseudomonas putida* strain genetically engineered for 1,2,3-Trichloropropane bioremediation. App Environ Microbiol 80(17):5467–5476. https://doi.org/10.1128/aem.01620-14

Samstag RW, Ducoste JJ, Griborio A, Nopens I, Batstone DJ, Wicks JD, Laurent J (2016) CFD for wastewater treatment: an overview. Water Sci Technol 74(3):549–563

Sandoval Á, Arias-Barrau E, Bermejo F, Cañedo L, Naharro G, Olivera ER, Luengo JM (2005) Production of 3-hydroxy-n-phenylalkanoic acids by a genetically engineered strain of pseudomonas putida. Appl Microbiol Biotechnol 67(1):97–105

Sharma PK, Balkwill DL, Frenkel, Vairavamurthy MA (2000) A new *Klebsiellaplanticola* strain (Cd-1) grows anaerobically at high cadmium concentrations and precipitates cadmium sulfide. App Environ Micro 66(7):3083–3087

Shuaib M, Azam N, Bahadur S, Romman M, Yu Q, Xuexiu C (2021) Variation and succession of microbial communities under the conditions of persistent heavy metal and their survival mechanism. Micro Patho 150:104713

Singh R (2013) Microbial waste management. Lap lambert Academic Publishing

Singh JS, Abhilash PC, Singh HB, Singh RP, Singh DP (2011) Genetically engineered bacteria: an emerging tool for environmental remediation and future research perspectives. Gene 480 (1–2):1–9

Singh N, Hui D, Singh R, Ahuja IPS, Feo L, Fraternali F (2017) Recycling of plastic solid waste: a state of art review and future applications. Compos Part B Eng 115:409–422

Sonune A, Ghate R (2004) Developments in wastewater treatment methods. Desalination 167:55–63

Sood N, Patle S, Lal B (2009) Bioremediation of acidic oily sludge-contaminated soil by the novel yeast strain Candida digboiensis TERI ASN6. Environ Sci Pollut Res 17(3):603–610. https://doi.org/10.1007/s11356-009-0239-9

Srivastava AK, Srivastava M, Kashyap PL, Srivastava AK (2016) Prospects of biocomposting in organic farming and environment management. In: Gupta RK, Singh SS (eds) Environmental biotechnology: a new approach. Environmental biotechnology: a new approach. Daya Publishing House, New Delhi, pp 147–178

Strong LC, McTavish H, Sadowsky MJ, Wackett LP (2000) Field-scale remediation of atrazine-contaminated soil using recombinant Escherichia coli expressing atrazine chlorohydrolase. Environ Microbiol 2(1):91–98

Sutherland DL, Ralph PJ (2019) Microalgal bioremediation of emerging contaminants – opportunities and challenges. Water Res 114921. https://doi.org/10.1016/j.watres.2019.114921

Suyal DC, Soni R, Singh DK, Goel R (2021) Microbiome change of agricultural soil under organic farming practices. Biologia 76:1315–1325

Tabatabaee A, Assadi MM, Noohi AA, Sajadian VA (2005) Isolation of biosurfactant producing bacteria from oil reservoirs. J Environ Health Sci Eng 2(1):6–12

Tang X, Huang Y, Li Y, Wang L, Pei X, Zhou D, He P, Hughes SS (2021) Study on detoxification and removal mechanisms of hexavalent chromium by microorganisms. Ecotoxicol Environ Safe 208:111699

Toribio J, Escalante AE, Soberón-Chávez G (2010) Rhamnolipids: production in bacteria other than Pseudomonas aeruginosa. Eur J Lipid Sci Technol 112(10):1082–1087

Tripathi N, Singh RS, Hills CD (2019) Microbial removal of sulphur from petroleum coke (petcoke). Fuel 235:1501–1505

Tyagi M, da Fonseca MMR, de Carvalho CCCR (2010) Bioaugmentation and biostimulation strategies to improve the effectiveness of bioremediation processes. Biodegradation 22 (2):231–241. https://doi.org/10.1007/s10532-010-9394-4

Ubando AT, Africa ADM, Maniquiz-Redillas MC, Culaba AB, Chen WH, Chang JS (2021) Microalgalbiosorption of heavy metals: a comprehensive bibliometric review. J Hazard Mt 402:123431

Varjani SJ (2017) Microbial degradation of petroleum hydrocarbons. Bioresour Technol 223:277–286

Vijayaraghavan K, Yun YS (2008) Bacterial biosorbents and biosorption. Biotech Adv 26 (3):266–291

Villaverde J, Láiz L, Lara-Moreno A, González-Pimentel JL, Morillo E (2019) Bioaugmentation of PAH-contaminated soils with novel specific degrader strains isolated from a contaminated industrial site effect of hydroxypropyl-β-cyclodextrin as PAH bioavailability enhancer. Front Microbiol 10. https://doi.org/10.3389/fmicb.2019.02588

Volke-Sepúlveda TL, Gutiérrez-Rojas M, Favela-Torres E (2003) Biodegradation of hexadecane in liquid and solid-state fermentations by *Aspergillus niger*. Bioresour Technol 87(1):81–86

Wang CL, Ozuna SC, Clark DS, Keasling JD (2002) A deep-sea hydrothermal vent isolate, *Pseudomonas aeruginosa* CW961, requires thiosulfate for Cd^{2+} tolerance and precipitation. Biotech Lett 24(8):637–641

Wang XB, Chi CQ, Nie Y, Tang YQ, Tan Y, Wu G (2011) Degradation of petroleum hydrocarbons (C6–C40) and crude oil by a novel Dietzia strain. Bioresour Technol 102:7755–7761

Whang LM, Liu PWG, Ma CC, Cheng SS (2008) Application of biosurfactants, rhamnolipid, and surfactin, for enhanced biodegradation of diesel-contaminated water and soil. J Hazard Mater 151(1):155–163

White DA, Hird LC, Ali ST (2013) Production and characterization of a trehalolipid biosurfactant produced by the novel marine bacterium Rhodococcus sp., strain PML026. J Appl Microbiol 115(3):744–755

Wu HK, Zhang X, Shao H, Chu L, Ruan C (2010) A critical review on the bio-removal of hazardous heavy metals from contaminated soils: issues, progress, eco-environmental concerns and opportunities. J Hazard Mat 174(1–3):1–8

Xu X, Liu W, Tian S, Wang W, Qi Q, Jiang P, Gao X, Li F, Li H, Yu H (2018) Petroleum hydrocarbon-degrading bacteria for the remediation of oil pollution under aerobic conditions: a perspective analysis. Front Microbiol 9. https://doi.org/10.3389/fmicb.2018.02885

Yang J, Liu R, Song W, Yang Y, Cui F, Qiao C (2012) Construction of a genetically engineered microorganism that simultaneously degrades organochlorine and organophosphate pesticides. Appl Biochem Biotechnol 166(3):590–598

Yang T, Chen M, Wang J (2015) Genetic and chemical modification of cells for selective separation and analysis of heavy metals of biological or environmental significance. Trend Analyt Chem 66:90–102

Youssef N, Simpson DR, Duncan KE, McInerney MJ, Folmsbee M, Fincher T, Knapp RM (2007) In situ biosurfactant production by bacillus strains injected into a limestone petroleum reservoir. Appl Environ Microbiol 73(4):1239–1247

Yuan X, Xue N, Han Z (2021) A meta-analysis of heavy metals pollution in farmland and urban soils in China over the past 20 years. J Environ Sci 101:217–226

Zeriouh H, Romero D, García-Gutiérrez L, Cazorla FM, de Vicente A, Pérez-García A (2011) Thc iturin-like lipopeptides are essential components in the biological control arsenal of *Bacillus subtilis* against bacterial diseases of cucurbits. Mol Plant-Microbe Interact 24(12):1540–1552

Zhang L, Zhang C, Cheng Z, Yao Y, Chen J (2013) Biodegradation of benzene, toluene, ethylbenzene, and o-xylene by the bacterium Mycobacteriumcosmeticumbyf-4. Chemosphere 90:1340–1347

Chapter 3
Microbial Fuel Cells for Wastewater Treatment

Prem Ranjan, Damini Maithani, Deep Chandra Suyal, Anup Kumar Singh, Krishna Giri, Vijay Kumar Sharma, and Ravindra Soni

3.1 Introduction

Waste generation became a global issue that needs to be managed through effective and eco-friendly approaches (Debbarma et al. 2017; Bhatt et al. 2021a, b). Microbial Fuel Cells (MFCs) offer a promising and testing innovation intending to energy and ecological issues such as bioremediation and wastewater treatment especially in far-off zones, furnished with biosensors, biohydrogen creation (Venkata Mohan et al. 2014; Gude 2016; Mercuri et al. 2016). Primarily, benefits that make microbial fuel cells more practical when executed in wastewater treatment include the immediate transformation of waste products energy to power, amenability to control and engineer, low carbon footprints, less resource consumption, no gas treatment, and no energy contribution for air circulation.

Prem Ranjan and Damini Maithani contributed equally with all other contributors.

P. Ranjan · A. K. Singh · R. Soni (✉)
Department of Agricultural Microbiology, College of Agriculture, Indira Gandhi Krishi Vishwavidyalaya, Raipur, Chhattisgarh, India

D. Maithani
Department of Microbiology, G.B. Pant University of Agriculture and Technology, Pantnagar, Uttarakhand, India

D. C. Suyal
Department of Microbiology, Akal College of Basic Sciences, Eternal University, Sirmaur, Himachal Pradesh, India

K. Giri
Rain Forest Research Institute, Jorhat, India

V. K. Sharma
Department of Post-harvest Science, Agricultural Research Organization – Volcani Centre, Rishon LeZion, Israel

© The Author(s), under exclusive license to Springer Nature Switzerland AG 2022

D. C. Suyal, R. Soni (eds.), *Bioremediation of Environmental Pollutants*,
https://doi.org/10.1007/978-3-030-86169-8_3

Although the use of microbial fuel cells is still in infancy, their concepts are continuously being targeted to develop modern methods and technologies that involve electron trading (Singh et al. 2010; Fraiwan and Choi 2016). Progress has been made in using MFCs to be used in environmental cleanup strategies including treatment of soil, wastewater, greenhouse gas emission, and environmental monitoring, etc. (Wu et al. 2020).

Although MFC offers an economical solution by converting waste into useful energy its implications on practical grounds are still limited to laboratories and research works. Thus, work is being carried out intensively on appropriate designing of MFCs to optimize its performance for environmental cleanup and power generation (Lovley 2006; Logan and Rabaey 2012; Kumar et al. 2012). The use of MFC at a commercial scale for wastewater reclamation and power generation is a primary stage and requires a lot more effort to expand their use beyond laboratories (Yang et al. 2011). Constraints like turbulence and non-uniform mixing in compartments, low power generation than the desired amount, cost-effectiveness, and detailed elucidation of electrode-electricigen interactions need to be looked into for achieving the aforementioned goals.

3.2 Characteristics of the MFCs

Based on its design MFCs can be a single chamber or two-chamber type (Pant et al. 2010). A two-chambered MFC consists of two electrode chambers separated via a proton exchange membrane (PEM). Electrons and protons are generated at anode via oxidation of organic substrates in presence of anaerobic microorganisms. Electrons are then traded towards the cathode through the external circuit, whereas, protons are conveyed to the cathode chamber via PEM. On the other hand, in a single chamber MFC there is no separating membrane in-between anode and cathode whereas the basic operating principle remains the same. The regular MFC is ordinarily half organic since just the anode chamber contains biocatalysts whereas the cathode chamber is abiotic, primarily consisting of terminal electron acceptors including, O_2, sulfate, ferricyanide, nitrate, and H_2O_2 which are reduced at the cathode. Oxygen is considered the most appropriate terminal electron acceptor (TEA) out of the aforementioned electron acceptors. Current generated in an MFC is measured either via a voltmeter or potentiostat attached to a computer (Logan 2008).

The air-cathode regularly comprises of impetus layer, terminal, and separator. Plus, separator assumes an inexorably fundamental part in MFCs as contrasted and synergist and anode materials, which builds the inward obstruction as well as diminishes the MFCs' exhibitions, along these lines seriously restrict the useful utilizations of MFCs. Also, numerous past endeavors have zeroed in for the most part on cutting edge abiotic impetuses materials for improving oxygen decrease responses, yet their application in MFCs is still exorbitant, generally coming about because the force and current densities in MFCs are moderately low in contrast to customary power devices. In correlation, biocathodes can be chosen as an appealing option instead of abiotic impetuses, where microbes are utilized on modest carbon cathodes. Although biocathodes are better than abiotic cathodes, very limited reports are available on their instrumentation and design. It has been observed that the application of MFCs has been increased significantly in the wastewater treatment process (Butti et al. 2016). Exhibition of MFCs with blended immunization would be superior to that with unadulterated culture vaccination in terms of force thickness and pollutants evacuation proficiency and would be effectively superior and advantageous for wastewater treatment with minimal expense and would enhance its commercialization as well.

3.3 Types of Microbial Fuel Cells

3.3.1 Single Chambered MFC or Air-Cathode MFC

It is an advanced single-chambered MFC that comprises five sections, *viz*., proton trade film, impetus layer, cathode, dispersion layer just as a defensive layer. An air-cathode MFC is outfitted with the anode and made up of carbon paper without wet sealing (Liu and Logan 2004). Further, cathodes are made dependent either on an independent unbending carbon paper lacking a proton layer, or a carbon proton trade/anode/film gathering created by holding the proton trade film straightforwardly onto an adaptable carbon-fabric terminal (Du et al. 2007). Liu et al. (2004) presented a solitary barrel-shaped Plexiglas chamber, made basically out of eight graphite poles as anode orchestrated in concentric circles around a solitary cathode. The air-permeable cathode that contains a platinum/carbon trade layer is incorporated inside a plastic tube. Further, Rabaey and Verstraete (2005) have shown that graphite-dependent cathode (external) and anode (internal) are used to make rounded MFC frameworks. Moreover, with the help of a fixed cathode and air-exposed anode, an SCMFC (solitary chambered microbial energy) unit was developed by Mohan et al. (2008).

3.3.2 Two-Chambered MFC

These are made by anode and cathode along with the particle trade layer around them. This layer does not affect the air circulation, energy reduction, and accumulation rate of the oxidants at the respective electrode. Although, MFCs are not as popularly used in business applications, but offer enormous scope for treating wastewater.

For instance, Min et al. (2005a) progressed a salt scaffold MFC to analyze power yield and demonstrated an expected use of MFCs for power generation from high-starch containing compounds and sewage. Similarly, a UMFC (up-stream microbial energy) unit was developed by He et al. (2006) through acquainting the cathode inside to accomplish improved execution of MFC for persistent treatment. It complements the UASB unit framework by fulfilling the requirement for a double chamber MFC. Also, the particle trade layer shared the greatest expense and decided its dependability for long-haul activity. Further, a special film less MFC was utilized by Jang et al. (2004) to diminish the construction expenses. Such designs were found more beneficial for environmental pollutant monitoring and analysis.

3.3.3 Benthic MFC (BMFC)

Benthic microbial power devices or residue microbial energy components are known as the frameworks to create power dependent on the electro potential contrasts existing between the oxic seawater and anoxic sediments (Li et al. 2008). During the time spent in BMFCs, microorganisms drive the electrons towards the anode either installed in or laid on top of the dregs. At the cathode, which is suspended in seawater, electrons are utilized for the reduction of free oxygen to water (Donovan et al. 2011). As of late, a large number of setups of BMFCs have been effectively proposed for their improved power generation capacity. Their unique plans for the most part comprise straightforward plates of graphite covered in the dregs along with water suspended cathode, yet discoveries have exhibited that such plans were exceptionally delicate and their connected force age was amazingly low. Further, a chamber-based BMFC was proposed by Nielsen et al. (2007) that has shown higher energy up to 3.8 W/m^3 (380 mW/m^2).

3.3.4 Stacked MFC (SFMC)

Voltage yields of an overall MFC are always less than hypothetical voltages, even in a specific circumstance of dismissing the inward misfortunes. However, an arrangement or equal SMFC is thought to be a productive answer for expanding the voltages. Further, the associated microbial energy unit can be arranged into two gatherings: arrangement and equal association. For example, multiple units containing SMFC were developed by Aelterman et al. (2006) that consist of graphite electrode terminals. Among these, one electrode was interfacing towards the outer electronic circuit. It leads to the conclusion that the improvement of SMFCs either in arrangement or equal could have an incredible capacity of creating higher force densities. In any case, overpotential wonder has likewise been noticed, what part of the way restricted the current thickness from expanding in the arrangement association. Moreover, Shimoyama et al. (2008) planned a novel exceptionally versatile MFC consisting of a bunch of tape terminals having two-sided air cathodes for forestalling water and/or air spillage. The cathode box was then covered using a proton film on each side. Stacked MFCs show the course type framework by sequentially associating non-compartmented microbial energy components

(NCMFCs) (Seo et al. 2009), which was utilized for taking out the need of air circulation and outside energy utilization. A sun-based cell was associated with NCMFCs for enacted power age (Zhuang and Zhou 2009). It was developed by using a PVC tube having equally distant pores and an anode chamber. Furthermore, a cylindrical layer cathode was built by using hot pressing carbon materials followed by folding over the external surface of the anode chamber. In the SCS-MFC framework, two extraordinary implanted highlights were thought about including a serpentine stream design that permitted a high anode surface territory concerning the reactor volume. The another contains various MFC units worked either independently or in stacked modes under similar conditions (Wang and Han 2009).

3.4 Principles of Waste Management by the Use of Microbial Fuel Cells

The type of substrate utilized in an MFC is a major factor for power generation. Thus, a variety of substrates ranging from pure substrates to a mixture of organic compounds are used in MFCs (Pant et al. 2010). Pollutants discharged from domestic, industrial, and agriculture sectors are rich in these organic compounds. Thus, utilizing them as substrate can serve a dual purpose i.e. energy generation and environmental cleanup (Singh et al. 2010).

An MFC behaves like a galvanic cell generating exergonic electrochemical responses having negative free-response energy and this returns precipitously with energy discharge. The standard free energy can without much of a stretch be changed over into a standard cell voltage (or electromotive power, emf) ΔE0 as demonstrated in this equation as

$$\Delta E^0 = -\left[\sum v_i \Delta G_i{}^0{}_{\text{, products}} - \sum v_i \Delta G_i{}^0\text{, reactants}\right]/nF = -\Delta G/nF$$

Here, the ΔG^0 addresses the free energies of arrangement reactants and other components (J per mol) and n (moles) addresses the stoichiometry variables of the redox response, and F stands for Faraday's steady (96,485.3 C per mol). ΔG^0 of a response estimates the greatest measure of valuable work that can be acquired from a response of thermodynamic framework. The hypothetical cell voltage or electromotive power (emf) of the general response decides whether the framework is equipped for power age, the equation can be presented as

$$\Delta \text{Ecell}^0 = \Delta E^0{}_{\text{cathode}} - \Delta E^0{}_{\text{anode}}$$

The negative free-response energy prompts a positive-standard voltage of the cell. It, therefore, differentiates an electrolysis cell from the galvanic cells. The prior cell type has positive free-response energy along with negative cell voltage. In an MFC, ΔG^0 of the response comes negative along with positive emf. It shows their potential for unconstrained power age from the response. For instance, if acetic acid is utilized as the natural substrate equation can be represented as:

$$\text{At Anode}: CH_3COO - +4H_2O \rightarrow 2HCO_3 - +9H^+ + 8e^-$$

$$\text{At Cathode}: 2O_2 + 8H^+ + 8e^- \rightarrow 4H_2O$$

Overall reaction: $CH_3COO^- + 2O_2 \rightarrow 2HCO_3^- + H^+$ ($\Delta G = -847.60$ kJ/mol; emf $= 1.094$ V) (Rozendal et al. 2008)

3.5 Advantages of MFCs

- MFCs offer a few energy (direct power age, energy investment funds by anaerobic treatment because of end of air circulation, low ooze yield, and unified and decentralized applications); environmental (water recovery, low carbon impression, lower muck volumes for removal); financial (income through energy and worth added items synthetics, low operational expenses, kill downstream cycles) and operational advantages (self-age of microorganisms, great protection from ecological pressure, and amiable to continuous checking and control) (Li et al. 2014).
- They produce clean power straightforwardly from wastewater with no requirement for detachment, purging, and transformation of the energy items. In the examination, methane and hydrogen can be created from anaerobic absorption measures which require division and decontamination before their utilization.
- MFCs are harmless to the ecosystem innovations since they can create clean power straightforwardly and work at gentle working conditions particularly at encompassing temperatures.
- These can make a maximum of 1430 Wh/m^3 from an essential ooze or 1800 Wh/m^3 from the treated profluent. MFCs devour just 24 W or 76 Wh/kg-COD in normal conditions (primarily for taking care of and blending in the reactor), around one significant degree not exactly initiated slime based vigorous cycles (~300 W or 600 Wh/kg-COD) (Ge et al. 2013; Zhang et al. 2013).
- MFCs devour just around 10% of the outside energy for their activity when contrasted and ordinary actuated slime measures showing extraordinary potential for energy investment funds as well as could be expected energy recuperation from sewage treatment processes.
- The yield of exo-electrogenic microbes (0.07e0.16 gVSS per gCOD) is considerably lesser than the actuated slop (0.35e0.45 gVSS per gCOD) and thus, diminishes slime creation fundamentally. Ooze the board is a major challenge for the sewage treatment plants. A few investigations have reported substitute ways to gather the energy substance of the sewage (Rittmann and McCarty 2001).
- Organic matter present in the wastewater is considered to be Glucose ($C_6H_{12}O_6$), net energy yield and ΔG^0 are presented in kJ per mol of the glucose molecule.

Furthermore, hydrogen, methane, ethanol, and others are used to steer the electrons and showed observable effects in the cell. Energy change misfortunes are prevalent in the burning and power modules innovations.

- MFCs recuperate a lot higher energy through power creation from different substrates. While thermodynamic examination presents an ideal viewpoint on MFCs, the real cycle execution and reasonable issues related to electron collecting components keep on being significant concerns that should be tended to in future MFC aim.

3.6 MFCs in Wastewater Management

MFCs can perform dual functions leading to the cleanup of wastewater and simultaneous power generation. Wastewaters from dairy, agriculture, dye industries, breweries, dairy, etc. are used as substrates for power generation in MFCs. Wastewater contains more energy than is required to treat it. Energy is locked in the form of thermal energy, organic compounds, and elements such as nitrogen and phosphorous (Gude 2016).

Civil or homegrown wastewater consists of feedstock and pollutants in a good amount which can be utilized in MFCs for power age given the diminished harmfulness (Ren et al. 2016). Also, the initiated slime from traditional high-impact treatment measures is incredibly huge (Tee et al. 2016). The costs for wastewater treatment normally share over half of the all-out administration ventures (Xiao et al. 2011). Therefore, expenses for wastewater treatment can be minimized by diminishing the muck treatment costs. Also, the slime frequently contains undeniable degrees of organic content (approximately, 66%) from wastewater treatment plants (Ting and Lee 2007).

Agriculture put more accentuation on animal excrements wastewater, which has been reported to be the essential component of the farm waste (Zornoza et al. 2016). Huge volumes of farm's waste are associated with domesticated animals. It is estimated that the US alone generates approximately, 5.8×10^7 t of composts yearly. Wastewater should be blessed to receive meet release guidelines for staying away from water defilement and smell issues (Lu et al. 2009). Water contamination likewise results from surface water eutrophication which causes high groupings of nitrates and phosphate in wastewater (He et al. 2010). In a study conducted by Zhang et al. (2020) mustard tuber wastewater with high salt concentration was treated in an MFC with a single chamber. Thus carbon, nitrogen removal, and power generation were achieved at the same time.

Wastewater from food processing units is different from that of municipal wastewater in that it is non-toxic and readily degradable. However, it has high amounts of suspended solids and high biological oxygen demand (BOD). Wastewater released from industries utilizing molasses as raw material is one of the most polluted wastewaters as it is high-strength water with high COD and is moderately acidic (Sharma and Malaviya 2016). Such water needs to be treated before discharge into the environment. In a study conducted by Zhang et al. (2009) an attempt was made for electricity generation from molasses wastewater and treating it simultaneously by combining MFC and conventional approaches. This led to a maximum power generation of 1410.2 mW/m^2.

On the off chance that high CODs were not thought about, wastewater from breweries with a lot of natural matter can be picked as a reasonable substrate for organisms (Lu et al. 2009). Another main food-based enterprise is the dairy industry. Wastewater produced from dairy industries is rich in organic matter including protein, complex sugars, lipids, etc. It has been estimated that if dairy wastewater produced annually is used for electricity generation, the revenue of around $\$300 \times 10^6$ per annum can be generated (Venkata Mohan et al. 2010).

Various mechanical wastewaters, for the most part from the color assembling and material industry, incorporate extensively engineered substance colors (Cheng et al. 2016). Endeavors have been made to examine the decolorization limit of MFCs for specific colors. By and by, scarcely any examination was applied into genuine color wastewater. A constructed-wetland MFC was built by Fang et al. (2015) to generate a power of about 0.852 W/m^3. Moreover, it had achieved up to 95.6% of decolorization of wastewater polluted with azo dyes. Similarly, Kalathil et al. 2011) had developed a granular carbon-based MFC to treat and explore the dye-containing wastewater for making bioelectricity.

Wastewater in cathode chamber was essentially avirulent and non-poisonous and the harmfulness of anode emanating was likewise significantly decreased when contrasted with unique color wastewater after a 48-h which generated force thickness of 0.0017 KW/m^3 and voltage of 0.45 V. Critical enhancements in water treatment potential and power thickness have been explored in recent years.

3.7 Redox Reactions in MFCs (Metabolic Mechanism)

Toxins in the wastewater, for example, natural compounds and metals can be explored to deliver the power directly and precisely. Power creation is the aftereffect of Oxidation-reduction responses in MFCs cause electron delivery, movement, and acknowledgment through electrical and/or biochemical reactions at both electrode terminals. Among these one acts as electron donor and another as electron acceptor. Oxidation-decrease responses present conceivable bio electro-compound responses in microbial power devices creating power using polluted water as an electron contributor and different contaminations like nitrates, phosphates, and others as electron acceptors.

3.7.1 *Metabolism in the Anode Chamber*

In the past investigations, exogenous go-betweens were frequently added to produce a specific measure of force, which restricts their applications in the MFC due to their harmfulness and precariousness. Notwithstanding, this obstruction has been penetrated lately. Discoveries have affirmed that a normally existing local area of microorganisms could create power without thought of exogenous arbiters. Likewise, several reports have revealed that the biodegradation power of the microorganisms can be explored for making movements of the electrons in MFC. The important microorganisms are *Rhodopseudomonas palustris*, *Geobacter* spp., *Ochrobactrum* sp., *Pseudomonas* spp., *Shewanella* spp., *Chlorobium limicola, Arcobacter* sp., *Clostridium* sp., *Rhodofoferax* sp. *Chlamydomonas reinhardtii* and *Rhodobacter sphaeroides*. Moreover, the mixed cultures have shown higher efficiency of electron trading. Many investigations of "photograph MFCs" containing photosynthetic microorganisms have shown their potential in electron movement among the electrodes (Azwar et al. 2014). Besides, the variety of electrochemically dynamic microorganisms made MFCs ready to manage diverse wastewaters, a lot of which can scarcely be treated by the set up natural treatment strategies. For it, monosaccharides, unstable unsaturated fats and glycerol, adaptable food materials, cellulosic materials, and other natural polymeric natural substances can be explored (Rozendal et al. 2008). Under this thought, syntrophic collaborations ought to be genuinely considered to separate complex natural matter into less difficult substrates utilized by exoelectrogen (Logan and Rabaey 2012). As per a new report, MFCs seemed, by all accounts, to be more suitable for the debasement of medium and low strength wastewaters as an outcome of the presence of high part of fermentative as well as particulate materials that can barely be utilized by the potential microorganisms (Li et al. 2014).

Some common reactions occurring at anode chambers are as follows:

(a) When Glucose is used as substrate: $C_6H_{12}O_6$+ $12H_2O \rightarrow 6HCO_3^-$ + $3OH^+$+ $24e^- E^0 = -0.429$ V
(b) When Glycerol is used as substrate: $C_3H_8O_3$+ $6H_2O \rightarrow 3HCO3^-$ + $17H^+$+ $14e^- E^0 = -0.289$ V
(c) When malate is used as substrate: $C_4H_5O_5^-$ + $7H_2O \rightarrow 4H_2CO_3$+ $11H^+$+ $12e^- E^0 = -0.289$ V
(d) sulfur compounds as substrates: $HS^- \rightarrow S^0$+ H^++ $2e^- E^0 = -0.230$ V

3.7.2 *Metabolism in Cathode Chamber*

3.7.2.1 Abiotic Cathodes

Continuous impetuses or fake electrons are required for speeding up the oxidation-decrease response rates in the cathode on the outside of anodes. Reduction reactions that take place at the cathode are a major determinant in the current generation. But MFCs suffer cathodic limitations that affect the overall performance of MFCs. These include activation losses that occur due to activation energy barrier, internal resistances, mass transport losses, etc. Platinum with superb synergist capacity is the most suggested impetus for oxygen decrease in MFCs (Sivasankaran et al. 2016), yet its application is restricted significantly because of its over-the-top expense and harmfulness (Matesanz et al. 2016). To improve the performance of cathode certain mediators such as ferricyanide, are used that have a faster rate of reduction in comparison to oxygen thus preventing activation losses. Numerous analysts found capital expense as well as harmfulness to be lessened using metal oxides (like manganese (IV) dioxide, lead oxide, etc.) directly incorporated in cathodes. Exceptional morphologies and arrangement methods of macrocyclic compounds of metals are professed to be huge elements on oxygen decrease action. The oxidation-decrease response rates movement of macrocyclic compounds is impacted by the middle metallic particle and the ligand type. The most noteworthy oxidation-decrease response rates are found when cobalt (Co) or iron (Fe) are explored as the central metallic particle (Kargi and Eker 2007). Internal resistance can be reduced by reducing the spacing between two electrodes. Carbon nanotube and nanofiber, fundamentally because of their special electrical and mechanical properties, and electro-conductive polymers are used. The natural steadiness and high conductivity of the polymers make them potent for electro-catalysis purposes and other related applications (Dallas and Georgakilas 2015).

3.7.2.2 Biocathodes

Biocathodes, unlike conventional abiotic cathodes, utilizes microorganisms that help to carry out reactions. These can be grouped into high-impact and anaerobic biocathodes as indicated by the states of the cathodic chamber. As far as vigorous biocathodes, oxygen ordinarily fills in as the TEA, and both iron and manganese are first abiotically reduced and then oxidized again by microorganisms. However, in anaerobic biocathodes, TEAs are reduced directly through electrons generated through microbial metabolism. The benefit of biocathode with an anaerobic environment generally lies in the disposal of oxygen dissemination into the anode by means of the proton trade layer, which can conveniently forestall the deficiency of electrons for oxygen molecules. The biocathodes are considered more advantageous than abiotic ones because of two possible reasons. From one viewpoint, the MFC's supportability could be successfully improved after its utilization and secondly, the microbial digestion in biocathodes might be used to create alluring items or eliminate undesirable mixtures. Furthermore, they are used for wastewater treatment and removing nitrates, the expulsion of weighty metals, and sulfates. With the exception of studies on metabolic instruments and biocathodes, some centered around progress metal mixtures utilized as electron go-between, like Mn or Fe. These electron arbiters are first and foremost decreased by tolerating electrons from the cathodic terminal and subsequently oxidized with the help of oxidizing microscopic organisms, for example, *Pseudomonas fluorescens, Leptothrix discophora, Chlorella vulgaris, Thiobacillus ferrooxidans.* The microorganisms with photosynthetic abilities have additionally been fundamentally researched in cathode because of their highlights of oxygen supply, carbon dioxide catch, important biomass creation, as well as wastewater treatment at the same time (Xiao and He 2014).

3.7.2.3 Reaction in Biocathode Chamber (Reduction Reaction)

$$O_2 + 4H^+ + 4e^- \rightarrow 2H_2O\ E^0 = +1.230\ V$$

$$O_2 + 2H^+ + 2e^- \rightarrow H^2O^2E^0 = +0.269\ V$$

$$NO_3^- + 2e^- + 2H^+ \rightarrow NO_2^- + H_2O\ E0 = +0.433\ V$$

$$NO_2^- + e^- + 2H^+ \rightarrow NO + H_2O\ E^0 = +0.350\ V$$

$$NO + e^- + H^+ \rightarrow 1/2N_2O + 1/2H_2O\ E^0 = +1.175\ V$$

$$1/2N_2O + e^- + H^+ \rightarrow 1/2N_2 + 1/2H_2O\ E^0 = +1.355\ V$$

$$2NO_3^- + 12H^+ + 10e^- \rightarrow N_2 + 6H_2O\ E^0 = +0.734\ V$$

$$Fe^{3+} + e^- + H^+ \rightarrow Fe^{2+} + 1/2H_2O\ E^0 = +0.773\ V$$

$$MnO_2 + 4H^+ + 3e^- \rightarrow Mn^{2+} + 2H_2O\ E^0 = +0.602\ V$$

3.7.2.4 Performance of Microbial Fuel Cells in Agriculture Wastewater Management

There are several reports on the use of microbial fuel cells for converting agro-waste into useful power generation. For instance:

- Min et al. (2005b) have shown the generation of the highest electrical density up to 45 mW/m^2 along with 86% COD removal efficiency and 83% ammonia production by using a single-chambered MFC.
- In a study, using the two-chambered MFC, the highest electrical density was observed at 0.32 mW/m^2 and the maximum level of nutrients was observed to be N (84%), P (70%), K (91%) (Yokoyama et al. 2006)
- In the single chamber, microbial fuel cells, 79.76%, and 99% ten chemicals odors and explosive natural acids were reduced in the agriculture waste management (Kim et al. 2008)
- In a study conducted by Du et al. (2011), 71%, 88%, and 44% of tCOD, sCOD, and NH^{+4} were diminished in agriculture wastewater treatment respectively and

22 mW/m^2 electrical density was released if human waste material were decomposed before poured in two-chambered MFC.

- In the single chamber microbial fuel cells, 55 mw electrical production in agriculture system with the help of platinum-based cathode was found to be greater than platinum-free cathode by 23 mW power density. In the urine energy recital of MFC in agriculture waste treatment up to 75% of COD, removals were gained 4 days later (Santoro et al. 2013).

3.8 Organic Substrate Removal by MFCs

MFCs are capable of utilizing a number of biodegradable compounds from agricultural, industrial, dairy, domestic waste, and landfill leachates, etc. Unlike earlier studies which were conducted using synthetic wastewaters in order to analyze MFC performances, to understand the mechanisms for increasing efficiency of organic waste removal and energy recovery. Recent studies have been conducted to analyze the feasibility and real-time potential of these MFCs towards the wastewater from different sources.

A more than 90% of carbon was removed by employing MFC in the wastewater treatment plants. Moreover, their performances have also been evaluated for treating the organic substrates such as acetate, glucose, sucrose, etc. in the anode chamber that are utilized by microorganisms (Pant et al. 2011). Acetate is the final product of a number of metabolic pathways. It is an easy, simple, and most preferred biodegradable carbon source for MFCs (Bilfinger et al. 2008). MFCs that utilize acetate as substrates have been reported to generate higher power densities than glucose and domestic wastewater. For instance, Liu et al. (2005) reported that using a single chamber MFC, the power generated with acetate (506 mW/m^2) as substrate was found to be 66% higher than that produced with butyrate (305 mW/m^2).

In another study, conducted by Liu et al. (2009), the performance of an electroactive consortium was analyzed in a two-chamber MFC using protein and acetate-rich water as substrates. It has been observed that acetate-based MFC was able to produce higher electric power in comparison to the protein-rich wastewater.

Another commonly utilized substrate in MFCs is glucose. Rabaey et al. (2003) conducted an experiment to optimize glucose dosage for a mixed bacterial culture for efficient power generation and it was found that at the dosage of 3–5 $g\,L^{-1}$, power up to 3.6 $W\,m^{-2}$ was generated. Hu (2008) made a comparison between anaerobic sludge and glucose as a fuel for MFC and found that anaerobic sludge generated a limited power of 0.3 mWm^{-2}. However, a maximum of 0.161 W/m^2 power was generated when glucose was used in the same system.

In a study conducted by Lee et al. (2008), energy conversion competence (ECE) of acetate and glucose as substrates in MFC have been compared. The ECE was found to be 42% with acetate, but only 3% with glucose which led to a low current and power density.

City wastewaters have BOD fixations under 0.3 g per L and thus, considered low energy thickness transporters for MFCs. MFCs are additionally fit for treating wastewaters with BOD focuses surpassing 2 g per L because of the anoxic environment within the anode terminal. The high-strength wastewaters are produced from brewer plants, dairy homesteads, the food industry, and other waste streams. The wastewater released from food industries is generally rich in biodegradable compounds viz. carbohydrates, proteins, lipids, etc. (Gohil and Nakhla 2006). MFCs can deliver power in the range of 2 KWh/ton to 260 kWh/ton of wastewater material utilized for preparing food items. Further, it can be contingent upon the biological oxygen demand and water volume utilized all the while. Further, a total of 46,000 KW of force can be generated from low BOD dairy wastewaters. Borole and Hamilton (2010) have shown that a maximum of 1960 MW electricity can be generated from high biological oxygen demand wastewater, released from US dairies. Furthermore, Hoffmann et al. (2008) have revealed that the domesticated animal's wastewater is frequently high in natural material substances. They may contain undeniable degrees of nitrogen-rich compounds like proteins. Moreover, they are harder to debase natural materials, like cellulose. Further, slaughterhouse waste may likewise incorporate lipids other than sugars, natural acids, and proteins. It is considered a good MFC substrate because of the presence of low ammonium content and high starch amount.

Feng et al. (2008) have explored air cathode MFC for treating brewery waste. They have used the greatest force thickness of 0.528 W/m^2 are accomplished when 0.05 M phosphate support was mixed to the polluted water. The maximum force delivered by distillery wastewater is, however, is lower than that accomplished utilizing homegrown wastewater.

3.9 Removal of Nitrogen and Phosphorous in MFCs

Nitrogen and Phosphorous are two major elements dominantly present in wastewaters which causes eutrophication, and human health issues. MFCs are reported to remove excess of these nutrients from wastewaters. Customary natural nitrogen expulsion through notable nitrification-denitrification response is intensive in terms of carbon, cost, and energy. This cycle includes vigorous oxidation of alkali materials to nitrate and nitrite, and a dissimilatory reduction of nitrate called denitrification, which ultimately produces nitrogen gas (Ahn and Logan 2010). Nitrate can acknowledge electrons from natural mixtures to be decreased to nitrogen gas like that in an ordinary denitrification measure. Such an electron-moving cycle makes it conceivable to utilize nitrate as a TEA. The decrease of nitrate can produce a positive electric capability of 0.98 V when utilizing natural mixtures as acetic acid derivatives as an electron source (Madigan et al. 2010). In an MFC, the power needed to give the vital lessening capacity to denitrification can be definitely decreased if microorganisms utilize the cathodic anode straightforwardly as the electron benefactor (Virdis et al. 2008). *Geobacter* species are equipped for utilizing a graphite-based anode as an immediate electron benefactor during the oxidation of nitrate to nitrite. Further, the H_2 less denitrification at the cathode is combined with oxidation of natural carbon-containing MFCs at the anodes (Gregory et al. 2004). The principal factors influencing the nitrogen evacuation are broken down oxygen (DO), pH, C/N proportion, and power age by the anode cycle. Denitrification is hindered by high pH and significant degrees of DO hinder the denitrification. As the cathode terminal is influenced by the number of electrons delivered by the anode interaction, the proportion of C/N proportion turns into a significant boundary for the cycle execution (Kelly and He 2014).

Phosphorous is normally eliminated from the wastewater as struvite. Furthermore, few methodologies such as compound expansion, electrolysis, and CO_2 stripping can be used for controlled struvite recuperation from wastewater. Further, several struvite recuperations contemplate having zeroed in on expanding arrangement pH by means of substance base expansion (Cusick and Logan 2012).

Ichihashi and Hirooka (2012) have reported that when swine wastewater was used in a single-chambered MFC, 70 to 82% of phosphorous was expelled from MFC and was precipitated on the cathode as struvite crystals. In view of the high centralization of magnesium ions, it is revealed that phosphate was eliminated as struvite in light. Because the pH close to the cathode terminal was higher in comparison to the anode. Phosphorus is effectively encouraged as struvite (300–900 mg/m^2h) within the mono-chamber unit along with 40% phosphate solvent eliminated at a higher rate. Further, the energy efficiencies and hydrogen creation rates recommended the energy requests of struvite creation with a MEC will be essentially balanced by energy recuperated as hydrogen. This can considerably bring down the operational expenses for struvite recuperation. The impediments for struvite recuperation in MFCs are important for high salt and phosphate-containing wastewaters that make this interaction reasonable for pollutants from dairy ranches, agrarian fields, and livestock activities. Furthermore, Mg supplies may create a problem with this interaction and standard acidic or alkaline conditions for struvite framing.

3.10 Removal of Heavy Metals by MFCs

Metal particles in wastewater are not biodegraded into innocuous finished results and hence require uncommon strategies for treatment. Additionally, high redox possibilities were observed in the metal-containing bunches, and thus, they can be used as electron acceptors (Wang and Ren 2014). Moreover, this technique can prepare MFCs for eliminating the metal particles from the polluted water (Mathuriya and Yakhmi 2014). It has been found that a decrease in dissimilatory metals is an interaction that is utilized by microorganisms to monitor energy by metabolizing inorganic or natural electron benefactors and diminishing the metals. During this process, microorganisms help in the movement of electrons and generate electrochemical energy. Metal diminishing miniature organic entities are turning into an examination center because of their capability to work with bioremediation in regions that are debased with weighty metals. Besides, these organic entities are indispensable for the advancement of microbial energy units (Lovley 1993). Huang et al. (2015) accounted recuperation Cobalt as Co $(OH)_2$ in oxygen lessening biocathodes. Further, the most noteworthy cobalt recuperation pace of 79 ± 1 μmol L^{-1} h^{-1} with a 0.24 mol yield on per mol of COD are gotten with synchronous power creation of 0.0015 KW/m^3 under the ideal states of oxygen of 31 μM, pH of 5.6, and an underlying Co(II) convergence of 508 μM. Conclusively, it is a promising methodology for recuperating cobalt hydroxide in a harmless way.

3.11 Current Difficulties in Microbial Fuel Cells

For the most part, the chief issues of microbial energy units possess lesser force densities, lower toxin conversion rates, lower air cathode execution, and expenses related to electrodes and their ecological sway in correlation with customary wastewater treatment measures.

3.11.1 COD Expulsion Rates

Carbon, supplements, and different toxins are taken out in MFCs at an extremely low rate thus requiring extra plans, and taking into account the executives' issues including costs. Clauwaert et al. (2008) have proposed the natural evacuation pace of 5e10 kg per m^3 of COD per day to meet the practical efficiency of MFC-based water treatment. Janicek et al. (2014) have accounted for a wide scope of natural expulsion rates for going somewhere in the range of 0.0053 and 5.57 gCOD per L per day along with various feedstock involving industrial and homegrown polluted water, essential effluents, leachates from landfills, and compost slurry.

3.11.2 Low Power Densities

The creation of Low force thickness is a wide basic condition in microbial power devices. A wide scope of force densities somewhere in the range of 0.0018 and 2 W/m^2 have been announced (Janicek et al. 2014). In certain investigations, high force densities (W/m^3) are accounted for. Yet, these outcomes ought to be confirmed since the outcomes depend on the terminal surface zones and volume of the reactor with vulnerabilities. Fan et al. (2012) have recorded a high force thickness of 2870 W/m^3. It is accomplished by another U-formed current gatherer and separator materials. Frequently framework setup, terminal circuit engineering, layers, and anode materials have been accounted for to be the reasons for power thickness enhancements.

3.11.3 Biocathode and Air Cathode Advancement

The motor impediment additionally applies to the execution of the cathode terminal where both OH group and proton transport comes in equilibrium along with the electron stream in the MFC. Desloover et al. (2012) have observed that pH lopsidedness essentially influences the force creation in MFCs. Low cathode decrease exercises were as of late ascribed to low electron transport to the mass fluid from the impetus. This electron move obstacle to the ORR systems could be overwhelmed by biocathodes which likewise wipe out the requirement for costly impetuses. In any case, keeping dynamic biocathodes turns into another issue.

3.11.4 Integration with Other Valuable Cycles

The existing wastewater treatment processes can be coupled with MFCs along with other natural remediation practices. After the anaerobic absorption step or essential treatment, MFCs can be utilized as a preparing unit (Xu et al. 2016). MFCs can be coordinated inside the current wastewater treatment and the executive frameworks from homegrown levels (decentralized frameworks) to a local area level (concentrated) just as frameworks in the modern areas to expand their energy and asset use efficiencies.

3.12 Cost Analysis in Microbial Fuel Cells

Not many examinations have been accounted on financial and natural effect investigation of MFCs. Considering a potential power creation limit of 1000 mW/m^2, Logan (2005) has accounted for some simple financial matters for the treatment of wastewater using MFC. He estimated that around 16.4 billion liters of wastewater will be generated in a year from a place having a population size of 100,000 individuals. It can provide 2.3 MW power through MFCs. Representing different obstructions, for example, reactor detainment times and energy recuperation impediments, about 0.5 MW of power can be effortlessly caught utilizing MFCs. The financial matters of such a framework for power creation are accounted for to be practically identical with other customary options. Rozendal et al. (2008) introduced a financial investigation contrasting wastewater treatment alternatives utilizing the traditional framework, MECs, and MFCs. The initiated ooze framework plainly doesn't have the income age ability. While the anaerobic processing frameworks are fit for delivering net positive income, the MFC innovation isn't prudent. The MFC alternative can meet half all-out costs while on the other hand, MECs appear cost-cutthroat as of now. Cost dispersion shows that materials that make up the electrode's anode contribute a significant bit. The others include current gatherer materials and accessories. A future investigation presents an alternate expense dispersion situation indicating the replacement of anode and cathode expenses with the MFCs (Wei et al. 2011). For this to occur, modest terminal materials ought to be considered for improving the force execution and financial goals of MFCs.

The biocathode generated force densities are 20.1 W/m^3 for semicoke and 0.024 KW/m^3 for actuated carbon individually, which are greater than that acquired by graphite, i.e., 14.1 W/m^3 and for carbon material cathodes it is 0.017 KW/m^3. Further, the actuated carbon and semicoke expenses for graphite electrodes are just 22.7 and 2.8%, respectively. Furthermore, Janicek et al. (2015) have compared the anodes made up of steel networks and carbon fabric. They have observed that carbon fabric had 34% more arriving at 1.72 Wm^{-2} than that of steel network. Moreover, they had a higher static pressing factor value of 1.9 m. Moreover, carbon fibers possess greater hydrostatic pressing factor resistance, electrochemical resilience, and erosion opposition that make them suitable for the respective application.

Flow monetary investigations propose that MFCs alone are not efficient and cost-effective to generate sufficient energy from the wastewater. Therefore, energy recuperation techniques ought to be differentiated to deliver CH_4, biomass rich in lipids and bioelectricity. With this high-level plan, the general financial matters of the interaction could be altogether improved. A reciprocal plan of consecutive MFC and AD measures improve the net energy and money-saving advantages (Pham et al. 2006). The financial practicality of yeast fermenters joined with photobioreactors produces correlative power creation (Powell et al. 2009). The economical surveys revealed the positive trends for long-term investment in the field of microbial-based processes and the products viz. bioethanol, bioelectricity, biomass generation, etc.

3.13 Conclusion and Future Perspectives

Fossil fuels have been a major source of energy supply supporting industrial and economic growth all around the globe. But they cannot sustain the future energy requirements for a longer period. MFCs are a self-sustaining system of energy generation that can be a better alternative for energy generation. To advance the improvement of MFCs in supportable wastewater treatment, the future examination should emphasize comprehending the metabolic component, upgrade the activity conditions using strategies such as the development of a conductive thick biofilm, plan and design the MFCs accordingly for their use in wastewater treatment.

MFC stacks, the voltage inversion, and ionic short out are as yet enormous hindrances in the method of reasonable application due to bio catalyzed cathode responses in the MFC. The improvement of force and assortment in the framework will speed up the business use of MFCs. Currently, lots of effort are required to achieve high-performance MFCs generating high power without any constraints. Synergistic impact with other wastewater treatment innovation, consolidated innovation may speed up the utilization of MFCs in wastewater treatment.

References

Aelterman P, Babaey K, Pham HT, Boon N, Verstraete W (2006) Continuous electricity generation at high voltages and currents using stacked microbial fuel cells. Environ Sci Technol 40:3388–3394

Ahn Y, Logan BE (2010) Effectiveness of domestic wastewater treatment using microbial fuel cells at ambient and mesophilic temperatures. Bioresour Technol 101(2):469–475

Azwar MY, Hussain MA, Abdul-Wahab AK (2014) Development of biohydrogen production by photobiological, fermentation and electrochemical processes: a review. Renew Sust Energ Rev 31:158–173

Bhatt P, Bhatt K, Sharma A, Zhang W, Mishra S, Chen S (2021a) Biotechnological basis of microbial consortia for the removal of pesticides from the environment. Crit Rev Biotechnol 41(3):317–338

Bhatt P, Joshi T, Bhatt K, Zhang W, Huang Y, Chen S (2021b) Binding interaction of glyphosate oxidoreductase and C-P lyase: molecular docking and molecular dynamics simulation studies. J Hazard Mater 409(5):124927

Bilfinger JC, Byrd JN, Dudley BL, Ringeisen BR (2008) Oxygen exposure promotes fuel diversity for Shewanella oneidensis microbial fuel cells. Biosens Bioelectron 23(6):820–826

Borole AP, Hamilton CY (2010) Energy production from food industry wastewaters using bioelectrochemical cells. In: Emerging environmental technologies, vol II. Springer, pp 97–113

Butti SK, Velvizhi G, Sulonen MLK, Haavisto JM, Koroglu EO, Cetinkaya AF (2016) Microbial electrochemical technologies with the perspective of harnessing bioenergy: maneuvering towards upscaling. Renew Sustain Energy Rev 53:462–476

Cheng X, He L, Lu HW, Chen YW, Ren LX (2016) Optimal water resources management and system benefit for the Marcellus shale-gas reservoir in Pennsylvania and West Virginia. J Hydrol 540:412–422

Clauwaert P, Aelterman P, De Schamphelaire L, Carballa M, Rabaey K, Verstraete W (2008) Minimizing losses in bio-electrochemical systems: the road to applications. Appl Microbiol Biotechnol 79(6):901–913

Cusick RD, Logan BE (2012) Phosphate recovery as struvite within a single-chamber microbial electrolysis cell. Bioresour Technol 107:110–115

Dallas P, Georgakilas V (2015) Interfacial polymerization of conductive polymers: generation of polymeric nanostructures in a 2-D space. Adv Colloid Inter 224:46–61

Debbarma P, Raghuwanshi S, Singh J, Suyal DC, Zaidi MGH, Goel R (2017) Comparative in situ biodegradation studies of polyhydroxybutyrate film composites. 3Biotech 7(178):1–9. https://doi.org/10.1007/s13205-017-0789-3

Desloover J, Arends JA, Hennebel T, Rabaey K (2012) Operational and technical considerations for microbial electrosynthesis. Biochem Soc Trans 40(6):1233

Donovan C, Dewan A, Peng H, Heo D, Beyenal H (2011) Power management system for a2.5 W remote sensor powered by a sediment microbial fuel cell. J Power Sources 196:1171–1177

Du ZW, Li HR, Gu TY (2007) A state of the art review on microbial fuel cells: a promising technology for wastewater treatment and bioenergy. Biotechnol Adv 25:464–482

Du FZ, Li ZL, Yang SQ, Xie BZ, Liu H (2011) Electricity generation directly using human feces wastewater for life support system. Acta Astronaut 68:1537–1547

Fan Y, Han SK, Liu H (2012) Improved performance of CEA microbial fuel cells with increased reactor size. Energy Environ Sci 5(8):8273–8280

Fang Z, Song HL, Cang N, Li XN (2015) Electricity production from azo dye wastewater using a microbial fuel cell coupled constructed wetland operating under different operating conditions. Biosens Bioelectron 68:135–141

Feng Y, Wang X, Logan BE, Lee H (2008) Brewery wastewater treatment using air-cathode microbial fuel cells. Appl Microbiol Biotechnol 78(5):873–880

Fraiwan A, Choi S (2016) A stackable, two-chambered, paper-based microbial fuel cell. Biosens Bioelectron 83:27–32

Ge Z, Zhang F, Grimaud J, Hurst J, He Z (2013) Long-term investigation of microbial fuel cells treating primary sludge or digested sludge. Bioresour Technol 136:509e514

Gohil A, Nakhla G (2006) Treatment of food industry waste by bench-scale up-flow anaerobic sludge blanket-anoxic-aerobic system. Water Environ Res 974–985

Gregory KB, Bond DR, Lovley DR (2004) Graphite electrodes as electron donors for anaerobic respiration. Environ Microbiol 6(6):596–604

Gude VG (2016) Wastewater treatment in microbial fuel cells–an overview. J Clean Prod 122:287–307

He Z, Wagenr N, Minteer SD, Angenent LT (2006) An upflow microbial fuel cell with an interior cathode: assessment of the internal resistance by impedance spectroscopy. Environ Sci Technol 40:5212–5217

He L, Huang GH, Lu HW (2010) A stochastic optimization model under modeling uncertainty and parameter certainty for groundwater remediation design-Part I. Model development. J Hazard Mater 176(1–3):521–526

Hoffmann RA, Garcia ML, Veskivar M, Karim K, Al-Dahhan MH, Angenent LT (2008) Effect of shear on performance and microbial ecology of continuously stirred anaerobic digesters treating animal manure. Biotechnol Bioeng 100(1):38–48

Hu Z (2008) Electricity generation by a baffle-chamber membrane-less microbial fuel cell. J Power Sources 179(1):27–33

Huang L, Liu Y, Yu L, Quan X, Chen G (2015) A new clean approach for production of cobalt dihydroxide from aqueous Co (II) using oxygen-reducing biocathode microbial fuel cells. J Clean Prod 86:441–446

Ichihashi O, Hirooka K (2012) Removal and recovery of phosphorus as struvite from swine wastewater using microbial fuel cell. Bioresour Technol 114:303–307

Jang JK, Pham TH, Chang IS, Kang KH, Moon H, Cho KS et al (2004) Construction and operation of a novel mediator- and membrane-less microbial fuel cell. Process Biochem 39:1007–1012

Janicek A, Fan Y, Liu H (2014) Design of microbial fuel cells for practical application: a review and analysis of scale-up studies. Biofuels 5(1):79–92

Janicek A, Fan Y, Liu H (2015) Performance and stability of different cathode base materials for use in microbial fuel cells. J Power Sources 280:159–165

Kalathil S, Lee J, Cho MH (2011) Granular activated carbon based microbial fuel cell for simultaneous decolorization of real dye wastewater and electricity generation. New Biotechnol 29:32–37

Kargi F, Eker S (2007) Electricity generation with simultaneous wastewater treatment by a microbial fuel cell (MFC) with Cu and Cu-Au electrodes. J Chem Technol Biotechnol 82:658–662

Kelly PT, He Z (2014) Nutrients removal and recovery in bioelectrochemical systems: a review. Bioresour Technol 153:351–360

Kim JR, Dec J, Bruns MA, Logan BE (2008) Removal of odors from swine wastewater by using microbial fuel cells. Appl Environ Microbiol 74:2540–2543

Kumar AK, Reddy MV, Chandrasekhar K, Srikanth S, Mohan SV (2012) Endocrine disruptive estrogens role in electron transfer: bio-electrochemical remediation with microbial mediated electrogenesis. Bioresour Technol 104:547e556

Lee HS, Parameswaran P, Kato-Marcus A, Torres CI, Rittmann BE (2008) Evaluation of energy-conversion efficiencies in microbial fuel cells (MFCs) utilizing fermentable and non-fermentable substrates. Water Res 42(6):1501–1510

Li ZL, Kong L, Liu H (2008) Electricity generation using a baffled microbial fuel cell convenient for stacking. Bioresour Technol 99:1650–1655

Li WW, Yu HQ, He Z (2014) Towards sustainable wastewater treatment by using microbial fuel cells-centered technologies. Energy Environ Sci 7:911–924

Liu H, Logan BE (2004) Electricity generation using an air-cathode single chamber microbial fuel cell in the presence and absence of a proton exchange membrane. Environ Sci Technol 38:4040–4046

Liu H, Ramnarayanan R, Logan BE (2004) Production of electricity during wastewater treatment using a single chamber microbial fuel cell. Environ Sci Technol 38:2281–2285

Liu H, Cheng S, Logan BE (2005) Power generation in fed-batch microbial fuel cells as a function of ionic strength, temperature, and reactor configuration. Environ Sci Technol 39 (14):5488–5493

Liu Z, Liu J, Zhang S, Su Z (2009) Study of operational performance and electrical response on mediator-less microbial fuel cells fed with carbon-and protein-rich substrates. Biochem Eng J 45 (3):185–191

Logan BE (2005) Simultaneous wastewater treatment and biological electricity generation. Water Sci Technol 52(1):31–37

Logan BE (2008) Microbial fuel cells. Wiley

Logan BE, Rabaey K (2012) Conversion of wastes into bioelectricity and chemicals by using microbial electrochemical technologies. Science 337:686–690

Lovley DR (1993) Dissimilatory metal reduction. Ann Rev Microbiol 47:263–290

Lovley DR (2006) Bug juice: harvesting electricity with microorganisms. Nat Rev Microbiol 4 (7):497e508

Lu N, Zhou B, Deng LF, Zhou SG, Ni JR (2009) Starch processing wastewater treatment using a continuous microbial fuel cell with MnO2 cathodic catalyst. J Basic Sci Eng 17:65–73

Madigan MT, Martinko JM, Stahl D, Parker J (2010) Brock biology of microorganisms, 13th edn. Benjamin Cummings, San Francisco

Matesanz AI, Albacete P, Souza P (2016) Synthesis and characterization of a new bioactive mono (thiosemicarbazone) ligand based on 3,5-diacetyl-1,2,4-triazoldiketone and its palladium and platinum complexes. Polyhedron 109:161–165

Mathuriya AS, Yakhmi JV (2014) Microbial fuel cells to recover heavy metals. Environ Chem Lett 12(4):483–494

Mercuri EGF, Kumata AYJ, Amaral EB, Vitule JRS (2016) Energy by microbial fuel cells: scientometric global synthesis and challenges. Renew Sustain Energ Rev 65:832–840

Min B, Cheng SA, Logan BE (2005a) Electricity generation using membrane and salt bridge microbial fuel cells. Water Res 39:1675–1686

Min B, Kim JR, Oh SE, Regan JM, Logan BE (2005b) Electricity generation from swine wastewater using microbial fuel cells. Water Res 39:4961–4968

Mohan SV, Mohanakrishna G, Sarma PN (2008) Effect of anodic metabolic function on bioelectricity generation and substrate degradation in single chambered microbial fuel cell. Environ Sci Technol 42:8088–8094

Mohan SV, Mohanakrishna G, Velvizhi G, Babu VL, Sarma PN (2010) Bio-catalyzed electrochemical treatment of real field dairy wastewater with simultaneous power generation. Biochem Eng J 51(1–2):32–39

Nielsen ME, Reimers C, Stecher HA (2007) Enhanced power from chambered benthic microbial fuel cells. Environ Sci Technol 41:7895–7900

Pant D, Van Bogaert G, Diels L, Vanbroekhoven K (2010) A review of the substrates used in microbial fuel cells (MFCs) for sustainable energy production. Bioresour Technol 101 (6):1533–1543

Pant D, Singh A, Van Bogaert G, Alvarez Gallego Y, Diels L, Vanbroekhoven K (2011) An introduction to the life cycle assessment (LCA) of bioelectrochemical systems (BES) for sustainable energy and product generation: relevance and key aspects. Renew Sustain Energy Rev 15:1305–1313

Pham TH, Rabaey K, Aelterman P, Clauwaert P, De Schamphelaire L, Boon N, Verstraete W (2006) Microbial fuel cells in relation to conventional anaerobic digestion technology. Eng Life Sci 6(3):285–292

Powell EE, Mapiour ML, Evitts RW, Hill GA (2009) Growth kinetics of *Chlorella vulgaris* and its use as a cathodic half cell. Bioresour Technol 100:269–274

Rabaey K, Verstraete W (2005) Microbial fuel cells: novel biotechnology for energy generation. Trends Microbiol 23:291–298

Rabaey K, Lissens G, Siciliano SD, Verstraete W (2003) A microbial fuel cell capable of converting glucose to electricity at high rate and efficiency. Biotechnol Lett 25(18):1531–1535

Ren LX, He L, Lu HW, Chen YZ (2016) Monte Carlo-based interval transformation analysis for multi-criteria decision analysis of groundwater management strategies under uncertain naphthalene concentrations and health risks. J Hydrol 539:468–477

Rittmann BE, McCarty PL (2001) Biotecnología del medioambiente: Principios yaplicaciones. McGraw-Hill Interamericana de Espa~na

Rozendal RA, Hamelers HVM, Rabaey K, Keller J, Buisman CJN (2008) Towards practical implementation of bioelectrochemical wastewater treatment. Trends Biotechnol 26:450–459

Santoro C, Ieropoulos I, Greenman J, Cristiani P, Vadas T, Mackay A (2013) Power generation and contaminant removal in single chamber microbial fuel cells (SCMFCs) treating human urine. Int J Hydrog Energ 38:11543–11551

Seo HN, Woo JL, Tae SH, Doo HP (2009) Electricity generation coupled with wastewater treatment using a microbial fuel cell composed of a modified cathode with aceramic membrane and cellulose acetate film. J Microbiol Biotechnol 19:1019–1027

Sharma A, Malaviya P (2016) Irrigational impact of untreated and treated brewery-distillery effluent on seed germination of marigold (*Tagetes erecta* L.). J Environ Biol 37(1):115

Shimoyama T, Komukai S, Yamazawa A, Ueno Y, Logan BE, Watanabe K (2008) Electricity generation from model organic wastewater in a cassette-electrode microbial fuel cell. Appl Microbiol Biotechnol 80:325–330

Singh A, Smyth BM, Murphy JD (2010) A biofuel strategy for Ireland with an emphasis on production of biomethane and minimization of land-take. Renew Sustain Energy Rev 14:277–288

Sivasankaran A, Sangeetha D, Ahn YH (2016) Nanocomposite membranes based on sulfonated polystyrene ethylene butylene polystyrene (SSEBS) and sulfonated SiO2 for microbial fuel cell application. Chem Eng J 289:442–451

Tee PF, Abdullah MO, Tan IAW, Rashid NKA, Amin MAM, Nolasco-Hipolito C (2016) Review on hybrid energy systems for wastewater treatment and bio-energy production. Renew Sust Energ Rev 54:235–246

Ting CH, Lee DJ (2007) Production of hydrogen and methane from wastewater sludge using anaerobic fermentation. Int J Hydrog Energy 32:677–682

Venkata Mohan S, Velvizhi G, Annie Modestra J, Srikanth S (2014) Microbial fuel cell: critical factors regulating bio-catalyzed electrochemical process and recent advancements. Renew Sustain Energ Rev 40:779–797

Virdis B, Rabaey K, Yuan Z, Keller J (2008) Microbial fuel cells for simultaneous carbon and nitrogen removal. Water Res 42(12):3013–3024

Wang B, Han JI (2009) A single chamber stackable microbial fuel cell with air cathode. Biotechnol Lett 31:387–393

Wang H, Ren ZJ (2014) Bioelectrochemical metal recovery from wastewater: a review. Water Res 66:219–232

Wei J, Liang P, Cao X, Huang X (2011) Use of inexpensive semicoke and activated carbon as biocathode in microbial fuel cells. Bioresour Technol 102(22):10431–10435

Wu Q, Jiao S, Ma M, Peng S (2020) Microbial fuel cell system: a promising technology for pollutant removal and environmental remediation. Environ Sci Pollut Res 27(7):6749–6764

Xiao L, He Z (2014) Applications and perspectives of phototrophic microorganisms for electricity generation from organic compounds in microbial fuel cells. Renew Sustain Energ Rev 37:550–559

Xiao B, Yang F, Liu J (2011) Enhancing simultaneous electricity production and reduction of sewage sludge in two-chamber MFC by aerobic sludge digestion and sludge pretreatments. J Hazard Mater 189:444–449

Xu L, Zhao Y, Doherty L, Hu Y, Hao X (2016) The integrated processes for wastewater treatment based on the principle of microbial fuel cells (MFCs): a review. Crit Rev Environ Sci Technol 46 (1):60e-91

Yang Y, Sun G, Xu M (2011) Microbial fuel cells come of age. J Chem Technol Biotechnol 86 (5):625–632

Yokoyama H, Ohmori H, Ishida M, Waki M, Tanaka Y (2006) Treatment of cow-waste slurry by a microbial fuel cell and the properties of the treated slurry as a liquid manure. Anim Sci J 77:634–638

Zhang B, Zhao H, Zhou S, Shi C, Wang C, Ni J (2009) A novel UASB–MFC–BAF integrated system for high strength molasses wastewater treatment and bioelectricity generation. Bioresour Technol 100(23):5687–5693

Zhang F, Ge Z, Grimaud J, Hurst J, He Z (2013) In situ investigation of tubular microbial fuel cells deployed in an aeration tank at a municipal wastewater treatment plant. Bioresour Technol 136:316e321

Zhang L, Wang J, Fu G, Zhang Z (2020) Simultaneous electricity generation and nitrogen and carbon removal in single-chamber microbial fuel cell for high-salinity wastewater treatment. J Clean Prod 276:123203

Zhuang L, Zhou S (2009) Substrate cross-conduction effect on the performance of serially connected microbial fuel cell stack. Electrochem Commun 11:937–940

Zornoza R, Moreno-Barriga F, Acosta JA, Muñoz MA, Faz A (2016) Stability, nutrient availability and hydrophobicity of biochars derived from manure, crop residues, and municipal solid waste for their use as soil amendments. Chemosphere 144:122–130

Chapter 4
Critical Process Parameters and Their Optimization Strategies for Enhanced Bioremediation

Jazel Sosa-Martínez, Nagamani Balagurusamy, Suresh Kumar Gadi, Julio Montañez, Juan Roberto Benavente-Valdés, and Lourdes Morales-Oyervides

4.1 Introduction

Presently, the world is facing a severe environmental crisis as a result of increasing overpopulation and industrialization, which generate a high volume of waste and pollutants contaminating soil, water and air. Such contaminants include heavy metals, petroleum derivatives, polycyclic aromatic hydrocarbons (PAHs), organic pollutants, dyes, chlorinated compounds, greenhouse gases, and others (Goel et al. 2020; Azubuike et al. 2016; Okino-Delgado et al. 2019). Among the remediation approaches, bioremediation has emerged as an innovative, cost-effective and eco-friendly alternative with a high potential to restore contaminated environments efficiently (Dixit et al. 2015; Debbarma et al. 2017). Due to the high number of pollutant types and contaminated sites, the number of bioremediation techniques is quite large (Azubuike et al. 2016; Bhatt et al. 2021). Generally, bioremediation has been described as a process where plants, microorganisms, or their metabolites such as enzymes are employed to reduce, degrade, detoxify or remove environmental pollutants (Giri et al. 2017a, b; Okino-Delgado et al. 2019). The most common

J. Sosa-Martínez · J. Montañez · J. R. Benavente-Valdés · L. Morales-Oyervides (✉)
Faculty of Chemical Sciences, Department of Chemical Engineering, Autonomous University of Coahuila, Saltillo, Coahuila, Mexico
e-mail: lourdesmorales@uadec.edu.mx

N. Balagurusamy
Bioremediation Laboratory, Faculty of Biological Sciences, Autonomous University of Coahuila, Torreón, Coahuila, Mexico

S. K. Gadi
Faculty of Mechanical and Electric Engineering, Autonomous University of Coahuila, Torreón, Coahuila, Mexico

© The Author(s), under exclusive license to Springer Nature Switzerland AG 2022

D. C. Suyal, R. Soni (eds.), *Bioremediation of Environmental Pollutants*,
https://doi.org/10.1007/978-3-030-86169-8_4

classification is based on the application site of the process; in-situ and ex-situ (Azubuike et al. 2016).

In-situ techniques encompass treating the contaminants at the site of contamination, while ex-situ techniques require transporting the contaminants to another treatment location. In general, in-situ bioremediation techniques are considered less expensive than the ex-situ techniques, without considering the costs of processing design, research, and development studies. In-situ techniques involve bioaugmentation, biostimulation and phytoremediation and ex-situ techniques include biopiles, windrows and bioreactor-based systems.

Due to the raised concerns regarding environmental pollutants during the last decades, researchers have invested significant efforts to optimize the bioremediation process so that it can meet the requirements of technical and economic feasibility and a viable alternative for implementation. Bioremediation encompasses numerous process types, each of which requires a different optimization approach. Figure 4.1 shows the structural framework of this chapter, where available information is divided according to in-situ (phytoremediation, bioaugmentation and biostimulation) and ex-situ bioremediation (bioreactors, biosorption and enzymatic). Further, the most applied optimization approaches, critical factors affecting each bioremediation process type, quality targets and process design are discussed.

This chapter intends to provide knowledge gained from the bioremediation optimization studies performed to treat the most common pollutants. We discussed the most applied optimization approaches, critical factors affecting each bioremediation process type, quality targets and process design. The gathered information will provide scientific and technical recommendations to assist in the bioremediation implementation and improvement.

4.2 Optimization Approaches in Environmental Bioremediation Processes

All bioremediation types have one common feature; the bioremediation mechanisms have a high number of variables with a significant non-linearity degree due to the vast biological, physical and chemical reactions involved. Thus, bioremediation process optimization is definitely not an easy target. Despite that, a feasible application requires that the process capabilities are evaluated in advance to eliminate process alternatives that do not qualify for the bioremediation of a contaminated site. Most studies reviewed in this chapter employed statistical techniques to evaluate the effects of process factors on a particular response (see Tables 4.1, 4.2, 4.3, 4.4, 4.5 and 4.6). The methods were from the simplest to the most complex, depending on the previous knowledge and bioremediation process types. The common methods used were the One-Factor-at-a-Time method (OFAT), Response Surface Methodology (RSM), orthogonal arrays (Taguchi method) and Artificial Neural Networks (ANNs).

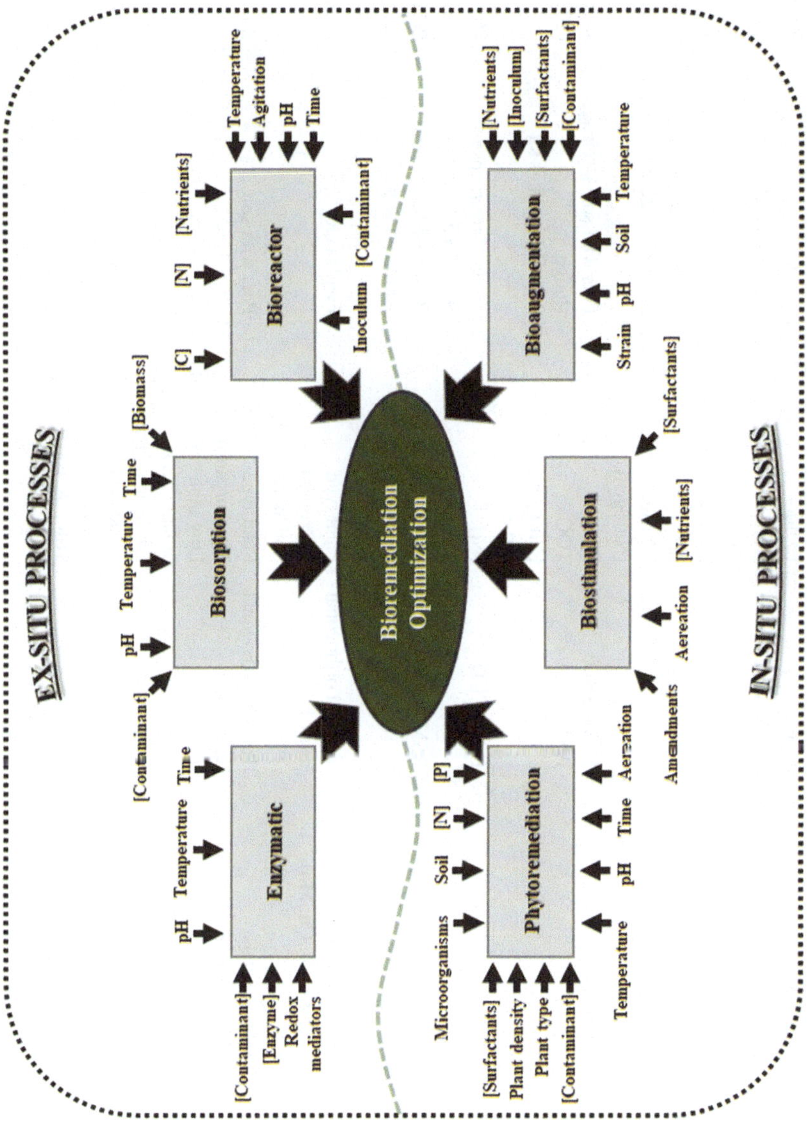

Fig. 4.1 Bioremediation processes types and their critical influencing parameters

Table 4.1 Summary of the optimization studies related to phytoremediation process

Plant	Pollutant	Optimization method	Factors	System	References
Ludwigia octovalvis	Arsenic	RSM–BBD ANN	[As] Time Aeration rate	Pilot reed bed	Titah et al. (2018)
Lemna valdiviana	Arsenic	RSM–CCD	pH [Phosphorous] [Nitrogen]	Polyethylene containers 30 × 30 × 22 cm	de Souza et al. (2019)
Sinapis alba L.	Cadmium	RSM–BBD ANN	[Cd] TOC [Nitrogen] Time	Pots with soil + sewage sludge 25 × 12 × 25, 4 kg	Jaskulak et al. (2020)
Alocasia puber	Nickel	RSM–CCD	[Ni] Time	Wetland microcosm 30 × 18.5 × 18 cm	Mohamad Thani et al. (2020)
Melastoma malabathricum L.	Lead	RSM–BBD	[Pb] Time Aeration rate	Pilot reed bed system 100 × 90 × 90 cm	Selamat et al. (2018)
Soybean Rainbow pink Kochia	Heavy metals (Cu, Zn, Pb, Cd, Mn)	Taguchi orthogonal array L_{16}	Plant type Microorganism Soil amendment	Field-scale	Li et al. (2019)
Maize Sorghum Sunflower	Heavy metals (Pb, Ni, Cd, Zn)	Taguchi orthogonal array L_{18}	Plant type Microorganism Low molecular weight organic acids Ethylene diamine tetra acetic acid Tween 80 Triton X-100	Pots with soil	Razmi et al. (2021)

Chrysopogon zizanioides (L.)	BOD/COD (Palm oil mill secondary effluent, POMSE)	RSM–CCD	[POMSE] Plant density Time	Wetland 30 × 30 × 50 cm	Darajeh et al. (2016)
Eichhornia crassipes *Pistia stratiotes*	COD Ammonium Nitrate Phosphorus (Wastewater)	Complete factorial design	Time Plant density [Phosphorus]	Polyethylene tanks 25 L wastewater	Ntakiyiruta et al. (2020)
Eichhornia crassipes	Organic matter (Wastewater)	Complete factorial design Multi-response optimization	Time Plant density [COD]	Mesocosm Containers 34 × 24 × 19 cm	Mahunon et al. (2018)
Macleaya cordata	Radionucleides (Contaminated soil)	Constrained optimization	[Chelating agent]	Pots with soil 25 × 15 × 20	Hu et al. (2021)
Chamaedorea elegans	Formaldehyde	RSM ANN	[Formaldehyde] Relative humidity Light intensity Total leaf area	Pilot scale chamber (357 L)	Teiri et al. (2020)

Table 4.2 Summary of the data on the optimization of bioaugmentation process

Microorganisms	Pollutants	Optimization method	Factors	Time	Systems	References
Agrobacterium tumefaciens, Cellulosimicrobium funkei, Shinella zoogloeoides, Bacillus aryabhattai	Chlorpyrifos [*O*, *O*-diethyl *O*-(3,5,6-trichloro-2-pyridinol) phosphorothioate]	RSM–CCD	Temperature [Chlorpyrifos] Inoculum size	30 days	Microcosm glass flask	Uniyal et al. (2021)
Sphingomonas sp.	Tricyclazole	OFAT	Temperature pH [Tricyclazole] C and N source	4 days	Microcosm glass flask	Wu et al. (2018)
Pandoraea sp. *Comamonas* sp. *Aspergillus* sp	Black liquor (Pulp and paper wastewater)	Taguchi orthogonal array L_{15}	Temperature pH Inoculum size COD Organic load rate	4 days	Microcosm glass flask	Zheng and Chai (2016)
Pseudomonas sp. *Rhodococcus* sp.	Benzene Toluene Styrene	OFAT	Microbial consortium ratio Initial sewage Sludge samples	3 days	Microcosm PVC containers	Feng et al. (2021)
Arthrobacter sp. ZXY-2	Atrazine	PBD RSM–CCD	Temperature pH Inoculum size [Sucrose] [$Na_2HPO_4 \cdot 12H_2O$]	60 days	Microcosm constructed wetland of polyvinylchloride	Zhao et al. (2020)
Bacillus sp. *Lysinibacillus* sp. *Rhodococcus* sp.	Heavy metals (Al, Cd, Cr, Fe, Ni, Pb and Zn)	OFAT	Microbial strain type	100 days	Microcosm plastic bag	Fauziah et al. (2017)
Fungal strain	Heavy metal	OFAT	Fungal consortia	100 days	Microcosm in poly bag	Hassan et al. (2019)

Sphingomonas haloaromaticaman, Pseudomonas putida Degrading bacteria consortium	Pesticides (Thiabendazole, imazalil, ortho-phenylphenol and diphenylamine)	OFAT	Bacterial strain	160 days	Pilot biobeds	Karas et al. (2016)
Achromobacter denitrificans strain PR1	Sulfamethoxazole	OFAT	Hydraulic retention time [Acetate] Inoculum size	36 days	Laboratory-scale membrane bioreactors	Nguyen et al. (2019)
Bacillus sp. GZT	2,4,6-tribromophenol	OFAT	Additives Co-substrates	50 days	Microcosm serum bottle	Xiong et al. (2017)
Bacterial consortium immobilized	Dichloromethane and n-hexane	Taguchi orthogonal array L_{16}	Inoculum Size Inoculation protocol Surfactant amount [Inorganic fertilizer]	60 days	Microcosm glass flask	Li et al. (2016)

Table 4.3 Summary of the optimization of biostimulation process

Microorganisms	Pollutants	Optimization method	Factors	Time	Systems	References
Not reported	Total Petroleum Hydrocarbon	RSM–CCD	[Nitrogen] [Phosphoru]	50 days	Microcosm cylindrical glass flasks	Martínez Álvarez et al. (2015)
Hydrocarbonoclastic microbial populations	Total Petroleum Hydro-carbon (Diesel)	OFAT	[Phosphoru]	15 days	Microcosm in plastic bags	Júlio et al. (2018)
Not reported	Polycyclic aromatic hydrocarbon	OFAT	Nutrient type (digestate, fraction of municipal solid waste, and a nutrient solution)	120 days	Microcosm plas-tic bottle	Bianco et al. (2020)
Not reported	Total Petroleum Hydro-carbon Hexavalent chromium	RSM–BBD	[Inorganic nutrient] [Surfactant] [Organic nutrient]	42 days	Microcosm in glass flask	Agarry and Ogunleye (2012)
Rhodovulum sp. DBP07	Di-butyl phthalate (DBP)	RSM–BBD	C source pH [NaCl]	5 days	Microcosm in glass flask	Baker et al. (2021)
Yeast-bacteria consortium	Total Petroleum Hydrocarbons	OFAT	Nutrient type (commercial fertilizer, humic substance, organic industrial waste)	90 days	Microcosm glass flask	Qiao et al. (2014)
Not reported	Total Petroleum Hydro-carbon (diesel-fuel)	OFAT	Nitrogen Oxygen rate	11 months	Microcosm in polyethylene columns	Kauppi et al. (2011)
Not reported	Hexavalent chromium	OFAT	Nutrient type Aireation	5 days	Microcosm in flask-type bottles	Carlos et al. (2016)
Pseudomonas sp. SDS-N2	Sodium dodecyl sulfate (SDS)	PBD RSM –CCD	[KNO_3] [Proline] [Folic acid] [Biotin] [Thiamine] [Sodium pyru-vate] [$FeSO_4$] [Citric acid]	12.5 h	Microcosm plas-tic bottle	Chen et al. (2019)

Microbial consortium	Total Petroleum Hydrocarbon	Taguchi orthogonal array L_{16}	[Molasses] [Whey] [Urea] [Tween 80]	60 days	Microcosm glass flasks	Khayati and Barati (2017)
Not reported	Crude oil	RSM –CCD	Biosurfactant Mineral fertilizer	30 days	Microcosm PVC rectangular recipients	Da Silva et al. (2009)
Not reported	Polycyclic aromatic hydrocarbons (Methyl-b-cyclodextrin)	OFAT	[Methyl-b-cyclodextrin]	13 months	Soil piles field	Simpanen et al. (2016)

Table 4.4 Summary of the optimization of bioprocesses for bioremediation

Microorganisms	Pollutants	Optimization method	Factors	References
Aspergillus flavus *Aspergillus fumigatus*, *Aspergillus niger*, *Aspergillus terreus*	Reactive Red 120	RSM–CCD	[Carbon] [Nitrogen] pH Temperature	Ameen and Alshehrei (2017)
Halomonas strain Gb	Toluidine Red	RSM–CCD	[Tolouidine Red] [NaCl] pH Temperature	Amini et al. (2019)
Serratia sp. and *Arthrobacter* sp.	Pyrene Chromium	Taguchi orthogonal array L_9	pH Temperature inoculum size	Ge et al. (2021)
Yeast consortium	Benzo[ghi] perylene	RSM–BBD	pH Temperature shaking speed inoculum size	Mandal et al. (2018)
Pseudomonas aeruginosa PA06 *Achromobacte*r sp	Pyrene	OFAT	Carbon source pH	Li et al. (2020)
Bacterial consortia	Propylene glycol	RSM–BBD	pH Salinity [PG] [Phosphate] [Nitrate]	Udaykumar et al. (2020)
Lysinibacillus macrolides DSM54T	4-Chlorobenzoic acid	RSM–CCD	pH Temperature [4-CBA] Inoculation size	Samadi et al. (2020)
Klebsiella variicola FH-1 *Arthrobacter* sp. NJ-1	Atrazine	RSM–BBD	[Carbon] [Nitrogen] Temperature	Gao et al. (2020a)
Candida tropicalis	Metalaxyl	OFAT	pH Temperature [Metalaxyl]	Derbalah et al. (2020)
Serratia ureilytica	Butachlor	PPB RSM–CCD	pH Temperature inoculum size	Mohanty and Jena (2019)
Chlorella sp. and *Scenedesmus* sp.	Carbon dioxide	OFAT	Flashing light continuous light	Martín-Girela et al. (2017)
Chlorella vulgaris and *Chlamydomonas reinhardtii*	Wastewater	OFAT	Hydraulic retention Time Photoperiod Light intensity [Nitrogen] [Phosphorus]	Ashok et al. (2019)

(continued)

Table 4.4 (continued)

Microorganisms	Pollutants	Optimization method	Factors	References
Spirulina sp.	Carbon dioxide	OFAT	Pore diameter Relative height CO_2 gas outlet flow	Guo et al. (2020)
Chlorella vulgaris	Chromium	OFAT	Hydraulic retention Time Solid retention time [Cr (VI)]	Lu et al. (2021)

The OFAT method considers the effect of a single factor by fixing other process variables at a constant value. This method was not the most preferred due to the neglection of the interactive effects between factors (Mahunon et al. 2018; Selamat et al. 2018; Jaskulak et al. 2020); an interactive effect is significant when the effect of a factor relies upon another factor's level. Since the bioremediation process is affected by many variables, this method is not the top choice. However, as discussed in the following sub-sections, the OFAT method is proper when it is desired to scan in detail (many levels) the effect of a single factor when the process itself does not allow other variables to have a wide range of working levels (Dauda and Erkurt 2020).

RSM was one of the most preferred methods precisely because it allows determining the effect of the evaluated factors and their interactive effects on a target response (see Tables 4.1, 4.2, 4.3, 4.4, 4.5 and 4.6). RSM consists of fitting a quadratic model to a set of data derived from an experimental design (full factorial or fractional factorial designs). Fractional factorial designs were mainly selected to reduce the number of experimental runs, such as Box-Benkhen Design (BBD) and Central Composite Design (CCD). Of course, RSM also has its disadvantages, principally the fact that it relies its predictions on a mathematical model. If there is a poor fit of the model to the experimental data, then there is a high probability of falling into wrong conclusions and non-optimal regions. Likewise, even though the model's error is acceptable, one could be misled to sub-optimal regions due to its parabolic shape (Sosa-Martínez et al. 2020).

On the other hand, the Taguchi method offers the advantage that it does not depend upon a mathematical model's fitting. It applies orthogonal arrays to identify the influence of individual factors and select the most appropriate combination of settings that improve performance (Deniz and Yildiz 2019). The method offers various advantages, mainly because it minimizes the required experimental runs by ensuring process consistency. Notwithstanding, one of the Taguchi method's main drawbacks is that it assumes that the interactive effects between factors are negligible, which could be counterproductive.

Regarding ANNs, these are computer-based systems highly used to model non-linear-systems and it is inspired by the nerve cells' learning process in the human brain. It has recently gained attention for bioremediation purposes due to its high prediction accuracy, even when it works with irrelevant and noisy data (Jaskulak

Table 4.5 Summary of the optimization studies on biosorption processes

Microorganisms	Pollutants	Optimizations	Factors	References
Codium vermilara	Copper	RSM-CCD	Algal dose pH [Copper] Time	Fawzy (2020)
Pseudomonas azotoformans	Cadmium Copper Lead	RSM-BBD	pH [Metal] [Biosorbent]	Choińska-Pulit et al. (2018)
Pseudomonas alcaligenes *Pseudomonas resinovorans* (immobilized)	Arsenic	PBD RSM-BBD	[Arsenic] Temperature Time Agitation Biomass dose [Matrix]	Banerjee et al. (2016)
Ganoderma lucidum	Lead Cadmium	OFAT	Pretreatments with NaOH H_2O_2 NaCl Time	Rozman et al. (2020)
Trichoderma viride	Lead Cadmium Copper	RSM–BBD	Temperature Biomass dose pH [Metal]	Singh et al. (2010)
Ceratocystis paradoxa MSR2	Chromium	RSM–BBD	pH Temperature Biomass dose [Metal]	Melvin et al. (2015)
Pleurotus mutilus	Nickel	OFAT Taguchi	pH [Metal] Time Particle size	Daoud and Selatnia (2019)
Cyanobacterial consortium	Chromium	RSM–CCD ANN	pH [Metal] Adsorbent dose	Sen et al. (2018)
Aphanizomenon ovalisporum	Copper Lead	Taguchi orthogonal array	[Metal] Algal dose pH Contact time	Flouty et al. (2019)
Chlorella pyrenoidosa	Copper	RSM	[Biosorbent] pH [Copper]	Moreira et al. (2019)
Chlorella kessleri	Lead	RSM–BBD Crow search algorithm	pH Temperature Biomass dose	Sultana et al. (2020)
Chlamydomonas sp.	Chromium	RSM	pH Biomass dose Contact time	Ayele et al. (2021)

Table 4.6 Summary of the optimization studies on enzymatic biodegradation

Enzymes	Pollutant groups	Statistical approach	Optimized factors	References
Peroxidase soybean	Sulfamethoxazole (SMX)	OFAT	Enzyme amount pH [H_2O_2] [SMX]	Al-Maqdi et al. (2018)
MnP, LiP	Synthetic dyes	RSM-CCD	Temperature Reaction time [H_2O_2]	Sosa-Martínez et al. (2020)
MnP	Textile dyes	OFAT	pH Temperature [H_2O_2]	Rekik et al. (2019)
MnP, Laccase	PAHs, dyes	OFAT	[Metal ions] [Organic solvents]	Zhang et al. (2016)
Laccase	Acetaminophen	OFAT	pH	Wang et al. (2018)
Laccase	Malachite green	RSM-CCD	Enzyme amount pH Reaction time [Malachite green]	Shanmugam et al. (2017)
Laccase	Malachite green	ANN	pH Temperature	Rashtbari and Dehghan (2021)
Laccase	Reactive blue 19	OFAT	Reaction time pH Temperature [Reactive blue 19]	Dauda and Erkurt (2020)
Laccase	2- and 4-chlorophenol	OFAT	Reaction time Enzyme amount Flow rate pH Temperature [2- and 4-chlorophenol]	Menale et al. (2012)
Chloroperoxidase	Lincomycin	RSM	pH Enzyme amount [H_2O_2] Reaction time	Zhu et al. (2020)
Chloroperoxidase	Acid Blue 113 Direct Black 38 Acid Black 10	OFAT	Enzyme amount pH Temperature [Acid Blue 113] [Direct Black 38] [Acid Black 10]	Jin et al. (2018)

(continued)

Table 4.6 (continued)

Enzymes	Pollutant groups	Statistical approach	Optimized factors	References
Crude extracts (2-HBP-3-monooxygenase, Catechol-2,3-dioxygenase, benzoate dioxygenase)	2-hydroxybiphenyl (2-HBP), catechol and benzoic acid	OFAT	pH Temperature Enzyme amount [Pollutants]	Younis et al. (2020)

et al. 2020; Teiri et al. 2020; Rashtbari and Dehghan 2021). However, like any method, ANN also has its share of disadvantages, like, its complexity for adapting the system, interpretation of the results and the need for a large amount of data to have accurate results. Furthermore, it is not always feasible for all bioremediation types (in-situ or ex-situ) and scales (laboratory, pilot, or on-site).

In addition, various remediation types have multiple responses or will be constrained (Mahunon et al. 2018; Ntakiyiruta et al. 2020; Hu et al. 2021). In such cases, applying a multi-response approach would be highly recommended to identify the best compromise (Zhao et al. 2020).

4.2.1 In-situ Bioremediation Processes Optimization

4.2.1.1 Phytoremediation

Phytoremediation utilizes plants and their interactions to aid in remediating polluted sites, including soil, water, air, sediments, sludge, among others. Such interactions could be physical, chemical, biochemical, or involve the synergistic action of microorganisms to degrade, remove or transform contaminants (Wang et al. 2015; Li et al. 2019).

This technique has been depicted as a less expensive technique than other bioremediation methods; however, its utilization is hindered by many considerations like low survival of plants, their adaptability to contaminated sites, accumulation of toxic products, likely volatilization of organic contaminants, etc. Indeed, considerable efforts have been made to overcome such limitations and gain more benefits from this technique. This chapter section focuses on the most recent optimization strategies, critical factors affecting the process, and the most commonly utilized target responses to improve the process efficiency.

A wide range of contaminants has been removed by phytoremediation like heavy metals, organic contaminants, petroleum hydrocarbon and radioactive materials. Depending on the pollutant type, the involved phytoremediation mechanisms can differ from phytoextraction, phytofiltration, phytostabilization, phytovolatilization and phytodegradation; thus, the variables influencing the process are numerous.

However, one factor will be crucial for the phytoremediation process despite the selected mechanism, which is the selection of plant species. The preferable characteristics are the tolerance to wide climatic conditions, high biomass yields, fast growth rates, low agricultural requirements and high survival rate in contaminated sites (Li et al. 2019). Usually, native species present such desired characteristics. Also, macrophytes, plants capable of hyper-accumulating contaminants like heavy metals, are commonly exploited. Apart from plant selection, environmental factors such as climate, pH, temperature, and contaminants characteristics play a crucial role in phytoremediation.

As many factors might influence this technique, establishing relationships between the plants, microbiome, surrounding conditions are crucial for its efficiency. In this regard, over the past years, particular emphasis has been made on avoiding one-factor-at-a-time traditional approaches and selecting modeling and optimization techniques that provide a better understanding of the biological, physical, chemical and biochemical processes involved and their interactions (Mahunon et al. 2018; Selamat et al. 2018; Jaskulak et al. 2020).

Table 4.1 summarizes the latest works on various pollutant type removal by phytoremediation. Reviewed process optimization studies were carried out under laboratory scale (Containers, tanks, pots), pilot systems (Wetland, reed bed, tanks) and also on site. Although the heterogeneity and variability of field conditions hinder the replicability of the optimized process under controlled conditions, applied optimization methods aid in predicting the phytoremediation process potential to a given contaminated scenario. Most of Table 4.1 studies optimized the process via response surface methodology, although other designs were also employed, such as the Taguchi method, ANN, complete factorial design (CFD) and even OFAT. Evaluated factors depended on the pollutant type, system and bioremediation mechanism. Such factors can directly impact the phytoremediation process, influence the pollutant's availability, affect the plant growth, or boost a series of synergistic activities from a biological perspective (plant–microorganism), all of which will translate into the desired target. Likewise, target optimization responses are also numerous, although contaminant removal efficiency (%) was undoubtedly the primary target among all the studies.

Regarding metals removal, the removal efficiency was the optimization objective either in terms of removal percentage or contaminant removed per plant biomass. However, other quality characteristics were quantified to evaluate the potential of plant species, such as the relative growth rate (RGR, mg/g.d) (de Souza et al. 2019), tolerance index (TI, %) (de Souza et al. 2019), translocation factor (TF, contaminant concentration in shoots/contaminant concentration in roots) (de Souza et al. 2019; Mohamad Thani et al. 2020; Razmi et al. 2021), bioaccumulation factor (BF, contaminant concentration in plant/contaminant concentration in the medium) (de Souza et al. 2019; Razmi et al. 2021), plant physico-chemical characteristics (biomass, length, chlorophyll content) (Li et al. 2019), shoot uptake (contaminant concentration in shoot/shoot dry matter) (Razmi et al. 2021), root uptake (contaminant concentration in root/shoot dry matter) (Razmi et al. 2021), phyto-extraction efficiency (shoot uptake/root dry matter) (Razmi et al. 2021), uptake efficiency (shoot + root uptake/root dry matter) (Razmi et al. 2021). Such responses will be of interest depending on the phytoremediation mechanisms such as phytoextraction, phytostabilization, phytovolatilization and rhizofiltration (Jaskulak et al. 2020; Razmi et al. 2021).

Regarding the evaluated factors, the most evaluated were contaminant concentration and time; other factors included nitrogen and phosphorous concentration, pH, temperature, aeration rate, plant type, and microorganism (see Table 4.1). Regarding the contaminant's concentration, RSM and ANN studies found an interactive effect between the concentration and other factors such as aeration rate (Titah et al. 2018), time (Selamat et al. 2018), nitrogen (Selamat et al. 2018) on the removal efficiency. Indeed, concentration-effect and optimum level might depend on the selected ranges of settings for the study, and one should consider such interactive effects. For example, for the removal of As and Pb, the sampling time's effect varied depending on the initial contaminant concentration and so did the optimum levels (Selamat et al. 2018). Environmental conditions such as pH and temperature are crucial for plant growth and optimum levels will depend upon the chosen plant type. The aeration rate was evaluated on the pilot reed bed, and the effects varied according to the pollutant type; while an increment in the aeration rate increased the arsenic removal (Titah et al. 2018), no aeration was required for Pb removal (Selamat et al. 2018).

Nitrogen and phosphorous composition in the medium are also crucial. Borges and collaborators demonstrated that nitrogen (in the form of nitrate) affected the medium's oxidation potential and, consequently, affects arsenic phytoremediation (de Souza et al. 2019). The same authors also demonstrated that phosphorous (in the form of phosphate) competes with arsenic for adsorption because they are transported into the plants using the same mechanisms (de Souza et al. 2019). Likewise, the addition of other chemicals such as chelating agents, surfactants, and low molecular weight organic acids to enhance the phytoremediation process is crucial and thus is convenient to evaluate such factors to improve the process (Razmi et al. 2021).

Other factors are related to improving the plants' survival in the contaminated site by supplementing the soil with soil amendments (Li et al. 2019; Jaskulak et al. 2020) or by boosting the phytoremediation process by the synergism between plants and microorganisms (Li et al. 2019; Razmi et al. 2021). The addition of sewage sludge was a significant factor to increase cadmium removal by *Sinapis alba L.* in soil (Jaskulak et al. 2020). Likewise, the combined addition of microorganisms and soil amendments resulted in a higher phytoremediation efficiency by soybean in terms of plant growth parameters and heavy metals removal (Li et al. 2019). The relevance of various biological and chemical associated factors on the phytoremediation process of contaminated soil (Pb, Ni, Cd, Zn) was evaluated via the Taguchi method and it was demonstrated that the most important factors were plant type and fungi inoculation (Razmi et al. 2021). Likewise, the multicriteria assessment (evaluation of multiple responses) allowed the authors to determine the phytoremediation mechanism associated to the plant type (phytoextraction or phytostabilization) (Razmi et al. 2021).

In the case of the phytoremediation process optimization for wastewater treatment, the removed contaminants include organic matter, ammonium, nitrate, phosphorus, petroleum hydrocarbons, among others (see Table 4.1). Studies were carried out in laboratory scale tanks and other pilot systems such as artificial wetlands, which were depicted as promising systems for wastewater treatment (Darajeh et al. 2016). Based on pollutants types, system and plant types, different mechanisms can take place like biodegradation and filtration, which, similar to the above-mentioned studies, are affected by many factors, including environmental and biological variables. Notwithstanding, other crucial factors influencing the process are plant density, biological oxygen demand (BOD) and chemical oxygen demand (COD). Various studies have agreed that increasing the plant density favors the wastewater treatment process efficiency, possibly due to increased surface area contact available for absorbing, filtering or degrading pollutants (Darajeh et al. 2016; Mahunon et al. 2018; Ntakiyiruta et al. 2020).

Furthermore, phytoremediation has also shown potential for removing volatile organic compounds (VOCs) from indoor air through stems and leaves absorption, plant metabolization and further degradation by rhizosphere microorganisms (Teiri et al. 2020). Recently, the phytoremediation process of formaldehyde by *Chamaedorea elegans* was optimized via artificial neural networks and response surface methodology, and it was reported that one of the most important factors in removing the pollutants was the unit leaf surface area (Teiri et al. 2020).

Phytoremediation is also promising for radionuclides-contaminated soil, where one crucial aspect is to improve the phytoavailability of radionuclides (Hu et al. 2021). The authors proposed using chelating agents to enhance the radionuclides phytoavailability and uptake; however, high doses of chelating agents could negatively affect plant growth. Therefore, the authors proposed a constrained optimization method where two responses were considered, plant growth and total bioaccumulation amount, and their constrained condition was to keep the biomass growth equal to that of the plant without using a chelating agent. Indeed, the effect of the factors has been evaluated for various responses related to contaminant removal and plant growth aspects as the ones presented in Table 4.1; however, in most of those optimization studies, authors only selected a single response as the optimization target. A useful approach would be to select a target that encompasses various responses such as RGR, TF, BF, plant physical and chemical characteristics when applicable. Examples of such multicriteria-optimization approaches can be referred to elsewhere (Mahunon et al. 2018; Ntakiyiruta et al. 2020).

It is evident that previous knowledge regarding the plant and pollutant type, removal mechanisms, ambient conditions and system characteristics (soil, water, or air) helps select the factors, the range levels and consequently the optimization method. Further, being the phytoremediation process efficiency quantified by a high number of responses, a multi-response optimization method would be useful to tackle the contamination problem in a holistic approach encompassing all the desired targets.

4.2.1.2 Bioaugmentation

Bioaugmentation is a bioremediation technique through the inoculation of microorganisms or activated sludge in soils affected by pollutants (Uniyal et al. 2021). For this purpose, pure strains or genetic material of bacteria, fungi, yeast, archaea, and algae are used; microbial consortia are also used, which allows using metabolic capabilities of several microorganisms.

Mainly bioaugmentation approach is used for pesticides, herbicides, fungicides, hydrocarbons, and pharmaceuticals' biodegradation (Karas et al. 2016; Uniyal et al. 2021). It is low cost compared to biostimulation, although it is also common to implement a bioaugmentation-biostimulation technique. Many environmental and microbial growth factors impact the efficiency of a bioaugmentation process. It is a complex system due to the existence of many microbial communities and the multiple generated reactions when adding a microorganism since the entire metabolism is interconnected.

The most recent bioaugmentation process optimization, applied methods, evaluated factors and other details are summarized in Table 4.2. It can be seen that bioaugmentation optimization can be approached by different routes: (1) To optimize the growth conditions of the degrading strains; (2) to optimize environmental conditions that influence the degradation capacity of degrading strains; (3) to optimize both growth and environmental parameters to determine ideal conditions for degradation. Evaluated process parameters will depend upon the route taken. Microbial-related factors included inoculum type, size and protocol, microbial consortium ratio, the addition of substrates and co-substrates. In comparison, environmental process parameters comprehended pH, temperature, soil type and moisture, salinity, and conductivity. Indeed, the number of factors affecting the process is high. Therefore, reviewed studies employed full factorial designs, Taguchi method, response surface methodology and screening designs to select the most significant variables, such as the Placket Burman Design (PBD) for optimization. A novel optimization scheme was proposed to combinedly optimize technical and environmental targets, where various techniques were integrated, including Life cycle assessment (LCA), PBD and RSM (Zhao et al. 2020).

One of the main challenges of bioaugmentation is the long adaptation time required by microorganisms to the pollutants and polluted sites. Therefore, selecting suitable degrading strains is a primary target. A common practice is to isolate a microorganism or a microbial consortium from a contaminated environment, then increasing the degrading capacity by subjecting it to a gradual increase of contaminant concentrations and even using the contaminant as the sole source of carbon (Nguyen et al. 2019). Moreover, researchers are inclined to isolate and utilize microorganisms from the contaminated place because they could adapt faster to the conditions and the stress caused due to the competition with other existing microorganisms (Di Gregorio et al. 2016; Li et al. 2016; Nguyen et al. 2019). Likewise, depending on the complexity of the pollutant and site conditions, it is recommended to use the bacterial consortia (Li et al. 2016).

Certainly, determining if a strain is ideal for bioaugmentation is more complex than looking only at the biodegradation rate due to the metabolic activity of the microbial communities that exist in the soil or sludge. The addition of a certain microorganism not only stimulates the pollutants elimination but also can affect the growth and degrading activity of other microorganisms. Studies have also been done to achieve biodegradation using immobilized or encapsulated inoculum, allowing greater stability and resistance in an extended period to the microbial cells (Mazumder et al. 2020). Of course, this alternative brings the necessity to optimize other process parameters such as optimal cell concentration, nature and quantity of encapsulating materials and immobilizer. In any case, coupling microorganisms to the environment's conditions is essential so that they do not lose activity (Xiong et al. 2017). This is why various authors agree that one key factor for successful bioaugmentation is the inoculum's optimal survival and colonization (Table 4.2).

On the other hand, for optimizing environmental process parameters, studies were carried out on a laboratory scale, simulated microcosm, pilot scale, and on-site (Table 4.2). Even though optimal conditions are determined in lab or microcosm studies, it is difficult to reproduce in actual polluted sites. Likewise, the presence of certain pollutants alters the characteristics of soil and water, such as electrical conductivity, pH, salinity, and others (Fauziah et al. 2017). From the environmental factors, studies have shown that pH will figure prominently. Even when the initial pH could be adjusted, as the metabolism progresses, organic acids and other synthesized compounds could alter the pH of the medium, triggering another series of metabolism by other groups of bacteria, such as acidophiles in some cases, which could result in solubilization of metal ions, and makes the metals more available and easily bio-accumulated (Hassan et al. 2019). Also, in various works where the partial bioremediation of contaminated soils was achieved, the pH was highly related to salinity and redox potential influence (Hassan et al. 2019). Although favorable results can be achieved initially, most of them agreed that they are not entirely effective at later stages, mainly because the inoculum fulfills its function in the first stage. However, synergistic action by other species is necessary to achieve overall efficiency (Fauziah et al. 2017).

Unlike other bioremediation techniques, time is not a critical parameter to optimize. Evidently, it is an important variable, but it is independently evaluated. Mainly because even when the selected strains' degradation capacity has already been tested, the time of acclimatization to the contaminant is a function of the pollutants type, genetic plasticity of the bacteria, resulting in faster degradation rates (days), or taking a longer duration of months (Zheng and Chai 2016; Wu et al. 2018; Feng et al. 2021).

It can be observed that there is no perfect route or optimization method, especially when there are many variables to deal with. By all means, identifying the factors that intervene positively and/negatively could provide an insight to choose the right steps. Additionally, knowledge of the contaminated site's contour, topography and behavior are to be considered to develop efficient bioremediation strategies.

4.2.1.3 Biostimulation

Biostimulation is another technique for bioremediation of contaminated soils and waters, which consists of modifying the physico-chemical conditions of the environment so that the microorganisms present in the site are activated and can biotransform or biodegrade pollutants (Martínez Álvarez et al. 2015). It is a simple method that does not require specialized equipment for its implementation.

In literature, it is common to find the combined application of biostimulation with bioaugmentation. Nevertheless, the optimization of the conditions considering a single strain or consortia underestimates the degradation capacity of endogenous strains.

The optimization of this technique presents two variants, optimizing environmental and growth parameters without considering a particular microorganism and optimizing conditions for a specific degrading microbial group detected in the area. As shown in Table 4.3, many compounds and factors can generate the desired effect on degrading microorganisms. Thus, it is necessary to know the microbial communities present, metabolic types, and the biodegradation capacity of endogenous microorganisms.

Experimental designs' common responses to determine the effects of biostimulation are microbial growth and degradation rate. It has also been reported that the measurement of metabolic activity through parameters such as ATP generation, elements of the electron transfer system, and activities of certain enzymes such as dehydrogenase give a clear idea about whether it is promoting metabolism or not (Chen et al. 2019).

Due to the low concentration of nutrients in the soil, most evaluated parameters are related to the microorganism growth stimulation by adding various nutrients (Li et al. 2016). Adding large amounts of nutrients could be detrimental by increasing the costs. Therefore, optimizing the process to ensure the highest biodegradation rate at a lower cost is critical. The addition of different carbon sources or adjusting the C/N ratio is proven to increase biodegradation. In general, two large groups of nutrients can be mentioned, agro-industrial residues and specific sugars.

Although it is desired that the microorganisms present in the contaminated areas use the pollutant as the only carbon source, it has been observed that the addition of co-substrates such as glucose, lactose, sucrose, etc. improves bioremediation. Biostimulation with glucose accelerated the elimination of a synthetic plasticizer by *Rhodovulum* sp. (Baker et al. 2021); however, biostimulation and bioaugmentation are not always applied simultaneously, so determining the appropriate carbon source for soil and water is crucial. In general, when the carbon source concentration is low, the removal of the pollutant decreases; when an optimal concentration is used, the degrading microbial communities increase in number and generates a higher biodegradation rate (Khayati and Barati 2017).

Using certain amendments such as humic substance, industrial fermentation waste, digestates, agro-industrial waste, manure from different livestock, municipal waste, molasses, whey, and others (Agarry and Ogunleye 2012; Qiao et al. 2014;

Khayati and Barati 2017; Bianco et al. 2020) aids in obtaining adequate nutrients, trace element, vitamins (Xiong et al. 2017) and optimum C/N ratio, increases the process efficiency and reduces the process cost.

The nutrient's physicochemical characteristics influence the growth of autochthonous microorganisms and the bioavailability of contaminating compounds. Studies on the use of different types of nutrients report that factors such as pH, C:N ratio, and redox potential have a strong influence on the biodegradation rate; and the type of influence, *viz.*, positive or negative, depends on the environment, microorganisms, pollutants (Hassan et al. 2019).

In some cases, inorganic nutrients such as calcium peroxide, urea, saltpeter, $FeSO_4^-$ are used as microbial stimulants (Simpanen et al. 2016; Khayati and Barati 2017; Chen et al. 2019) as they are homogeneous and unlike residues which are heterogeneous, and their elemental composition varies from one batch to another. Likewise, it has been reported that the use of fertilizers in contaminated soils favors the biostimulation process and facilitates the degradation of hydrocarbons, pesticides, herbicides and heavy metals (Da Silva et al. 2009).

The addition of surfactants in soils and bodies of water contaminated with oil and its derivatives has been shown to stimulate biodegradation; the most used is Tween 80 (Agarry and Ogunleye 2012; Khayati and Barati 2017), although various commercial surfactant brands are also reported (Da Silva et al. 2009). Although this amendment is not a nutrient source, it increases the bioavailability of oily compounds to microorganisms.

A strong correlation between bioremediation and N and P content has been reported and C:N:P ratio is one factor that favors or stimulates biodegradation (Carlos et al. 2016). The optimal C:N:P ratio has been widely studied and 100:10:1 is the reference value for the biodegradation of different pollutants (Júlio et al. 2018). However, each biostimulation test has different properties and the optimization of this parameter has led to different outcomes (Da Silva et al. 2009; Martínez Álvarez et al. 2017; Júlio et al. 2018).

With respect to removal of certain contaminants like hexavalent chromium (Carlos et al. 2016), total petroleum hydrocarbons (Qiao et al. 2014), diesel-fuel (Kauppi et al. 2011), biostimulation has been reported to be the best technique compared to bioaugmentation, as this process stimulates the growth and activity of innate microorganisms by providing optimum nutrient conditions.

4.2.2 Ex-situ Bioremediation Processes Optimization

4.2.2.1 Bioremediation in Bioreactor

Bioprocess at reactor level could be operated through different modes (batch, continuous, fed-batch, multistage) and by providing the optimum conditions that satisfy microorganisms growth and activity, the removal of contaminants is achieved (Azubuike et al. 2016). Many critical variables must be optimized and controlled, including process parameters, nutrient conditions and others to obtain high efficiency (Table 4.4).

Though fungi have been reported to tolerate less to high temperatures, they have the advantage of having enzymatic machinery with high hydrolytic capacity and affinity for many molecules. The optimum temperature range is broader in the case of bacterial strains. An optimal temperature of 38 °C has been reported for degradation of methylene blue by *Acinetobacter pittii* (Ogunlaja et al. 2020), while *Enterobacter* sp. degraded a textile effluent at an optimum temperature of 35 °C (Amini et al. 2019). It is suggested that high temperatures increase bacterial metabolic activity and thereby increases the substrate biodegradation rate. On the other hand, the optimum temperature for the bacterial degradation of various agrochemicals has been reported in the range of 30–33 °C (Gao et al. 2020a; Samadi et al. 2020). Regarding yeast, the strains *Rhodotorula* sp., *Debaryomyces hansenii* and *Hanseniaspora valbyensis* were able to degrade Benzo [ghe] perylene (polyaromatic hydrocarbons, PAH) at 30 °C with an efficiency of 63.83% (Mandal et al. 2018).

Like temperature, pH is another important factor in the function of a specific microorganism and its requirements. The pH can be adjusted to an initial value, but the microorganism's metabolism changes the pH due to product formation and substrate consumption. Authors could only evaluate the initial pH effect or its control to a specific range, but most experimental designs only consider the initial pH and measure its value at the end of the process. The final pH value is useful to explain the possible changes occurring during the biodegradation process. For pyrene degradation by *Pseudomonas aeruginosa*, the optimum pH was 5.5, but it increased to 8.5 at the end of the fermentation, and it was attributed to the optimum required level for the involved enzymes (Li et al. 2020).

Regarding time, the required duration of the process highly depends on microorganism and contaminant type. The optimal time it takes for a microorganism to reach a high biodegradation rate varies from few days up to a month (Ameen and Alshehrei 2017; Mohanty and Jena 2019; Derbalah et al. 2020; Li et al. 2020). Again, it depends on the genetic and metabolic capacity of the strain.

Agitation is another relevant parameter to process efficiency as it is related to the oxygen transfer rate. Likewise, for some microorganisms, the agitation can affect their morphology due to shear forces. Optimum reported levels (120–150 rpm) are similar for various microorganisms and contaminants types (Mandal et al. 2018; Gao et al. 2020b). However, understanding its effect on the process efficiency is crucial for the scale-up of the process since similar power input and oxygen levels will be needed in the scaled-up process.

As mentioned previously, the addition of co-substrates favors the process efficiency through co-metabolism. The most common source of co-substrate is glucose; it is cheap and easily metabolized. However, some microorganisms prefer another type of carbohydrate. For the degradation of Atrazine by *Klebsiella variicola*, different carbon sources were tested, sucrose, malt powder, soluble starch, corn flour and glucose, and it was observed that the highest growth and degradation rates were achieved with sucrose (Gao et al. 2020a). Thus, assessing the best carbon source before optimizing the process is a reasonable strategy like the one followed for the degradation of Reactive Red 120 by *Aspergillus flavus* (Ameen and Alshehrei 2017). The ideal carbon source was screened via PBD; then, the selected source concentration was optimized for the degradation process (Ameen and Alshehrei 2017). Regarding carbon source concentration, it has been reported that high concentrations might result in low biodegradation rates, due to acidification or as the membrane permeability is affected (Gao et al. 2020a; Seyedi et al. 2020).

Among the various nitrogen sources, yeast extract, peptone, ammonium sulfate, potassium nitrate, urea, malt and soybean meal are widely used. Some microorganisms prefer an inorganic nitrogen source, such as *Klebsiella variicola,* for Atrazine degradation (Gao et al. 2020a). In contrast, *Halomonas desiderata* preferred peptone as a nitrogen source to degrade Reactive Black 5 and Reactive Red 152 (Seyedi et al. 2020). Selecting an appropriate nitrogen source (organic or inorganic) is key to increase the microorganism biodegradation capacity. Similar to the carbon source, after selecting the nitrogen source, its concentration effect on the target response must be evaluated.

Similar to biostimulation, various polluting wastes can be used ex-situ like industrial or municipal water effluents. Those wastes are rich in organic and mineral compounds that serve perfectly as a source of carbon, nitrogen and provides the required minerals necessary for growth. If the microorganisms are capable of growing by using such raw materials could be favorable in terms of environmental and economic aspects. Even though the effluents could provide the carbon source for growth, it is necessary to optimize its concentration or the C/N ratio to achieve higher degradation levels.

Additionally, there are other variables related to the medium composition, such as the salt concentration; for various applications, the medium's salinity is relevant (Amini et al. 2019; Udaykumar et al. 2020). In this context, halotolerant microorganisms have a potential application for the treatment of textile effluents since the use of various types of salts is typical of the textile process (Seyedi et al. 2020).

Other important process variables are related to the microorganism state, such as inoculum type, size and age. It has been reported that the inoculum size is a critical parameter affecting biodegradation processes (Mandal et al. 2018; Samadi et al. 2020). Likewise, it has been proved that higher inoculum levels do not necessarily imply achieving a higher degradation level (Mohanty and Jena 2019).

In addition to the above bioprocesses, microalgae have stood out as a natural resource for bioremediation due to their high efficiency in treating polluted water, eliminating nutrients and toxic substances, such as pesticides, herbicides, and heavy metals (Table 4.4). This characteristic is due to its ability to accumulate significant toxic compounds without affecting its biological activity. Another benefit of microalgae is the ability to remove CO_2 and produce O_2 through photosynthesis, degrading organic compounds and generating high biomass density, which can be reused in other processes, adding value to bioremediation (Martín-Girela et al. 2017). Microalgae cultivation offers a potential strategy for treating high-nutrient wastewater and obtaining high-value-added microalgae. However, microalgae for removing contaminants face different challenges, such as dissolved solids and competing microorganisms that can affect the process. The use of culture systems for the production of microalgae during the bioremediation process requires ensuring controlled culture conditions, so the use of closed culture systems such as photobioreactors offers significant productivity advantages that include high production and efficiency of contaminant removal.

Hence, the development of new types of bioreactors has gained attention in recent years (Martín-Girela et al. 2017; Guo et al. 2020; Lu et al. 2021). The most studied factors in the design of reactors for bioremediation purposes using microalgae have been comparing the configuration, the size of membrane pore, the species of microalgae and the lighting conditions in the system. In this context, the OFAT method has served to compare the condition that favors removing the pollutant under study. The effect of flashing light and continuous light from red light-emitting diodes in a biofilm reactor using a microalgae consortium was evaluated for CO_2 fixation, observing that flashing light increases CO_2 uptake in bioreactors type (Martín-Girela et al. 2017). The use of biofilm reactors has also shown favorable results in removing phosphorus, nitrogen, and ammonia, showing that light intensity and retention time are the main factors that affect the ability of microalgae to remove pollutants (Ashok et al. 2019).

In membrane reactors, the pore size is a crucial factor to increase the efficiency of conversion or fixation of pollutants. This was demonstrated using a novel porous nickel-foam-filled CO_2 absorptive photobioreactor system to convert CO_2 to $NaHCO_3$ to improve photosynthesis of microalgal cells; increasing the pore size decreased the conversion capacity (Guo et al. 2020). The influence of pore size can be explained using the Ergun equation, where viscous resistance and inertial resistance are negatively correlated with the nickel foam pores' diameter, which blocks the passage of CO_2 and decreases the conversion reaction.

Membrane reactors have also been used to remove metals such as Cr (VI), where solid retention time and hydraulic retention time are the main factors to consider and can be controlled independently. Under a continuous culture system operation using *Chlorella vulgaris*, a reduction in Cr (VI) concentration of 50% was achieved at 3 days of hydraulic retention and 40 days of solids retention time (Lu et al. 2021).

In summary, a two-stage study would be an effective way to optimize the process performance at a bioreactor level due to the variables affecting the bioprocess include aspects related to the process conditions, medium factors, and inoculum characteristics. A first approach could be dedicated to selecting the significant factors affecting the target; then, depending on the number of factors, the design can be chosen to optimize the bioremediation process.

4.2.2.2 Biosorption

Biosorption is one of the most popular bioremediation techniques, which involves using natural materials to remove environmental pollutants (Banerjee et al. 2016; Daoud and Selatnia 2019). While biosorption via microorganisms has been successfully demonstrated for inorganic and organic materials, heavy metals removal has been focused mainly.

The technique involves various physical and chemical mechanisms (ion exchange, surface complexation, electrostatic attraction, diffusion); however, the process parameters are mostly optimized in terms of the adsorption principles. Unlike other bioremediation techniques, the number of factors influencing the process is not high. Table 4.5 summarizes the most recent works involving the biosorption process optimization utilizing microorganisms for removal of various contaminants, where it can be seen that the most evaluated factors were pH, temperature, residence time, biosorbent and pollutant concentration. Similar to previous techniques, the most popular optimization approach was applying multivariate designs (RSM, ANN, Taguchi Method), although an OFAT approach was also employed. The standard optimization target objective was the removal efficiency (%). In addition, other responses were quantified to assess the process potential and understand the interactions between the biosorbent and the pollutants and the nature of the process, including pollutant uptake (q_t, mg of contaminant per gram of biosorbent), equilibrium pollutant uptake (q_e, mg of contaminant per gram of biosorbent) and thermodynamic parameters (enthalpy change, entropy change and Gibb's free energy change). The prevailing method of addressing the biosorption optimization was to evaluate the process parameters effect followed by biosorption kinetics studies, kinetics modeling and thermodynamic studies under optimized conditions.

One of the most evaluated factors is the pH since it affects the pollutants' chemical speciation as well as the cell surface charges and hence affects the active sites on the biosorbent (Banerjee et al. 2016; Kumar et al. 2018; Fawzy 2020). Generally, a pH range between 3 and 6 is reported for metals biosorption by microorganisms due to highly acidic pH values (<3) could cause a positive charge on the biosorbent; thus, inhibiting metal adsorption (Banerjee et al. 2016; Fawzy 2020). For instance, the biosorption capacity by *Pseudomonas azotoformans* biomass was evaluated and optimized via RSM-BBD for the removal of copper, lead and cadmium (Choińska Pulit et al. 2018); while the pH effect was not significant for removal of cadmium in the pH range of 4–6, but was significant for copper (linear and quadratic) and lead (linear only). Additionally, the interactive effects between pH and other factors (biosorbent and metal concentration) were negligible for all the metals. Similarly, the use of RSM for the biosorption of Cu by *Chlorella pyrenoidosa* showed that the interactions of pH-biosorbent concentration and pH-copper concentration were not significant (Moreira et al. 2019). If the interactions between pH and other factors are assumed negligible, a wide range of pH could be evaluated by other methods such as OFAT or Taguchi method. An example of such an approach was the biosorption studies conducted for nickel removal by *Pleurotus mutilus*, where the process was first evaluated under a wide pH range (3–10) by keeping other parameters constant then further optimized through the Taguchi method (Daoud and Selatnia 2019). Similarly, an L16 orthogonal array was used for the biosorption of Pb (II) and Cu (II) with the cyanobacterium *Aphanizomenon ovalisporum* and pH had the greatest impact on the removal of Pb and Cu among the evaluated factors (metal concentration, algal dose, pH, and contact time) (Flouty et al. 2019). On the other hand, other authors have identified an interactive effect between pH and biomass concentration via RSM using microalgae biomass (Sultana et al. 2020; Ayele et al. 2021). In any case, the biosorbent's functional groups ionic state and pollutant characteristics are useful for selecting a suitable pH range for optimization studies,

Regarding the effect of biosorbent loading, the usual pattern is that the removal efficiency (%) increases by increasing biomass concentration due to a higher surface area availability (Kumar et al. 2018; Sen et al. 2018; Fawzy 2020). However, for the uptake capacity (mg of contaminant per gram of biosorbent), usually, a maximum is reached; that is, the biosorbent reaches a maximum capacity to adsorb a contaminant independently of increasing the biosorbent dosage (Melvin et al. 2015; Banerjee et al. 2016). It has also been recommended to consider biomass aggregation when increasing biosorbent loading (Melvin et al. 2015) and the particle size (Daoud and Selatnia 2019) due to it can affect the available surface area and consequently the removal efficiency. Likewise, various authors have reported significant interactive effects between the biosorbent loading and contaminants concentration (Flouty et al. 2019; Fawzy 2020). It has been observed that at low contaminants concentration, increasing the biomass loading does not affect the process; on the other hand, the removal efficiency is highly reliant on the biomass loading at high contaminants concentration levels attributed to the available sites for biosorption (Flouty et al. 2019; Fawzy 2020).

Although the temperature is not the most significant factor, evaluation of its effect aids in understanding the process thermodynamics (Singh et al. 2010; Banerjee et al. 2016). For instance, for the removal of copper by fungal biomass, the biosorption reduction by increasing temperature indicated that the process is exothermic (Singh et al. 2010). Similar results were reported for the biosorption of the dye reactive black B by *T. harzianum* where a negative enthalpy value indicated that the sorption process was exothermic. Furthermore, a removal efficiency reduction by increasing the process temperature has been attributed to a higher adsorbate's mobility causing desorption (Melvin et al. 2015).

As for residence time, its effect contribution evaluation is critical to establish the process kinetics and the required time to reach the equilibrium (maximum adsorption). Kinetic models are highly used to understand the mechanisms in play during the biosorption processes (Singh et al. 2010; Flouty et al. 2019; Fawzy 2020; Rozman et al. 2020). Also, the residence time might affect the distribution between the contaminants and the biosorbent surface in a solution (Flouty et al. 2019). For this reason, a prolonged time does not necessarily imply a higher removal efficiency and various authors have reported the positive effect of short residence time (Singh et al. 2010; Melvin et al. 2015; Flouty et al. 2019; Fawzy 2020; Rozman et al. 2020). That is, at the beginning of the process, the adsorption sites on the biomass surface are high, but as the process develops, the repulse forces between the free and adsorbed contaminants hinder the further adsorption of remained free contaminants (Flouty et al. 2019). Furthermore, the utilized system might affect the process kinetics.

Undoubtedly, all the efforts deployed provide guidance to clearly know which factors affect a biosorption process and are crucial to evaluate when new biomass sources are tested. Likewise, future biosorption optimization studies should focus on establishing the optimum conditions using samples from real polluted sites to understand the process mechanisms in a system containing a complex mixture of pollutants (both organic and inorganic).

4.2.2.3 Enzymatic Bioremediation

As previously discussed, microbial degradation offers an environment-friendly and cost-effective treatment for pollutant bioremediation. Nonetheless, it has been proposed to utilize the microbial enzymatic extracts rather than the microorganisms to overcome certain disadvantages such as the long adaptation time, inhibitory effect of the contaminants on growth and activity of the microorganisms, or biomass accumulation. Moreover, utilizing microbial isolated enzymes or crude enzymatic extracts could provide a more efficient and better-controlled process (Okino-Delgado et al. 2019).

The required enzyme for the bioremediation process will depend on the type of pollutant. For instance, oxidoreductases can neutralize pollutants rich in freed radicals, while hydrolases can help the decomposition of organic compounds (Bhandari et al. 2021). In this regard, oxidoreductases are the leading group of enzymes applied in bioremediation, followed by hydrolases and some lyases. Manganese peroxidase, lignin peroxidase, laccase, versatile peroxidase, from microbial sources are enzymes that can degrade a wide range of highly polluting and recalcitrant compounds such as phenol, dyes, pesticides, herbicides, fungicides, pharmaceuticals and personal care products. These so-called green catalysts are eco-friendly because their use considerably reduces the use of oxidizing chemical components, reducing the time and energy in degradation (Bhandari et al. 2021).

However, enzymes' application in bioremediation still faces considerable challenges even though these biomolecules can catalyze the breakdown of most contaminating compounds; the main disadvantage is the high cost of production and purification. For this reason, process optimization is of relevant importance both in the production of enzymes and in biodegradation, thus ensuring the use of resources to fulfill better the gaps that hinder the enzymes' application on an industrial scale.

Extensive interest has been placed on compiling the existing data on the promising enzymes for bioremediation applications, developed technologies, enzymatic degradation mechanisms by enzymes and the challenges yet to be met. Table 4.6 summarizes the most recent works on this subject, the selected process optimization method, pollutant types, utilized enzymes and evaluated factors. Unlike other bioremediation methods, the enzyme-assisted processes for pollutants removal/biodegradation are better controlled and the number of factors influencing their performance is not as high. Of course, the process still could be affected by numerous variables and those are the ones related to the enzyme's biochemical properties and their mode of action. Highly studied factors include temperature, pH, the concentration of enzymes, pollutants, mediators, and reaction time. The most common target response is the pollutant degradation/removal (%), although ensuring enzyme stability at the process conditions is also relevant (i.e. half-life time).

Temperature is one of the main evaluated factors in enzyme-assisted bioremediation processes. Undoubtedly, optimal temperature values can vary depending on the enzyme type, even in the same enzyme's isoform. Thermostable enzymes are desirable for bioremediation, especially on an industrial level. Laccase has been reported to exhibit high degradation rates at temperatures in the range of 50–60 °C (Zhang et al. 2016; Dauda and Erkurt 2020). Besides, it is recommended to have records of the enzyme-producing microorganisms environment to select appropriate temperature optimization (Menale et al. 2012). For example, the laccase produced by Trametes orientalis had a maximum activity at 80 °C; this fungus was isolated from a region with an extremely hot climate (Zheng et al. 2017). Nevertheless, in general, laccase above 60 °C loses catalytic capacity affecting stability and structure (Menale et al. 2012).

Likewise, the type of contaminant also influences the optimal degradation temperature. The use of a crude extract produced by *Phanerochaete chrysosporium* showed different optimal degradation temperatures for various dyes (Sosa-Martínez et al. 2020).

Generally, peroxidases exhibit broader temperature ranges for optimum degradation results. MnP managed to degrade dyes and polycyclic aromatic hydrocarbons at an optimal temperature of 70 °C (Zhang et al. 2016). Chloroperoxidase of *Caldariomyces fumago* was used to degrade antibiotics, and was reported to have an optimal temperature of 60 °C (Zhu et al. 2020). However, if the enzymes are highly affected by the temperature, it leads to the enzyme's denaturation and negatively affects the pollutant degradation efficiency. Enzymes' stability can be improved by the addition of cofactors or redox mediators, which in addition to enabling the oxidation-reduction reactions, they could also provide stability to the structure of the enzyme. In the absence of MnSO, the optimum temperature for textile dyes degradation by MnP was 40 °C, while with MnSO addition, the maximum degradation was obtained at 50 °C (Rekik et al. 2019).

Regarding the effect of pH, most enzyme-assisted biodegradation processes occur mainly in mildly acidic or neutral environments (Younis et al. 2020). Peroxidases are stable in pH ranges from 3 to 8; however, reported optimal values for degradation by MnP is in the range of 4–5 (Rekik et al. 2019) and 4 for chloroperoxidase (Jin et al. 2018).

Laccase has optimal pH values of 4 to 5 for the degradation of a wide range of contaminants (Dauda and Erkurt 2020). A study on the optimization of pH conditions revealed the importance of this parameter for laccase due to its conformation with four Cu ions in an active site; the acidic pH allows the adequate transfer of electrons among one Cu ions (Wang et al. 2018). In general, the optimum pH value helps to maintain the stability of the catalytic site. Moreover, through response surface methodology analysis, it was shown that interactive effects between the pH and enzyme concentration were significant for malachite green degradation by laccase (Shanmugam et al. 2017). Those authors reported that acidic conditions were useful to keep the catalytic site's conformational stability and thus improve biodegradation efficiency.

However, the type of contaminant did not affect the optimal pH value in a crude extract produced by *Corynebacterium variabilis*, obtaining similar degradation rates for 2-hydroxybiphenyl (2-HBP), catechol and benzoic acid (Younis et al. 2020). The above indicates that the pH is an important factor that directly influences the enzyme's performance despite the target contaminant.

Various authors have demonstrated that enzymes have greater activity in biodegradation with the addition of a redox mediator. In the degradation of sulfamethoxazole, it was reported that a redox-mediator (1-hydroxybenzotriazole) for the complete degradation of the contaminant (Al-Maqdi et al. 2018).

For peroxidase enzymes, the concentration of H_2O_2 is a critical factor, so its optimization is decisive to carry out successful bioremediation. Low H_2O_2 concentrations cause low catalytic activity and high concentrations could cause oxidation of amino acid residues and heme group in the enzyme and inactivate it. Optimum H_2O_2 concentrations for degradation of various dyes by MnP and LiP varied from 0.01 to 2 mM (Rekik et al. 2019; Sosa-Martínez et al. 2020).

Certainly, enzyme's and pollutant's concentration are important factors to optimize in terms of the process economics, which becomes more relevant when a purified enzyme is applied. Usually, increasing enzyme concentration might increase the pollutant's biodegradation; however, it is crucial to establish the minimum enzyme concentration to achieve the desired contaminants degradation (Zhu et al. 2020). For malachite green biodegradation, it was observed that a relatively low enzyme concentration sufficed to achieve a higher degradation level (>90%) (Shanmugam et al. 2017); however, the dye degradation at intermediate levels was already >90% (Constant coefficient) in that response optimization study. Indeed, the evaluated settings (factors levels) are crucial to withdrawn conclusions with confidence.

From the above discussed, one can conclude that the most commonly evaluated factors were the temperature, pH, enzyme and pollutant concentration and reaction time. Having fewer variables affecting the process might explain the reason for the preference of the one-factor-at-a-time method for evaluating their effect in most of the studies (Table 4.6). Although it is not the most recommended method (due to interactive effects), as previously mentioned, the OFAT design offers the possibility to scan in greater detail the sole influence of one factor (more studied levels) by keeping other factors constant, particularly if those factors require a narrow-range-working level due to the enzyme's stability. Of course, knowledge of the enzyme's mode of action and properties is useful to assist in selecting the factors and their levels for process optimization studies. One reasonable approach could be to evaluate the enzymes under factors affecting their performance despite the contaminant, such as pH and temperature. Once the process settings that ensure enzymes' stability are selected, it would be useful to cover a wide range of enzyme/pollutant concentration levels and establish the process kinetics to increase efficiency both in terms of degradation yields and economics of the process.

4.3 Concluding Remarks

In light of the discussions above, it is clear that it is essential to consider the higher number of parameters to develop an efficient bioremediation strategy. Undeniably, any optimization method will have its own advantages and disadvantages. The approach followed will depend on the available resources and also on the previous knowledge of the process.

In cases where multiple factors could influence the bioremediation process, but little is known about their effect, it would be advisable to follow a two-step approach. The first reasonable step is to conduct screening experiments. Two-levels designs are convenient for selecting the most significant factors influencing the process and dismissing the not-significant ones. They will be helpful to guide the selection of factors and their levels for process optimization. Then, depending on the number of significant factors, one can select the most appropriate optimization method as the second step. Additionally, defining all the process targets early on is a fundamental start. Indeed, the pollutants removal/degradation is an obvious target; however, as discussed, some bioremediation processes efficiency is measured in terms of various quality targets or imply certain constraints. A multi-response optimization approach would be highly recommended to address this environmental pollution problem from a holistic perspective encompassing all the long-desired objectives.

Lastly, despite the selected method and bioremediation process, it is imperative to run a validation test of the optimum settings to achieve reproducibility and confidence of the results. It is essential to acknowledge that most of the above-discussed optimization methods are black-box methods, so the conclusions will only be relevant to the evaluated scale and factor's levels. Nevertheless, such methods and developed models help to assess the bioremediation process potential for a particular polluted site, establish the most critical variables for control, and can offer recommendations for process design, scale-up, and implementation strategies.

References

Agarry SE, Ogunleye OO (2012) Box-Behnken design application to study enhanced bioremediation of soil artificially contaminated with spent engine oil using biostimulation strategy. Int J Energy Environ Eng 3:1–14. https://doi.org/10.1186/2251-6832-3-31

Al-Maqdi KA, Hisaindee S, Rauf MA, Ashraf SS (2018) Detoxification and degradation of sulfamethoxazole by soybean peroxidase and UV + H2O2 remediation approaches. Chem Eng J 352:450–458. https://doi.org/10.1016/j.cej.2018.07.036

Ameen F, Alshehrei F (2017) Biodegradation optimization and metabolite elucidation of reactive red 120 by four different *aspergillus* species isolated from soil contaminated with industrial effluent. Ann Microbiol 67:303–312. https://doi.org/10.1007/s13213-017-1259-1

Amini B, Otadi M, Partovinia A (2019) Statistical modeling and optimization of toluidine red biodegradation in a synthetic wastewater using *Halomonas* strain Gb. J Environ Health Sci Eng 17:319–330. https://doi.org/10.1007/s40201-019-00350-5

Ashok V, Shriwastav A, Bose P, Gupta SK (2019) Phycoremediation of wastewater using algal-bacterial photobioreactor: effect of nutrient load and light intensity. Bioresour Technol Rep 7:100205. https://doi.org/10.1016/j.biteb.2019.100205

Ayele A, Suresh A, Benor S, Konwarh R (2021) Optimization of chromium(VI) removal by indigenous microalga (*Chlamydomonas* sp.)-based biosorbent using response surface methodology. Water Environ Res wer.1510. https://doi.org/10.1002/wer.1510

Azubuike CC, Chikere CB, Okpokwasili GC (2016) Bioremediation techniques–classification based on site of application: principles, advantages, limitations and prospects. World J Microbiol Biotechnol 32:1–18. https://doi.org/10.1007/s11274-016-2137-x

Baker A, Ahmad B, Alarjani KM, et al (2021) Biostimulation of *Rhodovulum* sp., for enhanced degradation of di-n-butyl phthalate under optimum conditions. Chemosphere 266:128998. https://doi.org/10.1016/j.chemosphere.2020.128998

Banerjee A, Sarkar P, Banerjee S (2016) Application of statistical design of experiments for optimization of As(V) biosorption by immobilized bacterial biomass. Ecol Eng 86:13–23. https://doi.org/10.1016/j.ecoleng.2015.10.015

Bhandari S, Poudel DK, Marahatha R et al (2021) Microbial enzymes used in bioremediation. J Chem 2021:1–17. https://doi.org/10.1155/2021/8849512

Bhatt P, Bhatt K, Sharma A, Zhang W, Mishra S, Chen S (2021) Biotechnological basis of microbial consortia for the removal of pesticides from the environment. Crit Rev Biotechnol 41(3):317–338

Bianco F, Race M, Papirio S, Esposito G (2020) Removal of polycyclic aromatic hydrocarbons during anaerobic biostimulation of marine sediments. Sci Total Environ 709:136141. https://doi.org/10.1016/j.scitotenv.2019.136141

Carlos FS, Giovanella P, Bavaresco J et al (2016) A comparison of microbial bioaugmentation and biostimulation for hexavalent chromium removal from wastewater. Water Air Soil Pollut 227. https://doi.org/10.1007/s11270-016-2872-5

Chen Y, Wang L, Dai F et al (2019) Biostimulants application for bacterial metabolic activity promotion and sodium dodecyl sulfate degradation under copper stress. Chemosphere 226:736–743. https://doi.org/10.1016/j.chemosphere.2019.03.180

Choińska-Pulit A, Sobolczyk-Bednarek J, Łaba W (2018) Optimization of copper, lead and cadmium biosorption onto newly isolated bacterium using a Box-Behnken design. Ecotoxicol Environ Saf 149:275–283. https://doi.org/10.1016/j.ecoenv.2017.12.008

Da Silva AC, De Oliveira FJS, Bernardes DS, De França FP (2009) Bioremediation of marine sediments impacted by petroleum. Appl Biochem Biotechnol 153:58–66. https://doi.org/10.1007/s12010-008-8457-z

Daoud N, Selatnia A (2019) Taguchi optimization method for nickel removal from aqueous solutions using non-living *Pleurotus mutilus*. Arab J Sci Eng 44:10067–10077. https://doi.org/10.1007/s13369-019-04108-x

Darajeh N, Idris A, Fard Masoumi HR et al (2016) Modeling BOD and COD removal from palm oil mill secondary effluent in floating wetland by *Chrysopogon zizanioides* (L.) using response surface methodology. J Environ Manag 181:343–352. https://doi.org/10.1016/j.jenvman.2016.06.060

Dauda MY, Erkurt EA (2020) Investigation of reactive Blue 19 biodegradation and byproducts toxicity assessment using crude laccase extract from *Trametes versicolor*. J Hazard Mater 393:121555. https://doi.org/10.1016/j.jhazmat.2019.121555

de Souza TD, Borges AC, Braga AF et al (2019) Phytoremediation of arsenic-contaminated water by *Lemna Valdiviana*: An optimization study. Chemosphere 234:402–408. https://doi.org/10.1016/j.chemosphere.2019.06.004

Debbarma P, Raghuwanshi S, Singh J, Suyal DC, Zaidi MGH, Goel R (2017) Comparative in situ biodegradation studies of polyhydroxybutyrate film composites. 3Biotech 7(178):1–9. https://doi.org/10.1007/s13205-017-0789-3

Deniz F, Yildiz H (2019) Taguchi DoE methodology for modeling of synthetic dye biosorption from aqueous effluents: parametric and phenomenological studies. Int J Phytoremediation 21:1065–1071. https://doi.org/10.1080/15226514.2019.1594687

Derbalah ASH, El-Banna A, Saad Allah M (2020) Efficiency of *Candida tropicalis* for potential degradation of metalaxyl in the aqueous media. Curr Microbiol 77:2991–2999. https://doi.org/10.1007/s00284-020-02121-0

Di Gregorio S, Siracusa G, Becarelli S et al (2016) Isolation and characterization of a hydrocarbonoclastic bacterial enrichment from total petroleum hydrocarbon contaminated sediments: potential candidates for bioaugmentation in bio-based processes. Environ Sci Pollut Res 23:10587–10594. https://doi.org/10.1007/s11356-015-5944-y

Dixit R, Wasiullah MD et al (2015) Bioremediation of heavy metals from soil and aquatic environment: an overview of principles and criteria of fundamental processes. Sustain 7:2189–2212. https://doi.org/10.3390/su7022189

Fauziah SH, Agamuthu P, Hashim R et al (2017) Assessing the bioaugmentation potentials of individual isolates from landfill on metal-polluted soil. Environ Earth Sci 76:1–6. https://doi.org/10.1007/s12665-017-6739-x

Fawzy MA (2020) Biosorption of copper ions from aqueous solution by *Codium vermilara*: optimization, kinetic, isotherm and thermodynamic studies. Adv Powder Technol 31:3724–3735. https://doi.org/10.1016/j.apt.2020.07.014

Feng S, Gong L, Zhang Y et al (2021) Bioaugmentation potential evaluation of a bacterial consortium composed of isolated *Pseudomonas* and *Rhodococcus* for degrading benzene, toluene and styrene in sludge and sewage. Bioresour Technol 320:124329. https://doi.org/10.1016/j.biortech.2020.124329

Flouty R, El-Khoury J, Maatouk E, El-Samrani A (2019) Optimization of Cu and Pb biosorption by *Aphanizomenon ovalisporum* using Taguchi approach: kinetics and equilibrium modeling. Desalin Water Treat 155:259–271. https://doi.org/10.5004/dwt.2019.24005

Gao N, Zhang J, Pan Z et al (2020a) Biodegradation of atrazine by mixed bacteria of *Klebsiella variicola* strain FH-1 and *Arthrobacter* sp. NJ-1. Bull Environ Contam Toxicol 105:481–489. https://doi.org/10.1007/s00128-020-02966-y

Gao T, Qin D, Zuo S et al (2020b) Decolorization and detoxification of triphenylmethane dyes by isolated endophytic fungus, *Bjerkandera adusta* SWUSI4 under non-nutritive conditions. Bioresour Bioprocess 7. https://doi.org/10.1186/s40643-020-00340-8

Ge S, Gu J, Ai W, Dong X (2021) Biotreatment of pyrene and Cr(VI) combined water pollution by mixed bacteria. Sci Rep 11:1–7. https://doi.org/10.1038/s41598-020-80053-2

Giri K, Rai JPN, Pandey S, Mishra G, Kumar R, Suyal DC (2017a) Performance evaluation of isoproturon-degrading indigenous bacterial isolates in soil microcosm. Chem Ecol 33 (9):817–825. https://doi.org/10.1080/02757540.2017.1393535

Giri K, Suyal DC, Mishra G, Pandey S, Kumar R, Meena DK, Rai JPN (2017b) Biodegradation of isoproturon by *Bacillus pumilus* K1 isolated from foothill agroecosystem of north west Himalaya. Proc Natl Acad Sci India Sect B-Biol Sci 87(3):839–848. https://doi.org/10.1007/s40011-015-0667-x

Goel R, Debbarma P, Kumar P, Suyal DC, Kumar S, Mahapatra BS (2020) Assessment of soil chemical quality, soil microbial population and plant growth parameters under organic and conventional rice–wheat cropping system. Agric Res 10:193–204

Guo W, Cheng J, Liu S et al (2020) A novel porous nickel-foam filled CO2 absorptive photobioreactor system to promote CO2 conversion by microalgal biomass. Sci Total Environ 713:136593. https://doi.org/10.1016/j.scitotenv.2020.136593

Hassan A, Pariatamby A, Ahmed A et al (2019) Enhanced bioremediation of heavy metal contaminated landfill soil using filamentous fungi consortia: a demonstration of bioaugmentation potential. Water Air Soil Pollut 230. https://doi.org/10.1007/s11270-019-4227-5

Hu N, Chen S, Lang T et al (2021) A novel method for determining the adequate dose of a chelating agent for phytoremediation of radionulides contaminated soils by *M. cordata*. J Environ Radioact 227:106468. https://doi.org/10.1016/j.jenvrad.2020.106468

Jaskulak M, Grobelak A, Vandenbulcke F (2020) Modeling and optimizing the removal of cadmium by *Sinapis alba* L. from contaminated soil via response surface methodology and artificial neural networks during assisted phytoremediation with sewage sludge. Int J Phytoremediation 22:1321–1330. https://doi.org/10.1080/15226514.2020.1768513

Jin X, Li S, Long N, Zhang R (2018) Improved biodegradation of synthetic azo dye by anionic cross-linking of chloroperoxidase on ZnO/SiO2 nanocomposite support. Appl Biochem Biotechnol 184:1009–1023. https://doi.org/10.1007/s12010-017-2607-0

Júlio ADL, de Fernandes RCR, Costa MD et al (2018) A new biostimulation approach based on the concept of remaining P for soil bioremediation. J Environ Manag 207:417–422. https://doi.org/10.1016/j.jenvman.2017.11.061

Karas PA, Perruchon C, Karanasios E et al (2016) Integrated biodepuration of pesticide-contaminated wastewaters from the fruit-packaging industry using biobeds: bioaugmentation, risk assessment and optimized management. J Hazard Mater 320:635–644. https://doi.org/10.1016/j.jhazmat.2016.07.071

Kauppi S, Sinkkonen A, Romantschuk M (2011) Enhancing bioremediation of diesel-fuel-contaminated soil in a boreal climate: comparison of biostimulation and bioaugmentation. Int Biodeterior Biodegrad 65:359–368. https://doi.org/10.1016/j.ibiod.2010.10.011

Khayati G, Barati M (2017) Bioremediation of petroleum hydrocarbon contaminated soil: optimization strategy using Taguchi design of experimental (DOE) methodology. Environ Process 4:451–461. https://doi.org/10.1007/s40710-017-0244-9

Kumar V, Signh J, Kumar P (2018) Response surface methodology based optimization of cadmium and lead remediation from aqueous solution by water hyacinth (*Eichhornia crassipes* [Mart.] Solms) and its anatomical study. Arch Agric Environ Sci 3:163–173. https://doi.org/10.26832/24566632.2018.0302010

Li J, Guo C, Lu G et al (2016) Bioremediation of petroleum-contaminated acid soil by a constructed bacterial consortium immobilized on sawdust: Influences of multiple factors. Water Air Soil Pollut 227. https://doi.org/10.1007/s11270-016-3117-3

Li X, Wang X, Chen Y et al (2019) Optimization of combined phytoremediation for heavy metal contaminated mine tailings by a field-scale orthogonal experiment. Ecotoxicol Environ Saf 168:1–8. https://doi.org/10.1016/j.ecoenv.2018.10.012

Li J, Chen W, Zhou W et al (2020) Synergistic degradation of pyrene by *Pseudomonas aeruginosa* PA06 and *Achromobacter* sp. AC15 with sodium citrate as the co-metabolic carbon source. Ecotoxicology. https://doi.org/10.1007/s10646-020-02268-3

Lu MM, Gao F, Li C, Yang HL (2021) Response of microalgae *Chlorella vulgaris* to Cr stress and continuous Cr removal in a membrane photobioreactor. Chemosphere 262:128422. https://doi.org/10.1016/j.chemosphere.2020.128422

Mahunon SER, Aina MP, Akowanou AVO et al (2018) Optimization process of organic matter removal from wastewater by using *Eichhornia crassipes*. Environ Sci Pollut Res 25:29219–29226. https://doi.org/10.1007/s11356-018-2771-y

Mandal SK, Ojha N, Das N (2018) Process optimization of benzo[ghi]perylene biodegradation by yeast consortium in presence of ZnO nanoparticles and produced biosurfactant using Box-Behnken design. Front Biol (Beijing) 13:418–424. https://doi.org/10.1007/s11515-018-1523-1

Martínez Álvarez LM, Lo Balbo A, Mac Cormack WP, Ruberto LAM (2015) Bioremediation of a petroleum hydrocarbon-contaminated Antarctic soil: optimization of a biostimulation strategy using response-surface methodology (RSM). Cold Reg Sci Technol 119:61–67. https://doi.org/10.1016/j.coldregions.2015.07.005

Martínez Álvarez LM, Ruberto LAM, Lo Balbo A, Mac Cormack WP (2017) Bioremediation of hydrocarbon-contaminated soils in cold regions: development of a pre-optimized biostimulation biopile-scale field assay in Antarctica. Sci Total Environ 590–591:194–203. https://doi.org/10.1016/j.scitotenv.2017.02.204

Martín-Girela I, Curt MD, Fernández J (2017) Flashing light effects on CO2 absorption by microalgae grown on a biofilm photobioreactor. Algal Res 25:421–430. https://doi.org/10.1016/j.algal.2017.06.008

Mazumder A, Das S, Sen D, Bhattacharjee C (2020) Kinetic analysis and parametric optimization for bioaugmentation of oil from oily wastewater with hydrocarbonoclastic *Rhodococcus pyridinivorans* F5 strain. Environ Technol Innov 17:100630. https://doi.org/10.1016/j.eti.2020.100630

Melvin SS, EAA M, Chidambaram R (2015) Biosorption of Cr(VI) by *Ceratocystis paradoxa* MSR2 using isotherm modelling, kinetic study and optimization of batch parameters using response surface methodology. PLoS One 10:1–23. https://doi.org/10.1371/journal.pone.0118999

Menale C, Nicolucci C, Catapane M et al (2012) Optimization of operational conditions for biodegradation of chlorophenols by laccase-polyacrilonitrile beads system. J Mol Catal B Enzym 78:38–44. https://doi.org/10.1016/j.molcatb.2012.01.021

Mohamad Thani NS, Mohd Ghazi R, Abdul Wahab I et al (2020) Optimization of phytoremediation of nickel by *Alocasia puber* using response surface methodology. Water 12:2707. https://doi.org/10.3390/w12102707

Mohanty SS, Jena HM (2019) Evaluation of butachlor biodegradation efficacy of *Serratia ureilytica* strain AS1: a statistical optimization approach. Int J Environ Sci Technol 16:5807–5816. https://doi.org/10.1007/s13762-018-1958-6

Moreira VR, Lebron YAR, Freire SJ et al (2019) Biosorption of copper ions from aqueous solution using *Chlorella pyrenoidosa*: optimization, equilibrium and kinetics studies. Microchem J 145:119–129. https://doi.org/10.1016/j.microc.2018.10.027

Nguyen PY, Silva AF, Reis AC et al (2019) Bioaugmentation of membrane bioreactor with *Achromobacter denitrificans* strain PR1 for enhanced sulfamethoxazole removal in wastewater. Sci Total Environ 648:44–55. https://doi.org/10.1016/j.scitotenv.2018.08.100

Ntakiyiruta P, Briton BGH, Nsavyimana G, et al (2020) Optimization of the phytoremediation conditions of wastewater in post-treatment by *Eichhornia crassipes* and *Pistia stratiotes*: kinetic model for pollutants removal. Environ Technol (UK) 1–26. https://doi.org/10.1080/09593330.2020.1852445

Ogunlaja A, Nwankwo IN, Omaliko ME, Olukanni OD (2020) Biodegradation of methylene blue as an evidence of synthetic dyes mineralization during textile effluent biotreatment by *Acinetobacter pittii*. Environ Process 7:931–947. https://doi.org/10.1007/s40710-020-00443-6

Okino-Delgado CH, Zanutto-Elgui MR, do Prado DZ et al (2019) Enzymatic bioremediation: current status. Challeng Obtain Process Appl 79–101. https://doi.org/10.1007/978-981-13-7462-3_4

Qiao J, Zhang C, Luo S, Chen W (2014) Bioremediation of highly contaminated oilfield soil: bioaugmentation for enhancing aromatic compounds removal. Front Environ Sci Eng 8:293–304. https://doi.org/10.1007/s11783-013-0561-9

Rashtbari S, Dehghan G (2021) Biodegradation of malachite green by a novel laccase-mimicking multicopper BSA-Cu complex: performance optimization, intermediates identification and artificial neural network modeling. J Hazard Mater 406:124340. https://doi.org/10.1016/j.jhazmat.2020.124340

Razmi B, Ghasemi-Fasaei R, Ronaghi A, Mostowfizadeh-Ghalamfarsa R (2021) Investigation of factors affecting phytoremediation of multi-elements polluted calcareous soil using Taguchi optimization. Ecotoxicol Environ Saf 207:111315. https://doi.org/10.1016/j.ecoenv.2020.111315

Rekik H, Zaraî Jaouadi N, Bouacem K et al (2019) Physical and enzymatic properties of a new manganese peroxidase from the white-rot fungus *Trametes pubescens* strain i8 for lignin biodegradation and textile-dyes biodecolorization. Int J Biol Macromol 125:514–525. https://doi.org/10.1016/j.ijbiomac.2018.12.053

Rozman U, Kalčíková G, Marolt G et al (2020) Potential of waste fungal biomass for lead and cadmium removal: characterization, biosorption kinetic and isotherm studies. Environ Technol Innov 18:100742. https://doi.org/10.1016/j.eti.2020.100742

Samadi A, Sharifi H, Ghobadi Nejad Z et al (2020) Biodegradation of 4-Chlorobenzoic acid by *Lysinibacillus macrolides* DSM54T and determination of optimal conditions. Int J Environ Res 14:145–154. https://doi.org/10.1007/s41742-020-00247-4

Selamat SN, Halmi MIEB, Abdullah SRS et al (2018) Optimization of lead (Pb) bioaccumulation in *Melastoma malabathricum* L. by response surface methodology (RSM). Rend Lincei 29:43–51. https://doi.org/10.1007/s12210-017-0656-5

Sen S, Nandi S, Dutta S (2018) Application of RSM and ANN for optimization and modeling of biosorption of chromium (VI) using cyanobacterial biomass. Appl Water Sci 8:1–12. https://doi.org/10.1007/s13201-018-0790-y

Seyedi ZS, Zahraei Z, Jookar Kashi F (2020) Decolorization of reactive black 5 and reactive red 152 azo dyes by new haloalkaliphilic bacteria isolated from the textile wastewater. Curr Microbiol 77:2084–2092. https://doi.org/10.1007/s00284-020-02039-7

Shanmugam S, Ulaganathan P, Swaminathan K et al (2017) Enhanced biodegradation and detoxification of malachite green by *Trichoderma asperellum* laccase: degradation pathway and product analysis. Int Biodeterior Biodegrad 125:258–268. https://doi.org/10.1016/j.ibiod.2017.08.001

Simpanen S, Mäkelä R, Mikola J et al (2016) Bioremediation of creosote contaminated soil in both laboratory and field scale: investigating the ability of methyl-β-cyclodextrin to enhance biostimulation. Int Biodeterior Biodegrad 106:117–126. https://doi.org/10.1016/j.ibiod.2015.10.013

Singh R, Chadetrik R, Kumar R et al (2010) Biosorption optimization of lead(II), cadmium(II) and copper(II) using response surface methodology and applicability in isotherms and thermodynamics modeling. J Hazard Mater 174:623–634. https://doi.org/10.1016/j.jhazmat.2009.09.097

Sosa-Martínez JD, Balagurusamy N, Montañez J et al (2020) Synthetic dyes biodegradation by fungal ligninolytic enzymes: process optimization, metabolites evaluation and toxicity assessment. J Hazard Mater 400:123254. https://doi.org/10.1016/j.jhazmat.2020.123254

Sultana N, Hossain SMZ, Mohammed ME et al (2020) Experimental study and parameters optimization of microalgae based heavy metals removal process using a hybrid response surface methodology-crow search algorithm. Sci Rep 10:1–15. https://doi.org/10.1038/s41598-020-72236-8

Teiri H, Hajizadeh Y, Samaei MR et al (2020) Modelling the phytoremediation of formaldehyde from indoor air by *Chamaedorea Elegans* using artificial intelligence, genetic algorithm and response surface methodology. J Environ Chem Eng 8:103985. https://doi.org/10.1016/j.jece.2020.103985

Titah HS, Halmi MIEB, Abdullah SRS et al (2018) Statistical optimization of the phytoremediation of arsenic by Ludwigia octovalvis- in a pilot reed bed using response surface methodology (RSM) versus an artificial neural network (ANN). Int J Phytoremediation 20:721–729. https://doi.org/10.1080/15226514.2017.1413337

Udaykumar R, Srinivas NS, Jegadeesan GB (2020) Biodegradation of propylene glycol wastewater using bacterial consortia isolated from municipal wastewater treatment sludge–process kinetics and optimization. Water Air Soil Pollut 231. https://doi.org/10.1007/s11270-020-04657-0

Uniyal S, Sharma RK, Kondakal V (2021) New insights into the biodegradation of chlorpyrifos by a novel bacterial consortium: process optimization using general factorial experimental design. Ecotoxicol Environ Saf 209:111799. https://doi.org/10.1016/j.ecoenv.2020.111799

Wang Q, Gu M, Ma X et al (2015) Model optimization of cadmium and accumulation in switchgrass (*Panicum virgatum* L.): potential use for ecological phytoremediation in Cd-contaminated soils. Environ Sci Pollut Res 22:16758–16771. https://doi.org/10.1007/s11356-015-4878-8

Wang K, Huang K, Jiang G (2018) Enhanced removal of aqueous acetaminophen by a laccase-catalyzed oxidative coupling reaction under a dual-pH optimization strategy. Sci Total Environ 616–617:1270–1278. https://doi.org/10.1016/j.scitotenv.2017.10.191

Wu H, Shen J, Jiang X et al (2018) Bioaugmentation potential of a newly isolated strain Sphingomonas sp. NJUST37 for the treatment of wastewater containing highly toxic and recalcitrant tricyclazole. Bioresour Technol 264:98–105. https://doi.org/10.1016/j.biortech.2018.05.071

Xiong J, Li G, An T (2017) The microbial degradation of 2,4,6-tribromophenol (TBP) in water/sediments interface: investigating bioaugmentation using *Bacillus* sp. GZT. Sci Total Environ 575:573–580. https://doi.org/10.1016/j.scitotenv.2016.09.017

Younis SA, El-Gendy NS, Nassar HN (2020) Biokinetic aspects for biocatalytic remediation of xenobiotics polluted seawater. J Appl Microbiol 129:319–334. https://doi.org/10.1111/jam.14626

Zhang H, Zhang S, He F et al (2016) Characterization of a manganese peroxidase from white-rot fungus *Trametes* sp.48424 with strong ability of degrading different types of dyes and polycyclic aromatic hydrocarbons. J Hazard Mater 320:265–277. https://doi.org/10.1016/j.jhazmat.2016.07.065

Zhao X, Bai S, Tu Y et al (2020) Parameter optimization of environmental technologies using a LCA-based analysis scheme: a bioaugmentation case study. Sci Total Environ 737:140284. https://doi.org/10.1016/j.scitotenv.2020.140284

Zheng Y, Chai L-Y (2016) Combined alkali precipitation and bioaugmentation process for the treatment of black liquor under alkaline condition. Procedia Environ Sci 31:715–724. https://doi.org/10.1016/j.proenv.2016.02.056

Zheng F, An Q, Meng G et al (2017) A novel laccase from white rot fungus *Trametes orientalis*: purification, characterization, and application. Int J Biol Macromol 102:758–770. https://doi.org/10.1016/j.ijbiomac.2017.04.089

Zhu X, Wang X, Wang L et al (2020) Biodegradation of lincomycin in wastewater by two-level bio-treatment using chloroperoxidase and activated sludge: degradation route and eco-toxicity evaluation. Environ Technol Innov 20:101114. https://doi.org/10.1016/j.eti.2020.101114

Chapter 5
Microbial Biosensors for Real-Time Monitoring of the Bioremediation Processes

Seerpatham Divyasorubini, Shyami Menaka Kandage, Senal Liyanage, Charitha Rajapakse, and Gayathri N. Silva

5.1 Bioremediation: An Eco-friendly Tool for Environmental Rehabilitation

Environmental pollution is one of the most crucial and commonly discussed issues for decades worldwide, yet the establishment of appropriate solutions or remedial measures for this problem is still in its infancy. Despite the seminal advancement in science and technology, the world is currently experiencing diverse adverse impacts of environmental pollution. Deliberate and accidental discharge of contaminants on large scales has amplified the health risks and environmental degradation equally affecting both developing and developed nations. According to the World Bank reports, the largest environmental cause of disease and premature deaths is pollution, with more than nine million premature deaths worldwide (Pollution 2021). This is 15 times higher than the deaths caused by wars and other forms of violence (Pollution 2021). Diverse physical, chemical and thermal approaches to mitigate the pollution or to restore the contaminated sites have been brought into action over the years. However, inherent limitations and drawbacks in these methods (e.g., high costs and production of toxic intermediates) have led scientists to shift towards novel environmental remediation methods such as biodegradation or bioremediation. Bioremediation involves biological systems such as microorganisms, their products or plants for the rehabilitation of contaminated soil or water. This eco-friendly method employs the naturally occurring enzymatic processes of the microorganisms and plants (phytoremediation) or sometimes a combination of both (rhizoremediation) to transform toxic pollutants into an innocuous state (Arora 2018). Depending on the site of application, bioremediation is of two types; ex situ and in situ. Bioremediation techniques have been applied in various ecosystems

S. Divyasorubini · S. M. Kandage · S. Liyanage · C. Rajapakse · G. N. Silva (✉)
Department of Chemistry, University of Colombo, Colombo, Sri Lanka
e-mail: gayathris@chem.cmb.ac.lk

© The Author(s), under exclusive license to Springer Nature Switzerland AG 2022

D. C. Suyal, R. Soni (eds.), *Bioremediation of Environmental Pollutants*,
https://doi.org/10.1007/978-3-030-86169-8_5

such as cleaning up groundwater, lagoons, sludge, oil spills, water streams, agricultural sites, and reclamation of sites contaminated with heavy metals, radioactive elements, petroleum and hydrocarbons (Goel et al. 2008; Bhatnagar and Kumari 2013; Arora 2018; Bhatt et al. 2021).

The ultimate goal of managing polluted sites is to transform them into a non-hazardous, pollutant-free site that is environmentally acceptable to be utilized for future purposes. Accordingly, monitoring the bioremediation process has become an indispensable aspect in environment research to evaluate the overall performance and to predict its outcome. Even though bioremediation is highly appreciated as an environmentally friendly and cost-effective tool in cleaning up contaminated platforms, it is imperative to clearly demonstrate its efficiency, reliability, reproducibility and predictability. Therefore, an interdisciplinary, systemic conceptual framework of monitoring the bioremediation process at each level is required for the successful implementation of the bioremediation setup. Once an appropriate bioremediation technique is chosen, regular monitoring has to be done to provide sufficient information for the optimization of the bioremediation process and to evaluate the efficacy for further scaling up of the treatment procedure. Therefore, the development of stringent and accurate monitoring protocols must be tailored to provide comprehensive details of the efficacy of the treatment process. The monitoring techniques for bioremediation must be constructed in a way that could address the following critical questions.

1. Economical—Is the chosen bioremediation technique is economically competitive with other methods?
2. Chemical—Were the targeted endpoints of cleaning the polluted site is achieved?
3. Sustainability—Is the bioremediation method employed environmentally and economically sustainable and could be reproduced in the future?
4. Engineering—Can the treatment process used be engineered and optimized further to use in a different context? Can the process be used to establish predictive models to extrapolate outcomes in another application?
5. Eco-toxicological—Did the treatment process transformed the polluted site completely into a harmless site? Is there any toxic effect or threat posed by bioremediation technique to humans or local biodiversity?
6. Biological involvement—What proportion of the biotic and abiotic factors contributed during the treatment process?

Considering the abovementioned facts, monitoring the bioremediation process can be performed concerning three different aspects, namely: (1) determining the efficiency of the pollutant degradation process, (2) assessing the survival and activity of degradative microbes, and (3) eco-toxicity assessment. The techniques that are used in environmental remediation are anticipated to provide efficient and accurate measurements to determine the overall efficiency of the bioremediation process.

5.2 Quantification of Pollutant Degradation by Non-microbial Tools

Remediation efficacy is usually quantified by the time-dependent endpoint measurement of the complete disappearance of the pollutant. This method is aided by the advancement of different analytical techniques to obtain mechanistic details of the degradation of specific pollutants, especially under in situ conditions (Pandey et al. 2009). Such analytical techniques can be advantageous or disadvantageous depending on the type of environment (the content of organic matter or clay in soil), nature of the contaminant or mixture of contaminants and other factors (Fig. 5.1). For example, Gas Chromatography (GC)/Flame Ionization Detection (FID) is preferable for identifying environmental contaminants as FID shows a linear response to a wide range of concentrations of the target. However, the sensitivity of this method reduces in the presence of oxidizable carbon compounds. Similarly, luminescence techniques are recognized as highly sensitive and selective tools for detecting microbial activity and the content of aromatic compounds. Nevertheless, the applicability of luminescence technologies for detection purposes is limited by the field conditions and the presence of multispecies oil-degrading microbial communities (Andreoni and Gianfreda 2007). One of the most effective

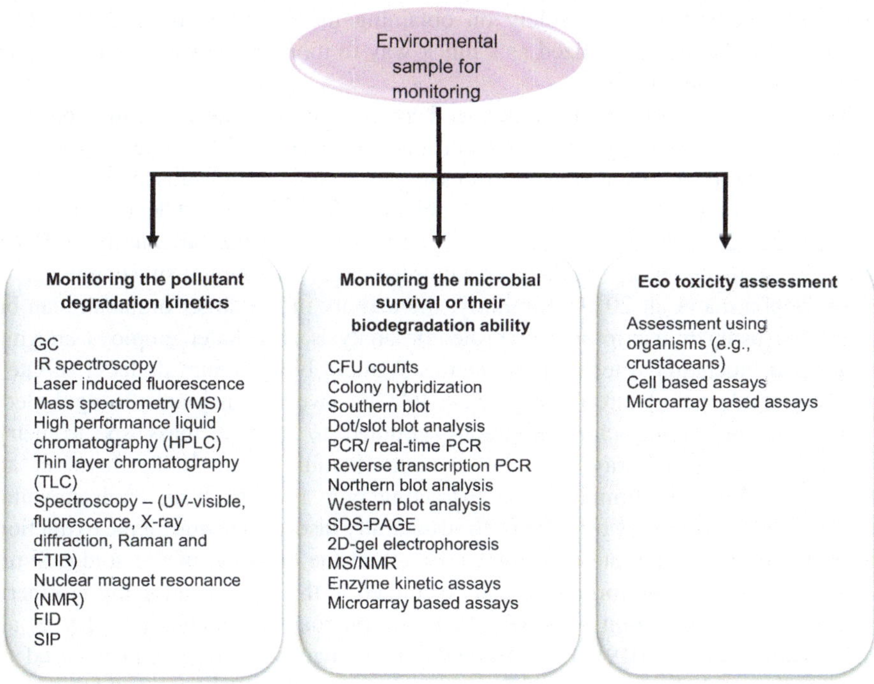

Fig. 5.1 Widely employed analytical techniques available to monitor the pollutant degradation process

techniques of monitoring the bioremediation process is the spectroscopic methods, among which UV-visible and Fourier-transform infrared spectroscopy (FTIR) methods are widely used. *Par excellence* of the spectroscopic analytical methods is due to their ability to rapidly monitor the degradation process while accurately and efficiently identifying the degradation intermediates (Pandey et al. 2009). Fluorescence spectroscopy is another analytical technique that is very sensitive in detecting aromatic pollutants, albeit the applicability of this method in complex mixtures is often minimal due to poor resolving features of the spectra (Gómez et al. 2004). To overcome the limitation of analyzing and quantifying complex mixtures, scientists have used a combinatorial approach involving infra-red (IR), fluorescence, synchronous luminescence spectrometry and GC techniques (Gómez et al. 2004).

Although the early analytical methods employed to monitor environmental remediation were largely depending on the kinetics of pollutant removal, the application of an eco-friendly analytical tool that utilizes survival and biodegradation abilities of microorganisms to quantify pollutant removal has shown a growing interest over the years. In this context, both the survival of the microorganism and the biodegradation ability as a function of time have been used to determine the effectiveness of the bioremediation process (Pandey et al. 2009). Cellular and molecular techniques are employed to assess microbial cell survival and their activity (Fig. 5.1). These techniques can be categorized into two broad sectors, namely: culture-dependent and culture-independent (Suyal et al. 2019a, b; Kumar et al. 2021). The culture-dependent method depends solely on obtaining the colony-forming unit (CFU) counts and is frequently utilized as a quick way to monitor the survival of a target microbe (Pandey et al. 2009).

Therefore, non-culturing technologies became more popular and equipped with modern molecular biology tools. To determine the survival of the target organism at the site of treatment, DNA isolated from samples is subjected to subsequent analysis (e.g., southern blot hybridization, dot/slot blot hybridization and PCR amplification, the latter being used for quantification of DNA). Positive amplification of the DNA sequence of interest indicates the survival of the target microorganism (Pandey et al. 2009; Debbarma et al. 2017). Similarly, the activity of the target organism can be monitored using transcriptome or proteome analysis. The latter employs enzyme assays and biochemical techniques such as Western blot, sodium dodecyl sulfate–polyacrylamide gel electrophoresis (SDS-PAGE), two dimensional (2D)-gel electrophoresis and liquid chromatography coupled to tandem mass spectrometry (MS/MS) via electrospray ionization source (Stenuit et al. 2008; Suyal et al. 2019a, b). However, limitations in extracting pure proteins from environmental samples and the tendency of proteins to structurally disorganize during the extraction process (protein denaturation) significantly affect the outcome of the study. Similarly, the activity of the organism can be studied at the RNA level using Northern blot, reverse transcription PCR (RT-PCR) or microarray analysis (Pandey et al. 2009; Kumar et al. 2019). High throughput microarray analyses can be used to monitor the catabolic potential of the targeted organism in real-time and have been used to monitor the bioremediation process in wastewater and other complex environments (Dennis et al. 2003). Although microarray-based assays provide the advantage of high throughput, comprehensive and quantitative characterization of microbial communities, further extensive research has to be conducted to validate the applicability of this technique in diverse environmental samples (Zhou and Thompson 2002).

5.3 Limitations Associated with Conventional Monitoring Techniques

Research conducted over the years has clearly demonstrated the positive impacts of using non-microbial analytical methods to decipher and monitor the degradation of pollutants. However, quantitation of pollutants using such methods greatly depend on the extraction method of the pollutant as different extraction methods performed under various conditions may give inconsistent results. Therefore, different extraction methods were employed primarily based on the physio-chemical characteristics of the pollutant to enhance the extraction efficiency. These methods are mainly categorized as exhaustive and non-exhaustive extraction methods (Pandey et al. 2009). Another major drawback in using analytical methods for pollutant degradation is the inability to distinguish between biological vs. non-biological degradation. The final quantitation of the pollutant may vary due to the non-biological phenomena such as photolysis, wash off, leaching, diffusion and adsorption to the substrate, and thus may not completely reflect the full potential of the bioremediation process (Strand et al. 2003). As a solution, stable isotope probing (SIP) technology that relies on the content of stable isotopes in the molecule of interest can be applied. In this technique, fluxes of specific chemicals are traced in microorganisms by introducing a heavier stable isotope such as C^{13} to the microbial community (Panigrahi et al. 2018). Incorporation of the isotope into microbial cellular components such as nucleic acids can be detected after separation by gradient centrifugation and can be used as biomarkers in SIP technology. Widely used informative biomarkers include DNA, RNA or phospholipid fatty acids (PLFA). Nonetheless, SIP technology is based on the assimilatory process of the microorganisms and therefore, non-assimilatory processes such as co-oxidation fall outside the applicable framework of SIP technology. Furthermore, the use of SIP technology is restricted due to the necessity of the substrate to be labeled close to 100% for a successful density-based separation. Such labeling is expensive and requires long incubation times, and may not be suitable for regular monitoring purposes. Also, the synthesis of isotope-labeled DNA is limited by the replication efficiency of the organism. In contrast, RNA-SIP is a better responsive marker as the rate of RNA synthesis occurs at a high rate in active cells and will be efficiently labeled. However, in this context, efficient RNA extraction methods must be employed for the broad application of this method (Andreoni and Gianfreda 2007). Even though culture-dependent techniques such as colony hybridization has high selectivity and efficacy for assessing microbial activity and survival, these techniques have their inherent limitations in culturing microbes, such as non-amenability for existing culturing protocols or the presence of viable but not culturable microbes (Pandey et al. 2009).

5.4 Eco-toxicity Assessment

The ultimate goal of the bioremediation process is to completely remove the hazardous waste or transform it into a harmless state such that the site could be reused in the future. Assessment of eco-toxicity in the pollutant site may not directly indicate the progress of the treatment method. However, eco-toxicity assessment paints the final picture of the bioremediation process by assessing whether the bioremediation process has positively impacted to reduce the toxicity imposed in the eco-system or not. At the end of an ideal bioremediation process, a significant reduction in the eco-toxicity of the target environment is anticipated. The most common eco-toxicity assessment assays employ luminescent marine bacteria, fungal biomass, shrimps, earthworms and crustaceans (Barajas-Aceves et al. 2002; Pandey et al. 2009). However, these assays require a longer time for evaluation as the toxicity is tested over a range of concentrations at different time points and may not be suitable for high throughput eco-toxicity assessment during the bioremediation process. Therefore, human cell line-based in vitro assays and microarray-based assays were developed. Furthermore, microarray-based analysis can be considered as a rapid, cost-effective and high throughput technique to study the catabolic responsiveness of an organism in the presence of a toxic chemical (Pandey et al. 2009).

Due to the limitations associated with conventional monitoring techniques, scientists have expanded the research to elucidate novel methods to assess environmental remediation. As a result, biosensors have emerged as a strong and versatile tool that could be engineered to get precise data in real-time.

5.5 Biosensors as a Powerful and Innovative Analytical Tool

A biosensor is defined as an analytical device that uses a biological molecule or living organism to sense the target molecule (e.g., sugars, proteins, hormones, pollutants, toxins, chemical compounds, antigens, enzymes, nucleic acids and microorganisms) (Hansen and Sørensen 2001; Bahadir and Sezgintürk 2015). The idea of using biological molecules for sensing the presence of the target molecule dates back to the 1960s where Clarks and Lyons used electrodes immobilized with glucose peroxidase to measure the levels of glucose in biological samples (Vigneshvar et al. 2016). Since then, numerous research efforts lead to the discovery of highly accurate and selective biosensors that could be utilized in various applications, including medical diagnosis, environmental monitoring, drug discovery, and food quality monitoring. In contrast to the conventional methods, improved detection limits, selectivity, accuracy, and sensitivity of biosensors have led to continuous development in this realm and resulted in remarkable advances to bring about highly sensitive biosensors for high-throughput real-time monitoring purposes. Biosensors have two basic components, namely: the biological component and the transducer (Mehrotra 2016). The biological material or biomimetic functions as the recognition molecule, which is either in intimate contact with the transducer or integrated within the physiochemical transducer or transducing microsystem (Korotkaya 2014). The transducer or the detector element transforms the physicochemical changes into processable signals (optical, piezoelectric, electrochemical, electro-chemiluminescence, etc.) that are proportional to the amount of target molecule–bioreceptor interactions (Bhalla et al. 2016). These signals are produced due to a change in proton concentration, emission of light, emission of heat, production and uptake of gases such as oxygen and ammonia and many other mechanisms which result from sensing an analyte. This process of energy transformation is called signalization. The selection of biological material to develop a biosensor depends mainly on the analyte to be detected. In addition, biochemical specificity, storage, operational and environmental stability also plays a vital role in this selection (Lim et al. 2015). Biosensors can be classified into two broad categories based on the sensing biological material and the type of transducer used (Fig. 5.2). An ideal biosensor must possess properties such as specificity, sensitivity, reliability, portability and the ability to function in optically opaque solutions, real-time analysis and simplicity in operation (Bhalla et al. 2016). Biosensors are typically comprised of an electronic system including a signal amplifier, a processor and a display in addition to the bio-recognition site and the bio-transducer components (Bhalla et al. 2016). The processor functions as a reader which analyses signal modifications, amplify and present the detecting signal. The fluctuation in the signal entirely depends on the analyte bioreceptor interactions and truly reflects inherent bio-sensitivity to the analyte (Bhalla et al. 2016). High selectivity and specificity of biological materials combined with the processing power of microelectronic devices give rise to versatile biosensors that could be employed in various aspects for accurate detection and monitoring purposes. However, it is necessary to ensure that the analyte reaches the site of reaction in the biological material. The biological component and the electrical signal can be manipulated further by amplification and processing to improve the quality of biosensors to meet the market requirements. The analytical capabilities of a biosensor can be further increased by miniaturization and improved processing systems (Bhalla et al. 2016).

Different types of biosensors have been introduced over time to monitor the bioremediation process to assess its effectiveness (Fig. 5.2). Biosensors have emerged as an attractive monitoring tool and have gained much attention in contrast to conventional techniques. Portability, high selectivity, rapid and on-site point of care monitoring favored the use of biosensors over other conventional methods for monitoring purposes. Among the bio-recognition elements indicated in Fig. 5.2, enzymes have been widely used for monitoring purposes. Nevertheless, the tedious

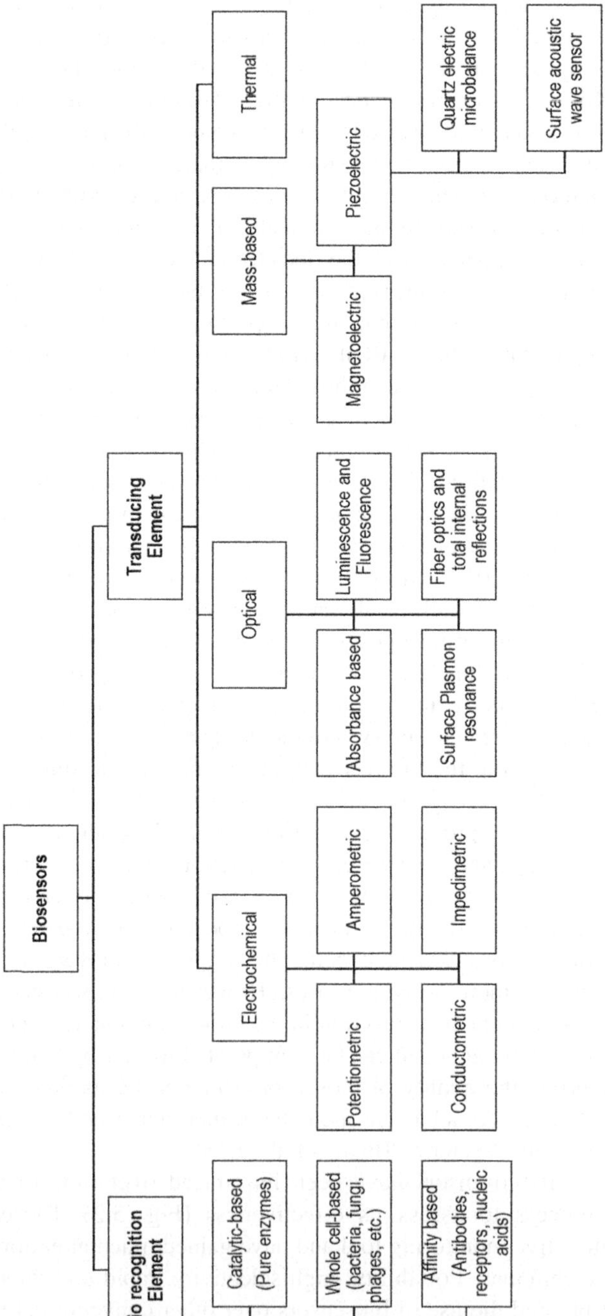

Fig. 5.2 Classification of biosensors

purification process, the necessity for co-enzymes or multi enzymes to give a detectable signal and the associated high costs in producing enzyme-based biosensors have led scientists to search for alternatives. Moreover, when considering enzymes and antibodies, maintaining their stability, specificity, activity and/or affinity in different environmental conditions is quite difficult (Park et al. 2013). Hence, scientists have presented microbial biosensors as convenient and cost-effective biosensors that could be utilized under a range of environmental conditions for monitoring the bioremediation process. Microbes can be cultivated on a large scale by following a simple cell culturing process. Besides, all the necessary co-factors for an enzymatic reaction reside inside the cell and comparably more stable and demonstrates more capacity to tolerability in harsh environmental conditions. Furthermore, microbes can be handled and manipulated easily in a way that can enhance the expression of a necessary pathway to augment the emitted signal or even can be manipulated to endure harsh environmental conditions.

This chapter is dedicated to provide a comprehensive overview of employing microbial biosensors for monitoring the bioremediation process and describe detailed mechanistic information about how synthetic biology, molecular biology, chemistry and engineering have interfaced to design a resourceful and ideal microbial biosensor.

5.5.1 Design and Fabrication of a Microbial Biosensor

Microbes have gained more research attention in developing biosensors as they have the potential to target a wide range of elements, and even they can be easily manipulated to enhance the specificity towards the substrates. Evidence from numerous research efforts suggests that genetic engineering of microbes is comparably much easier and seems to be better controlled and tailored to give the best-desired outcome than using plants or mammalian cells and other types of biosensors (Lim et al. 2015). There are two mechanisms by which microbial biosensors induce a specific reaction during the biosensing process; inhibition of cellular respiration or alteration of cell metabolism in the presence of an analyte of interest (Xu and Ying 2011). As a consequence of the change in microbial cellular metabolism induced by the targeted analyte, gene expression of the sensing elements is also changed. This change in gene expression is being detected and/or quantified using a microbial biosensor (Fig. 5.3). The sensing elements of a biosensor mainly consist of regulator genes and bio-recognition genes/reporter genes. The regulatory genes control the differential expression of the bio-recognition element based on the presence or absence of the target analyte. The bio-recognition gene or the reporter gene functions by converting the biological response into a detectable or measurable signal. Given the fact that the expression of the biorecognition gene is maneuvered by the regulator gene, the requirement of a promoter could be eliminated. Therefore, the design of a genetically engineered microbial biosensors may consist of only a regulator gene and a bio-recognition gene without a promoter. The resulting recombinant genes can

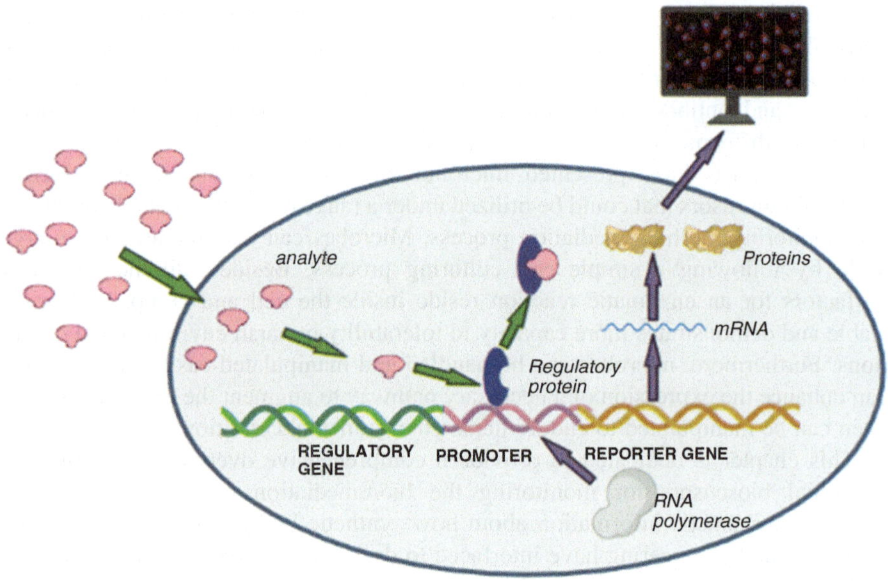

Fig. 5.3 Schematic illustrating the mechanism of a typical microbial biosensor. The diagram shows a negative regulatory mechanism where binding of the analyte to the regulatory protein frees the promoter region for RNA polymerase to express the reporter gene. Finally, the expressed reporter protein is detected by a transducer

then be cloned into the microbial host either by direct integration into the chromosome or by cloning into an appropriate plasmid vector and transforming it to the host. In the presence of the target compound or the analyte, the regulator stimulates the promoter, which in turn performs the transcription and translation of the reporter gene to a protein that can be detected as an electrochemical, chemiluminescent, colorimetric or fluorescent signal. The generated signal can be either a qualitative or a quantitative signal depending on the design of the biosensor. Based on the aforementioned facts, it is clear that the excitability of regulatory genes towards the analyte of interest is the key determinant of the specificity and sensitivity of a microbial biosensor (Bilal and Iqbal 2019). Therefore, proper selection of the regulatory genes, host strains and a suitable detection technique govern the successful utilization of a biosensor.

5.5.2 Host Strain

The selection of an appropriate bacterial strain is the most crucial step in designing a biosensor. The selected microbial strain must have the substrate specificity for the targeted analyte to detect the presence of it by eliciting a cellular response. Furthermore, there are several ways in which the microbial traits can be successfully translated into a routine assay format to fulfill different analytical and monitoring requirements, which include but are not limited to monitoring the bioremediation process and assessing the toxins and other chemicals in the environment. Microorganisms that have been widely utilized as biosensors generally do not have complex multicellular structures and are unicellular organisms. Immobilized cells can act as both sensing components and generators of the recognition signals. Bacteria and yeast are being widely used as the immobilized cell type in biosensors (Xu and Ying 2011).

Identification of the best-suited host strain for the biosensor improves its specificity, sensitivity and time-response (Gui et al. 2017). Selection of the host strain primarily depends on the target analyte, bio-sensing elements, sensitivity, specificity and the detection mechanism. A typical example of a highly sensitive biosensor is the biosensor hosted on *Pseudomonas putida* DOT-T1E, which was found to be the best strain for monitoring sites that are heavily contaminated with toxic organic compounds. *P. putida* DOT-T1E possesses this unique property as it is evolutionary optimized to survive in environments that are highly concentrated with such toxins whilst showing high substrate specificity to several contaminants such as antibiotics, toluene and flavonoids (Espinosa-Urgel et al. 2015; Gui et al. 2017). Another biosensor for the detection of Ni^{2+} in drinking water has been designed using the wild type *Escherichia coli* strain designated as *E. coli* TD2158. This strain has shown the highest level of sensitivity and activity for Ni^{2+} compared to other strains of *E. coli*, such as *E. coli* W3110-based biosensors that utilized the same mechanism of detection. Consequently, *E. coli* TD2158 was identified as the best host strain giving the highest sensitivity, to design the biosensor for detecting Ni^{2+} in drinking water (Gui et al. 2017). The selection of the host strain also depends on the type of analyte to be detected. For example, microbial biosensors have been constructed using *Acinetobacter baylyi* ADP1 for the detection of a broad range of alkanes and alkenes in water and soil. The use of *E. coli* as the host for this purpose has been restricted due to the poor accessibility and emulsifying capability of *E. coli* for oils. In contrast, *A. baylyi* ADP1is naturally adherent to the oil-water interface while emulsifying the minerals and oils and thus, it is an excellent strain for constructing a whole-cell biosensor for detecting a range of alkanes and alkenes in water, seawater or in oils (Gui et al. 2017).

Genetic approaches and gene modifications play a significant role in bringing successful microbial biosensors to execution. Genetic engineering provides solutions to most of the frequent drawbacks associated with the use of wild-type strains and helps to retain an improved selectivity and sensitivity within the biosensor. More commonly, bacteria are genetically engineered by incorporation of a specific reporter gene to respond to the chemicals present in the sample or physiological stress through synthesis of a reporter protein such as green fluorescence protein (GFP), luciferase, β-galactosidase, etc. (Xu and Ying 2011).

5.5.3 Reporter Genes

The performance of a microbial biosensor for the detection of environmental contaminants strongly depends on the reporter gene/s chosen to detect the genetic response for the contaminants and the type of regulatory protein associated with the promoter. A reporter gene is capable of converting its biological response into a measurable signal (e.g., electrical, optical, electrochemical, etc.) which determines the sensitivity and selectivity of the biosensor (Gui et al. 2017). Widely used reporter genes that can be successfully incorporated into microbial biosensors include *cat* (encodes bacterial chloramphenicol acetyltransferase), *cfp* (encodes coral fluorescence proteins), *lux* (encodes bacterial luciferase), *luc* (encodes firefly luciferase), *gfp* (encodes green fluorescent protein) and *lacZ* (encodes β-galactosidase) (Bullock and Gorman 2000; Hansen and Sørensen 2001; Chong and Ching 2016; Gui et al. 2017).

E. coli lacZ that encodes the protein β-galactosidase is one of the best-studied and most frequently used reporter genes in biosensors. *LacZ* exhibits unique advantages for analyte detection, such as employing convenient and sensitive colorimetric or fluorescent methods that utilize readily available chemiluminescent and electrochemical substrates. In addition, low detection limit (as low as 2 fg), ultra-high sensitivity, and an extensive dynamic detection range are some of the key advantages of using the *lacZ* reporter gene (Gui et al. 2017). The reporter gene bacterial luciferase (*lux*) catalyzes the oxidation of long-chain fatty aldehydes and reduces flavin mononucleotides to form the corresponding fatty acid and fructosamine in the presence of oxygen and produces bioluminescence as the reaction output (Xu and Ying 2011). *Lux* genes isolated from various bacterial strains (e.g., *luxCDABE* operon from *Vibrio fischeri*, *luxCDABFE* operon from *Photobacterium leiognathi*, and luciferase coding *luxAB* from *Vibrio harveyi*) have been widely used as reporter genes in biosensor constructs (Hansen and Sørensen 2001). Many researchers have used fiber optic technology to detect light emission from *lux* biosensors (Hansen and Sørensen 2001). Visualization of the light emission from the luciferase can be observed without disruption of the bacterial cell. However, *lux* is rarely used in mammalian cell-based biosensors due to thermal instability and protein dimerization which can lead to false interpretations (Gui et al. 2017). *Gfp* that encodes GFP has been widely used as a marker in many bacterial biosensor constructions. GFP is a very stable fluorescent protein that can be excited with UV or blue light and the fluorescence can be detected without bacterial cell lysis. Further, it does not require the addition of exogenous substrates or ATP to generate the signal (Xu and Ying 2011). Therefore, unlike *lacZ*, GFP is not limited by the accessibility of the substrate (Hansen and Sørensen 2001). Some variants of GFP have half-lives of more than 24 h, while the others have a half-life of ~40 min (Hansen and Sørensen 2001). This could be advantageous because GFP variants with an extended half-life can be produced even from weak promoters or in cells with low metabolic activity (Hansen and Sørensen 2001). Another advantage is the use of GFP variants with shorter half-life in transient (real-time/time-dependent) gene expression studies for the detection of various analytes (He et al. 2019). Therefore, to detect dynamics and to facilitate rapid degradation inside the microbial cells, these fluorescent reporters have been destabilized to shorten their half-lives. However, applications of this approach are limited by the very low signal intensities generated by the GFP variants with short half-life. A possible solution to overcome this challenge would be the development of a trans-timer using a destabilized GFP with another GFP variant called RFP (red fluorescent protein), which can detect the dynamics of gene expression in cells (He et al. 2019). When the target gene is 'on', the destabilized GFP expresses rapidly before the RFP expression turns on and when the target gene turns 'off', the GFP that was expressed will rapidly degrade leaving only the red signal by RFP. Therefore, in a biosensor, dynamic monitoring can be done using the ratio of green to red colour emitted by the GFP based trans-timer (He et al. 2019). Use of GFP has been limited due to the cost of equipment, such as flow cytometers and fluorimeters which are necessary to analyze the fluorescence signal. Another disadvantage is that the detection limit of GFP is higher than that of both β-galactosidase and luciferase (Hansen and Sørensen 2001).

Other reporter genes used include eukaryotic *luc* from the firefly *Photinus pyralis* (Hansen and Sørensen 2001). The firefly *luc* reporter gene has frequently been incorporated into mammalian and bacterial cells due to its high sensitivity and linearity over a broader range of analyte concentrations (Gui et al. 2017). At the same time, the use of *luc*, especially with mammalian cells, could overcome the thermally labile nature and dimeric protein interferences generally associated with bacterial *lux* (Gui et al. 2017). Another example is *crtA*, a gene involved in carotenoid biosynthesis. This reporter gene allows the detection of target analyte calorimetrically through the naked eye upon introducing it into a biosensor. When applied to a sample, *crtA*-based biosensors can change the color of the culture media from yellow to red without the addition of a supporting substrate and therefore, is considered a good choice for rapid detection of the target analytes in emergencies (Gui et al. 2017). A potential disadvantage associated with *crtA* is that the production of carotenoids is often affected by the metabolic fluxes of the host microorganism. This can interfere with the color intensity and the time required for color development when using the biosensor (Chong and Ching 2016). Therefore, a promising solution to produce an intense color development with little influence from the metabolic fluxes would be the use of *cfp* as the reporter gene (Chong and Ching 2016). It is considered a favorable candidate as the coral fluorescence protein not only enables a visible colorimetric change but also shows a minimal dependency on the amount of metabolites available (Chong and Ching 2016). Table 5.1 is a summarization of the reporter genes widely employed in microbial biosensors.

5.5.4 Regulatory Proteins

The gene regulatory proteins are one of the major components on which the performance of a biosensor depends (Gui et al. 2017). They are the proteins that

Table 5.1 Reporter genes that are frequently used in microbial biosensors (Chong and Ching 2016; Gui et al. 2017)

Gene	Detection method	Advantages	Disadvantages
lux	Luminescence	Easy measurement, rapid response	Thermal labiality, requirement for the substrate O_2
luc	Luminescence	High sensitivity, rapid response, thermal stability	Requirement for the substrates; O_2 and ATP, low permeability
gfp	Fluorescence	No substrate requirement, high stability	High cost of equipment, low sensitivity, lag-time for stable fluorescence, auto-fluorescence
lacZ	Luminescence, fluorescence, colourimetry, electrochemistry	High stability, wide variety of detection methods, detection by naked eyes	Substrate dependence (e.g., X-gal), low permeability
crtA	Colorimetry	Detection by naked eyes, no substrate requirement	Activity is affected by the metabolic fluxes of the host organism

influence the regions of a DNA molecule that are transcribed by RNA polymerase during transcription. These proteins, which include transcription factors, help control the synthesis of proteins in cells. They possess complex interactions with the target analytes or the contaminants of interest. These interactions are critical for the specificity and sensitivity of a biosensor (Gui et al. 2017). These regulatory proteins have been reported to respond to a wide array of compounds (e.g. sugars, vitamins, secondary metabolites, metal ions, amino acids and other lipid metabolites) and serves as a large reserve of biological components that can be utilized for designing in vivo biosensors (Shi et al. 2018).

Biosensors that have utilized the function of these regulatory proteins have shown higher selectivity, higher detection ranges and enhanced sensitivity when compared to conventional biosensors. The binding of the analyte to the regulatory protein induces a significant conformational change, thus activating/inhibiting the expression of the reporter gene (Raut et al. 2012). Therefore, the function of the regulatory protein can be either positive or negative in terms of the activation of the promoter for the expression of the reporter gene (Raut et al. 2012). In negative regulation, the regulatory protein is bound to the operator/promoter region, inhibiting the expression of the reporter gene. When the analyte is bound to the regulatory protein, it dissociates from the operator/promoter region, subsequently allowing RNA polymerase to carry out the downstream reporter gene expression (Fig. 5.2). In positive regulation, the analyte-regulatory protein complex binds to the operator/ promoter region facilitating binding of the RNA polymerase to carry out the expression of the reporter gene (Raut et al. 2012). For example, an *E. coli* biosensor for the detection of insecticide CPF (chlorpyrifos) has been designed by using a CPF inducible locus *chpAB* found in *Sinorhizobium meliloti*. CPF biosensors utilize a gene designated *chpR*, a cadC family transcription regulator, as a positive regulator for the *chpAB* operon, which is involved in the detection of CPF (Whangsuk et al. 2010). A brief comparison of a few microbial biosensors designed so far is given in Table 5.2.

Table 5.2 Widely used whole-cell biosensors for the detection of environmental pollutants (Gui et al. 2017)

Host strain	Reporter gene	Target analyte	Detection sensitivity
E. coli	*luxCDABE*	Arsenic	0.74–69 μg/l
E. coli	*lacZ*	Arsentate	<10 μg/l
D. radiodurans	*lacZ* *crtI*	Cadmium	1–10 mM 50 nM–1 nM
E. coli	*gap*	Chromate	100 nM
E. coli	*gfp*	Zinc Copper	16 μM 26 μM
E. coli	*luc*	Benzene, Toluene and Xylene	40 μM
E. coli	*luxAB*	Benzene, Toluene and Xylene	0.24 μM
P. putida	*luxAB*	Phenol	3 μM
B. sartisol	*luxAB*	Naphthalene and Phenanthrene	0.17 μM
E. coli	*luxAB*	C_6–C_{10} Alkanes	10 nM
E. coli	*luxCDABE*	Tetracycline	45 nM
S. typhimurium	*lacZ*	Single-stranded DNA	10 nM

5.5.5 Microbial Immobilization Techniques

A microbial biosensor is designed in such a way that host microorganisms are in close contact with the transducer. Therefore, the use of the proper immobilization technique is necessary to establish the required connectivity between the cell and the transducer which need to be placed in close proximity (Lei et al. 2006). An immobilization technique should preserve cell viability and functionality, and the immobilization matrix must provide mechanical stability to prevent cell leakage. Furthermore, such methods must ensure the efficient access of the analyte molecules into the cells (Lobsiger and Stark 2019). Chemical immobilization techniques such as covalent binding, cross-linking, and physical immobilization techniques such as adsorption, entrapment have been widely used in the fabrication of a microbial biosensor (Lobsiger and Stark 2019; Ganesan and Vasudevan 2021).

The covalent immobilization technique relies on the formation of strong covalent bonds between the different functional groups present on the microbial cell wall and the transducer. These functional groups include amine, carboxylic, sulfhydryl and tosyl groups. Harmful chemicals and harsh reaction conditions used in covalent binding have a negative impact on cell viability; thus covalent binding is rarely employed with viable microbial cells (Lei et al. 2006). Cross-linking is a process which involves the formation of a network of cells by interconnecting the functional groups in the outer membrane by chemicals such as glutaraldehyde and cyanuric chloride. Similar to covalent binding, cross-linking too may affect the cell viability. (Lei et al. 2006).

Physical immobilization techniques are in wide use when dealing with the viable cells, as it has the minimal interference with the native structure and function of the microorganism. Adsorption is considered to be the simplest form of physical immobilization technique. Microorganisms are immobilized due to different adsorptive interactions such as ionic, polar and hydrogen bonds (Lei et al. 2006). Entrapment is another widely used physical immobilization technique. During entrapment, cells are retained using dialysis, using a filter membrane or polymer (Lei et al. 2006). The polymers used in entrapment include hydrogels such as agarose, LB agar and alginate (Lobsiger and Stark 2019). However, low sensitivity limits the use of entrapment as the microbial immobilization technique (Lei et al. 2006). Lyophilization or freeze-drying is another physical technique used in the immobilization of microorganisms during the fabrication of a biosensor. This is a low-temperature dehydration process where water from a frozen sample is sublimed in a vacuum, thus preserving the integrity of the microbial cell (Lobsiger and Stark 2019).

5.6 Diversity of Microbial Biosensors

Tremendous research efforts equipped with the advancement in microbial biotechnology, micro-engineering and synthetic biology have led to the development of promising and more futuristic microbial biosensors with enhanced performance. This large pool of microbial biosensors can be categorized based on different criteria. In this chapter, the microbial biosensors are broadly categorized into three major categories in terms of their signal transducers utilized, namely: electrochemical, optical and microbial fuel-cell type biosensors.

5.6.1 Electrochemical Microbial Biosensors

Electrochemical microbial biosensors are the most widely available type of microbial biosensors and have been reported to have the highest sensitivity among all the available microbial biosensors. They mainly consist of a working electrode, microorganisms as a transducer layer for detection and a signal recording equipment (Lim et al. 2015). These types of biosensors exploit the respiratory electrochemical pathways of microorganisms. The analyte interacts with a component in the microorganism's respiratory pathway, which acts as an electron shuttle or a mediator. This interaction leads to an inhibition of the transmission of signals causing a change in the electrochemical potential, which is subsequently detected by the transducing mechanism (Ikeda and Kano 2001; Yang et al. 2018). To improve the sensitivity, externally supplied redox-active mediators which can get reduced in the cell can be used to amplify the signal via transferring electrons through the system (Gupta et al. 2019). As mentioned, electrochemical microbial biosensors are capable of providing specific quantitative or semi-quantitative analytical information with the use of biological recognition elements and can be further classified based on the mechanism used by the transducer to detect the signal (Xu and Ying 2011).

5.6.1.1 Types of Electrochemical Biosensors

(a) Voltametric Microbial Biosensors

This is the most versatile form of electrochemical biosensor type for the detection of chemical compounds. Each electric signal generated through a current and a voltage difference is recorded and correlates with a corresponding sample. Voltametric approaches can provide high selectivity and measurability via the position and density of the peak current signal. Low detection speeds and the requirement of complex components for the process are the potential limitations associated with these types of biosensors (Lim et al. 2015).

(b) Conductometric Microbial Biosensors

Conductometric microbial biosensors detect chemicals by the variation in conductivity of a sample solution caused by target analytes. Detection happens via the consumption or production of ions by the transducers. The conductance measurements are highly sensitive and can detect the target chemicals rapidly (Lim et al. 2015). In particular, they can easily be miniaturized as they do not require a reference electrode. Even with high sensitivity, the detection of solution conductance is considered to be nonspecific because the variation in conductivity can be affected by the electrical charge (Xu and Ying 2011). Microorganism-based conductometric biosensors are primarily being used in the detection of microbial toxicity in the dairy industry (Xu and Ying 2011).

(c) Amperometric Microbial Biosensors

Amperometric microbial biosensors monitor the concentration of the chemical by recording the current signal through the sample at a fixed potential with respect to a reference electrode (Lim et al. 2015). The corresponding current is obtained by the oxidation or reduction of electroactive species at the surface of the electrode. In particular, amperometric microbial biosensors have been recorded to provide greater sensitivity. Most of the biosensors designed to measure biological oxygen demand (BOD) belong to this category (Xu and Ying 2011).

(d) Potentiometric Microbial Biosensors

A Potentiometric microbial biosensor consists of either an ion-selective electrode or a gas-sensing electrode (Xu and Ying 2011). This approach uses the potential difference from a reference electrode and thus requires three electrodes as two working electrodes and a reference electrode. The need for a reference electrode for stable and accurate sensing is a limitation associated with potentiometric microbial biosensors. This type of biosensors shows a higher selectivity and sensitivity for the target chemical (Lim et al. 2015).

5.6.1.2 Application of Electrochemical Biosensors in Environmental Monitoring

Electrochemical microbial biosensors have proven their capability for the identification and analysis of different target compounds due to their simplicity, portability and cost-effectivity. Several attempts have been made to exploit the electrochemical sensors' potentialities to detect emerging contaminants in the environment, which include pesticides, antibiotics, heavy metals and perfluorinated compounds. Since electrochemical biosensors typically utilize the intrinsic electron transfer ability of microorganisms, the signal can be enhanced by simply supplementing electron mediators externally without the need for any genetic alterations (Gupta et al. 2019). A wide range of electrochemical microbial biosensors has been constructed to detect and monitor many environmental pollutants and parameters. These include BOD, toxins such as 3,5-dichlorophenol (DCP) and trichloroethylene, herbicides, pharmaceuticals, heavy metal ions such as Cu^{2+}, Cd^{2+}, Ni^{2+}, Pb^{2+}, As^{3+} and Zn^{2+}, and anions like sulfide (Table 5.3). Microorganisms used in these biosensors include *Shewanella oneidensis*, *Saccharomyces cerevisiae*, *Bacillus subtilis*, *E. coli*, *Chromobacterium violaceum*, *Thiobacillus thioparus* and *Pseudomonas* sp. (Pham et al. 2015; Yang et al. 2018; Gupta et al. 2019). A recently discovered group of bacteria called "exoelectrogens" are widely employed to develop electrochemical biosensors. Exoelectrogens have the ability to transfer electrons outside of the cell. Thus, their intracellular electrochemical pathways can be linked to an extracellular transducing mechanism (Yang et al. 2018).

Among the listed microorganisms, *E. coli* is identified as the most widely utilized organism in biosensors. A typical example includes the development of a cadmium sensing biosensor for on-site monitoring of water, seawater and soil samples. The biosensor has been designed by fusing the cadmium responsive promoter of *E. coli* to a promoterless *lacZ* that encodes the enzyme β-galactosidase. An electrochemical assay based on the activity of β-galactosidase is used to quantify the cadmium level. Enzymatic conversion of the substrate p-aminophenyl-β-D-galactopyranoside (ONPG) to p-aminophenol (ONP) generates a current signal which can be detected electrochemically. Under anaerobic conditions, this electrochemical biosensor is reported to have a sensitivity to Cd^{2+} concentrations as low as 25 nM in water and 5 μM in soil (Shin 2011). *S. oneidensis* MR-1 is an example of an exoelectrogen that can connect its internal electrochemical pathways to an external circuit. Here, the electric current flow is subjected to change based on the type and the concentration of the environmental pollutant. Yang et al. (2018) have developed an electrochemical bacterial biosensor exploiting this ability of *S. oneidensis* MR-1 for the detection of DCP levels in the water. In the presence of DCP, the produced electric current decreases in a concentration dependent manner (Yang et al. 2018). The cyanobacterium *Anabaena variabilis* provides a suitable biological system to detect the presence of photosynthesis-inhibiting herbicides. In photosynthetic organisms, photosystem II (PSII) harvests light energy via an electron transfer chain which ultimately oxidizes water to produce O_2. Photosynthesis-inhibiting herbicides

Table 5.3 Widely used electrochemical microbial biosensors used in environmental monitoring

Analyte/parameter monitored	Microorganisms utilized	Detecting/transducing mechanism
BOD	*B. subtilis*	Electrochemical (amperometry)
	C. violaceum R1	Electrochemical (amperometry)
	Mixed culture including *Geobacter* sp.	Bioelectrochemical
3,5-dichlorophenol (DCP)	*S. oneidensis* MR-1	Bioelectrochemical
	S. cerevisiae S288C	Electrochemical (amperometry)
	E. coli ATCC 25922, *B. subtilis* CGMCC 1.1086, *S. cerevisiae* S288C	Electrochemical (amperometry)
4-chlorophenol, phenol	*S. cerevisiae* S288C	Electrochemical (amperometry)
Trichloroethylene	*Pseudomonas* sp. ASA86	Electrochemical (potentiometry)
Herbicide (Diuron)	*A. variabilis*	Electrochemical (amperometry)
	Chlamydomonas Reinhardtii	Electrochemical (amperometry and potentiometry)
Herbicide (Atrazine)	*A. variabilis*	Electrochemical (amperometry)
Pesticides (Ametryn and Acephate)	*E. coli* ATCC 25922, *B. subtilis* CGMCC 1.1086, *S. cerevisiae* S288C	Electrochemical (amperometry)
Pharmaceuticals (Omeprazole, lansoprazole, naphthoflavone and methylcholanthrene)	*Arxula adeninivorans* G1212/YRC102	Electrochemical (amperometry)
Cu^{2+}	*S. cerevisiae* S288C	Electrochemical (amperometry)
	E. coli ATCC 25922	Electrochemical
	E. coli ATCC 25922, *B. subtilis* CGMCC 1.1086, *S. cerevisiae* S288C	Electrochemical (amperometry)
Cd^{2+}	*S. cerevisiae* S288C	Electrochemical (amperometry)
	E. coli ATCC 25922	Electrochemical
	E. coli ATCC 25922, *B. subtilis* CGMCC 1.1086, *S. cerevisiae* S288C	Electrochemical (amperometry)
Ni^{2+}	*S. cerevisiae* S288C	Electrochemical (amperometry)
Pb^{2+}	*S. cerevisiae* S288C	Electrochemical (amperometry)

(continued)

Table 5.3 (continued)

Analyte/parameter monitored	Microorganisms utilized	Detecting/transducing mechanism
	E. coli ATCC 25922	Electrochemical
Zn^{2+}	*E. coli* ATCC 25922	Electrochemical
As^{3+}	*S. oneidensis*	Bioelectrochemical
Sulfide	*E. coli* BL21	Voltammetry

competitively inhibit the electron transfer chain and thereby decrease the current generated (Fig. 5.4). Artificial redox mediators like quinone can be utilized to measure the electric current generated through this system (Tucci et al. 2020).

The concentration dependent inhibition of the current generation has been observed in the presence of two photosynthesis-inhibiting herbicides, diuron and atrazine. A three-electrode system, where the immobilized *A. variabilis* containing biosensor acts as the active electrode, has been utilized to obtain the electrochemical measurements. In the presence of atrazine, the electric current generated is decreased in a concentration dependent manner. In the presence of diuron, the electric current is completely inhibited due to diuron being a potent photosynthesis inhibitor (Tucci et al. 2020).

5.6.2 *Optical Microbial Biosensors*

These are the biosensor devices that make use of principles of optics for the transduction of a biochemical interaction into a detectable output signal (Xu and Ying 2011). Optical microbial biosensors are developed by coupling the ability of a microorganism to recognize a certain analyte (bio-recognition sensing element) with an optoelectronic transducer system (Gupta et al. 2019). The visual signal can be the result of bioluminescence, chemiluminescence, fluorescence or chromogenic detection (Axelrod et al. 2016; Bae et al. 2018). The two elements are frequency coupled via genetic modification by placing a reporter gene under the control of an analyte-specific promoter. Thus, the reporter gene is only expressed in the presence of the targeted analyte (Gupta et al. 2019). Optical microbial biosensors offer advantages such as flexibility and resistance to electrical noise. Optical fibers, as optical waveguides, have been largely used in optical microbial biosensors due to their low cost, small size and flexible geometry. The optical fiber-based microbial biosensors can be easily taken to the field for on-site monitoring (Xu and Ying 2011).

5.6.2.1 Bioluminescent Microbial Biosensors

Bioluminescent microbial biosensors measure the change in luminescence emitted by microorganisms (Su et al. 2011). They are mainly employed for risk assessment

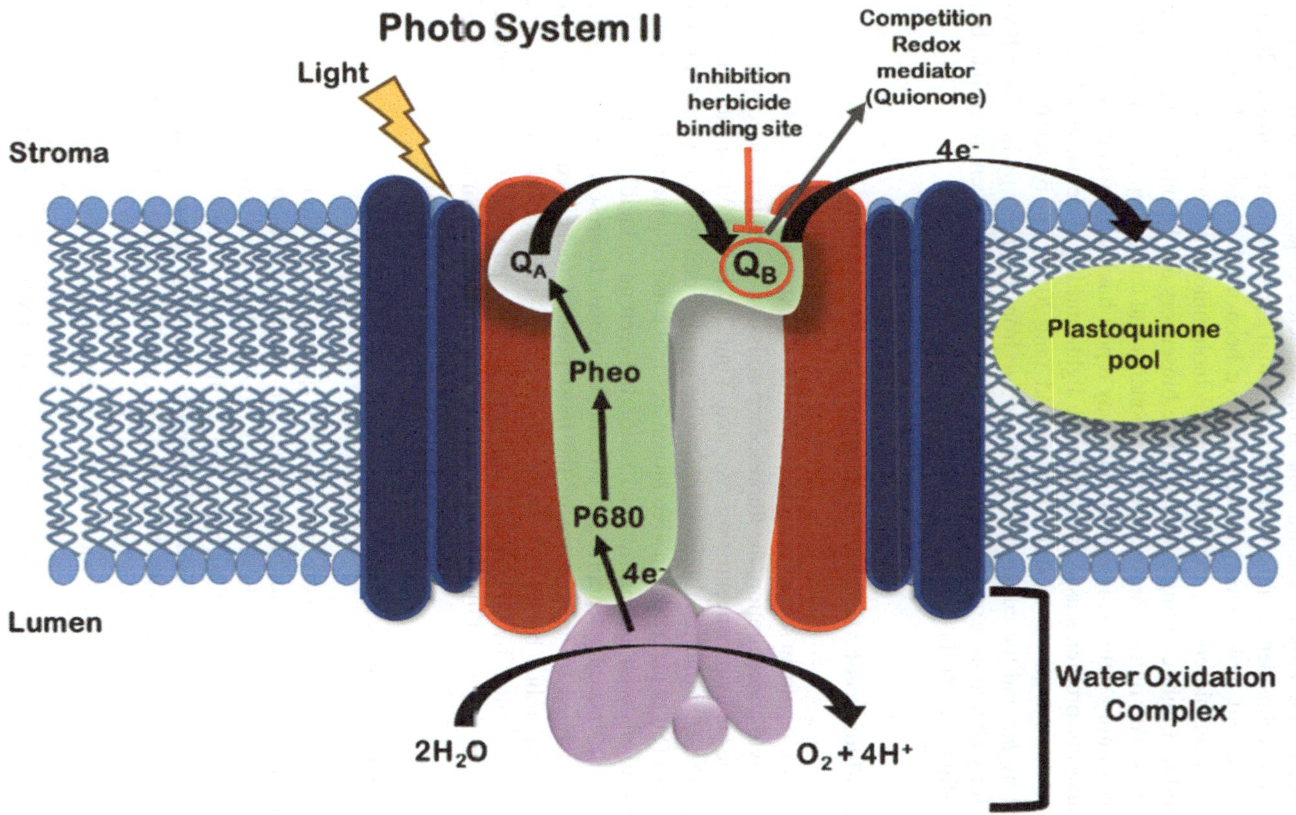

Fig. 5.4 Mechanism of action of photosynthesis-inhibiting herbicides in PSII. Electrons produced via the oxidation of water into O_2 are transported by the P680 protein. Naturally, the present electron acceptor plastoquinone accepts the electrons at the active site Q_B. An artificially provided redox mediator such as quinone can also competitively accept electrons through the same active site. Photosynthesis-inhibiting herbicides competitively bind the active site and inhibit the electron transfer chain

and environmental monitoring, especially for toxicity measurements and heavy metal detections. The emission of light by the microbial communities is mainly determined by the bioavailable fraction of the detected contaminant and therefore, this method is considered a reliable and efficient tool as a biosensor (Lim et al. 2015). Several forms of bioluminescent bacteria have been used in the development of bioluminescent biosensors, including natural bioluminescent bacteria and genetically modified light-emitting bacteria. These biosensors can be produced as single-chip, low-power, rugged, inexpensive components and can be deployed in a variety of non-laboratory settings. However, they may report having a lower efficiency due to the inherent problems associated with the light-emitting systems, which can be improved through genetic modification (Xu and Ying 2011). The bacterial luciferase encoded by *lux* and the eukaryotic luciferase *luc* from *Photinus pyralis* have been successfully served as reporter genes in a variety of bioluminescent microbial biosensors (Su et al. 2011).

5.6.2.2 Fluorescent Microbial Biosensors

Fluorescence occurs in some microbial cells when an external light source is applied. At low analyte concentration, the fluorescence emission intensity is directly proportional to the analyte concentration (Xu and Ying 2011). Based on the detection mode, fluorescent microbial biosensors can be divided into two categories, in vivo and in vitro (Su et al. 2011). In vivo biosensors make use of genetically engineered microorganisms with a transcriptional fusion between an inducible promoter and a reporter gene encoding a fluorescent protein. The Green fluorescent protein (GFP), which is encoded by *gfp*, is one of the most popular tools used in the in vivo fluorescent microbial biosensors due to its attractive stability and sensitivity. The fluorescence emitted by GFP can be conveniently detected with the use of modern optical equipment with little or no damage to the host system (Su et al. 2011).

5.6.2.3 Colourimetric Microbial Biosensors

Colorimetric microbial biosensors involve the generation of colored compounds that generates a signal which can be measured and correlated with the concentration of analytes (Su et al. 2011). These biosensors indicate the presence of the analyte by a visible color change of the microbial cell. Unlike in fluorescence or luminescence biosensors, colorimetric microbial biosensors generally do not need any special equipment for detection under light-shielded conditions and can be monitored by the naked eye both in the laboratory and the field (Fujimoto et al. 2006). The reporter genes which have been used widely in colorimetric biosensor include *lacZ*, *crtA* and *cfp* (Gui et al. 2017). *LacZ* from *E. coli* encodes the enzyme β-galactosidase, which splits its substrate X-gal into a blue colored product which can be detected by the naked eye. The colour intensity is proportional to the level of enzyme activity within a certain range thus, permits its use in a colorimetry based microbial biosensor to detect the analyte of interest (Shin 2011). Similarly, enzymes such as alkaline phosphatase and horseradish peroxidase are also being used to generate the detection signals in different colorimetric microbial biosensors (Shin 2011). Recently, carotenoid-based colourimetric biosensors have been developed with the use of reporter genes *crtA* and *crtI*, which have the additional advantage of substrate independent color developing ability or the detection of the target analyte (Shin 2011).

5.6.2.4 Optical Microbial Biosensors in Environmental Monitoring

Optical microbial biosensors have been optimized to monitor the quality and toxic levels in the water. These biosensors can assess BOD, heavy metals like Hg, Pd, As, Cu and Zn, organic pollutants such as formaldehyde and methyl parathion, as well as herbicides and pesticides (Table 5.4). The microorganisms that are being used for the development of such biosensors include *S. cerevisiae*, *E. coli*, *Sphingomonas* sp., *Chlamydomonas reinhardtii* and *Dictyosphaerium chlorelloides* (Axelrod et al. 2016; Bae et al. 2018; Gupta et al. 2019). For example, a simple microbial biosensor has been developed using the bacterium *E. coli* that employs naked-eye detection of color change for the on-site detection of phenolic compounds in water and soil. The bio-recognition is mediated by a plasmid harboring the *β-galactosidase* gene fused with the phenolic responsive *CapR* promoter. This biosensor has shown a significant sensitivity to phenolic compounds and can respond in concentrations range from 0.1 μM to 10 mM (Shin 2011). Another colorimetric biosensor for arsenic detection

Table 5.4 Widely used optical microbial biosensors for environmental monitoring

Analyte/parameter monitored	Microorganisms utilized	Transducing mechanism
BOD	*S. cerevisiae*	Chemiluminescence
Organic solvents including formaldehyde	*E. coli* TV1061	Bioluminescence
	E. coli DPD2794, DPD2544 and TV1061	Bioluminescence
Ammonium Hydroxide	*E. coli* TV1061	Bioluminescence
Endocrine destructive agents including 17-β-estradiol (E2), 17α-ethynylestradiol (EE2), diethylstilbestrol (DES) and estrone (E1)	*E. coli* DPD2794, DPD2544 and TV1061	Bioluminescence
	S. cerevisiae	Bioluminescence
Pesticides (Diuron, Simazine, Atrazine)	*C. reinhardtii*	Fluorescence
Herbicide (Simazine)	*Dictyosphaerium chlorelloides* Dc1M	Fiber optic–luminescent O_2 transducer
Mercury	*E. coli* TV1061	Bioluminescence
Pb^{2+}	*E. coli* DH5α	Fluorescence
Cu^{2+}	*S. cerevisiae*	Colorimetric

has been developed with the use of the reporter gene *crtA*, which expresses spheroiden monooxygenase (Shin 2011). The host strain used is an engineered strain of *Rhodovulum sulfidophilum* with a deleted *crtA* locus which appears in yellowish due to the accumulation of yellow spheroiden. The biosensor has been constructed by cloning the arsenite resistance operon from *E. coli* into a plasmid containing *crtA* and transforming it into the host strain. This operon consists of an operator/promoter region and a repressor gene *arsR*. The presence of AsO_2^- induces dissociation of the repressor protein from the operator, thus allowing the expression of spheroiden monooxygenase. This enzyme catalyzes the formation of reddish spheroiden from yellowish spheroiden. Therefore, in the presence of AsO_2^-, a color change from yellow to red could be observed by the naked eye. The reported limit of the detection of AsO_2^- by this biosensor is 5 μg/L (Su et al. 2011).

In addition to the colorimetric biosensors, fluorescent and bioluminescent microbial biosensors have played a significant role in monitoring metal contaminations. The bacterial luciferase or *lux* is a widely used reporter gene in the development of luminescent microbial biosensors. Expression of the *lux* genes in microorganisms can be controlled in a constitutive or inducible way. For example, in India and Vietnam, a luminescent-based bacterial biosensor has been developed and deployed in the field for the assessment of groundwater samples contaminated with arsenic. This biosensor can detect the analyte with more than 90% accuracy and has been applied on a large scale for the environmental monitoring of arsenic. This *E. coli* DH5α based biosensor has been developed with the *luxCDABE* reporter gene of *Vibrio fischeri* cloned with the arsenic resistant operon (*ars*) of a wild-type *E. coli*. The operator/promoter region and the *arsR* (negative regulator gene) of the *ars* operon are cloned with the reporter gene *lux*, which expresses bacterial luciferase only in the presence of the target analyte arsenic. This process generates a luminescent signal which can be detected quantitatively, within an arsenic concentration range of 0.74–60 μg/L (Sharma et al. 2013). Similarly, *V. fischeri* based bioluminescent microbial biosensor has been shown promising results for rapid determination of common environmental pollutants (Su et al. 2011). Another biosensor for the detection of heavy metal concentrations in wastewater was designed with the host organism *Acinetobacter sp.* employing the reporter genes *luxCDABE* (Su et al. 2011). Additionally, a bioluminescent biosensor with *Pseudomonas fluorescens* has been designed to detect the fraction of naphthalene present in the soil (Su et al. 2011).

Similarly, in vitro fluorescent whole-cell biosensors are designed and successfully applied to monitor environmental pollutants such as heavy metals and O_2 to assess the BOD levels in water (Su et al. 2011). For example, an Ag^+ and Cu^{2+} sensitive biosensor has been constructed using a two-component (plasmid) system consisting of an Ag^+ sensor and a regulator from bacterial *sil* operon coupled to a detector (Sharma et al. 2013). The membrane-bound protein SilS from *E. coli* J53 detects the Ag^+ ions, which then activates the secondary protein SilR by transphosphorylation. Phosphorylated SilR protein thus becomes an activator that activates the promoters of the *silE* and *silABC*. These two promoters have been separately cloned upstream of a promoterless *gfp* thus, creating two plasmids named *pRADEK.1* and *pRADEK.2*, respectively. These two plasmids containing one of the two promoters with *gfp* transformed into the host *E. coli* J53 can be used as a fluorescence biosensor to qualitatively and quantitatively detect Ag^+. Due to the close homology of the *sil* operon with the copper resistance genes, this biosensor may be successfully applied to detect Cu^{2+} in a similar way to Ag^+ (Sharma et al. 2013).

A microbial biosensor has been developed to monitor the levels of Cu^{2+} in water, utilizing a modified *S. cerevisiae* BY4742 strain. The modification involves two genetic changes in the AMP pathway of purine synthesis, which results in the production of a colored pigment in the presence of Cu^{2+} ions. The intensity of the color correlates with the concentration of Cu^{2+} in the medium. Here, the *ADE2*, which codes for AIR (5′-phosphoribosylaminoimidazole) carboxylase, is knocked out. In the absence of the enzyme, AIR gets accumulated in cells and subsequently oxidized into a red-colored pigment in the presence of O_2. Hence, the cells appear red in color. The second modification is to place the *ADE5,7*, which encodes GAR (5′-phosphoribosylglycinamide) synthase, under the control of the CUP1 promoter, which is induced in the presence of Cu^{2+}. GAR synthase catalysis the first step of the AMP pathway. Thus, the pathway is only initiated when Cu^{2+} is present. Consequently, the red color pigment is only produced in the presence of $Cu^{2+,}$ while in the absence of Cu^{2+}, the pathway is not initiated, and the cells remain white (Vopálenská and Palková 2015).

5.6.3 *Microbial Fuel-Cell Type Biosensors*

Microbial Fuel Cells are novel and promising tools in environmental biotechnology. They can be considered as devices that can convert chemical energy into electrical energy through catalytic reactions present in the electroactive microbes. Therefore, this type of biosensors can generate electricity as the original signal by bio-degradation of organic matter, i.e., the catalytic activity of microorganisms converts chemical energy to electric energy in response to the target analyte. While there are many applications of MFCs, they have been widely used in the construction of whole cell-based environmental biosensors.

The basic structure of an MFC consists of anodic and cathodic compartments separated by an ion-exchange membrane (Fig. 5.5). The anode and the cathode are connected via an external circuit. In addition to the anode, the anodic chamber contains a culture of electroactive bacteria in a medium rich with substrate organic compounds. The bacteria catalyze the oxidation of organic substrates and produce electrons and positively charged ions such as H^+, K^+ and Na^+. The electrons are captured by the anode, and the external circuit conducts them to the cathode. The ion exchange membrane facilitates the transfer of cations from the anodic chamber to the cathodic chamber to balance the charges. In the cathodic chamber, oxygen accepts the electrons and protons to produce water (Jung et al. 2007; Cui et al. 2019). The catalytic rate of the conversion of chemical energy to electric energy can be affected

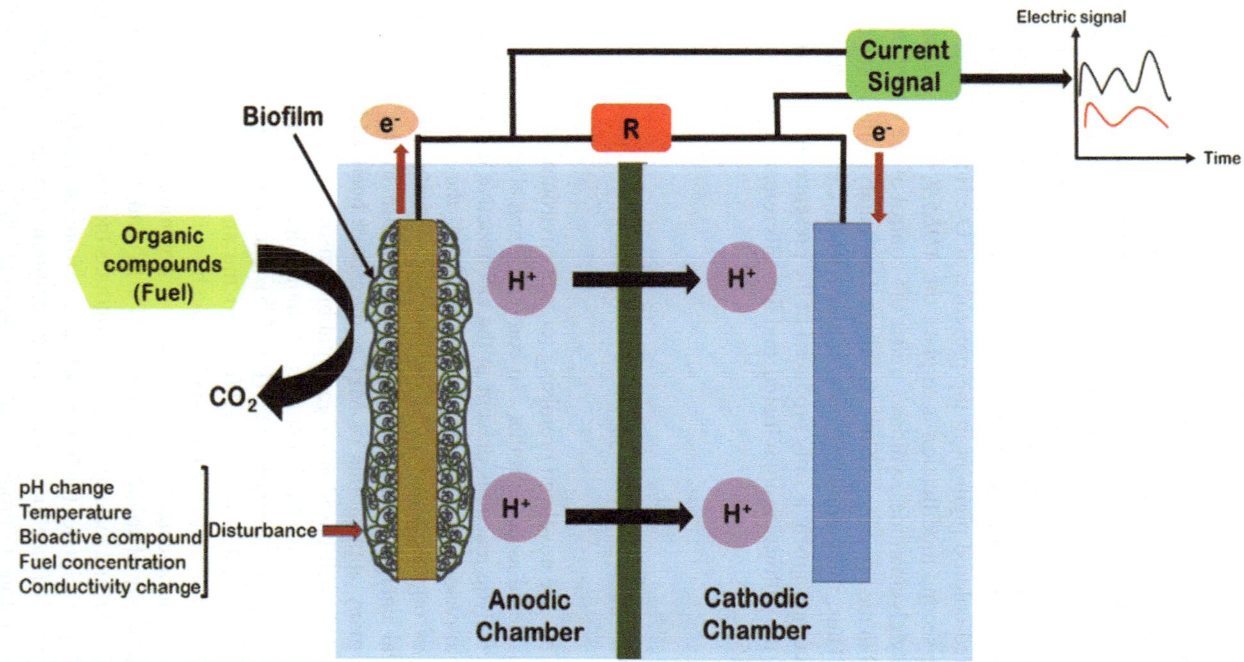

Fig. 5.5 Schematic of a microbial fuel cell. In the anodic chamber that contains microorganisms, the organic substrates are oxidized to produce electrons which are conducted to the cathodic chamber via an external circuit. Next, O_2 accepts the electrons to produce water. Oxidation of organic substrates in the anodic chamber also gives rise to cations which are transferred into the cathodic chamber across the ion exchange membrane

by various environmental parameters, which can lead to changes in the electron flow through the MFC, subsequently altering the electric current produced. Thus, the apparatus can be refined to construct a biosensor that measures such environmental parameters. In such a biosensor, the microorganisms in the anodic chamber would act as the receptor, which recognizes the changes in the environmental parameters. Instead of having a culture of bacteria, a biofilm containing bacteria can also be utilized. The anode which captures the electrons plays the role of the transducer. Typically the relationship between the electric current generated and the changes in the environmental parameters is considered to be linear (Cui et al. 2019).

5.6.3.1 Microbial Fuel-Cell Biosensors in Environmental Monitoring

Microbial biosensors have been developed using MFCs to detect a wide range of environmental parameters such as BOD, COD, heavy metal ions, and toxic compounds in water, as well as the activity of other microbes by utilizing biofilms, single bacterial cultures or mixed bacterial cultures (Table 5.5). The bacterial species that have been widely utilized in the development of MFC-based biosensors include, *Azospirillum*, *Acinetobacter*, *Ocillibacter*, *Shewanella loihica*, *Shewanella frigidimarina*, *Thermincola carboxydiphila*, *Pseudomonas aeruginosa*, *Ochrobactrum intermedium*, *Citrobacter freundii*, *Clostridium acetobutylicum* and *E. coli* (Cui et al. 2019; Gupta et al. 2019). For example, MFC-based biosensors have been used to detect toxic environmental agents. As mentioned, they utilize microbial metabolism as the driving force for the conversion of chemical energy into electrical energy. Therefore, the MFC output depends on the viability and activity of the bacterial cells (Sun et al. 2015). In contrast to the other microbial biosensor types, the MFC-based biosensor demonstrates long-term stability due to the self-healing property of the biofilm. In addition, the requirement of a transducer is eliminated in this biosensor as MFC is a self-powered device. However, MFC-type biosensors have certain limitations such as low substrate efficiency and low sensitivity due to the complicated biofilm (Fang et al. 2020).

Table 5.5 Frequently used microbial fuel-cell biosensors in environmental monitoring

Analyte/parameter monitored	Microorganisms utilized
BOD	*S. loihica* PV-4
	T. carboxydiphila, *P. aeruginosa*, *O. intermedium*, *S. frigidimarina*, *C. freundii*, *C. acetobutylicum*
Toxicity (avermectins (AVM), ivermectin (IVM), tetracyclines, heavy metals)	Mixed culture including *Azospirillum*, *Acinetobacter*, *Pseudomonas*, *Ocillibacter*
Cd^{2+}	*S. loihica* PV-4
Formaldehyde	*Shewanella oneidensis* MR-1

5.7 Advantages, Limitations and Future Challenges of Biosensors

Live cells offer the advantage of qualitative and quantitative analysis of a specific compound by emitting a signal as a response to their regular homeostasis process. Therefore, in contrast to conventional analytical techniques, microbial biosensor-based assays are undoubtedly much simpler and can be carried out without the need for expensive equipment. For example, hands-on demonstration carried out in public using the *E. coli* Arsr-LuxAB reporter assay provides evidence for the successful application of microbial bioreporter assays even by non-experts to get quality and accurate results in few hours (Van Der Meer and Belkin 2010). Moreover, unlike in enzyme-based biosensors, microbial biosensors do not exploit pure enzymes and hence, the stringent and costly purification steps can be excluded, and the entire reporter-sensor unit compacted in a self-replicating cell can be produced in a simple culturing step. In addition, enzymes and antibodies may subject to denaturation or inactivation during the extraction and purification process (Reshetilov et al. 2010). Such limitations can be overcome by the use of microbial biosensors. Modern advancements are more oriented to adapt a multi-well format for bioassays to achieve high throughput real-time monitoring. Therefore, microbes provide an ideal platform for miniaturizing the biosensors for high throughput monitoring whilst maintaining excellent accuracy in measurements required for better sample screening procedures. Furthermore, microbial biosensors come with the unique advantage of providing information pertinent to ecotoxicological safety endpoints of a particular contaminant in a site. This is because microbes elicit a specific response to the bioavailable fraction of the compound of interest in the sample (Van Der Meer and Belkin 2010). Given the strong analytical potential owned by microbes coupled with the cost-effectiveness, ease of handling, less technical hurdles, better stability in harsh environments and amenability for genetic manipulation to obtain pre-determined bioanalytical properties make microbial biosensors an ideal analytical tool to monitor environmental contaminants (Lim et al. 2015).

Although microbial biosensors provide many advantages in real-time monitoring of the bioremediation process, they do have a few drawbacks and limitations (discussed below) that need to be addressed in the near future.

5.7.1 Environmental Safety Concerns

Microbial biosensors utilize living microorganisms. Some of these microorganisms have been genetically modified to suit the needs of the monitoring process. Introducing a genetically modified microorganism may pose a risk to the environment as it can harm or disrupt the existing microbiome of the site, which may eventually disturb the ecological balance. Strict regulatory measures have been imposed restricting the use of genetically modified microbial biosensors in qualified laboratories and contained environments. Thus, precautions should be taken when microbial biosensors are utilized so as not to introduce any destructive microorganisms into the environment. Furthermore, the possible malignant effects of the microbes used in biosensors must be thoroughly researched.

5.7.2 Non-target Interaction and Poor Signal Quality

Non-target interaction is commonly associated with microbial biosensors, which utilize electrochemical detection and transducing mechanisms. This is because the microbial cells can interact with many chemical species such as ions and organic compounds found within the environment other than the targeted analyte. This can lead to false-positive measurements as well as high background noise leading to poor signal quality and reduced specificity of the biosensor (Gupta et al. 2019). Using appropriate controls such as multi strain assay or gas phase assays in the bioreporter assay may minimize the complications associated with chemical mixtures (Van Der Meer and Belkin 2010). The signal quality of an optical microbial biosensor depends on the expression of the reporter genes. Even in the same microorganism, the level of reporter gene expression may vary among the different cultures, leading to inconsistent sensitivities (Gupta et al. 2019). Furthermore, microorganisms are capable of rapid evolution in response to environmental changes. Thus, with prolonged usage, the sensitivity and selectivity of the biosensor might be altered (Cui et al. 2019). This might make microbial biosensors somewhat ill-suited for long-term monitoring of bioremediation.

5.7.3 Reliance on Genetic Manipulation When Designing

Most naturally occurring microorganisms are not ideal for the construction of biosensors. Therefore, certain characteristics of these microorganisms are required to be genetically manipulated to achieve the desired properties. Particularly the development of optical microbial biosensors heavily relies on the introduction of reporter genes via genetic engineering techniques (Gupta et al. 2019). This process introduces additional steps which can be time-consuming and expensive. Furthermore, it is challenging to achieve long-term genetic stability of the foreign gene expression.

5.7.4 The High Cost of Development and Maintenance

The development of microbial biosensors requires extensive research on the microorganisms as well as the instrumentations. This requires specialized facilities and human resources, which can be quite costly. Furthermore, the need for genetic alterations also increases the expense in the construction of microbial biosensors. The use of living microorganisms in microbial biosensors can also add to the maintenance cost of these instruments.

5.8 Future of Microbial Biosensors

The development and application of microbial biosensors have been on the rise during the past few years. One such development was a novel micro-chemostat platform that can be incorporated with microbial biosensors (Bae et al. 2018). Such designs allow the microbial culture to have a uniform environment that ensures long-term stability. In addition, other mechanisms to facilitate the continued stability of the microbial culture environment have also been proposed. These include remote monitoring of cellular and environmental parameters as well as self-stabilizing culture systems (Khire et al. 2020). Another ambitious prospect of microbial biosensing is to develop biosensors that can detect a wide range of signals in a well-coordinated manner. These biosensors may incorporate many microbial species and utilize genetically engineered microorganisms (Gupta et al. 2019).

This chapter mainly focuses on the use of transcriptional regulator/inducible promotor pairs in the design and fabrication of microbial biosensors for monitoring the bioremediation process. In addition, microbial biosensors are designed employing the riboswitch coupled to a reporter gene and the quorum sensing mechanism (Park et al. 2013). For example, an *E. coli* biosensor was designed linking thymidylate synthase with an anti-theophylline aptamer to monitor the theophylline concentration. The concentration of theophylline was monitored based on the dose-dependent repression of GFP expression by theophylline. Similarly, *E. coli* has been engineered to identify specific pathogens via sensing small diffusible molecules involved in the quorum-sensing through reporter gene expression (Park et al. 2013). Although both examples mentioned above are applied in medical diagnosis, it is clear that the underline principle of such biosensors can possibly be used to develop novel microbial biosensors to monitor the bioremediation process. Microfluidics and nanofabrication are another important aspect that has greatly contributed to the development of high throughput microbial biosensors. For example, microbial biosensors integrated with a centrifugal microfluidic platform reduce the time and resources required for analysis while retaining the analytical ability and enhancing portability. Incorporating nanofabrication into microfluidics allows further miniaturization of the microbial biosensor. For instance, microfluidic devices with separate and parallel channels could be designed using soft lithography techniques to screen different toxic compounds enabling high throughput assay on a chip (Lim et al. 2015). Further advances in microbial biosensors can be made by using a panel of bioreporter strains with the same output reporter protein but with different specificities (Van Der Meer and Belkin 2010). Built on this idea, a live cell array was designed in a silicon chip using *E. coli umuDp–lucFF* SOS reporter strain (Tani et al. 2007). Similarly, an array of *E. coli* heat shock responsive reporter cells was immobilized in micro-electrode chambers (100 nL each) to assess the water toxicity (Popovtzer et al. 2005). Here, silicon-based nano-bio chip, *E. coli* MC1061 that is genetically engineered to elicit a detectable electrochemical signal in the presence of the targeted toxicant was integrated into the nano volume electrochemical cells. The electric current generated by the cells was subsequently analyzed to trace the toxicant level in wastewater in less than 10 min (Popovtzer et al. 2005). In addition to microfluidics, another research group designed a paper-based biosensor to detect and monitor pathogenic bacteria using a quorum sensing mechanism (Brooks and Alper 2021). This paper-based biosensor can be presented as a convenient platform due to low production costs, portability and simplistic manufacturing process.

Microbial biosensors stand out as a promising tool for monitoring the efficacy of the bioremediation process. Further development in genomic and transcriptomic data, protein engineering, computational designing and simulation can accelerate the tailoring of microbial biosensors with better specificity and enhanced performance. Research and development in standardized and streamlined engineering methods will enable the design of more versatile genetic designs and immobilization techniques for real-time high throughput monitoring of the bioremediation process. We can anticipate increased adoption of better tailored, highly sensitive advanced microbial biosensors to effectively monitor the bioremediation process in near future.

References

Andreoni V, Gianfreda L (2007) Bioremediation and monitoring of aromatic-polluted habitats. Appl Microbiol Biotechnol 76(2):287–308. https://doi.org/10.1007/s00253-007-1018-5

Arora NK (2018) Bioremediation: a green approach for restoration of polluted ecosystems. Environ Sustain 1(4):305–307. https://doi.org/10.1007/s42398-018-00036-y

Axelrod T, Eltzov E, Marks RS (2016) Bioluminescent bioreporter pad biosensor for monitoring water toxicity. Talanta 149:290–297. https://doi.org/10.1016/j.talanta.2015.11.067

Bae J, Lim JW, Kim T (2018) Reusable and storable whole-cell microbial biosensors with a microchemostat platform for in situ on-demand heavy metal detection. Sensors Actuators B Chem 264:372–381. https://doi.org/10.1016/j.snb.2018.03.001

Bahadir EB, Sezgintürk MK (2015) Applications of commercial biosensors in clinical, food, environmental, and biothreat/biowarfare analyses. Anal Biochem 478:107–120. https://doi.org/10.1016/j.ab.2015.03.011

Barajas-Aceves M et al (2002) Effect of pollutants on the ergosterol content as indicator of fungal biomass. J Microbiol Methods 50(3):227–236. https://doi.org/10.1016/S0167-7012(02)00031-3

Bhalla N et al (2016) Introduction to biosensors. Essays Biochem 60(1):1–8. https://doi.org/10.1042/EBC20150001

Bhatnagar S, Kumari R (2013) Bioremediation: a sustainable tool for environmental management. Annu Rev Res Biol 3(4):974–993. https://www.journalarrb.com/index.php/ARRB/article/view/24983/46727. Accessed 27 Jan 2021

Bhatt P, Bhatt K, Sharma A, Zhang W, Mishra S, Chen S (2021) Biotechnological basis of microbial consortia for the removal of pesticides from the environment. Crit Rev Biotechnol 41(3):317–338

Bilal M, Iqbal HMN (2019) Microbial-derived biosensors for monitoring environmental contaminants: recent advances and future outlook. Process Saf Environ Prot 124:8–17. https://doi.org/10.1016/j.psep.2019.01.032

Brooks SM, Alper HS (2021) Applications, challenges, and needs for employing synthetic biology beyond the lab. Nat Commun 12(1):1–16. https://doi.org/10.1038/s41467-021-21740-0

Bullock C, Gorman C (2000) Fusions to chloramphenicol acetyltransferase as a reporter. Methods Enzymol 326(1967):202–221. https://doi.org/10.1016/s0076-6879(00)26056-4

Chong H, Ching CB (2016) Development of colorimetric-based whole-cell biosensor for organophosphorus compounds by engineering transcription regulator DmpR. ACS Synth Biol 5 (11):1290–1298. https://doi.org/10.1021/acssynbio.6b00061

Cui Y, Lai B, Tang X (2019) Microbial fuel cell-based biosensors. Biosensors 9(3):1–18. https://doi.org/10.3390/bios9030092

Debbarma P, Raghuwanshi S, Singh J, Suyal DC, Zaidi MGH, Goel R (2017) Comparative in situ biodegradation studies of polyhydroxybutyrate film composites. 3Biotech 7(178):1–9. https://doi.org/10.1007/s13205-017-0789-3

Dennis P et al (2003) Monitoring gene expression in mixed microbial communities by using DNA microarrays. Appl Environ Microbiol 69(2):769–778. https://doi.org/10.1128/AEM.69.2.769-778.2003

Espinosa-Urgel M et al (2015) Engineering biological approaches for detection of toxic compounds: a new microbial biosensor based on the *Pseudomonas putida* TtgR repressor. Mol Biotechnol 57(6):558–564. https://doi.org/10.1007/s12033-015-9849-2

Fang D et al (2020) Redox mediator-based microbial biosensors for acute water toxicity assessment: a critical review. Chem Electro Chem 7(12):2513–2526. https://doi.org/10.1002/celc.202000367

Fujimoto H et al (2006) Whole-cell arsenite biosensor using photosynthetic bacterium Rhodovulum sulfidophilum: Rhodovulum sulfidophilum as an arsenite biosensor. Appl Microbiol Biotechnol 73(2):332–338. https://doi.org/10.1007/s00253-006-0483-6

Ganesan S, Vasudevan N (2021) Genetically modified microbial biosensor for detection of pollutants in water samples. In: Gothandam KM et al (eds) Environmental biotechnology, vol 3. Springer, Cham, pp 85–103. https://doi.org/10.1007/978-3-030-48973-1_4

Goel R, Zaidi MGH, Soni R, Kusumlata SYS (2008) Implication of *Arthrobacter* and *Enterobacter* species for polycarbonate degradation. Int Biodeterior Biodegrad 61(2):167–172

Gómez RSG et al (2004) Spectroscopic determination of poly-aromatic compounds in petroleum contaminated soils. Water Air Soil Pollut 158(1):137–151. https://doi.org/10.1023/B:WATE.0000044829.26130.1c

Gui Q et al (2017) The application of whole cell-based biosensors for use in environmental analysis and in medical diagnostics. Sensors (Switzerland) 17(7):1–17. https://doi.org/10.3390/s17071623

Gupta N et al (2019) Cell-based biosensors: recent trends, challenges and future perspectives. Biosens Bioelectron 141:111435. https://doi.org/10.1016/j.bios.2019.111435

Hansen LH, Sørensen SJ (2001) The use of whole-cell biosensors to detect and quantify compounds or conditions affecting biological systems. Microb Ecol 42(4):483–494. https://doi.org/10.1007/s00248-001-0025-9

He L et al (2019) In vivo study of gene expression with an enhanced dual-color fluorescent transcriptional timer. eLife 8:1–20. https://doi.org/10.7554/eLife.46181

Ikeda T, Kano K (2001) An electrochemical approach to the studies of biological redox reactions and their applications to biosensors, bioreactors, and biofuel cells. J Biosci Bioeng 92(1):9–18. https://doi.org/10.1263/jbb.92.9

Jung RK et al (2007) Power generation using different cation, anion, and ultrafiltration membranes in microbial fuel cells. Environ Sci Technol 41(3):1004–1009. https://doi.org/10.1021/es062202m

Khire TS et al (2020) Microvascular mimetics for the study of leukocyte–endothelial interactions. Cell Mol Bioeng 13(2):125–139. https://doi.org/10.1007/s12195-020-00611-6

Korotkaya EV (2014) Biosensors: design, classification, and applications in the food industry. Foods Raw Mater 2(2):161–171. https://doi.org/10.12737/5476
Kumar S, Suyal DC, Yadav A, Shouche Y, Goel R (2019) Microbial diversity and soil physiochemical characteristic of higher altitude. PLoS One 14(3):e0213844
Kumar P, Dash B, Suyal DC, Gupta SB, Singh AK, Chowdhury T, Soni R (2021) Characterization of arsenic-resistant Klebsiella pneumoniae RnASA11 from contaminated soil and water samples and its bioremediation potential. Curr Microbiol. https://doi.org/10.1007/s00284-021-02602-w
Lei Y, Chen W, Mulchandani A (2006) Microbial biosensors. Anal Chim Acta 568(1–2):200–210. https://doi.org/10.1016/j.aca.2005.11.065
Lim JW et al (2015) Review of micro/nanotechnologies for microbial biosensors. Front Bioeng Biotechnol 3:1–13. https://doi.org/10.3389/fbioe.2015.00061
Lobsiger N, Stark WJ (2019) Strategies of immobilizing cells in whole-cell microbial biosensor devices targeted for analytical field applications. Anal Sci 35(8):839–847. https://doi.org/10.2116/analsci.19R004
Mehrotra P (2016) Biosensors and their applications – a review. J Oral Biol Craniofacial Res 6 (2):153–159. https://doi.org/10.1016/j.jobcr.2015.12.002
Pandey J, Chauhan A, Jain RK (2009) Integrative approaches for assessing the ecological sustainability of in situ bioremediation. FEMS Microbiol Rev 33(2):324–375. https://doi.org/10.1111/j.1574-6976.2008.00133.x
Panigrahi S, Velraj P, Subba Rao T (2018) Functional microbial diversity in contaminated environment and application in bioremediation. In: Microbial diversity in the genomic era. Elsevier, pp 359–385. https://doi.org/10.1016/B978-0-12-814849-5.00021-6
Park M, Tsai SL, Chen W (2013) Microbial biosensors: engineered microorganisms as the sensing machinery. Sensors (Switzerland) 13(5):5777–5795. https://doi.org/10.3390/s130505777
Pham HTM et al (2015) The determination of pharmaceuticals in wastewater using a recombinant Arxula adeninivorans whole cell biosensor. Sensors Actuators B Chem 211:439–448. https://doi.org/10.1016/j.snb.2015.01.107
Pollution (2021) The World Bank. Available at: https://www.worldbank.org/en/topic/pollution. Accessed 26 Jan 2021
Popovtzer R et al (2005) Novel integrated electrochemical nano-biochip for toxicity detection in water. Nano Lett 5(6):1023–1027. https://doi.org/10.1021/nl0503227
Raut N et al (2012) Engineered cells as biosensing systems in biomedical analysis. Anal Bioanal Chem 402(10):3147–3159. https://doi.org/10.1007/s00216-012-5756-6
Reshetilov, A.N., Iliasov, P.V., Reshetilova, T.A. (2010) The microbial cell based biosensors. In: Intelligent and biosensors. InTech. https://doi.org/10.5772/7159
Sharma P, Asad S, Ali A (2013) Bioluminescent bioreporter for assessment of arsenic contamination in water samples of India. J Biosci 38(2):251–258. https://doi.org/10.1007/s12038-013-9305-z
Shi S, Ang EL, Zhao H (2018) In vivo biosensors: mechanisms, development, and applications. J Ind Microbiol Biotechnol 45(7):491–516. https://doi.org/10.1007/s10295-018-2004-x
Shin HJ (2011) Genetically engineered microbial biosensors for in situ monitoring of environmental pollution. Appl Microbiol Biotechnol 89(4):867–877. https://doi.org/10.1007/s00253-010-2990-8
Stenuit B et al (2008) Emerging high-throughput approaches to analyze bioremediation of sites contaminated with hazardous and/or recalcitrant wastes. Biotechnol Adv 26(6):561–575. https://doi.org/10.1016/j.biotechadv.2008.07.004
Strand SE et al (2003) Results misinterpreted. Environ Sci Technol 37(19). https://doi.org/10.1021/es032577x
Su L et al (2011) Microbial biosensors: a review. Biosens Bioelectron 26(5):1788–1799. https://doi.org/10.1016/j.bios.2010.09.005
Sun JZ et al (2015) Microbial fuel cell-based biosensors for environmental monitoring: a review. Water Sci Technol 71(6):801–809. https://doi.org/10.2166/wst.2015.035

Suyal DC, Joshi D, Debbarma P, Soni R, Dash B, Goel R (2019a) Soil metagenomics: unculturable microbial diversity and its function. In: Varma A, Choudhary D (eds) Mycorrhizosphere and pedogenesis. Springer, Singapore, pp 355–362

Suyal DC, Joshi D, Kumar S, Soni R, Goel R (2019b) Differential protein profiling of soil diazotroph *Rhodococcus qingshengii* S10107 towards low-temperature and nitrogen deficiency. Sci Rep 9(20378):1–9

Tani H, Maehana K, Kamidate T (2007) On-chip bioassay using immobilized sensing bacteria in three-dimensional microfluidic network. Methods Mol Biol (Clifton, N.J.) 385:37–52. https://doi.org/10.1007/978-1-59745-426-1_4

Tucci M et al (2020) Microbial amperometric biosensor for online herbicide detection: photocurrent inhibition of *Anabaena variabilis*. Electrochim Acta 302(2019):102–108. https://doi.org/10.1016/j.electacta.2019.02.007

Van Der Meer JR, Belkin S (2010) Where microbiology meets microengineering: design and applications of reporter bacteria. Nat Rev Microbiol 8(7):511–522. https://doi.org/10.1038/nrmicro2392

Vigneshvar S et al (2016) Recent advances in biosensor technology for potential applications – an overview. Front Bioeng Biotechnol 4:11. https://doi.org/10.3389/fbioe.2016.00011

Vopálenská I, Palková Z (2015) New biosensor for detection of copper ions in water based on immobilized genetically modified yeast cells 72, pp 160–167. https://doi.org/10.1016/j.bios.2015.05.006

Whangsuk W et al (2010) ChpR is a chlorpyrifos-responsive transcription regulator in Sinorhizobium meliloti. J Mol Microbiol Biotechnol 18(3):141–147. https://doi.org/10.1159/000308514

Xu X, Ying Y (2011) Microbial biosensors for environmental monitoring and food analysis. Food Rev Intl 27(3):300–329. https://doi.org/10.1080/87559129.2011.563393

Yang Y et al (2018) Bioelectrochemical biosensor for water toxicity detection: generation of dual signals for electrochemical assay confirmation. Anal Bioanal Chem 410(4):1231–1236. https://doi.org/10.1007/s00216-017-0656-4

Zhou J, Thompson DK (2002) 'Challenges in applying microarrays to environmental studies. Curr Opin Biotechnol 204–207. https://doi.org/10.1016/S0958-1669(02)00319-1

Chapter 6
Recent Advancements in Mycoremediation

Ihsan Flayyih Hasan Al-Jawhari

6.1 Introduction

Human activities are continuously interfering with the environmental processes and making them polluted by adding heavy metals, radionuclides, hydrocarbons, and contaminants. Furthermore, the pollution of surface water, and groundwater as a result of urban and industrial growth has imposed negative consequences on both humans and the environment (EEA 2003; Kour et al. 2021). Therefore, the requirement for eco-friendly approaches is increasing day by day that can treat the pollutants in an effective way.

6.2 Impact of the Agrochemicals on the Ecosystem

Agriculture has a significant effect on terrestrial as well as aquatic ecosystems. The injudicious use of pesticides, plant hormones, fertilizers, and other agrochemicals has caused several environmental concerns (Debbarma et al. 2017; Bhatt et al. 2021a, b). Similarly, agriculture is the primary source of PO_4^{3-}, NO_3, and pesticide emissions, with the continuous increment over the last 35 years (Loebenstein and Thottappilly 2007; Önder et al. 2011). Further, unorganized land remediation policies have decreased the vegetation, soil organisms, and soil organic matter (Walls 2006; Arias-Estévez et al. 2008; FAO 2009).

I. F. H. Al-Jawhari (✉)
Department of Biology, College of Education for Pure Sciences, University of Thiqar, Al-Nasiriyah, Iraq

© The Author(s), under exclusive license to Springer Nature Switzerland AG 2022

D. C. Suyal, R. Soni (eds.), *Bioremediation of Environmental Pollutants*,
https://doi.org/10.1007/978-3-030-86169-8_6

6.2.1 Eutrophication and Algal Blooms in the Lakes

Blooming is the uncontrolled growth of the photosynthetic organisms in the lake due to the excessive accumulation of organic carbon, phosphate, and nitrates. Direct runoff, deforestation, and agrochemical spray drift transfer terrestrial phosphate and nitrate to the water environment and causes eutrophication (Fig. 6.1). Due to it, the population of algae and phytoplankton grows rapidly due to high concentrations of chemical nutrients that promote their growth. As these organisms grow, the O_2 is depleted, resulting in life stop in the profundal zone due to the lack of oxygen (Rial-Otero et al. 2003). It is greatly affected by heat, nutrients, flow rate, and other abiotic elements (OECD 2012).

Impact indicators:

- Large increase in microalgae and macroalgae growth on the epilimnion layer. The sunlight penetration gets decreased which leads to the loses of underwater aquatic vegetation.
- A nutrient imbalance can lead to a change in microalgae components, allowing toxic algal blooms to flourish.
- Reduce diversity and make a negative impact on the food web due to changes in benthic species composition.
- In the coastal and maritime environment, it results in reduced dissolved O_2 levels.

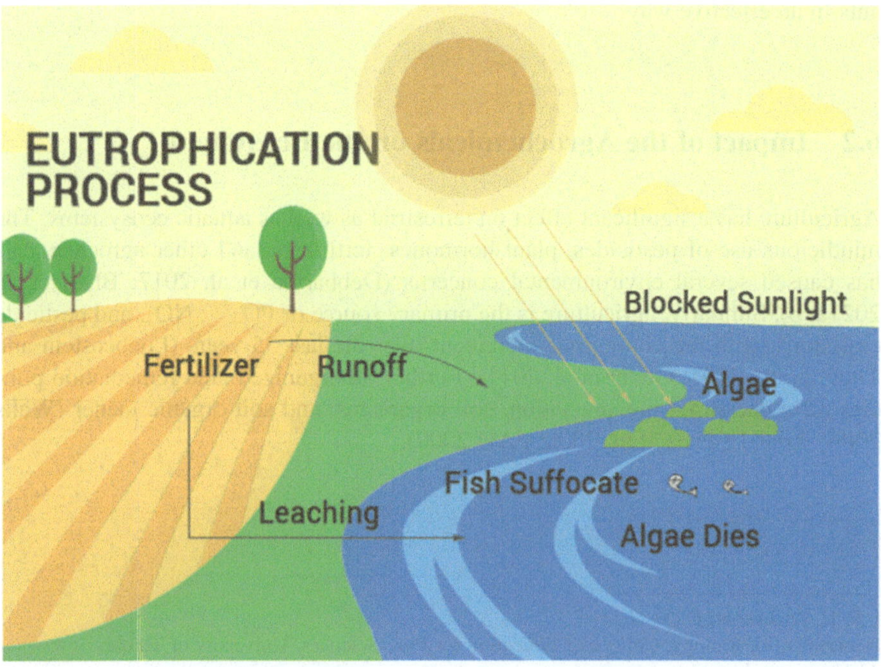

Fig. 6.1 Blooming processes in the lakes

Besides, phosphorus, nitrogen is another important factor that contributes towards eutrophication. Several algal species are able to fix atmospheric nitrogen into nitrate (Howarth 2006). It was observed that nitrogen load in the Yellow Sea (China) and the North Sea (Europe) was 10–15 times higher than the natural levels. Moreover, nitrogen loads to US coastal waters were on average 6 times higher than natural levels (Howarth 2006). In Europe and the United States, agriculture is now the primary source of nitrogen supply.

6.3 Mycoremediation of Pesticides

Pesticides have a significant economic effect by protecting and lowering pest-related agricultural sabotage and thereby, increasing productivity and yield. In recent years, pesticides used have been increased in agriculture. Due to their excessive use throughout the world, the 1940s and 1950s are known as the "pesticide era" (Graeme 2005). Aside from agricultural uses, pesticides are used in large quantities for urban plantations maintenance, sanitary handling and storage, vegetation control, and forest protection. Pesticides have also been used in the reduction of the disease *viz.* malaria and typhus fever. Every year, hundreds of thousands of tonnes of pesticides are used around the world to boost agricultural production by reducing pests (Liu and Xiong 2001).

Overuse of agrochemicals has imposed negative impacts on the environment (Giri et al. 2017a, b). They reach specific organisms in less than 5% of cases, with the majority of them leach into the subsoil and contaminate groundwater (Kookana et al. 1998). Alternatively, if it is immobile, it causes magnification by increasing chemicals levels in the aquatic ecosystem and harm native organisms (Amakiri 1982). Pesticide pollution can cause contamination of surface and sediments (Juhler et al. 2001; Bhatt et al. 2021a). Moreover, it has a negative impact on microbial diversity, especially bacterial, archaeal and fungal communities (Ahmed et al. 1998). Agrochemicals spray also affects non-vegetation. They can show deviation or volatility from treatment regions that have been processed to non-target plants, climate, and the soil. Low crop yields, the devastation of soil organisms even in microhabitats, and unwanted chemical residue concentrated in crops are the main outcomes of agrochemical overuse (Edwards 1986). Huge amounts of pesticide are used, within local soil habitat and thus cause pesticide spills (Gan and Koskinen 1998). Pesticides are poisonous and intransigent by the environment, causing pollution in ecosystems. As a result, health issues occur in the end-users. Carcinogenicity, mutagenicity, immunosuppression, hormonal imbalance, and different other health issues are all side effects of pesticides (Gupta 2004).

Chemical treatment, volatilization, and incineration are all physical and chemical processes for removing pesticide residues from the soil. Large quantities of acids and alkalis are produced as a result of chemical treatment and volatilization, which must be discharged. Incineration is a highly dependable physical–chemical process for pesticide removal; but, due to the risk of harmful pollution and high economic costs, it has a lot of public opposition. Further, excavating soil from a contaminated site and transporting it to other location for cleaning, is expensive and inefficient, and require a large storage region (Nerud et al. 2003). Therefore, there is a pressing need to evolved pesticide remediation techniques that are safe, convenient, and cost-effective (Zhang and Chiao 2002). As a result, many biological processes involving the bioremediation of organic substances by organisms have grown in popularity as a treatment option for pesticide residues in terrestrial and aquatic ecosystems.

Bioremediation is usually is less-costly than physical and chemical methods for removing contaminants without the need for excavation. This kind of process has the ability to treat pollutants onsite (Kearney and Wauchope 1998). Microorganisms are the main factor for the recirculation of biological materials and, they have developed a diverse set of enzymes, activation systems, and pathways for degrading and utilizing pesticides as a source of energy (Talaro and Talaro 2002; Goel et al. 2008; Kumar et al. 2021).

Mycodegradation and my deterioration are two approaches in which fungi degrade a wide range of contaminants (Singh et al. 2021). Mycoremediation is the use of fungi in the ecosystem to degrade wastes and contaminants. This term is derived from two words—“mukēs” and “remedium” representing the meaning “fungus” and “restoring balance”, respectively.

6.4 Mycoremediation of Herbicides

Many fungi, such as *Phanerochaete velutina, Coriolus versicolor,* and *Pleurotus ostreatus* have been the capacity to break down different herbicides like atrazine, and various abilities (Bending et al. 2001). Al-Jawhari and Al-Sead (2016) showed that *Aspergillus versicolor* was more efficient for biodegradation and the five-week was the optimum period for biodegradation of granstar (*Tribenuron methyl*) to different metabolites.

Zboniska et al. (1992) investigated the remediation of glyphosate by *Penicillium citrinum. Penicillium notatum* was found to use the herbicide as a source of P and use the CH_6NO_3P route to break it down (Bujacz et al. 1995). *Trichoderma harzianum*, and *Aspergillus niger* have shown studies to break down glyphosate (Krzysko-Lupicka et al. 1997). Klimeka et al. (2001) have shown *Penicillium chrysogenum* was the important microorganism to use glyphosate as a nitrogen source. Since the fungal cell is known to lack observable activity of the enzyme nitrogen reductase, this isolate was incapable of converting NO_3 to NH_4^+. Further, Lipok et al. (2003) have discovered a plant pathogen, *Alternaria alternata* that was able to use glyphosate as a sole source of nitrogen.

Mougin et al. (1994) confirmed that the white-rot fungus *Phanerochaete chrysosporium* biotransformed atrazine and resulted in the reduction of herbicide concentration by 48% under in vitro conditions within the early 4 days after treatment. The formation of hydroxylated and/or N-dealkylated metabolites was clearly demonstrated by the mineralization of the herbicide’s ethyl group. In fungi

N-Dealkylation is more common than hydroxylation, resulting in atrazine hydrolysis through direct or secondary metabolite (Mougin et al. 1994).

Bending et al. (2002) tested the ability of nine species of white-rot fungi to degrade different mono-aromatic pesticides. The breakdown of atrazine (20 μg g^{-1}) was measured in sterilization made up of dirt, wheat straw, and peat. Bastos and Magan (2009) have examined the ability of *T. versicolor* to break down atrazine in sterilized soil for 24 weeks. The fungal treatment was found to accelerate atrazine breakdown. Similarly, Elguetaa et al. (2016) have analyzed the atrazine degrading ability of white-rot fungi and found it effective.

6.5 Mycoremediation of Heavy Metals

Heavy metal refers to any metal that is relatively denser and more toxic in comparison to the other metals, even at its lower concentrations. Heavy metal pollution is most visible in mining regions and abandoned mine areas. Metals are released into the atmosphere during their extraction and processing processes (Lenntech Water Treatment 2004). When heavy metals are present in the environment, they are well known for their toxicity. From the standpoints of ecology, evolution, nutrition, and the environment, it is a major source of interest. Heavy metals accumulation in soils is a significant source of concern in agricultural production because it has negative consequences on food production, cereals growth lead to phyto-poisons, soil microbe, and human health (Dash et al. 2021). The biological toxicity impacts of heavy metals in human biochemistry are a major source of concern. These heavy metals are non-biodegradable and possess a longer half-life. Heavy metals can be absorbed by the crops growing in the polluted regions. The biodegradation method becomes hard if the soil has been polluted by heavy metals. Su (2014) has enumerated several properties of heavy metal pollution in the environment, including its global impact and distribution, latency and long-term damage, arduous biodegradation, and so on. Further, one of the major risk factors associated with heavy metals is an occupational hazard. Workers in mines and those engaged in the manufacture of heavy metals, as well as residents near industrial sites, have been exposed to suspended particulate matter (Ogwuegbu and Muhanga 2005).

6.5.1 Toxicity of Heavy Metals

Lead (Pb)

There are many different sources of lead emissions. Mainly it discharges to the ecosystem as a result of smelting and mining of mines. Lead poisoning occurs when a lead is ingested or inhaled through any means inside the body (Ferner 2001). Further, it causes severe health problems i.e.: hemoglobin synthesis, kidney dysfunctions, joint and reproductive system impairment, as well as heart dysfunction

(Ogwuegbu and Muhanga 2005). Also, ingestion may cause neurological problems including serious and irreversible brain damage. The brain development of children is also harmed (Udedi 2003).

Mercury

Eating polluted fish, amalgam dental fillings, coal burning, gold mining are some major sources of mercury exposure (Balaguru et al. 2016). In its inorganic form, mercury is extremely toxic. Oral conditions such as gingivitis and stomatitis, serious brain and CNS damage, and congenital malformations are linked with this pollution (Ferner 2001; Lenntech Water Treatment 2004).

Arsenic

Some sources of arsenic include long-term consumption of arsenic-contaminated water and consumption of oil additives and plants cultivated on arsenic-rich soils. Arsenic has been linked to an increased risk of cancer. Arsenic poisoning can cause cancer in a variety of organs, including the kidney, liver, heart, respiratory system, and skin (Ratnaike 2003; Kumar et al. 2021). Moreover, symptoms may worsen if acute toxicity occurs. It causes muscle pain and stomach pain followed by blood in urine, loose motions, vomiting, and convulsions. Arsenic poisoning typically impairs the functioning liver, kidney, skin, and lungs (Kapaj et al. 2006). Arsenic discouraged the production of (ATP) through breathing (INECAR 2000), and high levels of exposure can result in death (Ogwuegbu and Ijioma 2003).

Cadmium

Cadmium, as a contaminant, exists in phosphate fertilizers, industrial dyes, and other outputs as a pollutant. It's a highly poisonous and dangerous metal. Cadmium-containing goods are rarely recycled, but they are often thrown away with household waste. As a result, pollution of the surrounding environment occurs particularly when waste is incinerated. It is released as a by-product during the refining of zinc (and sometimes lead), (IARC 1993). Cigarette smoking, battery making, coatings, plastics making, and air from the smelting refineries are the main sources of cadmium poisoning that can harm the kidney and liver. Cadmium is a human carcinogen according to the World Health Organization (Il'yasova and Schwartz 2005).

6.5.2 Remediation of Heavy Metals by Fungi

Heavy metals biosorption is aided by exploring endophytic fungi (Yang et al. 2012). More specifically, *Penicillium* sp., and *Aspergillus niger* have been shown to possess a higher biosorption potential in removal metal-contaminant environments by adsorbing heavy metals from the mixed pollutants (Ahmad et al. 2006). The biological material's potential to absorb heavy metals from the contaminated site using metabolically influenced or physicochemical adsorption mechanisms is commonly referred to as biosorption (Fourest and Roux 1992). It entails species removing metal ions by separating organisms solid and liquid (Yang et al. 2012). Biosorption has received a lot of publicity because it is a very powerful and practical way to remove heavy metals (Cruz et al. 2004; Ting et al. 2008). According to an estimate, biomass, whether life or death, can serve as a bio-absorption and high-potency heavy metal ions (Bayramoglu et al. 2002).

Xiao et al. (2010) have obtained an endophytic fungus, *Microsphaeropsis* sp. LSE10, from the plant *Solanum nigrum*. It has shown higher Cd bio-absorption ability. Similarly, Deng et al. (2011) have demonstrated that an endophytic *Mucor* species isolated from *Brassica chinensis* was able to bioaccumulate and absorb heavy metals, especially lead and cadmium. Choo et al. (2015) have demonstrated that the fungal endophyte of *Nypa fruticans, Pestalotiopsis* sp., was able to grow luxuriously under the high concentration of chromium, lead, copper, and other heavy metals. Further, Zhang et al. (2008) have reported the heavy metal accumulating property in an endophytic *Exophiala pisciphila*. Li et al. (2012b, 2016) have observed that a higher abundance of metal-resistant endophytes in any host plant increases the ability of phytoremediation. Moreover, the soil heavy metal concentration affects the colonization rate of the endophytes and thus, bioaccumulation of the contaminants. Further, the dominant genera were more resistant to heavy metals than other genera (Li et al. 2016). In another study, Li et al. (2012a) have shown that regions contaminated with lead and zinc encouraged the colonization of the endophytes and thus promoted phytoremediation. Deng et al. (2014) have reported that *Portulaca* possesses high heavy metals aggregation power. It was correlated with the extensive mycelia producing capacity of this genus (Meena and Sarita 2017). Kratochvil and Volesky (1998) have summarized the differences between traditional and biological methods of biosorption. They have revealed that biological methods are cost-effective, highly efficient, easy to recover, and eco-friendly in comparison to the conventional alternatives.

6.5.3 Mycoremediation of Hydrocarbon Pollution

Hydrocarbons are the carbon and hydrogen-containing compounds that can be generated from fuels and other similar sources i.e.: diesel, oil, kerosene, benzene, ethylbenzene, toluene, and xylenes. Among these, aromatic hydrocarbons are more dangerous for the environment than aliphatic ones.

The presence of crude oils and similar contaminants in the soil, cause harmful influences on soil organisms (Scott 2003). Petroleum hydrocarbon pollution of soil causes several issues. Petroleum hydrocarbon composition is found in the atmosphere is harmful to the environment (Scott 2003). Owing to the accumulation of hydrocarbon pollutants in animal and plant tissues, wide harm influence by hydrocarbons pollutants to the local ecosystem can result in mutations and/or deaths (Alvarez and Vogel 1991). The health risks associated with carbon emissions derive mainly from direct contact with a polluted ecosystem of pollutants, as well as secondary contamination of water beneath the soil.

PAHs are Polycyclic aromatic hydrocarbons that are resistant to biodegradation and are often associated with oil pollution. Maliszewska-Kordybach (1999) found that contaminants can migrate long distances by air and precipitate on the earth's surface, plant vegetation, and water bodies (Cheema et al. 2009). The rhizosphere of a plant controls its microbiome through different types of interactions (Huang et al. 2004; Haritash and Kaushik 2009). Endophytes support their hosts by using relevant degradation pathways and metabolic capacities, resulting in increased phytodegradation of organic pollutants and decreased phyto-toxicity (Soleimani et al. 2010). Moreover, it also affects the evaporation and transpiration of volatile pollutants (Weyens et al. 2009).

6.6 Petroleum Biodegradation by Fungi

Many microorganisms have the efficiency to remove hydrocarbons in the atmosphere, water, and soil. Natural biodegradation takes a long time. As a result, rather than relying on a single organism, microbe from different genera may work together to expand the restricted range of degradation. Several types of microbes have been found in petroleum-contaminated soil, water, and surface (Surovtseva et al. 1997). For the clearance of PAH pollutants, ideas in field degradation of pollutants are more favored in today's environment (Ndimele 2010). Al-Hawash et al. (2019) found that the capacity of *Aspergillus* sp. RFC-1 to degrade petroleum hydrocarbons was assessed using surface adsorption, and the results showed that by day 7 of incubation, crude oil, naphthalene (NAP), phenanthrene (PHE), and pyrene (PYR) removal capacities had reached 60.3%, 97.4%, 84.9%, and 90.7%, respectively. While, crude oil, NAP, PHE, and PYR had biodegradation capacities of 51.8%, 84.6%, 50.3%, and 55.1%, respectively. Extracellular enzymes extracted from the crude-oil-polluted environment from six *Aspergillus* species efficiently break down crude oil, (Zhang et al. 2016). *Penicillium sp.* RMA1 and RMA2, two crude oil-degrading fungi were isolated from the Rumaila oilfield, that could grow in petroleum hydrocarbons composition media and removed 75% and 55% of hydrocarbons, respectively, suggesting that these fungi could efficiency catabolism petroleum hydrocarbons (Al-Hawash et al. 2017). The lipolytic enzyme of *Aspergillus niger* isolated from an oil-contaminated ecosystem can removal polyaromatic hydrocarbons in petroleum polluted environment (Mauti et al. 2016). Sandhu (2016) obtained that the high remediation capacity of *Aspergillus* sp. to kerosene because of its superior mycelial growth and extracellular enzymes activity. Kadri et al. (2017) fungi-produced enzymatic systems are important for transfer petroleum hydrocarbons to CO_2 or catabolism products. Extracellular enzyme production is the main mechanism for fungal hydrocarbon degradation (Zhang et al. 2016). Al-Jawhari (2016) found that the highest anthracene removal rates were achieved after 7-day incubation with a mixed pure culture of *Aspergillus niger* and *Penicillium funiculosum*. Michael et al. (2020) obtained the enzyme activity in two fungi *Aspergillus oryzae* and *Mucor irregularis* growing them on Bushnell Haas

(BH) mineral agar supplemented with the hydrocarbon at various concentrations, such as 5%, 10%, 15%, and 20%, with a dextrose power. After 15 days of incubation, hydrocarbon remediation potentials of these fungi were confirmed using GC/MS in BH broth culture filtrates pre-supplemented with 1% engine oil. The results indicated that *M. irregularis* only breakdown the long-chain hydrocarbons and BTEX. This study confirmed that *A. oryzae* and M. *irregularis* have the potential to be exploited in the bio-treatment and removal of hydrocarbons from polluted soils.

Exploration of endophytic fungal genera in phytoremediation appears to be a more promising feature of phytoremediation (Mohsenzadeh et al. 2010). By fostering pyrene accumulation in the roots of the host, inoculation of *Lewia* sp., resulted in a remarkably high removable of pyrene in comparison with non-inoculated ones (Cruz-Hernández et al. 2013). Further, a fungal endophyte *Phomopsis liquidambari*, exists to uses phenolic 4-hydroxybenzoic acid, as a sole carbon source and as a capacity breakdown of (PAH) in pure culture. One of the most important factors in the rhizospheric deterioration of petroleum-contaminated soils is the presence of endophytes. Mohsenzadeh et al. (2010) have analyzed seven plant species from hydrocarbon-polluted areas. Moreover, several endophytic *Fusarium* strains *viz. Fusarium acuminatum, Fusarium reticulatum,* and *Fusarium equiseti* are also reported to have hydrocarbon-degrading potential. This group has also proposed that combined use of endophytes and plants were more effective than their application. Similarly, basidiomycete, *Phanerochaete chrysosporium*, was discovered to degrade many organic contaminants in 1985 (Hammel et al. 1992). With the assistance of enzymes such as (MnP) (LiP), *Phanerochaete chrysosporium* is known to break down pyrene and anthracene by oxidation (Lei et al. 2007). It is worth noting that white-rot fungi account for one-third of the literature on mycoremediation (Singh 2006). Various strains of white-rot fungi have been identified to work on a wide variety of organic compounds that are resistant to others. *Phanerochaete chrysosporium*, the white-rot fungus, is one of the best fungi for a breakdown of poisons and insoluble wastes throughout the setting (Singh 2006). Figure 6.2 showed the degraded Phenanthrene by dioxygenase enzyme *phn*Ac/*phn*Ad. In recent years, the significance of white-rot fungi for fungal hydrocarbon bioremediation has also received scientific attention (Rahman et al. 2014).

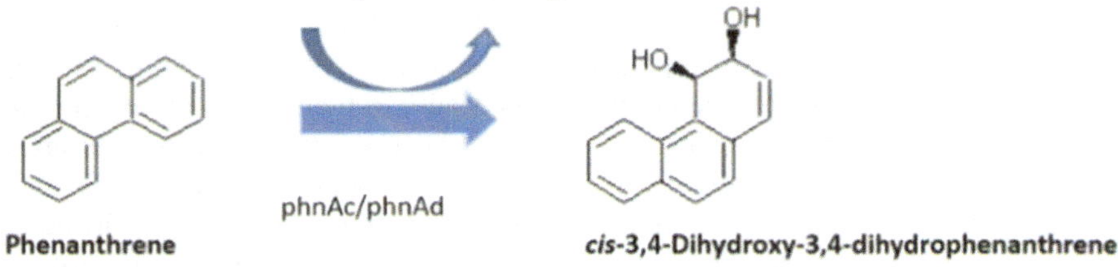

Fig. 6.2 Reaction catalyzed by the hydroxylating dioxygenase *phn*Ac/*phn*Ad as the first step of the phenanthrene degradation pathway

6.7 Myconanoparticles

Nanotechnology is a branch of science that deals with metal nanoparticles in various forms. A nanoparticle (10^9 m) is a small piece that functions as a catalyst in terms of its transportation and assets as a whole (Gholami-Shabani et al. 2016a, b, c). Fungi are integral microorganisms from the catalog of microorganisms used to synthesize metal nanoparticles and are superior to different microorganisms in several ways. They shape a mycelial mesh, which aids in their development withstand the flow pressure, agitation, and other bioreactor associated stresses. Further, the potential of fungal strains to grow and reproduce on readily available, low-cost substrates, reflects their capacity to produce a wide variety of industrially important compounds (Dhanasekar et al. 2015). The use of fungi in nanotechnology for the making of nanoparticles is known as mycosynthesis of metal nanoparticles or myconanotechnology (Honary et al. 2012; Prasad 2016). Many fungal species, such as *Aspergillus*, *Fusarium*, *Penicillium*, and *Verticillium* have been used as promising tools for metal nanoparticles. Different fungal species are potential candidates for metal nanoparticle making. Mycosynthesis of silver, gold, gold-silver alloy, platinum, selenium nanoparticles has already been studied. Further, *Trichoderma viride* was used for green synthesis for gold nanoparticles using para-nitro phenol and amino phenol (Mishra et al. 2014).

6.8 Biodegradation in Soils by Fungi

Heavy metal ions are believed to be immobilized by microorganisms by binding them to their cell walls (Vankar and Bajpai 2008). Moreover, they can convert certain pollutants into soluble substances and use them as a source of nutrients and energy (Kumar et al. 2008). The presence of microorganisms that enhance phytostimulation or rhizodegradation can speed up the phytoremediation method (Kavamura and Esposito 2010).

Fungi are beneficial for the biodegradation of heavy metal-polluted sites due to they possess high biomass content (Mann 1990). Because of their negative charge, the cell walls of fungi can act as a cation exchanger (Muraleedharan et al. 1991; Fomina et al. 2007). Furthermore, because of their low-cost processing, fungal biomasses may act as bio-absorption compounds (Maurya et al. 2006). Further, fungi play a very important role on the Earth because of their ability to decompose, turn, and cycle nutrients (Archana and Jaitly 2015). The capacity of fungi to break down anthropogenic substances was recorded for the first time in a study by Wunch et al. (1999). *Marasmiellus troyanus* degraded PAH (benzo[α] pyrene) in liquid culture, to the researchers. After the publication of this study, various writers and research groups have been tasked with gathering more proof to resolve this problem (Anastasi et al. 2013; Rhodes et al. 2014; Singh et al. 2015).

6.9 Conclusions

Future research should be concentrated on developing real-time biodegradation assays, and large-scale bioremediation experiments. They should use potential native strains as well as genetically engineered microorganisms. More work is required today to gain a deeper understanding of how agricultural soils function as a whole, and a deeper understanding of the dynamics of these heterogeneous ecosystems, as well as the interactions of different microorganisms. Much deeper insight into the associated mechanisms is needed. It is possible to develop technologies to improve degradation efficiency through studying the process of degradation. Through a better understanding of the mechanisms, such as cell immobilization in various systems and the development and use for waste removal in the field, we will improve the efficiency of degradation. Further, fungi in the soil can be used to make metal nanoparticles and as green remediators to investigate the ability of fungi in heavy metal biosorption and extraction from polluted areas for the benefit of the industry. Fungi may be used as a factory for a variety of purposes, including investigating nanoremediation, soil fertility, and ecosystem balance. Nanoparticles may also minimize pesticide usage by enabling native fungi to biosynthesize the nanoparticles, which is emerging as a cutting–edge technology for mankind. With the rapid production of environmentally friendly fungal synthesis procedures, the field of green nanotechnology must blossom in order to make the earth greener and safer.

References

Ahmad I, Ansari MI, Aqil F (2006) Biosorption of Ni, Cr and Cd by metal tolerant *Aspergillus niger* and *Penicillium* sp. using single and multi-metal solution. Indian J Exp Biol 44:73–76

Ahmed M, Ismail S, Mabrouk S (1998) Residues of some chlorinated hydrocarbon pesticides in rain water, soil and ground water, and their influence on some soil microorganisms. Environ Int 24:665–670

AI-Jawhari IF (2016) Bioremediation of anthracene by *aspergillus Niger* and *Penicillium funiculosm*. Int Res J Bio Sci 5(6):1–11

AI-Jawhari IF, AI-Sead KG (2016) Fate of herbicide Granstar (Tribenuron methyl) in wheat field in AI-Nasiriya governorate. Int Res J Bio Sci 5(8):22–37

Al-Hawash AB, Alkooranee JT, Abbood HA, Zhang J, Sun J, Zhang X, Ma F (2017) Isolation and characterization of two crude oil-degrading fungi strains from Rumaila oil field, Iraq. Biotechnol Rep 17:104–109. https://doi.org/10.1016/j.btre.2017.12.006

Al-Hawash AB, Zhang X, Ma F(2019) Removal and biodegradation of different petroleum hydrocarbons using the filamentous fungus *Aspergillus* sp. RFC-1. Microbiol 8(1):1–14. https://doi.org/10.1002/mbo3.619

Alvarez PJ, Vogel TM (1991) Substrate interactions of benzene, toluene, and para-xylene during microbial degradation by pure cultures and mixed culture aquifer slurries. Appl Environ Microbiol 57(10):2981–2985. https://doi.org/10.1128/aem.57.10.2981-2985.1991

Amakiri MA (1982) Microbial degradation of soil applied herbicides. Nij J Microbiol 2:17–21

Anastasi EM, Wohlsen TD, Stratton HM, Katouli M (2013) Survival of Escherichia coli in two sewage treatment plants using UV irradiation and chlorination for disinfection. Water Res 47 (17):6670–6679

Archana A, Jaitly AK (2015) Role of myco-communities in the field of heavy metal remediation. Biolife 3(1):77–108

Arias-Estévez M, López-Periago E, Martínez-Carballo E, Simal-Gándara J, Mejuto JC, García-Río L (2008) The mobility and degradation of pesticides in soils and the pollution of ground-water resources. Agric Ecosyst Environ 123(4):247–260

Balaguru P, Hariharan V, Manivel R, Trakroo M (2016) Measuring respiratory pressures with mercury manometer in low economic health care settings–an analytical study. J Clin Diagnostic Res 10:CC12

Bastos AC, Magan N (2009) *Trametes versicolor*: potential for atrazine bioremediation in calcareous clay soil, under low water availability conditions. Int Biodeterior Biodegrad 63:389–394

Bayramoglu G, Denizli A, Bektas S, Arica MY (2002) Entrapment of Lentinus sajor-cajuinto Ca-alginate gel beads for removal of Cd (II) ions from aqueous solution: preparation and bio-sorption kinetics analysis. Microchem J 72:63–76

Bending MP, Anderron A, Ander P, Stenström J, Torstensson L (2001) Establishment of white-rot fungus *Phanerochaete chrysosporium* on unsterile straw in solid substrate fermentation system intended for degradation of pesticides. World J Microbiol Biotechnol 17:627–633

Bending G, Friloux M, Walker A (2002) Degradation of contrasting pesticides by white rot fungi and its relationship with ligninolytic potential. FEMS Microbiol Lett 212:59–63

Bhatt P, Bhatt K, Sharma A, Zhang W, Mishra S, Chen S (2021a) Biotechnological basis of microbial consortia for the removal of pesticides from the environment. Crit Rev Biotechnol 41(3):317–338

Bhatt P, Joshi T, Bhatt K, Zhang W, Huang Y, Chen S (2021b) Binding interaction of glyphosate oxidoreductase and C-P lyase: molecular docking and molecular dynamics simulation studies. J Hazard Mater 5:409:124927

Bujacz B, Wieczorek P, Krzysko-lupicka T, Golab Z, Lejczak B, Kavfarski P (1995) Organophosphonate utilization by the wild-type strain of Penicillium notatum. Appl Environ Microbiol 61(8):2905–2910

Cheema SA, Khan MI, Tang X, Zhang C, Shen C, Malik Z, Ali S, Yang J, Shen K, Chen X (2009) Enhancement of phenanthrene and pyrene degradation in rhizosphere of tall fescue (*Festuca arundinacea*). J Hazard Mater 166:1226–1231

Choo J, Sabri NBM, Tan D, Mujahid A, Müller M (2015) Heavy metal resistant endophytic fungi isolated from *Nypa fruticans*. Ocean Sci J 50:445–453

Cruz CCV, da Costa ACA, Henriques CA, Luna AS (2004) Kinetic modeling and equilibrium studies during cadmium biosorption by dead Sargassum sp. biomass. Bioresour Technol 91:249–257

Cruz-Hernández A, Tomasini-Campocosio A, Pérez-Flores LJ, Fernández-Perrino FJ, Gutiérrez-Rojas M (2013) Inoculation of seed-borne fungus in the rhizosphere of *Festuca arundinacea* promotes hydrocarbon removal and pyrene accumulation in roots. Plant Soil 362:261–270

Dash B, Sahu N, Singh AK, Gupta SB, Soni R (2021) Arsenic efflux in Enterobacter cloacae RSN3 isolated from arsenic-rich soil. Folia Microbiol 66:189–196

Debbarma P, Raghuwanshi S, Singh J, Suyal DC, Zaidi MGH, Goel R (2017) Comparative in situ biodegradation studies of polyhydroxybutyrate film composites. 3Biotech 7(178):1–9. https://doi.org/10.1007/s13205-017-0789-3

Deng Z, Cao L, Huang H, Jiang X, Wang W, Shi Y, Zhang R (2011) Characterization of Cd-and Pb-resistant fungal endophyte *Mucor* sp. CBRF59 isolated from rapes (*Brassica chinensis*) in a metal-contaminated soil. J Hazard Mater 185:717–724

Deng Q, Ramsköld D, Reinius B, Sandberg R (2014) Single-cell RNA-Seq reveals dynamic, random monoallelic gene expression in mammalian cells. Science 343(193):193–196

Dhanasekar NN, Rahul G, Narayanan KB, Raman G, Sakthivel N (2015) Green chemistry approach for the synthesis of gold nanoparticles using the fungus Alternaria sp. J Microbiol Biotechnol 25:1129–1135. https://doi.org/10.4014/jmb.1410.10036

Edwards CA (1986) In: Van Hofsten B, Eckstrom G (eds) Agrochemicals as environmental-pollutants. In control of pesticide applications and residues in food. A guide and directory. Swedish Science Press, Uppsala

EEA (European Environment Agency) (2003) Europe's environment: the third assessment. State of Environment report No 1/2003. Copenhagen

Elguetaa S, Santosa C, Limab N, Diezc MC (2016) Atrazine dissipation in a biobed system inoculated with immobilized white-rot fungi. Arch Agron Soil Sci 62:1451–1461

FAO (Food and Agriculture Organization of the United Nations) (2009) How to feed the world in 2050. FAO, Rome. http://www.fao.org/fileadmin/templates/wsfs/docs/expert_paper/How_to_Feed_the_World_in_2050.pdf

Ferner DJ (2001) Toxicity, heavy metals. Med J 2:1

Fomina M, Charnock JM, Hillie S, Alvarez R, Gadd GM (2007) Fungal transformations of uranium oxides. Environ Microbiol 9(7):1696–1710

Fourest E, Roux J-C (1992) Heavy metal biosorption by fungal mycelial by-products: mechanisms and influence of pH. Appl Microbiol Biotechnol 37:399–403

Gan J, Koskinen WC (1998) Pesticide fate and behaviour in soil at elevated concentrations. In: Kearney PC (ed) Pesticide remediation in soils and water. Wiley, Chichester, pp 59–84

Gholami-Shabani M, Imani A, Shams-Ghahfarokhi M, Gholami-Shabani Z, Pazooki A, Akbarzadeh A, Riazi G, Razzaghi-Abyaneh M (2016a) Bioinspired synthesis, characterization and antifungal activity of enzyme-mediated gold nanoparticles using a fungal oxidoreductase. J Iran Chem Soc 9:1–10. https://doi.org/10.1007/s13738-016-0923-x

Gholami-Shabani M, Shams-Ghahfarokhi M, Gholami-Shabani Z, Akbarzadeh A, Riazi G, Razzaghi-Abyaneh M (2016b) Biogenic approach using sheep milk for the synthesis of platinum nanoparticles: the role of milk protein in platinum reduction and stabilization. Int J Nanosci Nanotechnol 12:199–206

Gholami-Shabani M, Shams-Ghahfarokhi M, Gholami-Shabani Z, Razaghi-abyaneh M (2016c) Microbial enzymes: current features and potential applications in nanobiotechnology. In: Prasad R (ed) Advances and applications through fungal nanobiotechnology. Springer, Heidelberg. https://doi.org/10.1007/978-3-319-42990-8_5

Giri K, Rai JPN, Pandey S, Mishra G, Kumar R, Suyal DC (2017a) Performance evaluation of isoproturon-degrading indigenous bacterial isolates in soil microcosm. Chem Ecol 33(9):817–825. https://doi.org/10.1080/02757540.2017.1393535

Giri K, Suyal DC, Mishra G, Pandey S, Kumar R, Meena DK, Rai JPN (2017b) Biodegradation of isoproturon by *Bacillus pumilus* K1 isolated from foothill agroecosystem of north west Himalaya. Proc Natl Acad Sci India Sect B-Biol Sci 87(3):839–848. https://doi.org/10.1007/s40011-015-0667-x

Goel R, Zaidi MGH, Soni R, Kusumlata SYS (2008) Implication of *Arthrobacter* and *Enterobacter* species for polycarbonate degradation. Int Biodeterior Biodegrad 61(2):167–172

Graeme M (2005) Resistance management – pesticide rotation. Ontario Ministry of Agriculture, Food and Rural Affairs

Gupta P (2004) Pesticide exposure-Indian scene. J Dent Technol 198:118–119

Hammel KE, Gai WZ, Green B, Moen MA (1992) Oxidative degradation of phenanthrene by the ligninolytic fungus *Phanerochaete chrysosporium.* Appl Environ Microbiol 58:1832–1838

Haritash AK, Kaushik CP (2009) Biodegradation aspects of polycyclic aromatic hydrocarbons (PAHs): a review. J Hazard Mater 169:1–15

Honary S, Barabadi H, Gharaei-Fathabad NF (2012) Green synthesis of copper oxide nanoparticles using *Penicillium aurantiogriseum*, *Penicillium citrinum* and *Penicillium waksmanii.* Dig J Nanomater Bios 7:999–1005

Howarth RW (2006) Atmospheric deposition and nitrogen pollution in coastal marine ecosystems. In: Visgilio GR, Whitelaw DM (eds) Acid in the environment: lessons learned and future prospects. Springer, New York, pp 97–116

Huang X-D, El-Alawi Y, Penrose DM, Glick BR, Greenberg BM (2004) A multi-process phytoremediation system for removal of polycyclic aromatic hydrocarbons from contaminated soils. Environ Pollut 130:465–476

IARC (1993) Beryllium, cadmium, mercury, and exposures in the glass manufacturing industry. World Health Organization, International Agency for Research on Cancer. https://monographs.iarc.fr/ENG/Monographs/vol58/mono58.pdf

Il'yasova D, Schwartz GG (2005) Cadmium and renal cancer. Toxicol Appl Pharmacol 207:179–186

INECAR (2000) Institute of Environmental Conservation and Research. Position paper against mining in Rapu-Rapu, Published by INECAR, Ateneo de Naga University, Philippines. www.adnu.edu.ph/Institutes/Inecar/pospaper1.asp

Juhler R, Sorensen S, Larsen L (2001) Analysing transformation products of herbicide residues in environmental samples. Water Res 35:1371–1378

Kadri T, Rouissi T, Brar SK, Cledon M, Sarma S, Verma M (2017) Biodegradation of polycyclic aromatic hydrocarbons (PAHs) by fungal enzymes: a review. J Environ Sci 51:52–74. https://doi.org/10.1016/j.jes.2016.08.023

Kapaj S, Peterson H, Liber K, Bhattacharya P (2006) Human health effects from chronic arsenic poisoning–a review. J Environ Sci Heal Part A 41:2399–2428

Kavamura VN, Esposito E (2010) Biotechnological strategies applied to the decontamination of soils polluted with heavy metals. Biotechnol Adv 28:61–69

Kearney P, Wauchope R (1998) Disposal options based on properties of pesticides in soil and water. In: Kearney P, Roberts T (eds) Pesticide remediation in soils and water, Wiley series in agrochemicals and plant protection. Kluwer Academic, Dordrecht

Klimeka S, Marini L, Hofmann M, Isselstein J (2001) Additive partitioning of plant diversity with respect to grassland management regime, fertilisation and abiotic factors. Basic Appl Ecol 9:626–634

Kookana RS, Di HJ, Aylmore L (1998) Degradation rates of eight pesticides in surface and subsurface soils under laboratory and field conditions. Soil Sci 163:404–411

Kour D, Kaur T, Devi R, Yadav A, Singh M et al (2021) Beneficial microbiomes for bioremediation of diverse contaminated environments for environmental sustainability: present status and future challenges. Environ Sci Pollut Res 28:24917–24939

Kratochvil D, Volesky B (1998) Advances in the biosorption of heavy metals. Trends Biotechnol 16:291–300

Krzysko-Lupicka T, Stroff W, Kubs K, Skorupa M, Wieczorek P, Lejczak B, Kafarski P (1997) The ability of soil borne fungi to degrade organophosphonate carbon-to-phosphorus bonds. Appl Environ Microbiol 48:549–552

Kumar P, Dash B, Suyal DC, Gupta SB, Singh AK, Chowdhury T, Soni R (2021) Characterization of arsenic-resistant Klebsiella pneumoniae RnASA11 from contaminated soil and water samples and its bioremediation potential. Curr Microbiol. https://doi.org/10.1007/s00284-021-02602-w

Kumar M, Bhatt G, Duffy CJ (2008) An efficient domain decomposition framework for accurate representation of geodata in distributed hydrologic models. Int J Geogr Inf Sci 23(12):1569–1596

Lei A-P, Hu Z-L, Wong Y-S, Tam NF-Y (2007) Removal of fluoranthene and pyrene by different microalgal species. Bioresour Technol 98:273–280

Lenntech Water Treatment (2004) Lenntech water treatment and air purification. Water Treatment, Lenntech, Rotterdamseweg. www.excelwater.com/thp/filters/water-purification.htm

Li H-Y, Li D-W, He C-M, Zhou Z-P, Mei T, Xu H-M (2012a) Diversity and heavy metal tolerance of endophytic fungi from six dominant plant species in a Pb–Zn mine wasteland in China. Fungal Ecol 5:309–315

Li H-Y, Wei D-Q, Shen M, Zhou Z-P (2012b) Endophytes and their role in phytoremediation. Fungal Divers 54:11–18

Li X, Li W, Chu L, White JF Jr, Xiong Z, Li H (2016) Diversity and heavy metal tolerance of endophytic fungi from Dysphania ambrosioides, a hyperaccumulator from Pb–Zn contaminated soils. J Plant Interact 11:186–192

Lipok J, Dombrovska L, Wieczorek P, Kafarski P (2003) The ability of fungi isolated from stored carrot seeds to degrade organophosphonate herbicides. In: Del Re AAM, Capri E, Padovani L, Trevisan M (eds) Pesticide in air, plant, soil and water system, proceeding of the XII symposium pesticide chemistry. Piacenza

Liu YY, Xiong Y (2001) Purification and characterization of a dimethoate-degrading enzyme of *Aspergillus niger* ZHY256 isolated from sewage. Appl Environ Microbiol 67:3746–3749 (Martens R (1976) Degradation of endosulfan by soil microorganisms. Appl Environ Microbiol 31:853–858)

Loebenstein G, Thottappilly G (2007) Agricultural research management. Springer, Dordrecht MA (Millennium Ecosystem Assessment) (2005) Ecosystem services and human well-being: wetlands and water synthesis. World Resources Institute, Washington, DC, 68 pp. Web site: http://www.millenniumassessment.org/en/index.aspx)

Maliszewska-Kordybach B (1999) Sources, concentrations, fate and effects of polycyclic aromatic hydrocarbons (PAHs) in the environment. Part a: PAHs in air. Polish J Environ Stud 8:131–136

Mann H (1990) Biosorption of heavy metals by bacterial biomass. In: Volesky B (ed) Biosorption of heavy metals. CRC Press, Boca Raton, FL, pp 93–138

Maurya NS, Mital AK, Cornel P, Rother E (2006) Biosorption of dyes using dead macro-fungi: effect of dye structure, ionic strength and pH. Bioresour Technol 97:512–521

Mauti GO, Onguso J, Kowanga DK, Mauti EM (2016) Biodegradation activity of *Aspergillus niger* lipase isolates from a tropical country garage. J Sci Innov Res 5(1):15–18

Meena K, Sarita S (2017) Mycoremediation potential of Pleurotus species for heavy metals: a review. Bioresour Bioprocess 4(32):1–9. https://doi.org/10.1186/s40643-017-0162-8

Michael DA, Solveig T, Chiara D, Xiao W, Shihan X, Mario AM, Wenyuan G, Segun GJ, Lorenozo P (2020) Hydrocarbon degradation and enzyme activities of *Aspergillus oryzae* and *Mucor irregularis* isolated from Nigerian crude oil-polluted sites. Micro 8(1912):1–19. https://doi.org/10.3390/microorganisms8121912

Mishra A, Kumari M, Pandey S, Chaudhry V, Gupta KC, Nautiyal CS (2014) Biocatalytic and antimicrobial activities of gold nanoparticles synthesized by *Trichoderma* sp. Bioresour Technol 166:235–242. https://doi.org/10.1016/j.biortech.2014.04.085

Mohsenzadeh F, Nasseri S, Mesdaghinia A, Nabizadeh R, Zafari D, Khodakaramian G, Chehregani A (2010) Phytoremediation of petroleum-polluted soils: application of Polygonum aviculare and its root-associated (penetrated) fungal strains for bioremediation of petroleum-polluted soils. Ecotoxicol Environ Saf 73:613–619

Mougin C, Laugero C, Asther M, Dubroca J, Frasse P, Asther M (1994) Biotransformation of the herbicide atrazine by the white rot fungus *Phanerochaete chrysosporium*. Appl Environ Microbiol 60:705–708

Muraleedharan TR, Leela L, Venkobachar C (1991) Biosorption: an attractive alternative for metal removal and recovery. Curr Sci 61:379–385

Ndimele PE (2010) A review on the phytoremediation of petroleum hydrocarbon. Pak J Biol Sci 13:715

Nerud F, Baldrian J, Gabriel J, Ogbeifun D (2003) Nonenzymic degradation and decolorization of recalcitrant compounds. In: Sasek V et al (eds) The utilization of bioremediation to reduce soil contamination: problems and solutions. Kluwer Academic Publishers, Dordrecht, pp 29–48

OECD (2012) Water quality and agriculture – meeting the policy challenge. OECD studies on water. OECD Publishing, Paris

Ogwuegbu MO, Ijioma MA (2003) Effects of certain heavy metals on the population due to mineral exploitation. In: International conference on scientific and environmental issues. University of Ado Ekiti, Ekiti State, pp 8–10

Ogwuegbu MOC, Muhanga W (2005) Investigation of lead concentration in the blood of people in the Copperbelt Province of Zambia. J Environ 1:66–67

Önder M, Ceyhan E, Kahraman A (2011) Effects of agricultural practices on environmental. In: International conference on biology, environment and chemistry (ICBEC), vol 24 © IACSIT Press, Singapore

Prasad R (2016) Advances and applications through fungal nanobiotechnology. Springer, Cham

Rahman MM, Jusoh I, Husaini A, Seman IA, Sing NN (2014) Biodegradation and ligninolytic enzymes profiles of the newly synthesized organotin (IV)-treated non durable tropical wood species. J Biochem Technol 5:743–750

Ratnaike RN (2003) Acute and chronic arsenic toxicity. Postgrad Med J 79:391–396

Rial-Otero R, Cancho-Grande B, Arias-Estévez M, López-Periago E, Simal-Gándara J (2003) Procedure for the measurement of soil inputs of plant-protection agents washed off through vineyard canopy by rainfalls. J Agric Food Chem 51(17):5041–5046

Rhodes A, Skea J, Hannon M (2014) The global surge in energy innovation. Energies 7(9):5601–5623. https://doi.org/10.3390/en7095601

Sandhu SS, Shakya M, Deshmukh L, Aharwal RP, Kumar S (2016) Determination of hydrocarbon degrading potentiality of indigenous fungal isolates. Int J Environ Sci 6(6):1163–1172

Scott SL (2003) Biodegradability and toxicity of total petroleum hydrocarbon leachate from land treatment units. M.S. California Polytechnic State University

Singh H (2006) Mycoremediation: fungal bioremediation. Wiley, Hoboken

Singh SK, Srivastava PK, Singh D, Han D, Gautam SK, Pandey AC (2015) Modeling groundwater quality over a humid subtropical region using numerical indices, earth observation datasets, and X-ray diffraction technique: a case study of Allahabad district, India. Environ Geochem Health 37(1):157–180

Singh M, Singh D, Rai P, Suyal DC, Saurabh S, Soni R, Giri K, Yadav AN (2021) Fungi in remediation of hazardous wastes: current status and future. In: Yadav AN (ed) Recent trends in mycological research, fungal biology. Springer Nature, Cham

Soleimani M, Afyuni M, Hajabbasi MA, Nourbakhsh F, Sabzalian MR, Christensen JH (2010) Phytoremediation of an aged petroleum contaminated soil using endophyte infected and non-infected grasses. Chemosphere 81:1084–1090

Su C (2014) A review on heavy metal contamination in the soil worldwide: situation, impact and remediation techniques. Environ Skept Critics 3:24

Surovtseva EG, Ivoilov VS, Belyaev SS (1997) Degradation of the aromatic fraction of oil by an association of gram-positive and gram-negative bacteria. Microbiology 66:65–69

Talaro KP, Talaro A (2002) Foundations in microbiology, 4th edn. McGraw Hill, New York

Ting FAN, Yunguo L, Baoying F, Guangming Z, Chunping Y, Ming Z, Haizhou Z, Zhenfeng T, Xin W (2008) Biosorption of cadmium (II), zinc (II) and lead (II) by *Penicillium simplicissimum*: isotherms, kinetics and thermodynamics. J Hazard Mater 160:655–661

Udedi SS (2003) From Guinea worm scourge to metal toxicity in Ebonyi state. Chem Niger New Millenn Unfold 2:13–14

Vankar PS, Bajpai D (2008) Phyto-remediationof chrome-VI of tannery effluent by Trichoderma species. Desalination 222:255–262

Walls M (2006) Agriculture and environment. The standing committee on agricultural research (SCAR) foresight group. 22 strani. http://ec.europa.eu/research/agriculture/scar/pdf/scar_foresight_environment_en.pdf

Weyens N, van der Lelie D, Taghavi S, Newman L, Vangronsveld J (2009) Exploiting plant–microbe partnerships to improve biomass production and remediation. Trends Biotechnol 27:591–598

Wunch KG, Alworth WL, Bennet JW (1999) Mineralization of benzo[a]pyrene by Marasmiellus troyanus, a mushroom isolated from a toxic waste site. Microbiol Res 154:75–79

Xiao X, Luo S, Zeng G, Wei W, Wan Y, Chen L, Guo H, Cao Z, Yang L, Chen J (2010) Biosorption of cadmium by endophytic fungus (EF) Microsphaeropsis sp. LSE10 isolated from cadmium hyperaccumulator *Solanum nigrum* L. Bioresour Technol 101:1668–1674

Yang HB, Tan N, Wu FJ, Liu HJ, Sun M, She ZG, Lin YC (2012) Biosorption of uranium (VI) by a mangrove endophytic fungus fusarium sp.# ZZF51 from the South China Sea. J Radioanal Nucl Chem 292:1011–1016

Zboniska E, Lejczak B, Kafarski (1992) Organophosphonate utilization by the wild - type strain of Pseudomonas fluorescens. Appl Environ Microbiol 58(9):2993–2999

Zhang J, Chiao C (2002) Novel approaches for remediation of pesticide pollutants. Int Environ Pol 18:423–433

Zhang D, Hui D, Luo Y, Zhou G (2008) Rates of litter decomposition in terrestrial ecosystems: global patterns and controlling factors. J Plant Ecol 1(2):85–93

Zhang JH, Xue QH, Gao H, Ma X, Wang P (2016) Degradation of crude oil by fungal enzyme preparations from *Aspergillus* spp. for potential use in enhanced oil recovery. J Chem Technol Biotechnol 91(4):865–875

Chapter 7
Genetically Modified Organisms for Bioremediation: Current Research and Advancements

Inoka C. Perera and Erandika Harshani Hemamali

7.1 Introduction

Release of petroleum by-products, crude oil, xenobiotic compounds, heavy metals, radioactive substances, and emission of greenhouse gases into the environment has exponentially increased over the last two decades owing to global industrialization, urbanization, unsafe agricultural practices, mining activities, population growth, and civil war. It causes lasting impacts on the environment and human health depending on the pollutant and degree of pollution (Bai et al. 2017; Power et al. 2018; Wang 2018). Many different strategies were employed over time to restore these degraded landscapes and to minimize the impact of effluents emitted to the environment. Genetically modified organisms hold a small fraction in these interventions, but holds a great promise for the future.

As an industry, crude oil holds the largest market share where they are used in industries, transportation, and household activities as a major source of energy. It is a mixture of naturally occurring hydrocarbons and the process of extraction can cause small to large-scale petroleum contaminations that are regularly reported worldwide during exploration, refining, transportation, and storage. Air pollution occurs via hydrocarbon gas liquids and soil and water pollution occur via liquid and solid hydrocarbons. Hydrocarbon contamination is greatly affected in particular ecosystems (Wang et al. 2011). However, the estimation of the impact on aquatic and terrestrial environments and on human health is not feasible. Soil, fresh water bodies, and plants are affected mainly through industrial effluents and seepage of oil storage tanks (Moubasher et al. 2015). Marine environments are greatly affected owing to

I. C. Perera (✉) · E. H. Hemamali
Faculty of Science, Department of Zoology and Environment Sciences, Synthetic Biology Laboratory, University of Colombo, Colombo, Sri Lanka
e-mail: icperera@sci.cmb.ac.lk

© The Author(s), under exclusive license to Springer Nature Switzerland AG 2022
D. C. Suyal, R. Soni (eds.), *Bioremediation of Environmental Pollutants*,
https://doi.org/10.1007/978-3-030-86169-8_7

Fig. 7.1 Deepwater Horizon in flames 20th April 2010

accidental and intentional oil spills during transportation, mining and in periods of civil war (Linden et al. 2004; Saadoun 2016) (Fig. 7.1).

Accumulation of xenobiotic compounds in ecosystems with the release of synthetic compounds from medicine, food, agriculture, textile, and other products has been increasing with technological advancements (Giri et al. 2017a, b; Atashgahi et al. 2018a, b). Xenobiotic compounds persist in ecosystems due to less bioavailability and create toxic effects upon exposure to humans and other organisms (Atashgahi et al. 2018a, b).

A large amount of heavy metals are being unearthed and dispersed with uncontrolled applications in pesticides, fertilizers, and industrial practices. However, heavy metals have received special emphasis due to their high toxicity even at trace amounts. They can accumulate in the human body through food chains and direct absorption through the skin since those are not metabolized in the body. In addition, reduction in crop production and food quality is the other emerging problem behind heavy metal pollution (Tarekegn et al. 2020).

Radioactive waste always reminds us of high energy radiation is discharged through nuclear explosions, testing of nuclear weapons, accidents at nuclear power plants, volcanic eruption, and mining of radioactive ores (Nies 2018). Since radioactive substances emit ionizing radiation, they cause great impact on human health and the surrounding environment. Exposure to radiation causes skin irritation, genetic defects, cancer, and fatal effects to humans. Furthermore, they can destroy wildlife and marine life, and cause soil infertility (Pollution 2014; Nies 2018). The emerging problems are with the radioactive substances can trace to being ubiquitous with the technological developments and they produce a low level of radiation. The effect of such long-term exposure to low level radiation may not be visible for many generations to come.

It is a known fact that the greenhouse gas emission have increased exponentially after late 1700s owing to the industrial revolution. This has caused many impacts on ecosystems and human health. World health organization reports show that seven million deaths occur every year due to exposure to ambient and household air pollution (WHO 2021). In addition, it was estimated that half-of the world's population will be living in water stressed areas by 2025. However, human and ecological health are increasingly threatened and it is a great concern worldwide (Albert et al. 2018).

Restoring ecological health and improving human health is a mainstream discussion in every forum. Turning to nature to find solutions to this problem by harnessing what nature has to offer us and responsible utilization becomes the theme of this chapter.

7.2 Bioremediation

The perspective of an environmental scientist for sustainable environmental remediation is the implementation of rapid restoration strategies at the contaminated sites in an eco-friendly and economically viable framework. Because of the exorbitant costs involved and adverse effects of physiochemical methods such as addition of chemicals and coagulants, filtration, ion exchange methods, reverse osmosis, physical removal etc., they are of limited usage.

Bioremediation on the other hand employs the use of bacteria, fungus, plants, and enzymes for removing and detoxifying pollutants (Dash et al. 2021; Kumar et al. 2021; Kour et al. 2021). This is considered a better alternative owing to its capabilities of complete mineralization, lower liability, low cost, feasibility in in situ applications, and eco-friendliness. However, the rate of biodegradation depends on the concentration of the contaminant and the number of catalytic organisms as well as the number of catalytic enzymes produced by each organism, amount of oxygen, amount of nutrients, pH, temperature, moisture, etc. Therefore, natural degradation is a slow process compared to pollution accretion.

Depending on the state of the contaminant to be remediated, two bioremediation techniques named; in-situ and ex-situ are used. The application of non-pathogenic microorganisms that has the capability of converting toxic pollutants into less toxic compounds is called in-situ bioremediation. This is widely used for the remediation of petroleum contaminated sites. However, supplement of nutrients such as nitrogen, phosphorus, sulphur and oxygen is required for the growth and survival of bacterial species. This a comparatively less expensive and effective method. In addition, the chemotactic ability of the microorganisms is another important factor that aids efficient in-situ bioremediation.

Ex-situ bioremediation refers to treating pollutants after excavating the contaminated site. Slurry phase bioremediation and solid phase bioremediation are the two main types of ex-situ bioremediation. The slurry phase bioremediation is done in a bioreactor providing the optimal growth conditions to the microorganisms. Here, water and other nutrients are mixed with contaminated soil. After the treatment, water is separated from the soil.

Solid phase bioremediation is mainly used for soil contaminated with toxic compounds. Here, the contaminated soil is treated in above–ground treatment areas taking care not to allow pollutants to escape. Conditions required for microbial growth are properly maintained.

Biostimulation and bioaugmentation are other general methods that are widely used for the remediation of contaminated sites. Biostimulation is providing suitable conditions for the growth of indigenous microorganisms present in the contaminated sites. Bioaugmentation is an application of microbial cultures that are cultivated in a laboratory into a contaminated site or a bioreactor.

Even though bioremediation is more advantageous compared to physiochemical methods, there are limitations and disadvantages in bioremediation. Lack of microorganisms evolved to degrade many xenobiotic compounds, biological reactions are specific, bi-products generated are highly toxic than initial compound, there is no acceptable endpoint for bioremediation, treatment time is longer, and difficulties in controlling volatile compounds are the key problems associated with bioremediation. To mitigate these issues, knowledge, innovative ideas, and novel techniques are essential.

With the advances in biotechnology, the use of genetically modified organisms for bioremediation is a rapidly developing field. Thus, organisms that can persist in many different stress conditions owing to pollutant accretion have been developed by using recombinant DNA technology in many laboratories throughout the world. The development of novel genetic engineering methods, identification of extremophiles, and understanding of their genetics, physiology, and metabolic pathways are of utmost importance in developing effective bioremediation strategies (Bhatt et al. 2021). This chapter summarizes recent advances and current research on bioremediation with genetically modified organisms.

7.3 Bioremediation by Extremophilic Organisms

Organisms that have evolved to live in extreme environments such as high or low pH, high temperature, cold conditions, acidic conditions, high salinity, high pressure, nutrient deficiencies, etc. are considered extremophiles. This group of organisms can be categorized as extremophilic organisms and extremotolerant. Organisms in need of one or more extreme conditions to thrive are named extremophilic and the organisms are named as extremotolerant when they grow well in both normal and extreme conditions. Moreover, extremophiles can be categorized further as acidophilic (organisms live at low pH), alkalophilic (organisms live at high pH), thermophilic (organisms live at high temperatures), psychrophilic (organisms live at low temperatures), barophilic (organisms live at high pressure), oligotrophic (organisms live at nutrient deficiencies), etc. according to the particular extreme conditions that they can thrive (Sani and Nawanietha Krishanraj 2017). With the advancement of genetic engineering techniques, extremophiles have acquired special attention owing to their various biotechnological applications. Understanding the metabolic pathways and characterization of proteins that give advantages in extreme environmental conditions are trending topics among environmental scientists (Dumorné et al. 2017).

Deinococcus geothermalis and *Deinococcus radiodurans* are applied for bioremediation of radioactive wastes at heated environments owing to their capabilities of reducing Cr^{+6}, Hg^{+2}, Tc^{+7}, Fe^{+3}, Mn^{+3}, Mn^{+4}, and U^{+6}. In addition, xenobiotics such as toluene, diethyl sulphate, and ethyl acetated can be degraded by *D. geothermalis* strain T27. Removal of $^{99}Tc^{+7}$ from the contaminated environments is very difficult via chemical methods due to its capability to form complexes. Thermophiles, *Thermoterrabacterium ferrireducens, Thermoproteus usoniensis, Pyrobaculum islandicum,* has been reported to reduce $^{99}Tc^{+7}$ into insoluble $^{99}Tc^{+4}$. *Shewanella putrefaciens* acquired capability to reduce U^{+6} via c-type cytochrome in the periplasm.

Penicillium chrysogenum, *Pseudomonas* sp., Acinetobacter sp., Mycobacterium sp., and *Basillus subtilis* have acquired the ability to utilize many different compounds present in petroleum hydrocarbons as their sole source of carbon and energy. *Pseudomonas fluorescens, Pseudomonas aeruginosa, Microbacterium profundi strain Shh49T, Saccharomyces cerevisiae*, *Geobacter* sp., *Rhodopseudomonas palustris*, *Acromonas* sp., and *Trichoerma* sp. possess the capability to utilize heavy metals. *Bacillus firmus, Bacillus macerans, Exiguobacterium indium, Acinetobactor baumanii, Penicillium ochrochlorum, and Tremetes trogii* have been isolated in industrial effluents with dye compounds and *Arthrobactor sp.*, *Enterobacter* sp., *Photobacterium* sp., *Bacillus* sp., *Staphylococcus* sp., and *Pseudomonas* sp. persist on pesticides contaminated sites (Peeples 2014; Shukla and Singh 2020).

7.4 Designer Organisms for a Cleaner Tomorrow

Even though many organisms have been evolving in degrading or converting pollutants, the rate of biodegradation depends on many environmental and biological factors. Accumulation of pollutants affects adversely for almost all organisms including humans. Some of the pollutants (Xenobiotic compounds) are resistant to natural degradation owing to the catabolic pathways that are yet to be resolved. With the rapid development of molecular engineering technology, molecular biologists are trying to construct genetically modified organisms that have the capability in biodegradation. Enhancement of enzyme specificities, substrate affinities, cellular localization, expression, and genetic stability occurs mainly via mutations, the introduction of novel genes, pathways, and regulatory mechanisms via cloning, fusion proteins, into heterologous hosts.

Gene editing is a DNA manipulation technique by using engineered nucleases having a growing popularity among genetic engineers due to the potential advantages. These engineered nucleases are named molecular scissors. Gene editing is used for the manipulation of genes that can convert toxic chemicals into less toxic compounds and for the remediation of xenobiotic compounds (Butt et al. 2018; Hussain et al. 2018). The main gene editing tools are CRISPR-Cas9, Transcription activator-like effector nucleases, and Zinc finger nucleases (Waryah et al. 2018).

7.4.1 Gene Transfer

Molecular cloning, homologous recombination, horizontal gene transfer are used to insert or delete or replace genes. Thus, particular organisms can be engineered to grow on pollutants. Figure 7.2 indicates the types of probable genes that are currently used for the genetic manipulations to enhance biodegradation. However, identification and characterization of novel gene clusters that can improve bioremediation through genetic alterations provide basis for the novel bioremediation strategies.

Owing to anthropogenic activities, agricultural practices, and effluents, arsenic (As) is added into the soil and water. Arsenic adversely affects human health due to its genotoxic and mutagenic effects. Bio volatilization is a process that involve a series of methylation reactions catalysed by enzymes. Soil microbes have evolved to

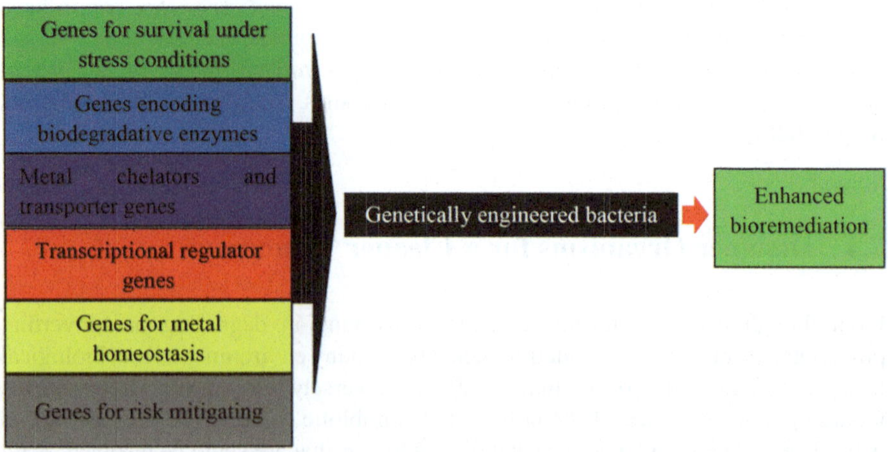

Fig. 7.2 Genes that are currently manipulated for the construction of genetically modified bacterial strains

form volatile arsenic (Kumar et al. 2021). The enzyme arsenite S-adenosylmethionine methyltransferase which is encoded by *arsM* gene can convert arsenicals into volatile arsenic. Liu et al. have cloned the *arsM* gene in *Rhodopseudomonas palustris* in the expression vector pET28a and transformed it into the *Sphingomonas desiccabilis* and *Bacillus idriensis*. They have found that about 2.2–4.5% of arsenic has been removed from the soil after 30 days of incubation with these genetically modified strains (Liu et al. 2011).

Accumulation of heavy metals in *E. coli* was enhanced by introducing *Arabidopsis thaliana* phytochelatin synthase (AtPCS) gene via pBluscript plasmid. Phytochelatin is a metal binding cysteine rich peptide synthesized from glutathione. Here, for the easy detection of AtPCS, it is fused to a C-terminal flag epitope and expressed under the control of the T5 Promoter/lac operator/repressor system. They have shown that intracellular metal content is increased when this genetically modified *E. coli* strain is grown in a medium supplemented with Cd, Cu, and As form 20, 5, and 50 folds respectively. However, no significant intercellular accumulation of Ag, Zn, and Hg was observed (Sauge-merle et al. 2003).

Sun et al. (2019) identified yeast as an organism that can be improved well for heavy metal remediation. Since metal transporters, storage components, and chelators play a pivotal role in the hyperaccumulation of metal ions in cells, genes that express metal transporters through cell membranes, vacuole and an oxidoreductase (ZRT1, ZRT2, CTR1, CTR3, FTR1, FTE4, SMF1, SMF2, ZRT3, CTR2, SMF3, FET3) have been cloned into 2 μ plasmids which contain GAL1 promoter and were transformed to constructed engineered yeast strains. ZRT1 and 2 selectively transport Zn ions and CTR1 and 3 are selectively transport Cu. The oxidoreductase (FET3) membrane transporter SMF1 were not shown with such selectivity. The natural degradation of SMF1 was reduced by mutating the ubiquitination site by site-directed mutagenesis and SMF1 expression was enhanced by deleting the BSD2 ubiquitin ligase (Sun et al. 2019).

Vitreoscilla stercoraria is a gram-negative bacterium that can live in oxygen limited environments. The first bacterial hemoglobin protein (VHb) was isolated from *Vitreoscilla stercoraria* (Webster and Hackett 1966). VHb is produced at oxygen stress conditions and plays a key role in binding oxygen at low concentrations and delivering directly to the terminal respiratory oxidases. In addition, they pass the signal to transcriptional regulators in response to oxygen concentrations and control the expression of many genes (Stark et al. 2012). The expression of VHb in different bacterial species enhances cell density, oxidative metabolism, and bioremediation at oxygen stress conditions; i.e. in *Burkholderia*, the degradation of 2,4,-DNT was increased.

Today, many environments are co-contaminated with different pollutants. Chlorpyrifos is an organophosphate and γ-Hexachlorocyclohexane is an organochlorine found in insecticides. Co-contamination of these compounds and heavy metals in nature contributes to potentiate toxicity. Yang et al. has constructed a genetically engineered *Sphingobium japonicum* UT26 strain for the remediation of Cd^{+2}, Chlorpyrifos, and γ-Hexachlorocyclohexane (Yang et al. 2015). *Sphingobium japonicum* UT26 is a gram-negative bacterium isolated from the γ-Hexachlorocyclohexane contaminated site in Japan (Nagata et al. 2010). The capability of degrading γ-Hexachlorocyclohexane in *Sphingobium japonicum* UT26 is well characterized. In general, phytochelatins are used for metal remediation since they have metal binding moieties (Alkorta et al. 2004). Methyl parathion hydroxylase hydroxylates chlorpyrifos. The access of metals and chlorpyrifos into phytochelatins and methyl parathion hydroxylase is greatly reduced as these proteins are produced in the cytoplasm. Therefore, they have chemically synthesized a fusion gene that contains *mpd* gene that encodes methyl parathion hydroxylase in *Stenotrophomonas* sp. (Yang et al. 2006) and a synthetic phytochelatin. This fusion protein is coupled to a truncated ice nucleation protein so that fusion proteins can transport to the cell surface. The fusion gene is cloned to the surface expression vector pVINPEM using *E. coli–Pseudomonas* shuttle vector pVLT33 (Yang et al. 2008, 2015). The engineered *Sphingobium japonicum* UT26 strain can be used to simultaneously detoxify heavy metal and pesticide contaminated sites (Cao et al. 2013; Yang et al. 2015).

7.4.2 Genetic Mutations Improve Biodegradation of Pollutants

Gene mutations are involved in genetic variations and contribute to both the survival and the extinction of organisms. Many different environmental and biological factors such as exposure to radiation and chemicals, viral infections, mating patterns, etc. are associated with mutations. Bacteria attribute to high mutation frequencies. Mutations induced owing to pollutants imply organisms mounting a survival strategy to tolerate harmful conditions. Taking cues from nature, organisms are engineered by mutating the genes to improve their capabilities of degrading or detoxifying pollutants according to requirements. Strategies such as random mutagenesis by exposure to mutagens, site directed mutagenesis, PCR based methods, TA strategy are currently used for the construction of genetically modified organisms.

Phenol is an aromatic hydrocarbon that contains a hydroxyl group. Phenolic compounds are widely used in industries to synthesis many different organic chemicals. Phenolic compounds are further grouped into simple phenols, bi-phenols, and poly-phenols depending on the number of phenol rings of a compound. Many industries such as textile, dyes, explosives, oil, gas, and coal utilize phenol as a raw material and phenolic compounds are discharged into the environment through effluents. Phenolic compounds are considered as pollutants owing to long term and short term severe effects on humans and animals. Therefore, phenols are enlisted in the United States Environment Protection Agency (USEPA) and European Union (EU) under harmful chemical categories.

Ozonation, use of UV light, and filtration methods are applied to remove phenols from effluents (Anku et al. 2019). In addition, microorganisms that have the ability to degrade phenolic compounds (*Pseudomonas pseudoalcaligenes* (Kurzbaum et al. 2010), *Acinetobacter* sp. (Liu et al. 2009), *Candida* sp. (Basak et al. 2013), *Sphingomonas* sp. (Liu et al. 2009) have been reported).

Mao et al. (2015) isolated *Pseudochlorobactrum* sp. that can degrade 1800 mg/l phenols from active sludge from a wastewater treatment plant. *Pseudochlorobactrum* sp. strain that has enhanced phenol degradation capability was screened upon subjecting to UV light for 120 s for random mutagenesis. The tolerance to pH and temperature variations and faster phenol degradation had been observed in mutated strain, *Pseudochlorobactrum* sp. XF1-UV (Mao et al. 2015).

Today, plastics have invaded earth owing to over consumption and low biodegradability. Plastics of various forms became a necessity for the day today life and over 380 million tons of plastics are produced per year where 5% of them are single use plastics (Plastics – the Facts 2019). About 90% of the plastic waste is accumulated in the ocean without a proper waste management strategy (Schmidt et al. 2017). Bacteria and fungi species that have the capability to degrade polythene have evolved and their underlining mechanisms are to be uncovered (Debbarma et al. 2017; Singh et al. 2021). Attachment of microbes on the plastic surfaces and biofilm formation are initial steps in plastic biodegradation (Goel et al. 2008). *Pseudomonas aeruginosa* (Ogunbayo et al. 2019), *Aspergillus niger* (Raaman et al. 2012; Ogunbayo et al. 2019), *Fusarium lini, Bradyrhizobium japonicum, Pseudomonas sp. AKS2* can efficiently get attached to the plastic surfaces and are identified as plastic degrading organisms (Łabużek et al. 2004). It has been reported that mutations in O-antigen which plays a fundamental role in the attachment of gram-negative bacteria on hydrophilic surfaces can enhance the affinity towards hydrophobic surfaces (Bogino et al. 2013). In addition, by expressing cellulose binding domain (CBD) of the cellobiohydrolase I (CBHI) in *Trichoderma reesei* in the cell surface of yeast (*Saccharomyces cerevisiae*) and mycoremediation of lignin have been enhanced by increasing binding affinity to cellulose (Jafari et al. 2013).

Pseudomonas putida strains are utilizing terephthalic acid for growth as a sole source of carbon and for accumulation of medium chain length polyhydroxyalkanoate at limited nitrogen conditions. However, they are not degrading ethylene glycol that is produced in the terephthalic acid degradation as a carbon source. A recent study has demonstrated that a mutated strain of *Pseudomonas putida* KT2440 that overexpresses glyoxylate carboligase operon and glycolate oxidase operon has acquired the capability of utilizing ethylene glycol as the sole source of carbon (Ann et al. 2018; Li et al. 2019). A mutated *P. putida* strain constructed by DNA restructuring had also improved the degradation of low density polyethylene (Anantharam and Talkad 2018).

The extracellular enzyme laccase contributes to the oxidation of many xenobiotic compounds such as synthetic dyes, aromatic ammines, polyphenols. Mutant strains constructed by mutating bacillus sp. by substituting the Glutamic acid 188 residues with Lysine, Arginine, and Alanine were exhibited to have thermal stability and thermal activation compared to the wild type. Moreover, mutants constructed by substituting Glutamic acid 188 with Alanine, Leucine, Isoleucine, Valine, Lysine, and Arginine have exhibited tolerance to organic solvents (Rasekh et al. 2014).

Cytochrome P450 monooxygenase in *Bacillus megaterium 3* oxidizes aromatic compounds. Mutants generated by error-prone PCR in the gene; cytochrome P450 monooxygenase showed an improved hydroxylation capability towards chrysene and pyrene (Santos et al. 2019).

7.4.3 Genetically Engineered Enzymes in Bioremediation

Enzymes are macromolecules that function as catalysts in biological reactions. In bioremediation, enzymes produced by bacteria, fungi, and plants are utilized as biocatalysts instead of using the whole microorganisms. This technique has been identified as an advantageous method compared to the use of microorganisms in bioremediation. For the reasons that purified or partially purified enzymes can be applied to the contaminated sites. Enzymes can be applied to poor nutrient conditions, enzymatic reactions are specific and generation of toxic by-products is prevented.

Phanerochaete chysosporium, *Trametes versicolor*, *Bjerkandera adjusta* and *Ceriporiopsis subvermispora* are white rot fungi that produce ligninolytic enzymes (laccases, peroxidases). These enzymes convert toxic organic pollutants into nontoxic compounds. Laccases that have low substrate specificities cleave aromatic rings of cyclic hydrocarbons. Peroxidases generate reactive oxygen species initiating the oxidation of pollutants. However, the degradation of pollutants into simpler forms by biocatalysts is achieved through further genetic modifications. The genetically engineered biocatalysts are grown in bioreactors for efficient removal of pollutants (Akhtar and Mannan 2020).

The enzymes that are predominantly involved in the bioremediation include cytochrome P450s, laccases, hydrolases, dehalogenases, dehydrogenases, proteases, and lipases. These enzymes are involved in degradation of polymers, aromatic hydrocarbons, halogenated compounds, dyes, detergents, agrochemical compounds etc. with the use of various mechanisms such as oxidation, reduction, elimination, and ring-opening (Bhandari et al. 2021).

Under natural conditions, the production of these enzymes from its native host is considerably low. Therefore genetic approaches such as isolating and transferring the coding genes of the enzyme into a better expression host, ensures that the enzyme production is enhanced such that biotransformation of compounds occurs more effectively (Sharma et al. 2018).

Using DNA technology and genetic engineering approaches that involves change or modification in the basic amino acid structure of the enzyme, recombinant enzymes can be produced with higher activity and stability, with enhanced shelf-life, substrate range, pH, temperature stability and stress tolerance. So that the engineered enzyme would have a higher capacity to degrade the contaminant under defined environmental conditions. (Sharma et al. 2018).

Nitrobenzene 1,2-dioxygenase is an enzyme that catalyses the conversion of nitrobenzene into catechol and nitrite. It also oxidizes the aromatic rings of mononitotolunes and dinitrotoluenes at the nitro-substituted carbon. The important features of regiospecificity with nitroarene and enantiospecificity with naphthalene is determined by the residues at positions 258, 293, and 350 in the α subunit. Phenylalanine amino acid residue at position 293, near the active site of nitrobenzene 1,2-dioxygenase is modified/substituted with Glutamine such that it results in an increase of 2.5-fold oxidation rate against 2,6 dinitrotoluene (Ju and Parales 2006). Similarly, 2-nitrotoluene dioxygenase which is responsible for the conversion of nitrotoluene to 3-methyl catechol and nitrite was modified at position 258 by site directed mutagenesis (Singh et al. 2008).

Polyethyleneimine (PEI) can bind onto the surface of Horseradish peroxidase (HRP) mainly via hydrophobic interaction and van der Waals interactions. The complex formation between HRP and PEI induces a more compact conformation of the enzyme and thus the hydrophobicity of the microenvironment surrounding the heme pocket is enhanced. The non-planarity of the porphyrin ring in the heme group contributes to an increased degree of exposure to the active centre, significantly enhancing the catalytic efficiency of HRP in the presence of high molecular weight PEIs (Huang et al. 2018).

The technique of site-directed mutagenesis is used to alter genes thereby protein sequences and to explore the structure-function relationship of enzymes. The basis of rational protein design is the combined use of protein crystallography to provide detailed knowledge of the three-dimensional structures of proteins, and elucidating the residues that are of significance to the enzyme catalysis and substrate specificity (Taylor et al. 2011).

Halogenated organic compounds, such as 1,2,3-trichloropropane (TCP), are stable and chemically inert molecules. As a result, their natural degradation is difficult in the environment. Studies conducted using random mutagenesis of haloalkane dehalogenase from *Rhodococcus* sp. m15–3 (DhaA) has yielded an enzyme eightfold more efficient than the initial protein, significantly improving the degradation of TCP (Bosma et al. 2002).

The key residues function in modulating entrance into the active site buried in haloalkane dehalogenase were determined by error-prone PCR followed by computer modelling, site saturation mutagenesis. The combined rational design and directed evolution strategy employed in the study yielded a mutant haloalkane dehalogenase. Use of mutated haloalkane dehalogenase achieved a 30-fold improvement of TCP degradation and also the resulting substitutions promote the formation of an activated reaction complex. Thus, solvent accessibility to the active site of the mutated haloalkane dehydrogenase decreased. This study demonstrates the substitutions in the "access tunnels" not in the catalytic site can also be critical for engineering proteins to enhance their catalysis (Pavlova et al. 2009).

Studies have applied a directed evolution strategy to the arsenic resistance (*ars*) operon from *Staphylococcus aureus*. The *ars* operon encodes a repressor (*arsR*), a membrane-efflux protein (*arsB*), and arsenate reductase (*arsC*). *E. coli* is resistant to arsenate and arsenite when *ars* operon is expressed. A mutant operon that has a 40-fold improvement of arsenate resistance was isolated with three rounds of DNA shuffling; random fragmentation of the operon, followed by mutagenic PCR, and library generation and screening. Interestingly, they have observed ten coding mutations in the arsB protein in final evolved version of the operon. Though there are no mutations in arsC, the activity of arsenate reductase has significantly improved (more than tenfold). This study clearly elicits that directed evolution is capable of improving the function of pathways (Carameri et al. 1997).

Biphenyl oxygenases (BOs) catalyse the oxygenation followed by degradation of polychlorinated biphenyls (PCBs). Biochemical properties of BOs are improved by directed evolution. The gene shuffling experiments confirmed large subunit of the biphenyl oxygenases present in *Pseudomonas pseudoalcaligenes* KF707 and *Burkholderia cepacia* LB400 plays a pivotal role in improving degradation properties for PCB and single aromatic compounds such as benzene, toluene, etc. (Suenaga and Furukawa 2006).

Toluene-ortho monooxygenase (TOM) has the ability to modify its substrate specificity via direct evolution. Different mutant proteins that effectively oxidize chlorinated ethane and naphthalene have been reported and they form dihydroxy and trihydroxy compounds from benzene and toluene, to enhance the oxidation of nitroaromatics, and the degradation of chlorinated aliphatic compounds (Canada et al. 2002; Ryu and Wood 2005; Leungsakul et al. 2006).

Saturation and random mutagenesis of aniline dioxygenase isolated from *Acinetobacter* sp. strain YAA have developed a mutated enzyme that harbour the capability to metabolize aniline, 2,4-dimethylaniline (a carcinogenic aromatic amine), and 2-isopropylaniline. Thereby the hydroxylation of aromatic amines is significantly improved (Ang et al. 2009).

The 1,2,3-trichloropropane (TCP) is highly toxic, recalcitrant in the environment and is a ground water pollutant which is frequently detected at sites where chemical waste has been inappropriately disposed. Interestingly, no natural microorganisms are known to degrade TCP under oxic conditions, even though anaerobic and oxidative biotransformations are possible. Studies have used a combination of protein and metabolic engineering to construct bacteria that use TCP as a growth substrate to facilitate biodegradation.

The haloalkane dehalogenase called DhaA from *Rhodococcus* was used as the starting enzyme which has very low but detectable activity with TCP. Researchers have determined its protein structure and have developed enzyme variants through directed evolution and particularly by inserting mutations in the substrate access/product exit tunnel that connects solvent and active site to have increased activity towards TCP with much more thermal stability (Janssen and Stucki 2020).

7.4.4 Genetically Modified Plants for Improved Phytoremediation

Plants have been used in bioremediation for centuries. Historical records show that the man-made irrigation systems in ancient Sri Lanka promoted growing selected trees around them and are known to be a measure taken for improving water quality. With growing amount of pollutants released, the natural capacity of these plants would not be sufficient to perform this. Hence, genetically modified, fast growing, and high biomass plants are called for phytoremediation (Nedjimi 2021). Phytoextraction (removal of contaminants by using plant roots), phytovolatilization (conversion of non-volatile compounds into volatile compounds by using plants), phytofiltration (absorb or adsorb pollutants by using plant roots or seedlings), phytostabilization (conversation of toxic compounds into less toxic compounds by using plants), phytodegradation (degradation of organic compounds by using plants), and rhizosphere bioremediation (removal of contaminants with the help of rhizosphere microorganisms) are the different types of phytoremediation explored for this purpose (Li et al. 2019; Nedjimi 2021). In addition, soil microbes invaded the roots are activated to degrade or detoxify pollutants. Phytoremediation contributes to the removal of organic pollutant, and metals and prevent soil leaching and it is an attractive strategy today.

Heavy metals are released to the environments mainly due to anthropogenic activities. Since heavy metals are non-degradable, they persist in soil for a long period. Therefore it is a long-term threat to ecosystems (Demarco et al. 2019). In general, heavy metals such as Zn, Cu, Mn, Ni are required for the form and function of plants and are named as essential heavy metals. However, Cd, Hg, Pb, and As are highly toxic and are not required for the physiological and biochemical process of plants (Chaffai and Koyama 2011). These non-essential heavy metals contribute to biomagnification and they are a major threat to human and animal health. Removal of heavy metals from contaminated sites is of utmost importance to prevent them from entering the food chains. Root systems of plants form the rhizosphere ecosystem so that they can increase the bioavailability of metal ions by root exudates and stabilize the soil fertility (Mishra et al. 2017). More than 400 plant species that can be utilized for phytoremediation have been identified (Sciences et al. 2008). These plants are slow growing and require strict nutritional and environmental conditions. To ensure the efficient removal of heavy metals, plant traits and environmental factors that limit the metal remediation should be improved. Recently, chemical assisted phytoremediation by applying synthetic cheaters such as ethylene di-ammine tetra acetic acid (EDTA), Ethylene glycol tetra acetic acids, or organic acids such as acetic acid, citric acid into the contaminated sites to form the metal complexes to enhance the bioavailability has shown promising results (Agnello et al. 2013). Thereby the metal uptake of low accumulator plants is increased. Biochar assisted phytoremediation has also shown promising results (Sun et al. 2018). Microbial assisted phytoremediation can improve heavy metal uptake by altering the soil pH and plant exudates composition. Plant growth promoting bacteria increase the plant growth by alleviating heavy metal stress and toxic effects. Use of transgenic plants i.e. enhancing the phytoremediation by improving plant traits such as pollutant uptake, translocation, sequestration, and tolerance by genetic manipulations, have moved into mainstream research with encouraging results (Buhari et al. 2019; Cherian and Oliveira 2005; Khan 2006; Suman et al. 2018).

However, a better understanding of molecular mechanisms of degradation of pollutants is crucial.

For phytoremediation of metals, metal tolerance is a prerequisite. This is achieved either by screening clones in greenhouses or selecting plant species that are naturally grown in contaminated sites. *Arabidopsis* has been identified as a model organism for genetic manipulations to improve plant traits associated with phytoremediation (Koornneef et al. 2010).

ZIP proteins are metal ion transporters. Novel investigations have been focused on using genes encoding ZIP proteins for phytoremediation. It has been reported in a model study, the uptake of Cd and Zn was improved by 150% upon the overexpression of AtIRT1 gene in *Arabidopsis thaliana* (Connolly et al. 2002). *Noccaea cearulescens* is a metal hyperaccumulating plant belongs to *Brassicaceae* family used to recover metal contaminated sites. Introducing the gene, NcZNT1 that encodes Zn transporter of *N. cearulescens* into *A. thaliana* improved the accumulation of Zn and Cd (Lin et al. 2016).

Metal tolerance is increased by inserting metal transporter genes such as ABC transporters improving metal sequestration in the vacuole. Overexpression of ABC genes improves the translocation of heavy metals. *Nicotiana tabacum* (Tobacco) plant was genetically modified by expressing yeast metallothionein gene to improve cadmium tolerance (Huang et al. 2012; Chen et al. 2015).

Overexpression of genes of cation diffusion facilitator (CDF) family proteins facilitates hyperaccumulation of metals. In addition, proteins encoded by the CDF family are metal tolerance/transport proteins that contribute to maintaining the level of ions in the cytoplasm. It has been reported that the gene OsMTP1 of *Oryza sativa* L. cv. IR64 in tobacco improved the accumulation of Cd and tolerance to As (Das et al. 2016).

Phytochelatins are metal chelators produced by plants and enzymes where phytochelatin synthase and c-glutamyl cysteine synthetase are involved in the synthesis of phytochelatins. In many transgenic plants, metal tolerance has been enhanced by introducing genes that encode these enzymes. For an instance, Cd and Pb accumulation and Cd tolerance were improved in *Nicotina glauca* and *Nicotina tabacum* by introducing phytochelatin synthase gene. However, a recent study reported that the uptake of Cd is improved by overexpressing both genes; phytochelatin synthase and c-glutamyl cysteine gene compared to single gene transformants and wild type (Chen et al. 2015).

Metallothioneins are cysteine rich proteins associated with the homeostasis of essential metals such as Zn, Cu, etc. Overexpression of metallothioneins in transgenic plants has improved metal tolerance. For an instance, tolerance to Cu and Cd of transgenic *A. thaliana* was improved by introducing and overexpressing metallothionine gene IlMt2a of *Iris lcatae var. chinensis* into *A. thaliana* (Gu et al. 2014). Overexpression of SbMt2 gene of *Solicornia brachiate* in tobacco has improved its metal tolerance as well as Zn translocation (Chaturvedi et al. 2014).

Recent studies have demonstrated that overexpression of genes responsible for transcription and DNA repair enhances metal tolerance (Fae et al. 2014; Charfeddine et al. 2016). Underlining mechanisms attributed are upregulation of genes encoding antioxidant enzymes, metallothioneins, metal transporters, and enzymes of DNA repair systems. Accumulation of reactive oxygen species under stress conditions damages the particular plant species. Improvement of phytoremediation properties by optimizing the activity of antioxidant enzymes and reactive oxygen species (ROS) scavengers is a special concern. However, plant responses observed in transgenes and wild type upon heavy metal stresses are debatable. For an instance, overexpression of superoxide dismutase, ascorbate peroxidase, and catalase generates a lower amount of ROS compared to wild type plants when exposed to Cu, Cd, and As. However, Iannone, Groppa and Benavides, 2014 has reported that catalase is not playing a pivotal role in tobacco plants upon Cd toxicity. Reduction of catalase activity by genetic modifications enhances the expression of alternative defence mechanisms mediated by guaiacol peroxidase and ascorbate peroxidase to Cd stress. Genetic factors affecting the rate of glutathione and phytochelatin production have been determined by introducing *gshl* gene of *E. coli* to *Brassica juncea*. The γ-ECS transgenic seedlings with increased production of phytochelatins, glutathione, non-protein thiols, γ-GluCys were tolerant to Cd. The genes; AtNramps, AtPcrs, CDA in *Arabidopsis thaliana* and *gshI*, *gshII,* PCS cDNA clone in *Brassica juncea* contribute to metal uptake and translocation (de Farias and Chaves 2011).

Arabidopsis thaliana harbouring *MerA* and *MerB* genes have shown increased tolerance to Hg. In addition, increased metal tolerance was achieved by altering oxidative stress responsive enzymes such as glutathione-S-transferase, peroxidase, and 1-aminocyclopropane-1-carboxylic acid deaminase (de Farias and Chaves 2011). Transgenic Alfalfa plant constructed by co-expressing GST gene of *Trichoderma vires* and cytochrome P450 2E1 gene in humans has shown improved efficiency in metal accumulation, tolerance to heavy metals; Cd, Hg, and xenobiotic; trichloroethylene (Zhang and Liu 2011).

Transplastomic plants are developed through genetic modifications in plastid DNA instead of genomic DNA. This is achieved mainly by homologous recombination. Thus genetic defects such as gene silencing can be eliminated by precisely modifying the specific sites (Bock 2015; Daniell et al. 2016). In addition, the following advantages are attributed to genetic modifications in plastid DNA.

1. Expression of bacterial genes is not difficult because codon optimization is not required in transgenes (Quesada-vargas et al. 2005)
2. Low risk in transient gene transfer into the environment through pollen (Maliga 2002)

Successful mercury phytoremediation was achieved by using transplastomic tobacco plants constructed by introducing bacterial merA and merB genes into chloroplast genomes. The expression of merA and merB genes induces plant tolerance to a high concentration of phenylmercuric acetate (Ruiz et al. 2003; Hussein et al. 2007). Expression of murine metallothionein gene (MT1) enhances the accumulation of Hg in tobacco cells and translocated actively in leaves.

Successful insertion of three genes that encode antioxidant enzymes; dehydroxy ascorbate reductase (DHAR), glutathione S-transferase (GST), and glutathione reductase (GR) into chloroplast genome contributes to the stimulation of antioxidant system of tobacco plant under Cd and Zn treatment (Martret et al. 2011). However, screening of the most suitable gene is critical for transformation into plastid DNA because some genes are more effective when they are transformed into chromosomes than chloroplast DNA (Martret et al. 2011).

Cadmium (Cd) detoxification by transgenic *Arabidopsis* sp. was achieved by successful transformation of BrMT1 gene which encodes type 1-metallothionien of *Brassica rapa* into chloroplasts. In addition, the expression of BrMT1 gene significantly reduces the chlorosis and accumulation of H_2O_2 (Kim et al. 2007).

Gene silencing is the other strategy used for the enhancement of metal remediation. Here, the small RNA inhibits the translation of target mRNA (Saurabh and Vidyarthi 2014). Arsenic reductase converts arsenate into arsenite in roots. Silencing of arsenate reductase gene via RNA interference in *Arabidopsis* improved the As translocation to shoot compared to wild type (Dhankher et al. 2006). In rice plants, OsNRAMP5 transporter contributes to uptake Cd, Fe, and Mn. Silencing of the gene *OsNRAMP5* increases the translocation of Cd to shoot, however, the content of Cd to root to shoot has reduced. Genetically modified OsNRAMP5 RNAi plant was constructed by using natural Cd accumulating cultivar Anjana Dhan (A5i). Even though the biomass of A5i plant with silenced OsNRAMP5 gene remains unchanged, the Cd accumulation in shoot was twice higher than the wild type plant (Sugimoto et al. 2014).

Use of a combination of transgenic plants and rhizobacteria for pollutant remediation has achieved significant advantages in the removal of pollutants. In a recent study it was reported that the transgenic *Arabidopsis thaliana* plant constructed by substituting the bacterial gene N-demethylase PdmAB into chloroplast has shown improved tolerance to isoproturon (IPU) while efficiently translocating to leaves. IPU is demethylated to 3-(-4-isoprophylphenyl)-1-methylurea (MDIPU) and can be released to outside. Rhizobacteria *Sphingobium* sp. *strain 1017-1* can further mineralize MDIPU in the rhizosphere. Thereby complete removal of IPU is achieved (Yan et al. 2018). The use of combination of transgenic plants and microorganisms is determined as one of the more efficient approaches to remediate contaminated sites with organic pollutants over individual phytoremediation (Dhankher et al. 2011).

To date, a plethora of genetically modified organisms/biocatalysts have been developed and have successfully applied for the bioremediation. We have summarized few newly modified organisms/biocatalysts in Table 7.1 in addition to the above-mentioned examples for perusal of readers interested in genetic engineering.

7.5 Impact and Challenges of Using Genetically Modified Organisms for Bioremediation

Even though the efficiency of bioremediation can be improved using genetically modified organisms, a prior risk assessment is essential before applying to the environment. Genetic manipulations of a particular organism can alter the pattern

Table 7.1 Genetically modified organisms/biocatalysts for efficient bioremediation

Organism/ Biocatalyst	Bioremediation	Modifications done	Phenotypic improvement	References
Deinococcus radiodurance	Radioactive substances	Incorporation of a new ion transporter gene; MerH from *M. marinum*	Ionic Hg degradation	Gupta and Walther (2016)
Bacillus pumilus	Dyes	Mutations in laccase gene	High thermo tolerance and increased the efficiency of decolorization	Luo et al. (2017)
Alcaligenes sp.	Xenobiotic compounds	Overexpression of DMFase gene	Enhancement of phytoremediation and microbial activity	Hussain et al. (2018)
Pseudomonas sp.	Xenobiotic compounds	Overexpression of SDS degrading gene	Enhancement of pentafluorosulfanyl-substituted aminophenols	Saccomanno et al. (2018)
Escherichia coli	Heavy metal	Overexpression of phytochelatin synthase gene of *schizosaccharymyces* and Cd transporter gene mntA	Removal of Cd	Kang et al. (2007)
Rhodococcus RHA1	Organic pollutants	Transformation of pPC3 plasmid containing *fbc* operon from *A. globifirmis*	Degrades 4-chlorobenzene	Microbiol et al. (2015)
Alfalfa	Phytoremediation of heavy metals	Overexpression of Arabidopsis ATP sulfurlase gene	Enhanced heavy metal accumulation	Kumar et al. (2019)
Mesorhizobium huakuii	Heavy metal	Overexpression of phytochelatins of *A. thaliana*	Accumulation and degradation of Cd^{2+}	Porter et al. (2017)

of gene expression, and copies of a particular gene of interest will integrate into the genome. Thus, potential risks associated with genetically engineered organisms cannot be fully ascertained. According to IUCN, the World Conservation Union has identified following environmental risks related to the release of GMOs into the environment (IUCN 2004).

1. Genetic contamination/inbreeding
2. Competition with natural species
3. Increases selection pressure on target and non-target organisms
4. Ecosystem damage and destruction due to novel species
5. Horizontal transfer of recombinant genes to other organisms
6. Adverse effects on human health (enhanced pathogenicity, emergence of new diseases, etc.)

However, unidentified and indirect effects can be observed and those vary according to species, characteristics of particular organisms, environmental factors, interactions among other species, etc. Major challenge associated with the development of genetically modified organisms for the bioremediation is mitigating potential risks associated with particular organisms (Saxena et al. 2020).

7.6 Future Perspectives

Continuous understanding of the genetic and physiological characteristics of microorganisms, plants, and other organisms is essential to improve organisms towards bioremediation. Identification, characterization, and isolation of novel species and biocatalysts are important for the development of effective bioremediation strategies. The process of rapid identification, characterization, isolation, purification, and genetic manipulations is directly linked to the current advancements in proteomics, metabolomics, genomics, transcriptomics, and phenomics. Studying the regulatory mechanisms of these pathways would turn a new leaf in bioengineering of these organisms. The strategies for which would be a subject for another review. However, synthetic biological approaches with proper regulatory mechanisms may find solutions for some of the drawbacks in engineered organisms. For effective bioremediation, the development of environmentally-safe genetically modified organisms for synergetic approaches is highly recommended.

References

Agnello AC et al (2013) Enhanced phytoremediation: a review of low molecular weight organic acids and surfactants used as amendments. Crit Rev Environ Sci Technol 44:2531–2576. https://doi.org/10.1080/10643389.2013.829764

Akhtar N, Mannan MA (2020) Mycoremediation: expunging environmental pollutants. Biotechnol Rep 26:e00452. https://doi.org/10.1016/j.btre.2020.e00452
Albert ON, Amaratunga D, Haigh RP (2018) Evaluation of the impacts of oil spill disaster on communities and its influence on restiveness in Niger Delta, Nigeria. Procedia Eng 212:1054–1061. https://doi.org/10.1016/j.proeng.2018.01.136
Alkorta I, Hernandez-allica J, Amezaga I (2004) Recent findings on the phytoremediation of soils contaminated with environmentally toxic heavy metals and metalloids such as zinc, cadmium, lead, and arsenic. Rev Environ Sci Bio/Technol 3. 10.1023/B
Anantharam H, Talkad MS (2018) Plastic (LDPE) degradation by induced mutations in *Pseudomonas putida* plastic (LDPE) degradation by induced mutations in *Pseudomonas putida*. J Environ Sci 12(9):34–40. https://doi.org/10.9790/2402-1209033440
Ang EL, Obbard JP, Zhao H (2009) Directed evolution of aniline dioxygenase for enhanced bioremediation of aromatic amines. Appl Microbiol Biotechnol 1063–1070. https://doi.org/10.1007/s00253-008-1710-0
Anku WW, Mamo MA, Govender P (2019) Phenolic compounds in water: sources, reactivity, toxicity and treatment methods. Natural sources importance and applications https://doi.org/10.5772/66927
Ann M et al (2018) 'Engineering *Pseudomonas putida* KT2440 for efficient ethylene glycol utilization. Metab Eng 48:197–207. https://doi.org/10.1016/j.ymben.2018.06.003
Atashgahi S, Shetty SA et al (2018a) Flux, impact, and fate of halogenated xenobiotic compounds in the gut. Front Physiol 9:1–18. https://doi.org/10.3389/fphys.2018.00888
Atashgahi S, Sánchez-Andrea I et al (2018b) Prospects for harnessing biocide resistance for bioremediation and detoxification. Science 746:743–746
Bai X et al (2017) Linking urbanization and the environment: conceptual and empirical advances. Annu Rev Environ Resour 1–26
Basak B, Bhunia B, Dutta S (2013) Kinetics of phenol biodegradation at high concentration by a metabolically versatile isolated yeast *Candida tropicalis*. Environ Sci Pollut Res. https://doi.org/10.1007/s11356-013-2040-z
Bhandari S et al (2021) Microbial enzymes used in bioremediation. J Chem 1.17. https://doi.org/10.1155/2021/8849512
Bhatt P, Bhatt K, Sharma A, Zhang W, Mishra S, Chen S (2021) Biotechnological basis of microbial consortia for the removal of pesticides from the environment. Crit Rev Biotechnol 41(3):317–338
Bock R (2015) Engineering plastid genomes: methods, tools, and applications in basic research and biotechnology. Annu Rev Plant Biol 66:211–241. https://doi.org/10.1146/annurev-arplant-050213-040212
Bogino PC et al (2013) The role of bacterial biofilms and surface components in plant-bacterial associations. Int J Mol Sci 14:15838–15859. https://doi.org/10.3390/ijms140815838
Bosma T, Stucki G, Janssen DB (2002) Biodegradation of 1, 2, 3-trichloropropane through directed evolution and heterologous expression of a haloalkane dehalogenase gene. Appl Environ Microbiol 68(7):3582–3587. https://doi.org/10.1128/AEM.68.7.3582
Buhari ML et al (2019) Role of biotechnology in phytoremediation. J Biotechnol Phytochem 3(1)
Butt H et al (2018) Engineering plant architecture via CRISPR / Cas9-mediated alteration of strigolactone biosynthesis. BMC Plant Biol 1–9
Canada KA et al (2002) Directed evolution of toluene ortho -monooxygenase for enhanced 1-naphthol synthesis and chlorinated ethene degradation. J Bacteriol 184(2):344–349. https://doi.org/10.1128/JB.184.2.344
Cao X, Yang C, Liu R (2013) Simultaneous degradation of organophosphate and organochlorine pesticides by *Sphingobium japonicum* UT26 with surface-displayed organophosphorus hydrolase. Biodegradation 24:295–303. https://doi.org/10.1007/s10532-012-9587-0
Carameri A, Dawes G, Rodriguez E (1997) Molecular evolution of an arsenate detoxification pathway by DNA shuffling. Nat Biotechnol 15:436–438. https://doi.org/10.1038/nbt0597-436

Chaffai R, Koyama H (2011) Heavy metal tolerance in *Arabidopsis thaliana* research. Advances in botanical research, 1st edn. Elsevier. https://doi.org/10.1016/B978-0-12-385851-1.00001-9

Charfeddine M et al (2016) The effect of cadmium on transgenic potato (*Solanum tuberosum*) plants overexpressing the StDREB transcription factors. Plant Cell, Tissue Organ Cult (PCTOC) 0(0):0. https://doi.org/10.1007/s11240-016-1130-2

Chaturvedi AK et al (2014) The SbMT-2 gene from a halophyte confers abiotic stress tolerance and modulates ROS scavenging in transgenic tobacco. PLoS One 9(10). https://doi.org/10.1371/journal.pone.0111379

Chen Y, Liu Y, Ding Y (2015) Overexpression of PtPCS enhances cadmium tolerance and cadmium accumulation in tobacco. Plant Cell Tissue Organ Cult (PCTOC). https://doi.org/10.1007/s11240-015-0710-x

Cherian S, Oliveira MM (2005) Transgenic plants in phytoremediation: recent advances and new possibilities. Environ Sci Technol 39(24):9377–9390

Connolly EL, Fett JP, Guerinot ML (2002) Expression of the IRT1 metal transporter is controlled by metals at the levels of transcript and protein accumulation. Plant Cell 14:1347–1357. https://doi.org/10.1105/tpc.001263.1348

Daniell H et al (2016) Chloroplast genomes: diversity, evolution, and applications in genetic engineering. Genome Biol 17:1–29. https://doi.org/10.1186/s13059-016-1004-2

Das N, Bhattacharya S, Maiti MK (2016) Enhanced cadmium accumulation and tolerance in transgenic tobacco overexpressing rice metal tolerance protein gene OsMTP1 is promising for phytoremediation. Plant Physiol Biochem 105:297–309. https://doi.org/10.1016/j.plaphy.2016.04.049

Dash B, Sahu N, Singh AK, Gupta SB, Soni R (2021) Arsenic efflux in *Enterobacter cloacae* RSN3 isolated from arsenic-rich soil. Folia Microbiol 66:189–196

de Farias PCM, Chaves ALS (2011) Transgenic plants for enhanced phytoremediation – physiological studies. In: Genetic transformation. https://doi.org/10.5772/24355

Debbarma P, Raghuwanshi S, Singh J, Suyal DC, Zaidi MGH, Goel R (2017) Comparative in situ biodegradation studies of polyhydroxybutyrate film composites. 3Biotech 7(178):1–9. https://doi.org/10.1007/s13205-017-0789-3

Demarco CF, Afonso TF, Pieniz S (2019) Phytoremediation of heavy metals and nutrients by the *Sagittaria montevidensis* into an anthropogenic contaminated site at southern of Brazil. Int J Phytoremed 0(0):1–8. https://doi.org/10.1080/15226514.2019.1612843

Dhankher OP et al (2006) Hyperaccumulation of arsenic in the shoots of Arabidopsis silenced for arsenate reductase (ACR2). PNAS 103(14):5413–5418

Dhankher OP et al (2011) Biotechnological approaches for phytoremediation. In: Plant biotechnology and agriculture, pp 309–328. https://doi.org/10.1016/B978-0-12-381466-1.00020-1

Dumorné K et al (2017) Extremozymes: a potential source for industrial applications. J Microb Biotechnol 27(4):649–659

Fae M et al (2014) Copper-mediated genotoxic stress is attenuated by the overexpression of the DNA repair gene MtTdp2 a (tyrosyl-DNA phosphodiesterase 2) in Medicago truncatula plants. Plant Cell Rep 33(7):1071–1080. https://doi.org/10.1007/s00299-014-1595-6

Giri K, Rai JPN, Pandey S, Mishra G, Kumar R, Suyal DC (2017a) Performance evaluation of isoproturon-degrading indigenous bacterial isolates in soil microcosm. Chem Ecol 33(9):817–825. https://doi.org/10.1080/02757540.2017.1393535

Giri K, Suyal DC, Mishra G, Pandey S, Kumar R, Meena DK, Rai JPN (2017b) Biodegradation of isoproturon by *Bacillus pumilus* K1 isolated from foothill agroecosystem of north west Himalaya. Proc Natl Acad Sci India Sect B-Biol Sci 87(3):839–848. https://doi.org/10.1007/s40011-015-0667-x

Goel R, Zaidi MGH, Soni R, Kusumlata SYS (2008) Implication of A*rthrobacter* and *Enterobacter* species for polycarbonate degradation. Int Biodeterior Biodegrad 61(2):167–172

Gu C et al (2014) Overexpression of Iris. lactea var. chinensis metallothionein IlMT2a enhances cadmium tolerance in *Arabidopsis thaliana*. Ecotoxicol Environ Safety 105:22–28. https://doi.org/10.1016/j.ecoenv.2014.04.002

Gupta DK, Walther C (2016) Radionuclide contamination and remediation through plants. Springer, Cham

Huang J et al (2012) Fission yeast HMT1 lowers seed cadmium through phytochelatin-dependent vacuolar sequestration. Plant Physiol 158(April):1779–1788. https://doi.org/10.1104/pp.111.192872

Huang A et al (2018) Conformation and activity alteration of horseradish peroxidase induced by the interaction with gene carrier polyethyleneimines. Spectrochim Acta Part A: Mol Biomol Spectrosc 188:90–98. https://doi.org/10.1016/j.saa.2017.06.046

Hussain I et al (2018) Microbe and plant assisted-remediation of organic xenobiotics and its enhancement by genetically modified organisms and recombinant technology: a review. Sci Total Environ 628–629:1582–1599. https://doi.org/10.1016/j.scitotenv.2018.02.037

Hussein HS et al (2007) Phytoremediation of mercury and organomercurials in chloroplast transgenic plants: enhanced root uptake, translocation to shoot, and volatilization. Environ Sci Technol 41(24):8439–8446

Iannone MF, Groppa MD, Benavides MP (2014) Cadmium induces different biochemical responses in wild type and catalase-deficient tobacco plants. Environ Exp Bot. https://doi.org/10.1016/j.envexpbot.2014.07.008

IUCN (2004) Genetically modified organisms and biosafety

Jafari M et al (2013) Bioremediation and genetically modified organisms. In: Fungi as bioremediators, pp 433–451. https://doi.org/10.1007/978-3-642-33811-3

Janssen DB, Stucki G (2020) Perspectives of genetically engineered microbes for groundwater bioremediation. Environ Sci Process Impacts R Soc Chem 487–499. https://doi.org/10.1039/c9em00601j

Ju K, Parales RE (2006) Control of substrate specificity by active-site residues in nitrobenzene dioxygenase. Appl Environ Microbiol 72(3):1817–1824. https://doi.org/10.1128/AEM.72.3.1817

Kang SH et al (2007) Bacteria metabolically engineered for enhanced phytochelatin production and cadmium accumulation. Appl Environ Microbiol 73(19):6317–6320. https://doi.org/10.1128/AEM.01237-07

Khan AG (2006) Mycorrhizoremediation—an enhanced form of phytoremediation. J Zhejiang Univ Sci 7(7):503–514. https://doi.org/10.1631/jzus.2006.B0503

Kim SH et al (2007) Chloropiasto-targeted BrMT1 (Brassica rapa type-1 MetMIothionein) enhances resistance to cadmium and ROS in transgenic Arabidopsis plants. J Plant Biol 50:1–7

Koornneef M et al (2010) The development of Arabidopsis as a model plant. Plant J 61:909–921. https://doi.org/10.1111/j.1365-313X.2009.04086.x

Kour D, Kaur T, Devi R, Yadav A, Singh M et al (2021) Beneficial microbiomes for bioremediation of diverse contaminated environments for environmental sustainability: present status and future challenges. Environ Sci Pollut Res 28:24917–24939

Kumar V et al (2019) Enhancement of heavy metal tolerance and accumulation efficiency by expressing Arabidopsis ATP sulfurylase gene in alfalfa. Int J Phytoremed 0(0):1–10. https://doi.org/10.1080/15226514.2019.1606784

Kumar P, Dash B, Suyal DC, Gupta SB, Singh AK, Chowdhury T, Soni R (2021) Characterization of arsenic-resistant Klebsiella pneumoniae RnASA11 from contaminated soil and water samples and its bioremediation potential. Curr Microbiol. https://doi.org/10.1007/s00284-021-02602-w

Kurzbaum E et al (2010) Efficiency of phenol biodegradation by planktonic *Pseudomonas pseudoalcaligenes* (a constructed wetland isolate) vs. root and gravel biofilm. Water Res 44(17):5021–5031. https://doi.org/10.1016/j.watres.2010.07.020

Łabużek S, Nowak B, Pająk J (2004) The susceptibility of polyethylene modified with bionolle to biodegradation by filamentous fungi. Pol J Environ Stud 13(1):59–68

Leungsakul T, Johnson GR, Wood TK (2006) Protein engineering of the 4-methyl-5-nitrocatechol monooxygenase from Burkholderia sp. strain DNT for enhanced degradation of nitroaromatics. Appl Environ Microbiol 72(6):3933–3939. https://doi.org/10.1128/AEM.02966-05

Li W et al (2019) Laboratory evolution reveals the metabolic and regulatory basis of ethylene glycol metabolism by pseudomonas putida KT2440. Environ Microbiol 00. https://doi.org/10.1111/1462-2920.14703

Lin Y et al (2016) Expression of the ZNT1 zinc transporter from the metal hyperaccumulator *Noccaea caerulescens* confers enhanced zinc and cadmium tolerance and accumulation to *Arabidopsis thaliana*. PLoS One 11(3):1–30. https://doi.org/10.1371/journal.pone.0149750

Linden O, Jernelöv A, Egerup J (2004) The environmental impacts of the Gulf War 1991, IR-04-019/April, Forestry, (April)

Liu YJ, Zhang AN, Wang XC (2009) Biodegradation of phenol by using free and immobilized cells of Acinetobacter sp. XA05 and Sphingomonas sp. FG03. Biochem Eng J 44(13):187–192. https://doi.org/10.1016/j.bej.2008.12.001

Liu S et al (2011) 'Arsenic removal from contaminated soil via biovolatilization by genetically engineered bacteria under laboratory conditions. J Environ Sci 23(9):1544–1550. https://doi.org/10.1016/S1001-0742(10)60570-0

Luo Q et al (2017) Functional expression enhancement of *Bacillus pumilus* CotA-laccase mutant WLF through site-directed mutagenesis. Enzym Microb Technol 0–1. https://doi.org/10.1016/j.enzmictec.2017.07.013

Maliga P (2002) Engineering the plastid genome of higher plants Pal Maliga. Plant Biotechnol 5:164–172

Mao Z, Yu C, Xin L (2015) Enhancement of phenol biodegradation by pseudochrobactrum sp. through ultraviolet-induced mutation. Int J Mol Sci 16:7320–7333. https://doi.org/10.3390/ijms16047320

Martret L et al (2011) Tobacco chloroplast transformants expressing genes encoding dehydroascorbate reductase, glutathione reductase, and glutathione-S-transferase, exhibit altered anti-oxidant metabolism and improved abiotic stress tolerance. Plant Biotechnol J 9:661–673. https://doi.org/10.1111/j.1467-7652.2011.00611.x

Microbiol A et al (2015) Characterization of a protocatechuate catabolic gene cluster in *Rhodococcus ruber* OA1 involved in naphthalene degradation. Ann Microbiol (12). https://doi.org/10.1007/s13213-015-1132-z

Mishra J, Singh R, Arora NK (2017) Alleviation of heavy metal stress in plants and remediation of soil by rhizosphere microorganisms. Front Microbiol 8:1–7. https://doi.org/10.3389/fmicb.2017.01706

Moubasher HA et al (2015) Phytoremediation of soils polluted with crude petroleum oil using Bassia scoparia and its associated rhizosphere microorganisms. Int Biodeterior Biodegrad 98:113–120. https://doi.org/10.1016/j.ibiod.2014.11.019

Nagata Y et al (2010) Complete genome sequence of the representative -Hexachlorocyclohexane-degrading bacterium *Sphingobium japonicum* UT26. J Bacteriol 192(21):5852–5853. https://doi.org/10.1128/JB.00961-10

Nedjimi B (2021) Phytoremediation: a sustainable environmental technology for heavy metals decontamination. SN Appl Sci 3(286). https://doi.org/10.1007/s42452-021-04301-4

Nies H (2018) Pollution with radioactive substances. In: Handbook on marine environment protection, pp 413–427

Ogunbayo AO, Olanipekun OO, Adamu IA (2019) Preliminary studies on the microbial degradation of plastic waste using *Aspergillus niger* and *Pseudomonas* sp. J Environ Prot 10:625–631. https://doi.org/10.4236/jep.2019.105037

Pavlova M et al (2009) Redesigning dehalogenase access tunnels as a strategy for degrading an anthropogenic substrate. Nat Chem Biol 5(10):727–733. https://doi.org/10.1038/nchembio.205

Peeples TL (2014) Bioremediation using extremophiles. Microb Biodegrad Bioremed 2014:251–268. https://doi.org/10.1016/B978-0-12-800021-2.00010-8

Plastics – the Facts 2019 (2019) An analysis of European plastics production, demand and waste data (Plastics Europe 2018)

Pollution R (2014) Radioactive pollution. In: Methods of measuring environmental parameters, pp 380–384

Porter SS et al (2017) Association mapping reveals novel serpentine adaptation gene clusters in a population of symbiotic Mesorhizobium. ISME J Nat Publ Group 11(1):248–262. https://doi.org/10.1038/ismej.2016.88

Power AL et al (2018) Monitoring impacts of urbanisation and industrialisation on air quality in the Anthropocene using urban pond sediments. Front Earth Sci 6:1–18. https://doi.org/10.3389/feart.2018.00131

Quesada-vargas T, Ruiz ON, Daniell H (2005) Characterization of heterologous multigene operons in transgenic chloroplasts. Transcription, processing, and translation 1. Plant Physiol 138:1746–1762. https://doi.org/10.1104/pp.105.063040.Nickelsen

Raaman N et al (2012) Biodegradation of plastic by Aspergillus spp. isolated from polythene polluted sites around Chennai, July 2019

Rasekh B et al (2014) Protein engineering of laccase to enhance its activity and stability in the presence of organic solvents. Eng Life Sci 0:1–7. https://doi.org/10.1002/elsc.201300042

Ruiz ON et al (2003) Phytoremediation of organomercurial compounds via chloroplast genetic engineering 1. Plant Physiol 132:1344–1352. https://doi.org/10.1104/pp.103.020958.electron

Ryu K, Wood TK (2005) Protein engineering of toluene-o-xylene monooxygenase from *Pseudomonas stutzeri* OX1 for oxidizing nitrobenzene to 3-nitrocatechol, 4-nitrocatechol, and nitrohydroquinone. J Bacteriol 115:145–156. https://doi.org/10.1016/j.jbiotec.2004.08.008

Saadoun IMK (2016) Impact of oil spills on marine life. Intech Open, pp 76–103. https://doi.org/10.5772/57353

Saccomanno M et al (2018) Biodegradation of pentafluorosulfanyl-substituted aminophenol in Pseudomonas spp. Biodegradation. Springer, Cham. https://doi.org/10.1007/s10532-018-9827-z

Sani RK, Nawanietha Krishanraj R (2017) Extremophilic enzymatic processing of lignocellulosic feedstocks to bioenergy. Environmental science and pollution research. Springer International Publishing, Cham. https://doi.org/10.1007/978-3-319-54684-1_1

Santos GDA et al (2019) Directed evolution of P450 BM3 towards functionalization of aromatic O-heterocycles. Int J Mol Sci 20(3353):1–24

Sauge-merle S et al (2003) Enhanced toxic metal accumulation in engineered bacterial cells expressing *Arabidopsis thaliana* phytochelatin synthase. Appl Environ Microbiol 69 (1):490–494. https://doi.org/10.1128/AEM.69.1.490

Saurabh S, Vidyarthi AS (2014) RNA interference: concept to reality in crop improvement. Planta 239:543–564. https://doi.org/10.1007/s00425-013-2019-5

Saxena G et al (2020) Genetically modified organisms (GMOs) and their potential in environmental management: constraints, prospects, and challenges. In: Bioremediation of industrial waste for environmental safety, pp 1–19

Schmidt C, Krauth T, Wagner S (2017) Export of plastic debris by rivers into the sea. Environ Sci Technol 51:12246–12253. https://doi.org/10.1021/acs.est.7b02368

Sciences A, Pierce F, Health E (2008) Phytoremediation of heavy metal polluted soils and water. J Zhejiang Univ Sci 9(3):210–220. https://doi.org/10.1631/jzus.B0710633

Sharma B, Dangi AK, Shukla P (2018) Contemporary enzyme based technologies for bioremediation: a review. J Environ Manage 210:10–22. https://doi.org/10.1016/j.jenvman.2017.12.075

Shukla AK, Singh AK (2020) Exploitation of potential extremophiles for bioremediation of xenobiotics compounds: a biotechnological approach, current genomics. Bentham Science, vol 21, Number 3, pp 161–167(7). https://doi.org/10.2174/1389202921999200422122253

Singh S et al (2008) Bioremediation: environmental clean-up through pathway engineering. Curr Opin Biotechnol 437–444. https://doi.org/10.1016/j.copbio.2008.07.012

Singh M, Singh D, Rai P, Suyal DC, Saurabh S, Soni R, Giri K, Yadav AN (2021) Fungi in remediation of hazardous wastes: current status and future. In: Yadav AN (ed) Recent trends in mycological research, fungal biology. Springer Nature, Cham

Stark BC, Dikshit KL, Pagilla KR (2012) The biochemistry of Vitreoscilla hemoglobin mutants of VHb: structure-function studies other functions of VHb in addition to the apparent direct binding of VHb to terminal. Comput Struct Biotechnol J 3(4)

Suenaga H, Furukawa K (2006) Active-site engineering of biphenyl dioxygenase: effect of substituted amino acids on substrate specificity and regiospecificity. Appl Microbiol Biotechnol 168–176. https://doi.org/10.1007/s00253-005-0135-2

Sugimoto K et al (2014) From laboratory to field: OsNRAMP5-knockdown rice is a promising candidate for Cd phytoremediation in paddy fields. PLoS One 9(6):1–7. https://doi.org/10.1371/journal.pone.0098816

Suman J et al (2018) Phytoextraction of heavy metals: a promising tool for clean-up of polluted environment? Front Plant Sci. https://doi.org/10.3389/fpls.2018.01476

Sun W, Zhang S, Su C (2018) Impact of biochar on the bioremediation and phytoremediation of heavy metal (loid)s in soil. In: Advances in bioremediation and phytoremediation, pp 149–168. https://doi.org/10.5772/intechopen.70349

Sun GL, Reynolds EE, Belcher AM (2019) Designing yeast as plant-like hyperaccumulators for heavy metals. Nat Commun 1–12. https://doi.org/10.1038/s41467-019-13093-6

Tarekegn MM et al (2020) Microbes used as a tool for bioremediation of heavy metal from the environment. Cogent Food Agric 6(1):1–19. https://doi.org/10.1080/23311932.2020.1783174

Taylor P, Jez JM, Jez JM (2011) Toward protein engineering for phytoremediation: possibilities and challenges. Int J Phytoremediation 37–41. https://doi.org/10.1080/15226514.2011.568537

Wang Q (2018) Urbanization and global health: the role of air pollution. Iran J Pub Health 47 (11):1644–1652

Wang Q et al (2011) Potential approaches to improving biodegradation of hydrocarbons for bioremediation of crude oil pollution. J Environ Prot 02(01):47–55. https://doi.org/10.4236/jep.2011.21005

Waryah CB, Moses C, Arooj M (2018) Zinc fingers, TALEs, and CRISPR systems: a comparison. In: Epigenome editing: methods and protocols, methods in molecuar biology, pp 19–63

Webster DA, Hackett DP (1966) The purification and properties of cytochrome o from Vitreoscilla. J Biol Chem. © 1966 ASBMB. Currently published by Elsevier Inc; originally published by American Society for Biochemistry and Molecular Biology 241(14):3308–3315. https://doi.org/10.1016/S0021-9258(18)96464-1

WHO (2021) https://www.who.int/news-room/fact-sheets/detail/ambient-(outdoor)-air-quality-and-health

Yan X et al (2018) Enhanced and complete removal of Phenylurea herbicides by. Appl Environ Microbiol 84(14):1–16

Yang C et al (2006) Cloning of mpd gene from a chlorpyrifos-degrading bacterium and use of this strain in bioremediation of contaminated soil. FEMS Microbiol Lett 265:118–125. https://doi.org/10.1111/j.1574-6968.2006.00478.x

Yang C et al (2008) Surface display of MPH on pseudomonas putida JS444 using ice nucleation protein and its application in detoxification of organophosphates. Biotechnol Bioeng 99 (1):30–37. https://doi.org/10.1002/bit

Yang C et al (2015) An engineered microorganism can simultaneously detoxify cadmium, chlorpyrifos, and g -hexachlorocyclohexane. J Basic Microbiol 55:1–7. https://doi.org/10.1002/jobm.201500559

Zhang Y, Liu J (2011) Transgenic alfalfa plants co-expressing glutathione S-transferase (GST) and human CYP2E1 show enhanced resistance to mixed contaminates of heavy metals and organic pollutants. J Hazard Mater 189(1–2):357–362. https://doi.org/10.1016/j.jhazmat.2011.02.042

Chapter 8
Understanding the Role of Genetic and Protein Networking Involved in Microbial Bioremediation

Upasana Jhariya, Shweta Srivastava, Sanchita Das, Sakina Bombaywala, Sejal Mahallea, and Nishant A. Dafale

8.1 Introduction

Rapid urbanization and industrialization in the current era have led to the deposition of contaminants in the environment which directly-indirectly affects the terrestrial and aquatic health. Industries play prominent role in development of national economy, simultaneously they are the considered as major environmental polluters. Chemical manufacturing hub, pharmaceutical companies and more importantly crude oil processing industries are the major source of xenobiotic discharge into the environment (Bharagava et al. 2018). The flow of these discharges to various environmental niches causes increased pollutant dispersal which has created high risk to the 80% of world population and is considered as most vulnerable factor for all forms of life. The increasing rate of soil pollution has also compelled the attention towards increasing risk. As the rate of pollutants deposition increasing, it become necessary to remove these contaminants from existing highly polluted site to prevent their dispersal into the environment. Treatment of pollutants by using physico-chemical methods such as incineration, solidification, evaporation, chemical precipitation and oxidation reduction has been used from decades. Nevertheless, high chemical input, high processing cost and generation of harmful byproducts bound the application of these methods (Dangi et al. 2019).

U. Jhariya · S. Das · S. Mahallea
Environmental Biotechnology and Genomics Division, CSIR-National Environmental Engineering Research Institute (CSIR-NEERI), Nagpur, India

S. Srivastava · S. Bombaywala · N. A. Dafale (✉)
Environmental Biotechnology and Genomics Division, CSIR-National Environmental Engineering Research Institute (CSIR-NEERI), Nagpur, India

Academy of Scientific and Innovative Research (AcSIR), Ghaziabad, India
e-mail: na_dafale@neeri.res.in

© The Author(s), under exclusive license to Springer Nature Switzerland AG 2022
D. C. Suyal, R. Soni (eds.), *Bioremediation of Environmental Pollutants*,
https://doi.org/10.1007/978-3-030-86169-8_8

Bioremediation is an eco-friendly and reliable approach that employ microbes for their natural ability to neutralize or remove the toxic pollutants from the environment. Various, bacteria, algae and fungi have been reported for their high degradation capacity for environmental pollutants. Bioremediation through microbes is more reliable as they have unique enzymatic machinery, for broad substrate utilization capability (Azubuike et al. 2016; Singh et al. 2021). Microorganisms have evolved mechanisms for hydrocarbon degradation, generating metabolic intermediates that funnel to central metabolic pathways. They use different metabolic pathways for degradation of different pollutants, which involves different sets of genes and enzymes (Dafale et al. 2008; Bhatt et al. 2020). These genes and enzymes involved in pathways control the overall system of microbial action, also profound information about these factors allows to understand the bioremediation processes. Detailed understanding and identification of substrate specific organism, their interactions, catabolic gene structures and proteins involved in bioremediation can be gained by using advance molecular techniques and bioinformatic approaches (Fuentes et al. 2014).

Insights of bioremediation by microorganisms can be reflected either by culturable approach or unculturable approach. In culturable approach pure cultures are identified and genome sequencing reveals the information of promoters, genes and proteins involved in degradation pathways which can be reconstructed for improving bioremediation processes. To understand the full functional classification and taxonomic diversity, metagenomic approaches are popularly utilized (Kumar et al. 2016). Metagenome provides the insights of regulatory pathways, key metabolic genes and important set of microorganisms thriving in the contaminated sites. Recent advancement in microbial ecological DNA microarray has provided a platform for simultaneously analyzing the thousands of functional genes for characterizing the microbial communities participating in bioremediation of contaminated site (Techtmann and Hazen 2016). Technologies such as transcriptomics, metagenomics and metabolomics are used to analyze the physiology of microbes and also predict the pollutants biodegradation pathway. These advance technologies lead us to create a link in between genes and proteins involved in bioremediation approaches (Jennings et al. 2009).

The present book chapter elucidates the usage of computational tools for studying the interaction of genes and proteins involved in bioremediation. It sheds light on various approaches for exploring different culturable and non-culturable microbial community. This book chapter emphasizes on various OMICs and metaOMICs tools to produce myriads of intricate datasets and the analysis of these datasets introduce unprecedented analytical data for improving bioremediation processes. Furthermore, the application of gene editing tools as well as synthetic biology for microbial genome editing and pathways reconstruction have also been briefly discussed.

8.2 Unraveling the Potential of Culturable Microorganism

8.2.1 Genomic Approaches

Microorganisms are popularly known to thrive in adverse environment. They adapt to the environment by changing their genetic makeup and expressing the gene which are critical for their survival. These days the incessantly increasing pollution level requires an eco-friendly approach to treat the recalcitrant pollutants and maintain balance in ecosystem. Tiny little microorganisms degrade the organic contaminants of the environment and release harmless or non-toxic products mainly water, CO_2, and cellular biomass (Bharagava et al. 2018).

The metabolism of micro-organism is very complex and interconnectedness of the gene and their expression make the cellular process even more complicated. Genes playing key role in catabolic pathway are either chromosomal or extra-chromosomal, and are expressed with different physiologies and metabolic requirements (Rosanti et al. 2020). Advances in science and technology has enabled us to understand the genomic potential, genomic contexts of catabolic genes, mobility of genes, along with underlying physiology, its regulation and metabolism of the cell. Various approaches of genetics may also provide the biodegradation potential of micro-organism for different pollutants, and their metabolic differences among various biodegradation strains (Aguiar-Pulido et al. 2016; Debbarma et al. 2017).

Different molecular techniques such as 16S rRNA sequencing, whole genome sequencing, metagenomic analysis, transcriptomics analysis and other omics approaches have been explored and can be counted as a potent tool for the detoxification, eradication, and degradation, of highly toxic pollutants from the environment (Rawat and Rangarajan 2019).

8.2.1.1 Understanding Phylogeny Through 16S rRNA

The finding of sequences highly conserved in all the micro-organisms that is 16S rRNA, was the significant advancement in the field of microbial study. The 16S rRNA genes has delivered the very important area of phylogenetic characterization of the microorganisms which composes the microbial community. The 1600 base pairs long 16S rRNA gene has nine hypervariable regions of varying conservation namely from V1 to V9. Higher ranking taxa are often determined by most conservative regions while regions evolving more quickly help in identification of genus or species (Srivastava et al. 2020a).

The taxonomic identification revealed by 16S rRNA approach may determine the phylogenetic status of the organisms involved in critical steps of bioremediation (Saxena et al. 2021). Kalaimurugan et al. (2020) have isolated Chromium degrading bacteria from soil and characterized them through 16S rRNA sequencing. The 16S rRNA identified *Pseudomonas fluorescens and Bacillus safensis* have shown the 84% and 72% chromium degrading capacity (Kalaimurugan et al. 2020). Similarly,

Deng et al. (2020) have characterized the isolated crude oil degrading bacteria *Bacillus halotolerans* 1–1 and *Bacillus cereus* T-04 through 16S rRNA sequencing approach. Further they employed 16S rRNA sequencing method to study the impact of exogenous bacterial causing shift in microbial community of the crude-oil-contaminated soil under bioremediation processes (Deng et al. 2020).

The16S rRNA gene analysis has transformed the area of studying the microbial diversity in the environment by culture independent as well as culture-dependent approaches (Kumar et al. 2019; Suyal et al. 2021). PCR-amplified 16S sequences have been clustered together on the basis of similarity, for generating operational taxonomic units (OTUs) for inferring the taxonomy (Lovley 2003). The major drawback of applying 16S rRNA is that it fairly predicts the phylogeny of the organisms but it lacks the prediction of important aspects of their metabolic physiology related with bioremediation (Johnson et al. 2019). The organisms can be closely related with respect to the 16S rRNA sequences but their degradation potential for compounds may vary. For example, the TCE-degrader *D. ethanogenes* may cluster together on the basis of 16S rRNA but they differ in the chlorinated compounds degradation and the prediction of the compound which an uncultured organism can degrade may not be predicted from analysis of 16S rRNA sequence (Holmes et al. 2006). Thus, inferring the physiology behind the bioremediation processes is even more difficult if closely related organisms are not available in pure culture.

8.2.1.2 Mining the Physiology Through Whole Genome Sequencing

Since 16S rRNA cannot predict the whole physiology of organism therefore genome sequencing approach is being implemented to predict the full potential of organism. The pure culture isolated from the contaminated sites is often useful for bioremediation of organic pollutants. But applying genomics to bioremediation has metamorphosed the process. With high throughput sequencing approaches, it become possible to sequence whole genome of pure cultures which provide crucial information for bioremediation (Srivastava et al. 2021). Genome sequencing is most appropriate tool in understanding the physiology and extremely useful in providing the better insights of bioremediation-relevant microorganisms (Oyewusi et al. 2021). Hong et al. (2017) have isolated marine bacterium *Achromobacter* sp. HZo1 for crude-oil contamination degradation. Further genome sequencing revealed that the bacterium contains complete enzyme system for n-alkane degradation following the terminal oxidation pathway. Genome analysis also revealed the presence of gene related to the biosurfactant production. Therefore, genome data provided the better understanding of the hydrocarbon degrading potential and futuristic design strategies for bioremediating the polluted environment (Hong et al. 2017). Similarly, Jadeja et al. (2019) have implemented the genome sequencing on biosurfactant producing isolates *Acinetobacter* sp. AKBS16 and *Bacillus* sp. AKBS9. The sequencing data has identified the *NPRS* and *sfp* gene in the bacillibactin biosynthetic gene cluster of AKBS9 strain and gene cluster for emulsan biosynthetic in AKBS16 strain

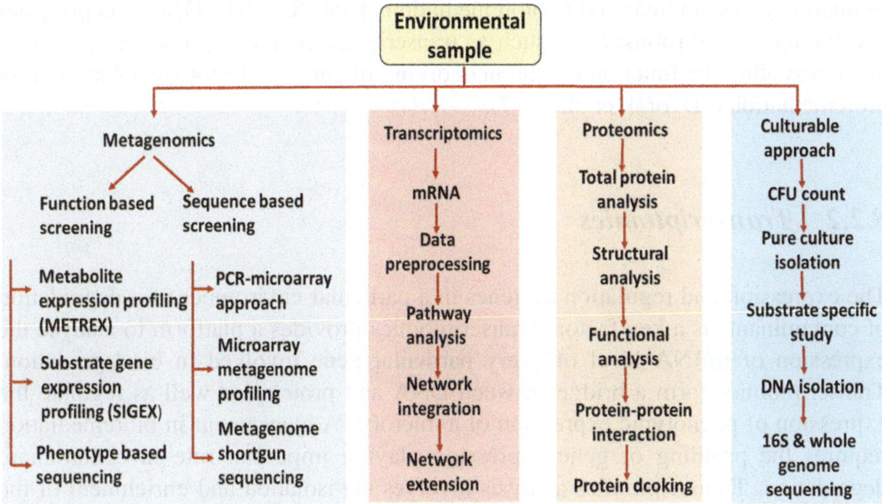

Fig. 8.1 Bioinformatic tools and strategies for characterization of genes and proteins involved in bioremediation

responsible for producing biosurfactant. Apart from this, several other gene involved in degradation of multiple aromatic compounds were also annotated in suggesting that the strains may play several other roles in bioremediation (Jadeja et al. 2019).

Advancement in bioinformatic tools has also spurred the interest in the genomics of micro-organism. Various tools such as Rapid annotation using subsystem technology (RAST) annotates archaeal and bacterial genomes for the protein encoding tRNA and rRNA genes, functional assignments of genes, predicting the subsystems in the genome, uses the genetic information to reconstruct the metabolic network and provides the easily downloadable output for the users (Aziz et al. 2008). Similarly, Prokka was introduced by *Seemann* in 2014 which uses command line and within 10 min annotates the draft bacterial genome. Utilization of bioinformatics tools unravels the hidden characteristics and provides potential for their application in various fields (Seemann 2014) (Fig. 8.1).

8.2.1.3 Applying Omics in Bioremediation

Bioremediation using different microorganism is one of best way to remediate and restore the contaminated environment. However, for enhanced bioremediation the growth and metabolism regulation of these microbes needs deep knowledge for their implementation. The scientific study of past few years has developed a platform embedded with different tools specially OMICs tools to study the physiological and functional factors of microbial system in a more integrated way involved in contaminant biodegradation (Rodríguez et al. 2020). OMICs approaches involve multidisciplinary and integrative research field for exploring the complex processes of microbial system involved in bioremediation (Table 8.1). The OMICs approaches involve the use of robust tools such as transcriptomics, proteomics, metabolomics for unrevealing the functioning and networking of genes and proteins taking part in bioremediations (Dvořák et al. 2017).

8.2.2 Transcriptomics

The expression and regulation of genes in a particular environment for degradation of contaminants is a key factor. Transcriptomics provides a platform to analyse the expression of mRNA level of every particular gene involved in biodegradation. Transcriptomics form a bridge between DNA and proteins as well as regulate the expression of phenotypic expression of a microbe. Advancement in bioremediation requires the profiling of gene expression playing important role in contaminant degradation. Transcriptomics analysis involves (1) isolation and enrichment of the total cellular mRNA, (2) using total mRNA synthesis of (3) either complete cDNA sequencing or microarrays analysis for cDNA hybridization (Singh 2006). DNA microarray is one of the powerful tool of transcriptomics helps to explore and analyse the various anabolic and catabolic gene expression of microbial culture. Whole genome sequencing of microbes made possible to recognize the potent gene involved in biodegradation (Pandey et al. 2019). Vilchez-Vargas et al. (2013) has developed a custom array system with a novel normalization strategy revealing the presence or abundance of catabolic, undetected catabolic groups and precisely their expression under pollutant environment (Vilchez-Vargas et al. 2013). Qiao et al. (2019) showed the application of transcriptomics for analysing the mode of action of perfluorooctane sulfonate on *Phanerochaete chrysosporium. They showed that* PFOS contamination can results into the change in membrane regulating gene structure, cell transportation, cellular redox process, synthesis and metabolism by causing mainly oxidative and membrane damage (Qiao et al. 2019).

8.2.3 Proteomics

The capability of living organisms to adapt change in environment originates from their scope to alter their molecular composition. Microorganisms can accomplish this phenomenon through copious mechanisms that modulate the diversity of molecular archive and their quantitative constitution (Azubuike et al. 2016). Molecular diversity is significantly expressed with respect to proteomes, thereby various proteoforms acquired from the same gene could collectively form distinct protein complexes along with their subsequent expression in the cell. In recent years, the survey of molecular and protein modular divergence and their retaliation towards altered environment has become feasible through novel omics approach i.e. proteomics and interactomics (Dangi et al. 2019).

Table 8.1 Mapping and modifying genes and proteins in microbes for improved bioremediation

Sr. no.	Organism	Approach	Contaminant	Experiment	References
1	*Sphingomonas desiccabilis* and *Bacillus idriensis*	Genomics	Arsenic	Over expression of arsM gene	Liu et al. (2011)
2	*Methylococcus capsulatus*		Chromium (VI)	CrR genes for Cr (VI) reductase activity	Hasin et al. (2010)
3	*B. subtilis* BR151 (pTOO24)		Cadmium	Luminescent cadmium sensors insertion	Ivask et al. (2011)
4	*Brevibacterium epidermidis* EZ-K02		Benzoate, p-hydroxybenzoate, acetophenone, catechol, gentisate	Draft genome sequencing	Ziganshina et al. (2018)
5	*Irpex lacteus*		Aromatic dye	Overexpression of MnP gene	Sun et al. (2016)
6	*Pseudomonas aeruginosa* N6P6		Lead	Over expression of bmtA gene	Kumari and Das (2019)
7	*Sphingomonas desiccabilis* and *Bacillus idriensis*		Arsenic	*arsM* gene shuffling	Liu et al. (2011)
8	*Naegleria*, *Vorticella*, *Arabidopsis*, *Asarum* and *Populus*	Metagenomics	Hydrocarbons	Amplicons sequencing using Illumina's MiSeq platform	Kachienga et al. (2018)
9	*Bacillus*, *Pseudoxanthomonas and Alcanivorax* sp.		Total petroleum hydrocarbons	DGGE and Illumina 16S metagenomic analyses	Koshlaf et al. (2016)
10	Community profiling		Diesel-contaminated microcosms	Metagenomic sequencing and analysis using STAMP, MG-RAST and M5RNA database	Jung et al. (2016)
11	γ- and δ-Proteobacteria and Euryarchaeota		Total petroleum hydrocarbon degradation	Metagenome sequencing Followed by metabolomics analysis	Sarkar et al. (2016)
12	*Thauera* sp., *Pseudoxanthomonas* sp., *Desulfomicrobium* sp., *Ottowia* sp.,		Azo-dye decolorization and degradation	16S rDNA metagenomic libraries construction	Dafale et al. (2010)

(continued)

Table 8.1 (continued)

Sr. no.	Organism	Approach	Contaminant	Experiment	References
	Acidovorax sp., and *Bacteriodetes bacterium* sp.				
13	Antibiotic resistance community		ARGs	ResFinder 3.2 and BLASTn program	Bombaywala et al. (2021)
14	*Geobacter* community	Proteomics	Uranium	Identification of uranium degrading proteins through metaproteomics	Wilkins et al. (2009)
15	*Geobacter* sp.		Uranium	c-type heme-containing protein analysis and overexpression of GscA gene	Yun et al. (2016)
16	*Miscanthus sinensis*		Antimony	Comparative proteomic analysis	Xue et al. (2015)
17	*Pseudomonas aeruginosa*		Cadmium	Shotgun proteomics approach	Izrael-Živković et al. (2018)
18	*Stenotrophomonas rhizophila*, *Xanthomonas retroflexus*, *Microbacterium oxydans* and *Paenibacillus amylolyticus*	Metaproteomics	Enhanced biofilm formation	Protein analysis and cross-feeding on specific amino acids	Herschend et al. (2017)
19	*Pseudomonas putida*, *Sphingobium* spp.	Transcriptomics	Organophosphates, pyrethroids and carbamates	RNA sequencing and DNA chip designing	Bolyen et al. (2019)
20	*Staphylococcus aureus* 29213		*Litsea cubeba* L. essential oil synthesis	Comparative transcriptomic analysis	Yang et al. (2020)
21	*Pseudomonas*, *Methylophaga*, *Pseudidiomarina, Thalassospira and Alcanivorax*		PAH	16S rRNA gene amplicon sequencing	Muangchinda et al. (2018)
22	*Pseudomonas putida* KT2440	Fluxomics	Hydrocarbons	Fluxomic applied for NADPH	Nikel and Chavarría (2015)

The term 'proteomics' was coined in the year 1995, which is considered to be a key post-genomic attribute that appeared through the development of complex datasets by genome sequencing. The analysis carried out in proteomics studies is of relevance as the phenotypes are a result of the protein activity rather genomic sequences. Generally, proteomics is based on efficient protein separation methods using 2-D Polyacrylamide Gel Electrophoresis (PAGE or 2-DE) novel bioinformatic approach along with Mass spectrometry (MS) (Pande et al. 2020). Nevertheless, 2-D PAGE is usually thought of as restricted system which constitutes of membrane proteins (hydrophobic) in proteomics. The process of bioremediation includes, proteome that comprises membrane proteins is of relevance, especially in biodegradation of pollutants, wherein any changes in niche specific microbes affect cell-surface protein receptors. For use in compartmental proteomics, new improvements in 2-DE have been designed by instituting a different approach for Multidimensional Protein Identification Technology (MudPIT) (Aslam et al. 2017).

MS has transformed environmental proteomics approach wherein investigation of compact units of proteins and peptides that in-turn enhanced the responsiveness in protein detection with minimized time and less complicated procedure. The recent developments in this technology along with advancements in other databases have over tuned the conventional proteomics study (Singh et al. 2016). MALDI-TOF-MS has been the frequently used MS system in identification of proteins that are extracted from gel electrophoresis, with the help of mass peptide fingerprinting. SELDI-TOF-MS is termed as the union of fractioned samples loaded on a chip fused with MALDI-TOF-MS analyser. A heterogeneity of characteristically showed targeted peptides were examined with SELDI in *Mytilus edulis* exposed to heavy metals and PAHs (Yadav et al. 2016).

Liquid-chromatography MS technique unlocks a novel analytical aperture for quick detection and characterization of pollutants in water. Also, the by-products of degradation have been addressed to evaluate the predestination of organic pollutants such as PAHs, surfactants, pesticides, cyanobacterial and algal toxins, disinfectants and pharmaceuticals in the ecological niches (Li and Zhu 2020).

8.2.3.1 Tracing the Developments on Bioremediation Using Proteomics Tools

In an organism, with differing environmental conditions the cellular expression of proteins varies. Along with changing external stimuli, the physiological response might change due to the adaptive response in organisms with respect to toxic agents present in the environment. Emergence of proteomic tools has enabled a substantial scrutiny of global switch in the constitution or abundance of proteins also, helped in targeting crucial proteins that are involved with the stimuli of various microorganisms in a differing environmental condition. Several studies have reported that, a set of proteins are responsible for the gene regulation in microbes with respect to the detection of contaminants (Naqvi et al. 2019).

PAHs and other omnipresent environmental contaminants are immensely crucial to eliminate from the ecological niche. Bioremediation of certain PAH contaminants have been partly attained by employing natural as well as microorganisms that have been genetically altered, as discussed by Herschend et al. (2017). By utilizing a proteomics technique, the observed physiological alterations in microbes at the time of bioremediation process provides better understanding for genes related to bioremediation and their regulatory pathways. In a reported study, a 81-kDa peptide that is generally activated during pyrene exposure in *Mycobacterium* sp. was retrieved using 2-DE (Techtmann and Hazen 2016).

8.2.4 Interactomics

Genomic mRNA profiling falls short to give insights into the working, arrangement and final segregation of the proteins or expressed genes. Novel proteomic techniques, have proved to be successful to provide all relevant information on protein activities. In addition, cellular structures are organized and maintained via a protein-protein interaction nexus (Jha et al. 2019), with several proteins participating in aggregation of multicomponent proteins. The detection and evaluation of these protein clusters is termed as 'Interactomics', it is generally based on affinity tag, pull down or MS-MS techniques on a protein level. Various experiments on protein networking, interaction and complex supermolecular structures display functional proteomics and second generation proteomics approach (Kumavath and Deverapalli 2013).

The increasing demand of omics-based applications for gene and protein functional analysis from a perspective of combined bioremediation are escalating the demand for microarray based experiments extensively. Previously, protein microarray technologies have been successfully used for targeting, analysis and quantification of protein functional genes in applied proteome research (Sharma and Shukla 2020). Apart from DNA chip technology, a vast diversity of protein-microarray based techniques have confirmed that interactomic techniques are competent to fill the gaps connecting transcriptomics to proteomics. Nevertheless, in bioremediation, microarray based peptide networking studies are yet to develop in order to recognize the pathways of chemotaxis for any niche specific bacterium contributing to degradation of environmental contaminants. The cellular structures form a vast nexus of DNA-protein or protein-protein interactions (Jhariya et al. 2021). Various proteins, usually from one pathway, form complex networks, interactomics analysis is hence targeted towards the role of such interconnections and their benefaction to cell metabolism changes (Imam et al. 2019). These interactomic products are either analyzed through computational softwares for protein interactions determination or via proteomic based approaches in order to extract proteins and identification with MS methods. The networking among DNA and proteins might regulate gene expression, hence changing metabolic and regulatory pathways. The general approach to understand the interactome nexus projects large scale gene interaction scanning through phenotypic analysis of mutants obtained via gene deletion. Interactomics based studies have showed the occurrence of novel pathways that include uncharacterized gene functions and well characterized gene clusters (Bludau and Aebersold 2020). Conjunctively, these new techniques have improved our knowledge about gene regulatory functions and their networking in various prokaryotes. Analysis based on interactomics of stress induced response systems have the capacity to address several drawbacks of omics approaches discussed above. The protein-DNA interactome analysis data also defines the fundamental gene regulatory functioning in bacterial cellular components. Understanding the networking structure is essential to know the alterations that take place during the transcriptional stage of cell cycle among environmental stress conditions. Furthermore, extensive research focused on understanding the gene interaction networks of bacteria such as *E. coli* and *H. pylori* have been conducted (Kim et al. 2015; Hauser et al. 2014). Additional efforts have constantly been made in order to make high throughput small-molecule screening for identification of molecules interacting and interfering with protein networking (Mabonga and Kappo 2019).

8.3 Metaomics for Unraveling the Unculturable Microorganism

8.3.1 Metagenomics

The complex networking among microbial communities involved in biodegradation of pollutants can be better understood through metagenomic analysis. Metagenomics is referred as community or environmental genomics or eco-genomics and is used for in-depth study of microbial diversity and their functions during biodegradation of polluted environment. The advancements in in field of metagenomics imparts features of key genes/enzymes involved in the biodegradation and detoxification process (Goodwin et al. 2016). Metagenomics utilizes several genome level approaches to characterize and unfold diversity of culturable and importantly unculturable microbes. It also reveals phylogenetically and taxonomically diverse gene operons (Uhlik et al. 2013).

Metagenomic approach often involves two methods such as targeted metagenomics and shotgun metagenomics. Targeted metagenomics involves probing a single gene with the goal to identify all the sequence of that particular gene present in a given environment. The phylogenetic diversity and abundance of selected genes is investigated in an environmental sample (Techtmann and Hazen 2016). Thus, targeted metagenomics registers a in depth shift in diversity and structure of a microbial community prior to and after the impact of environmental perturbation from contaminants (Parmar et al. 2018).

The total genomic content of an environment microbial community is subjected to sequencing in shotgun metagenomics. It seeks to provide a strong correlation between taxonomic classification and functional genes. There are two methods for screening of novel proteins and enzymes based on either gene sequence or function. In the method of sequence based screening, the sequences of probes and primers are based on the sequences of pre-determined coding genes for enzymes or other essential bioactive molecules (Kachienga et al. 2018). The PCR amplification is the used to determine these genes in the sample metagenome and consequently sequenced. For further cross-verification and over expression of these genes, they can be cloned into different gene expression models. The aim of function-based screening is to identify or discover genes with the ability to encode a novel compound that was previously unknown. In function-based screening method, the complete DNA is chopped into smaller fragments, which are ligated to vectors for the purpose of generating libraries. These clones are subsequently screening on the basis of specific phenotypic activity that enables detection of compounds with the required functional activity. Further, the sequence of 16 s RNA from a metagenome helps to identify diversity of microbial population in the sample. The study of comprehensive metagenomic 16 s RNA leads to deep understanding of cultivable as well as uncultivable microbial diversity and its interactions among themselves and with their particular habitat (Simon and Daniel 2011).

8.3.1.1 Case Studies of Metagenomic Approaches for Bioremediation

Metagenomics furthers our knowledge on biodegradation and detoxification mechanisms employed by microbes at the contaminated sites. Prospective microbial degraders or novel catabolic gene families responsible for bioremediation of organic or inorganic xenobiotic can also be identified through metagenomic approaches. A comparative metagenomic approach (Bohra et al. 2019) seeks to determine difference in functional diversity of microbial population at various contaminated sites impacted by a same specific pollutant. Advancement in sequencing technologies like NGS have enabled investigation into deeper layers microbial population and is also crucial for presenting an unbiased insight of taxonomic diversity, abundance and functions of microbial community present in the environment (Zwolinski 2007).

Martín et al. (2006) used function base screening method to unfold metabolic and ecological role of microorganisms through construction of metagenomic libraries. The study identified a dominant *Candidatus Accsumulibacter phosphatis*, which is a polyphosphate accumulating organism (PAO) involved in elevated phosphate removal systems (Martín et al. 2006). Metatranscriptomics has been used by Jennings et al. (2009) to identify upregulated genes in *Polaromonas* sp. JS666 strain during assimilation process of cis-dichloroethane (cDCE) (Jennings et al. 2009). A combination of metagenomic and proteomic approach was used to analyze an aromatic hydrocarbon biodegradation pathway in *Pseudomonas putida* KT 2440. In the study proteomic pipeline utilized 2 dimensional MS and removable isotope encoded affinity tag (Kim et al. 2006).

The study by Wang et al. (2015) emphasized the application of comparative and predictive metagenomics for identifying pollution biomarkers. Software called

MetaBoot was utilized for identification of hydrocarbon pollution biomarkers by comparison of metagenomic datasets retrieved from hydrocarbon contaminated habitats of 255 taxa and 414 functional modules (Wang et al. 2015). An extensive array of hydrocarbon degrading genes discovered from soil and water includes genes that degrade a variety of hydrocarbons including alkanes, naphthalene, methylphenanthrene, naphthalene, phenanthrene and several other aromatics (Dellagnezze et al. 2014). The bioremediation properties of several enzymes have been identified using metagenomic approaches. These enzymes have been identified from metagenomic datasets from soil, cow rumen, artificially polluted soils, chlorinated biphenyl polluted soils, wastewater influent and effluents, activated sludge, compost and oil contaminated water. The identified enzymes are laccases, esterases, dioxygenases, monooxygenases, phenol and polyaromatic hydrocarbon degrading enzymes, trichlorophenol catabolic enzymes, and hexadecane and alkane hydrolyzing enzymes. These enzymes are known to degrade insecticides, pesticides, dyes, plastics, dyes and fumigants (Ufarté et al. 2015).

Novel genes for biosurfactant production in marine bacteria and various strategies for increasing production of biosurfactants can be revealed by metagenomic analysis. For instance, novel biosynthesis pathways of N-acyl amino acids and palmitoyl putrescine have been discovered using metagenomics approach (Jackson et al. 2015; Williams and Trindade 2017).

The screening of microbes from radionucleotide and heavy metal polluted environment have paved way for discovery of novel radionucleotide and heavy metal bioremediating bacteria by the metagenomic analysis. For instance, the predominant microbial species survived in uranium polluted sites was discovered to be *Geobacter* species. Apart from this, several genes conferring resistance to heavy metals and radionucleotides have been identified by employing metagenomics approaches on varied environmental matrices (Hoyos-Hernandez et al. 2019; Xavier et al. 2019).

8.3.1.2 Tools Involved in Metagenomic Approach

A range of in silico softwares, internet resources, bioinformatics applications, and algorithms have been used to find correlation and interpret –omics and molecular data (Desai et al. 2010). Nevertheless, resources solely committed to microbial bioremediation are still limited. The pioneer bioinformatic tool that compiles bioremediation pathway data is a database developed by University of Minnesota called Biocatalysis/Biodegradation Database, which is an inventory of 187 pathways, 1287 reactions, 1195 compounds, 833 enzymes, 491 microorganism entries, and 259 biotransformation rules (http://umbbd.msi.umn.edu/). Another system known as MetaRouter maintains wide-ranging information on biodegradation and bioremediation in a system that permits updating of modification in query (Desai et al. 2010). MetaRouter is an integrated network that stores the information on compounds, chemical reactions, enzymes and microorganisms, thereby forming a relational database. It allows searching for pathways between two sets of chemical compounds. The visualization of possible pathways can be filtered on the basis of length of the pathway, particular enzyme present in a specified organism, intermediate chemical species having a number of values for a given chemical property, for example, solubility, melting point, etc. The web server of MetaRouter is freely accessible and can be administrated and accessed using a web interface (www.pdg.cnb.uam.es/biodeg_net/MetaRouter/) (Pazos et al. 2005). Meta genomc Analyzer (MEGAN) helps to analyze large metagenomic data efficiently. This tool also analyzes large chunk of metatranscriptomic data, functionally and taxonomically. The software program reads onto NCBI datasets and maps reads to SEED, COG and KEGG classifications. The software also uses wide range of visualization schemes such as co-occurrence plots and harbors in-built statistical tools like, PCoA (Principle Coordinate Analysis) and clustering tools (Huson et al. 2016).

Simple Metagenomic Analysis Shell for microbial communities (SmashCommunity) is a known as an annotation and analytical pipeline for metagenomic data. It is feasible for Sanger data and some other 454 sequencing techniques. It readily helps in metagenomic analysis such as assembly, gene prediction, quantitative phylogenetic analysis. It gives optimized parameters through Arachne and Celera for assembly of metagenomic sequences. It also provides GeneMark and MetaGene in order to predict protein coding sequences in metagenomic data. This software uses interactive Tree of Life (iTOL) web tool as visualization software. This software also compares multiple metagenomic data by clustering the profiles based on relative-entropy measure which compares quantitative profiles (Arumugam et al. 2010). Community Cyberinfrastructure for Advanced Microbial Ecology Research and Analysis (CAMERA) is a single platform for depositing, locating, analyzing, visualizing and data sharing about microbial biology via advanced web based analytical portal. CAMERA portably analyzes huge environmental metagenomic data, their environmental parameters and harbors tools for cross-analysis of environmental samples. CAMERA is different in a way as it collects and links metadata of environmental metagenomic data along with annotation which allows users with expressive queries to the database. Also, this allows data submission tools to share and forward data to other metagenomic sites and community archives (Sun et al. 2011).

8.3.2 Metatranscriptomics

Metatranscriptome provides knowledge about the expression and regulation profiling of composite microbial community found in the environment. In metatranscriptomics, extracted mRNA is converted to cDNA and then sequenced on NGS platform. It is an inventory of expressed genes in an environmental sample studied in a defined time and condition by looking at the total mRNA (Aguiar-Pulido et al. 2016). The modern metatranscriptomics techniques involves the use of stable isotope probing (SIP), which can be used for reclaiming the transcriptomes of any sample (Lueders et al. 2016). MetaVelvet, Trans-Abyss, AbySS, Oases, Cufflinks and Scripture are some tools used in metatranscriptomics. The challenges during metatranscriptomic in-situ analysis includes revival of high-quality mRNA, rapid degradation of mRNA and successful partitioning of mRNA and RNA species. These drawbacks can be evaded by direct applications of sequencing of cDNA, also feasibility of direct transcripts quantification can be employed (Simon and Daniel 2011).

8.3.3 Metaproteomics

Metaproteomics involves retrieval and analysis of protein from environmental matrices. It employs high resolution mass spectrometry followed by enzymatic digestion of protein and liquid chromatography (Hettich et al. 2013). Post transitional alterations and other modifications that may impact the protein activity during bioremediation processes can be studied by metaproteomics that provides protein complement found in an environment. It gives a correlation between taxonomic diversity and functional activities of microbes and their impact of functions of ecosystem. The difficulties during analysis of metatrascriptomics includes varying distribution of species, broad range of expression profiles of proteins and immense heterogeneity within the microorganisms (Pandey et al. 2019).

8.4 Metabolomics and Fluxomics

Metabolomics employs the identification and quantification of metabolites released in immediate environment of an organism (Aguiar-Pulido et al. 2016). The transformation of nutrients along with pollutants and health of environment can be grasped through metabolomic studies. It has application in analysis of catabolic pathways and role of microorganism in biodegradation of pollutants. Metabolomics does not depend on sequencing rather on numerous techniques like gas and liquid chromatography, and detection techniques like mass spectrometry and nuclear magnetic resonance that provide a spectrum of peak patterns that helps to identify and quantify metabolic molecules (Pandey et al. 2019). Similarly, fluxomics refers to set of techniques that try to determine rate of reaction inside a biological body. Thus, fluxomics seeks to capture the dynamics of phenotypes and functional interaction between the genome and the environment (Datta et al. 2020).

8.5 Prediction and Reconstruction of Bioremediation Pathways

Microorganism degrade contaminants by expressing genes or proteins in a group, and simultaneously activating the next gene or protein. In microorganisms, genes involved in microbial pathways are usually found in clusters, and during biodegradation their expression is regulated by special regulators which acts either by activation or repression. These gene clusters activated by microbes when they needed (Pande et al. 2020). Advancement in high throughput screening, whole-genome sequencing and information available in large repositories/databases from decades of research have framed a platform to predict the metabolic pathways. These technologies have also been open a way to reconstruct the original pathways or plot robust pathways for conversion of harmful pollutants into less harmful from, accumulated in the environment. There are mainly two approaches used for metabolic pathway reconstruction. (1) In-silico application in which various tools and computational algorithms are used for constructing and designing pathways. (2) Experimental approach which is used for verification of In-silico designed pathways by using various molecular techniques (Park and Choi 2020) (Fig. 8.2).

8.6 In-Silico Tools Used for Pathway Reconstruction

The whole genome sequencing and information related to natural metabolic pathway provides steps to modify them for better bioremediation process. This approach includes the enzymes of different steps and assembly of genes encoding these enzymes to reconstruct metabolic pathways. KEGG (Kanehisa 2017), MetaCyc (Caspi et al. 2018), Rhea (Morgat et al. 2019), and BRENDA (Schomburg et al. 2017) are few databases used for enzyme catalysed biochemical reaction prediction that render an inventory of enzymes for the building of pathways that can be modified with supplementary reactions to bridge in the gaps. Pathways built with these databases only represent chemical metamorphosis and do not take into account the physiological factors of microbes encoding pathways. BLAST (Chen et al. 2015) is a sequence alignment programme that can be employed to find statistically significant similarities among query sequences (nucleotides or proteins) and database sequence collections. This method, on the premise that homologous sequences are likely to encode similar proteins functionally within known and unknown networks.

The automated novel genome-scale metabolic model (GSMM) has also been used for reconstruction of metabolic pathways for enhanced bioremediation processes. GSMM aid the predication of phenotypic feature of microorganisms from their genotypes. GSMM model utilizes genomic information of a particular microorganism with the help of software and data repository for construction of metabolic pathway (Thiele and Palsson 2010). Some software such as KEGG and BioCyc

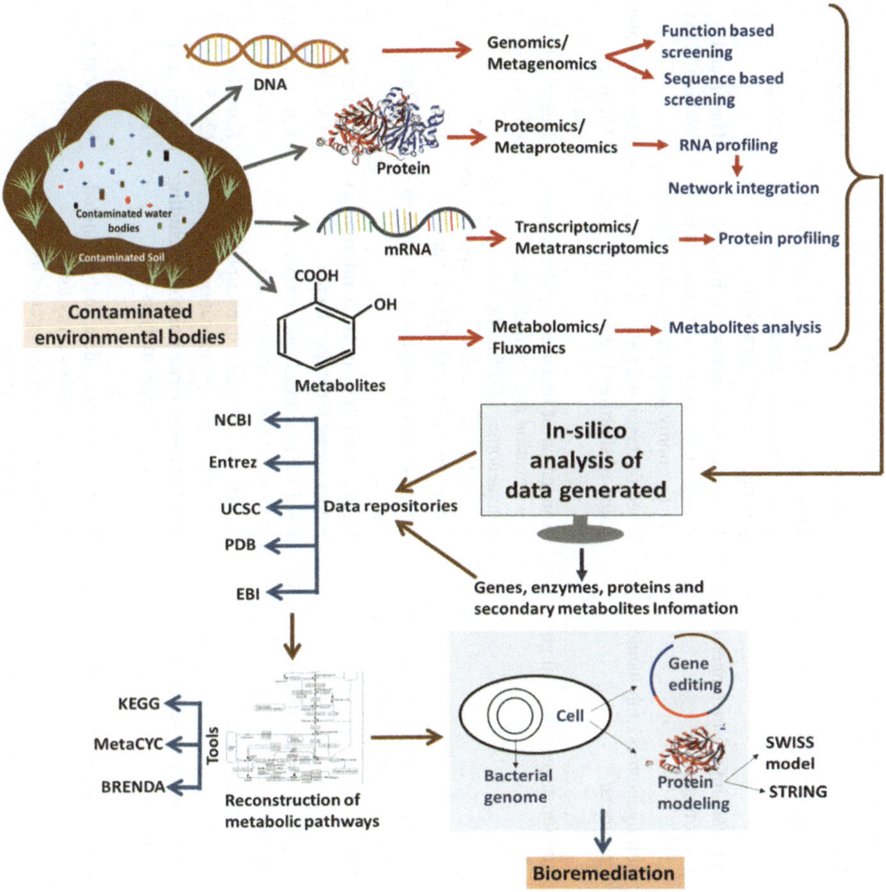

Fig. 8.2 Schematic representation of tools applied for exploring the possible pathways and genes involved in bioremediation

furnish pathways for particular microbe of complete genome (Caspi et al. 2016), and MetCyc reconstruct these metabolic networks on the basis of specific metabolites production. Similarly, KASS has also been used for reconstruction of microbial metabolic pathways. MG-RAST and MEGAN have also been used for assisting high-efficiency pathway reconstruction. Moreover, COBRA (Constraint-Based Reconstruction and Analysis) speculates optimum genomic modification for improving the rate and yield of metabolite production in bioremediation (Heirendt et al. 2019) (Table 8.2).

Advancement in technologies has also provided the way to De novo reconstruct the pathways by using bioinformatic tools (Fig. 8.3). De novo metabolic pathways can be reconstructed by predicting the biochemical reaction with the help of metabolites chemical structures used in the network (Kanehisa 2017). For example, PathPred and UMPPS (University of Minnesota Pathway Prediction System) has

Table 8.2 Tools used for bioinformatic analysis to improve bioremediation processes

Sr. no.	Tools	Approach	URL	Description	References
1	QIIME	Metagenomics	http://qiime.org/	Publically available tool for raw read data, taxonomic annotation and phylogenitic reconstruction	Caporaso et al. (2010)
2	MEGAN		http://ab.inf.uni-tuebingen.de/software/megan5/	Uses graphical interface to BLAST metagenome against available databases for taxonomy and functional analysis	Huson and Weber (2013)
3	Genometa		http://genomics1.mh-hannover.de/genometa/	Used for taxonomic and functional annotation of short-reads metagenomic data	Davenport and Tümmler (2013)
4	MG-RAST		http://metagenomics.anl.gov/	Annotation of functional and taxonomic classification and compares the metagenomes of different sites by using barcharts, trees, tables, heatmaps, circular recruitment plot, and KEGG maps	Keegan et al. (2016)
5	RAMMCAP		http://weizhong-lab.ucsd.edu/rammcap/cgi-bin/rammcap.cgi	Identifies raw read clusters and protein clusters	Li (2009)
6	GEPAS (**ge**ne ex**p**ression pattern **a**nalysis **s**uite)	Transcriptomics	http://www.gepas.org	**Used for analysis of DNA microarray gene expression experiments**	Vaquerizas et al. (2005)
7	TRAPID		http://bioinformatics.psb.ugent.be/webtools/trapid/	Used for functional and comparative analysis of transcriptome data	Van Bel et al. (2013)
8	TRUFA (**TR**anscriptome **u**ser-**f**riendly **a**nalysis)		https://trufa.ifca.es	Annotates raw reads, transcript assembly and expression quantification	Kornobis et al. (2015)
9	**M**icro**A**rray Genome**I**maging and **C**lustering **Tool** (MAGIC tool)		http://www.bio.davidson.edu/MAGIC	**Use for microarray analysis including clustering, image analysis etc.**	Heyer et al. (2005)
10	NIA Array analysis tool		http://lgsun.grc.nia.nih.gov/ANOVA	Use for microarray data analysis including two colour microarray using ANOVA, clustering of genes using three-dimensional graphics	Sharov et al. (2005)

11	AESOP	Proteomics	https://github.com/rohithmohan/aesop	Examine electrostatic interactions in protein, quantitatively compare the similarity based on electrostatic potentials	Harrison et al. (2017)
12	CAB-align (contact area-based alignment)		http://www.pharm.kitasato-u.ac.jp/bmd/bmd/Publications.html	Use to align protein structures, identify homologous sequences at residue level within related similar regions	Terashi and Takeda-Shitaka (2015)
13	MetaProteome analyzer (MPA)			Analysis and interpretation of metaproteomics and proteomics data including shotgun proteomics data analysis, analysis of taxa, ontologies, pathways and enzymes etc.	Muth et al. (2018)
14	SLiMSearch 4 (short linear motifs search)		http://bioware.ucd.ie/slimsearch.html	Identify and annotate functional modules, discover novel motifs	Davey et al. (2010)
15	d-Omix		http://www.biotec.or.th/isl/Domix	Construct protein phylogenetic trees and compare the similarity of proteins based on domain architecture and alignments	Wichadakul et al. (2009)
16	MeltDB	Metabolomics	http://meltdb.cebitec.uni-bielefeld.de	Use statistical analysis and annotation of datasets from metabolomics experiments and data visualization	Neuweger et al. (2008)
17	MetaboAnalyst		https://www.metaboanalyst.ca	It includes statistical metaanalysis, biomarker analysis, pathway analysis, two factor analysis, functional meta analysis etc.	Chong et al. (2018)
18	SECIM tools (south east Center for Integrated Metabolomics			Use python applications to show hierarchical cluster heatmap, principal component analysis, modular modularity clustering, basic statistical analysis methods	Kirpich et al. (2018)
19	MetaMapR		https://metamap.nlm.nih.gov	Web-based local tool for generating biochemical interaction (KEGG), structural similarity etc.	Grapov et al. (2015)

(continued)

Table 8.2 (continued)

Sr. no.	Tools	Approach	URL	Description	References
20	KEGG		https://www.genome.jp/kegg/	It is a database resource for biological system including biochemical pathways, chemical substances etc.	Kanehisa et al. (2016)
21	CGView server	Genomics	http://stothard.afns.ualberta.ca/cgview_server/	Analyze and generates graphical maps for circular genomes including any bacterial, plasmid, chloroplast or mitochondrial genome	Grant and Stothard (2008)
22	GSDraw (gene structure draw server)			Use to draw gene structure schematic diagrams on the basis of gene structure, phylogenetic tree, protein motif etc.	Wang et al. (2013)
23	MiGA (**mi**crobial **g**enomes **a**tlas)		http://microbial-genomes.org/	Use to classify an unknown query genomic sequence, taxonomic classification with available genome sequences	Rodriguez-R et al. (2018)
24	CoreGenes		http://pumpkins.ib3.gmu.edu:8080/CoreGenes or http://www.bif.atcc.org/CoreGenes	Use to compare two genomes to examine the total number of genes in common, percent value of genes in common, identify the unique genes contained in a pair of proteomes	Turner et al. (2013)
25	MTGIpick		http://bioinfo.zstu.edu.cn/MTGI or https://github.com/bioinfo0706/MTGIpick	Identify genomic islands from a single genome which is not annotated	Dai et al. (2016)
26	**G**enome **a**nnotation **t**ransfer **u**tility (GATU)		https://4virology.net/virology-ca-tools/gatu/	Annotates closely related reference genome. Reference genome database are BLASTed against the genome to be annotated	Tcherepanov et al. (2006)

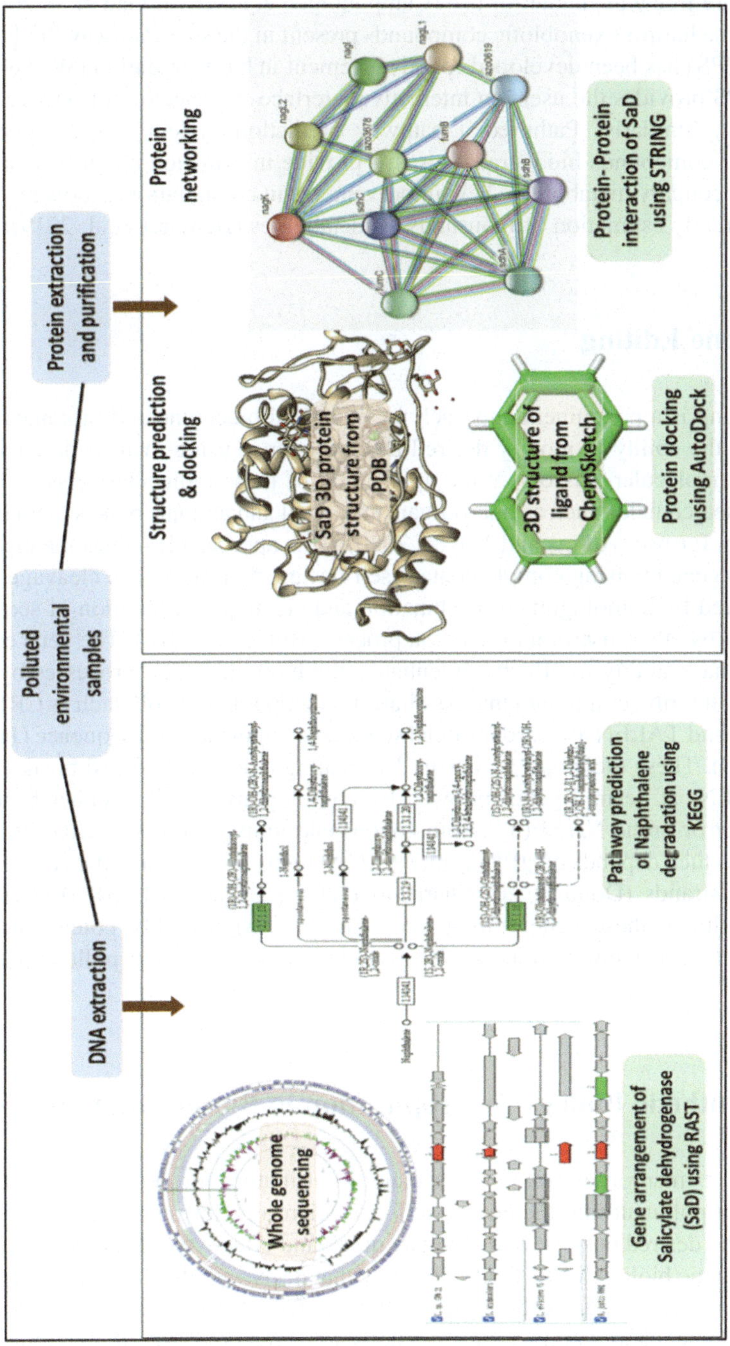

Fig. 8.3 Detailed characterization of Salicylate dehydrogenase using genomics and proteomics tools

been used for predicting pathways by using the chemical structure of metabolites and the tools are freely available for use (Campodonico et al. 2014; Gao et al. 2009).

For some harmful xenobiotic compounds present in the soil, Pathway Predication System (PPS) has been developed for advancement in biodegradation (Wicker et al. 2016). PPS provides the users an interactive interface for selection of only reaction of interest. Similarly, Pathpred a pathway prediction system inform about the xenobiotic compounds biodegradation also provide information about biosynthesis of plant secondary metabolites. Few improvements in computational tools may help in concurrently estimation of thousands of metabolites (Heirendt et al. 2019).

8.7 Gene Editing

Gene editing is a preeminent approach for microbial bioengineered bioremediation field with the ability to modify desired DNA sequence using man-made nucleases known as molecular scissors. Wide application of these molecular scissors in the area of research fields such as microorganisms, plant and animals makes them highly convenient for use (Atkins and Voytas 2020). This approach involves the cleavage of desired gene by using complementary self-designed primers. The cleavage made was repaired by homologous recombination and creating modification of sequence on interest by ether insertion or deletion process (Butt et al. 2018). The gene editing approach have ability to effectively enhance the bioremediation processes by conversion by harmful compound into less-harmful compounds. Tools such as CRISPR-Cas, ZFN and TALEN are used artificial scissors to cut the desired sequence (Jaiswal et al. 2019). These tools basically work by creating a double stranded break (ZFNs and TALEN) at desired gene followed by homologous repairing either by direct repair or error-prone NHEJ (non-homologous end joining). Further, CRISPR-Cas9 approach which depends on gRNA (guide RNA) helps to create specific cuts on both the DNA strands (Doudna and Charpentier 2014; Liang et al. 2015). The high compatibility of these gene editing tools aims to design highly potent microbes with modified metabolic networks for degradation of recalcitrant pollutants (Stein et al. 2018).

8.8 Synthetic Biology for Application in Bioremediation

Enzymes in microorganisms are the major contributors for degrading the highly recalcitrant pollutant into the non-toxic form. Micro-organism can be recircuited to produce the desired enzyme and facilitate quick and effective degradation of pollutants. Synthetic biology is the field of biology that deals with the designing of genetic circuits for performing the biological function (Rylott and Bruce 2020). Various biotechnological and computational methods are being employed to manipulate the host genome and turn them into microbial factories for the desired products. Recent computational methods have increasingly helped in designing the superior biocatalysts for biotransformation and biodegradation of widely spread pollutants. Altering the expression of degradation enzyme or fusing it with other metabolic protein is the most promising approach of synthetic biology (Naqvi et al. 2019). Tay et al. has combined MerR transcriptional regulator with a mercury absorbing operon encoding extracellular protein nanofiber or curli that help in biofilm formation. When mercury ions bind MerR regulator, mercury absorbing protein are expressed along with biofilm formation. The circuit operates only when there is relevant concentration of mercury (Tay et al. 2017). Similar approaches may help in designing the microbial cells that act according to the contaminants present in environment and metabolize them efficiently. In the field of bioremediation, synthetic biology technologies are still under development and previously exciting possibilities offers the use of bioengineered microbes to provide a safe, clean environment. Protein designing and systems biology uses mathematical techniques along with computational tools to model complex biological systems (Mackenzie and Grigoryan 2017). The combination of these techniques with gene editing allows single-base changes to an organism's DNA and expressing the desires protein or enzymes causing degradation of organic pollutants allows the retuning of microbial biological systems. The futuristic approach may involve the re-programming of genetic code by inclusion of unnatural amino acid which will open the way to reprogram the biological systems in order to synthesize the wide range of desired proteins (Rylott and Bruce 2020).

With recent advancements in computational power and underlying algorithms, protein modelling and molecular dynamics simulation is becoming a fundamental tool for investigating the bio-molecular assemblies at molecular level (Srivastava et al. 2020b). Molecular protein docking and molecular dynamics simulation are key resources of computer assisted bioremediation which gives the quick idea about the catalysis simulation of pollutants with desired enzymes on a computer system thus providing the catalytic potential of particular enzymes (Mackenzie and Grigoryan 2017). The molecular docking methods are used for exploring and estimating the binding features of the protein with ligand. In docking, the protein with catalytic center binds to the substrate to form a stable complex with suitable binding conformations. The algorithms of docking analysis, generates binding energy of suitable complexes formed from protein and ligand. Computational bioremediation uses the docking method to screen the pollutants for catalysis interaction, binding affinity, and predict the score for determining the appropriate degrading enzyme (Naqvi et al. 2019).

The pollutants are either organic or inorganic compounds whose degradation in microbial cells vary; and therefore, it is clear that single bioremediation pathway cannot work alone. Identifying the metabolic degradation pathways for overexpression of critical degradation enzymes and application of systems biology to the microorganism is very crucial. This approach will refine the knowledge of existing metabolic pathways for tuning it with efficient and increased microbial degradation of recalcitrant pollutants (Naqvi et al. 2019).

8.9 Future Prospects

The interplay of genes and proteins is crucial factor that need major consideration for effective bioremediation through microbes. For achieving the success in the area of bioremediation to clean-up the environment, the theoretical approaches for regulating the genetic and protein networking need to be epitomized for practical application. The computational tools have been widely explored for understanding the microbial shift and underlying physiology of community inhabiting the contaminated environment. But the knowledge generated by these tools are often complicated and it becomes difficult to infer the meaningful conclusions. The simplified algorithms may help the researchers to fully exploit the in-silico tools for their application in bioremediation. Moreover, there are still many technical loopholes which can be fixed by integrating the advanced methodology. The resources available need considerable progress to enhance the biodegradation capacity by illustrating the degradation pathways, protein-protein interaction or catalytic activity more deeply. The output generated by these data can further be synchronized with web-based algorithms shedding more light on enhanced degradation. The development of new technology for reconstructing the pathway comprises the combinations of various molecular biological tools like gene editing, shuffling and genome editing for the promising results contributing to the bioremediation of toxic contaminants.

8.10 Conclusion

The environment has been exposed to many contaminants, which are mostly chemically derived due to anthropogenic activities. Bioremediation techniques have recently progressed in terms of enzyme systems with degradative abilities However, definite information on bioremediation is lacking from a perspective of current bioremediation practices. Advancement in computational or bioinformatics technology is emerging as an alternating technique, which saves the scientific cost of labor and their precious time. Predictions of biodegradation pathways, molecular docking, and molecular dynamics simulation are the fundamental part of selecting appropriate degradative enzymes involved in bioremediation. Further with the help of pathway prediction tools the enzymes and its regulators can be altered in the desired manner. Developing the mathematical models along with fluxomics application may hold the future for effective bioremediation.

Acknowledgement Authors would like to thank Director, CSIR-NEERI, Nagpur for providing the support and facilities. The manuscript has been checked for plagiarism by Knowledge Resource Centre, CSIR-NEERI, Nagpur, India, and assigned KRC No.: CSIR-NEERI/KRC/2021/April/EBGD/2.

References

Aguiar-Pulido V, Huang W, Suarez-Ulloa V, Cickovski T, Mathee K, Narasimhan G (2016) Metagenomics, metatranscriptomics, and metabolomics approaches for microbiome analysis. Evol Bioinforma 12s1:EBO.S36436. https://doi.org/10.4137/EBO.S36436

Arumugam M, Harrington ED, Foerstner KU, Raes J, Bork P (2010) SmashCommunity: a metagenomic annotation and analysis tool: fig. 1. Bioinformatics 26:2977–2978. https://doi.org/10.1093/bioinformatics/btq536

Aslam B, Basit M, Nisar MA, Khurshid M, Rasool MH (2017) Proteomics: technologies and their applications. J Chromatogr Sci. https://doi.org/10.1093/chromsci/bmw167

Atkins PA, Voytas DF (2020) Overcoming bottlenecks in plant gene editing. Curr Opin Plant Biol. https://doi.org/10.1016/j.pbi.2020.01.002

Aziz RK, Bartels D, Best A, DeJongh M, Disz T, Edwards RA, Formsma K, Gerdes S, Glass EM, Kubal M, Meyer F, Olsen GJ, Olson R, Osterman AL, Overbeek RA, McNeil LK, Paarmann D, Paczian T, Parrello B, Pusch GD, Reich C, Stevens R, Vassieva O, Vonstein V, Wilke A, Zagnitko O (2008) The RAST server: rapid annotations using subsystems technology. BMC Genomics 9:75. https://doi.org/10.1186/1471-2164-9-75

Azubuike CC, Chikere CB, Okpokwasili GC (2016) Bioremediation techniques–classification based on site of application: principles, advantages, limitations and prospects. World J Microbiol Biotechnol. https://doi.org/10.1007/s11274-016-2137-x

Bharagava RN, Purchase D, Saxena G, Mulla SI (2018) Applications of metagenomics in microbial bioremediation of pollutants: from genomics to environmental cleanup. From genomics to environmental cleanup. In: Microbial diversity in the genomic era. Elsevier, pp 459–477. https://doi.org/10.1016/B978-0-12-814849-5.00026-5

Bhatt P, Rene ER, Kumar AJ, Kumar AJ, Zhang W, Chen S (2020) Binding interaction of allethrin with esterase: bioremediation potential and mechanism. Bioresour Technol 315:13845

Bludau I, Aebersold R (2020) Proteomic and interactomic insights into the molecular basis of cell functional diversity. Nat Rev Mol Cell Biol 21:327–340. https://doi.org/10.1038/s41580-020-0231-2

Bohra V, Dafale NA, Purohit HJ (2019) Understanding the alteration in rumen microbiome and CAZymes profile with diet and host through comparative metagenomic approach. Arch Microbiol 201:1385–1397. https://doi.org/10.1007/s00203-019-01706-z

Bolyen E, Rideout JR, Dillon MR et al (2019) Reproducible, interactive, scalable and extensible microbiome data science using QIIME 2. Nat Biotechnol 37:852–857. https://doi.org/10.1038/s41587-019-0209-9

Bombaywala S, Dafale NA, Jha V, Bajaj A, Purohit HJ (2021) Study of indiscriminate distribution of restrained antimicrobial resistome of different environmental niches. Environ Sci Pollut Res 28:10780–10790. https://doi.org/10.1007/s11356-020-11318-6

Butt H, Jamil M, Wang JY, Al-Babili S, Mahfouz M (2018) Engineering plant architecture via CRISPR/Cas9-mediated alteration of strigolactone biosynthesis. BMC Plant Biol 18:174. https://doi.org/10.1186/s12870-018-1387-1

Campodonico MA, Andrews BA, Asenjo JA, Palsson BO, Feist AM (2014) Generation of an atlas for commodity chemical production in Escherichia coli and a novel pathway prediction algorithm. GEM-Path Metab Eng 25:140–158. https://doi.org/10.1016/j.ymben.2014.07.009

Caporaso JG, Kuczynski J, Stombaugh J, Bittinger K, Bushman FD, Costello EK, Fierer N, P?a AG, Goodrich JK, Gordon JI, Huttley GA, Kelley ST, Knights D, Koenig JE, Ley RE, Lozupone CA, McDonald D, Muegge BD, Pirrung M, Reeder J, Sevinsky JR, Turnbaugh PJ, Walters WA, Widmann J, Yatsunenko T, Zaneveld J, Knight R (2010). QIIME allows analysis of high-throughput community sequencing data. Nat Methods. https://doi.org/10.1038/nmeth.f.303

Caspi R, Billington R, Ferrer L, Foerster H, Fulcher CA, Keseler IM, Kothari A, Krummenacker M, Latendresse M, Mueller LA, Ong Q, Paley S, Subhraveti P, Weaver DS, Karp PD (2016) The

MetaCyc database of metabolic pathways and enzymes and the BioCyc collection of pathway/genome databases. Nucleic Acids Res 44:D471–D480. https://doi.org/10.1093/nar/gkv1164
Caspi R, Billington R, Fulcher CA, Keseler IM, Kothari A, Krummenacker M, Latendresse M, Midford PE, Ong Q, Ong WK, Paley S, Subhraveti P, Karp PD (2018) The MetaCyc database of metabolic pathways and enzymes. Nucleic Acids Res 46:D633–D639. https://doi.org/10.1093/nar/gkx935
Chen Y, Ye W, Zhang Y, Xu Y (2015) High speed BLASTN: an accelerated MegaBLAST search tool. Nucleic Acids Res 43:7762–7768. https://doi.org/10.1093/nar/gkv784
Chong J, Soufan O, Li C, Caraus I, Li S, Bourque G, Wishart DS, Xia J (2018) MetaboAnalyst 4.0: towards more transparent and integrative metabolomics analysis. Nucleic Acids Res 46:W486–W494. https://doi.org/10.1093/nar/gky310
Dafale N, Wate S, Meshram S, Nandy T (2008) Kinetic study approach of remazol black-B use for the development of two-stage anoxic-oxic reactor for decolorization/biodegradation of azo dyes by activated bacterial consortium. J Hazard Mater 159:319–328. https://doi.org/10.1016/j.jhazmat.2008.02.058
Dafale N, Agrawal L, Kapley A, Meshram S, Purohit H, Wate S (2010) Selection of indicator bacteria based on screening of 16S rDNA metagenomic library from a two-stage anoxic-oxic bioreactor system degrading azo dyes. Bioresour Technol 101:476–484. https://doi.org/10.1016/j.biortech.2009.08.006
Dai Q, Bao C, Hai Y, Ma S, Zhou T, Wang C, Wang Y, Huo W, Liu X, Yao Y, Xuan Z, Chen M, Zhang MQ (2016) MTGIpick allows robust identification of genomic islands from a single genome. Brief. Bioinform 19:bbw118. https://doi.org/10.1093/bib/bbw118
Dangi AK, Sharma B, Hill RT, Shukla P (2019) Bioremediation through microbes: systems biology and metabolic engineering approach. Crit Rev Biotechnol 39:79–98. https://doi.org/10.1080/07388551.2018.1500997
Datta S, Rajnish KN, Samuel MS, Pugazlendhi A, Selvarajan E (2020) Metagenomic applications in microbial diversity, bioremediation, pollution monitoring, enzyme and drug discovery. A review. Environ Chem Lett. https://doi.org/10.1007/s10311-020-01010-z
Davenport CF, Tümmler B (2013) Advances in computational analysis of metagenome sequences. Environ Microbiol 15:1–5. https://doi.org/10.1111/j.1462-2920.2012.02843.x
Davey NE, Haslam NJ, Shields DC, Edwards RJ (2010) SLiMFinder: a web server to find novel, significantly over-represented, short protein motifs. Nucleic Acids Res 38:W534–W539. https://doi.org/10.1093/nar/gkq440
Debbarma P, Raghuwanshi S, Singh J, Suyal DC, Zaidi MGH, Goel R (2017) Comparative in situ biodegradation studies of polyhydroxybutyrate film composites. 3Biotech. 7(178):1–9. https://doi.org/10.1007/s13205-017-0789-3
Dellagnezze BM, de Sousa GV, Martins LL, Domingos DF, Limache EEG, de Vasconcellos SP, da Cruz GF, de Oliveira VM (2014) Bioremediation potential of microorganisms derived from petroleum reservoirs. Mar Pollut Bull 89:191–200. https://doi.org/10.1016/j.marpolbul.2014.10.003
Deng Z, Jiang Y, Chen K, Gao F, Liu X (2020) Petroleum depletion property and microbial community shift after bioremediation using bacillus halotolerans T-04 and Bacillus cereus 1-1. Front Microbiol 11:353. https://doi.org/10.3389/fmicb.2020.00353
Desai C, Pathak H, Madamwar D (2010) Advances in molecular and "-omics" technologies to gauge microbial communities and bioremediation at xenobiotic/anthropogen contaminated sites. Bioresour Technol. https://doi.org/10.1016/j.biortech.2009.10.080
Doudna JA, Charpentier E (2014) The new frontier of genome engineering with CRISPR-Cas9. Science (80) 346:1258096. https://doi.org/10.1126/science.1258096
Dvořák P, Nikel PI, Damborský J, de Lorenzo V (2017) Bioremediation 3.0: engineering pollutant-removing bacteria in the times of systemic biology. Biotechnol Adv. https://doi.org/10.1016/j.biotechadv.2017.08.001

Fuentes S, Méndez V, Aguila P, Seeger M (2014) Bioremediation of petroleum hydrocarbons: catabolic genes, microbial communities, and applications. Appl Microbiol Biotechnol. https://doi.org/10.1007/s00253-014-5684-9

Gao J, Ellis LBM, Wackett LP (2009) The University of Minnesota Biocatalysis/biodegradation database: improving public access. Nucleic Acids Res 38:D488–D491. https://doi.org/10.1093/nar/gkp771

Goodwin S, McPherson JD, McCombie WR (2016) Coming of age: ten years of next-generation sequencing technologies. Nat Rev Genet. https://doi.org/10.1038/nrg.2016.49

Grant JR, Stothard P (2008) The CGView server: a comparative genomics tool for circular genomes. Nucleic Acids Res 36:181–184. https://doi.org/10.1093/nar/gkn179

Grapov D, Wanichthanarak K, Fiehn O (2015) MetaMapR: pathway independent metabolomic network analysis incorporating unknowns: fig. 1. Bioinformatics 31:2757–2760. https://doi.org/10.1093/bioinformatics/btv194

Harrison RES, Mohan RR, Gorham RD, Kieslich CA, Morikis D (2017) AESOP: a python library for investigating electrostatics in protein interactions. Biophys J 112:1761–1766. https://doi.org/10.1016/j.bpj.2017.04.005

Hasin AAL, Gurman SJ, Murphy LM, Perry A, Smith TJ, Gardiner PHE (2010) Remediation of chromium(VI) by a methane-oxidizing bacterium. Environ Sci Technol 44:400–405. https://doi.org/10.1021/es901723c

Hauser R, Ceol A, Rajagopala SV, Mosca R, Siszler G, Wermke N, Sikorski P, Schwarz F, Schick M, Wuchty S, Aloy P, Uetz P (2014) A second-generation protein-protein interaction network of helicobacter pylori. Mol Cell Proteomics 13:1318–1329. https://doi.org/10.1074/mcp.O113.033571

Heirendt L, Arreckx S, Pfau T et al (2019) Creation and analysis of biochemical constraint-based models using the COBRA toolbox v.3.0. Nat Protoc 14:639–702. https://doi.org/10.1038/s41596-018-0098-2

Herschend J, Damholt ZBV, Marquard AM, Svensson B, Sørensen SJ, Hägglund P, Burmølle M (2017) A meta-proteomics approach to study the interspecies interactions affecting microbial biofilm development in a model community. Sci Rep 7:1–13. https://doi.org/10.1038/s41598-017-16633-6

Hettich RL, Pan C, Chourey K, Giannone RJ (2013) Metaproteomics: harnessing the power of high performance mass spectrometry to identify the suite of proteins that control metabolic activities in microbial communities. Anal Chem 85:4203–4214. https://doi.org/10.1021/ac303053e

Heyer LJ, Moskowitz DZ, Abele JA, Karnik P, Choi D, Campbell AM, Oldham EE, Akin BK (2005) MAGIC tool: integrated microarray data analysis. Bioinformatics 21:2114–2115. https://doi.org/10.1093/bioinformatics/bti247

Holmes VF, He J, Lee PKH, Alvarez-Cohen L (2006) Discrimination of multiple Dehalococcoides strains in a trichloroethene enrichment by quantification of their reductive dehalogenase genes. Appl Environ Microbiol 72:5877–5883. https://doi.org/10.1128/AEM.00516-06

Hong YH, Ye CC, Zhou QZ, Wu XY, Yuan JP, Peng J, Deng H, Wang JH (2017) Genome sequencing reveals the potential of Achromobacter sp. HZ01 for bioremediation. Front Microbiol 8:1–14. https://doi.org/10.3389/fmicb.2017.01507

Hoyos-Hernandez C, Courbert C, Simonucci C, David S, Vogel TM, Larose C (2019) Community structure and functional genes in radionuclide contaminated soils in Chernobyl and Fukushima. FEMS Microbiol Lett 366:180. https://doi.org/10.1093/femsle/fnz180

Huson DH, Weber N (2013) Microbial community analysis using MEGAN. In: Methods in enzymology. Academic, pp 465–485. https://doi.org/10.1016/B978-0-12-407863-5.00021-6

Huson DH, Beier S, Flade I, Górska A, El-Hadidi M, Mitra S, Ruscheweyh H-J, Tappu R (2016) MEGAN Community edition - interactive exploration and analysis of large-scale microbiome sequencing data. PLoS Comput Biol 12:e1004957. https://doi.org/10.1371/journal.pcbi.1004957

Imam A, Suman SK, Ghosh D, Kanaujia PK (2019) Analytical approaches used in monitoring the bioremediation of hydrocarbons in petroleum-contaminated soil and sludge. TrAC – Trends Anal Chem. https://doi.org/10.1016/j.trac.2019.05.023

Ivask A, Dubourguier H-C, Lee P, Kahru A (2011) SOILS, SEC 1 • SOIL ORGANIC MATTER DYNAMICS AND NUTRIENT CYCLING • bioavailability of cd in 110 polluted topsoils to recombinant bioluminescent sensor bacteria: effect of soil particulate matter. J Soils Sediments 11:231–237. https://doi.org/10.1007/s11368-010-0292-5

Izrael-Živković L, Rikalović M, Gojgić-Cvijović G, Kazazić S, Vrvić M, Brčeski I, Beškoski V, Lončarević B, Gopčević K, Karadžić I (2018) Cadmium specific proteomic responses of a highly resistant: *Pseudomonas aeruginosa* san ai. RSC Adv 8:10549–10560. https://doi.org/10.1039/c8ra00371h

Jackson SA, Borchert E, O'Gara F, Dobson ADW (2015) Metagenomics for the discovery of novel biosurfactants of environmental interest from marine ecosystems. Curr Opin Biotechnol. https://doi.org/10.1016/j.copbio.2015.03.004

Jadeja NB, Moharir P, Kapley A (2019) Genome sequencing and analysis of strains bacillus sp. AKBS9 and Acinetobacter sp. AKBS16 for biosurfactant production and bioremediation. Appl Biochem Biotechnol 187:518–530. https://doi.org/10.1007/s12010-018-2828-x

Jaiswal S, Singh DK, Shukla P (2019) Gene editing and systems biology tools for pesticide bioremediation: a review. Front Microbiol 10. https://doi.org/10.3389/fmicb.2019.00087

Jennings LK, Chartrand MMG, Lacrampe-Couloume G, Lollar BS, Spain JC, Gossett JM (2009) Proteomic and transcriptomic analyses reveal genes upregulated by cis-Dichloroethene in Polaromonas sp. strain JS666. Appl Environ Microbiol 75:3733–3744. https://doi.org/10.1128/AEM.00031-09

Jha V, Dafale NA, Purohit HJ (2019) Regulatory rewiring through global gene regulations by PhoB and alarmone (p) ppGpp under various stress conditions. Microbiol Res 227:126309. https://doi.org/10.1016/j.micres.2019.126309

Jhariya U, Dafale NA, Srivastava S, Bhende RS, Kapley A, Purohit HJ (2021) Understanding ethanol tolerance mechanism in Saccharomyces cerevisiae to enhance the bioethanol. Production: current and future prospects. Bioenergy Res. https://doi.org/10.1007/s12155-020-10228-2

Johnson JS, Spakowicz DJ, Hong BY, Petersen LM, Demkowicz P, Chen L, Leopold SR, Hanson BM, Agresta HO, Gerstein M, Sodergren E, Weinstock GM (2019) Evaluation of 16S rRNA gene sequencing for species and strain-level microbiome analysis. Nat Commun 10:1–12. https://doi.org/10.1038/s41467-019-13036-1

Jung J, Philippot L, Park W (2016) Metagenomic and functional analyses of the consequences of reduction of bacterial diversity on soil functions and bioremediation in diesel-contaminated microcosms. Sci Rep 6:1–10. https://doi.org/10.1038/srep23012

Kachienga L, Jitendra K, Momba M (2018) Metagenomic profiling for assessing microbial diversity and microbial adaptation to degradation of hydrocarbons in two south African petroleum-contaminated water aquifers /631/80 /631/326 /704/172 /13 /13/89 /82 /82/58 /82/75 /82/80 / 82/83 /82/16 article. Sci Rep 8:7564. https://doi.org/10.1038/s41598-018-25961-0

Kalaimurugan D, Balamuralikrishnan B, Durairaj K, Vasudhevan P, Shivakumar MS, Kaul T, Chang SW, Ravindran B, Venkatesan S (2020) Isolation and characterization of heavy-metal-resistant bacteria and their applications in environmental bioremediation. Int J Environ Sci Technol 17:1455–1462. https://doi.org/10.1007/s13762-019-02563-5

Kanehisa M (2017) Enzyme annotation and metabolic reconstruction using KEGG. In: Methods in molecular biology. Humana Press, pp 135–145. https://doi.org/10.1007/978-1-4939-7015-5_11

Kanehisa M, Sato Y, Kawashima M, Furumichi M, Tanabe M (2016) KEGG as a reference resource for gene and protein annotation. Nucleic Acids Res 44:D457–D462. https://doi.org/10.1093/nar/gkv1070

Keegan KP, Glass EM, Meyer F (2016) MG-RAST, a metagenomics service for analysis of microbial community structure and function. In: Methods in molecular biology. Humana Press, pp 207–233. https://doi.org/10.1007/978-1-4939-3369-3_13

Kim YH, Cho K, Yun S-H, Kim JY, Kwon K-H, Yoo JS, Kim SI (2006) Analysis of aromatic catabolic pathways in *Pseudomonas putida* KT 2440 using a combined proteomic approach: 2-DE/MS and cleavable isotope-coded affinity tag analysis. Proteomics 6:1301–1318. https://doi.org/10.1002/pmic.200500329

Kim H, Shim JE, Shin J, Lee I (2015) EcoliNet: a database of cofunctional gene network for *Escherichia coli*. Database 2015. https://doi.org/10.1093/database/bav001

Kirpich AS, Ibarra M, Moskalenko O, Fear JM, Gerken J, Mi X, Ashrafi A, Morse AM, McIntyre LM (2018) SECIMTools: a suite of metabolomics data analysis tools. BMC Bioinformatics 19:151. https://doi.org/10.1186/s12859-018-2134-1

Kornobis E, Cabellos L, Aguilar F, Frías-López C, Rozas J, Marco J, Zardoya R (2015) TRUFA: a user-friendly web server for *de novo* RNA-seq analysis using cluster computing. Evol Bioinforma 11:EBO.S23873. https://doi.org/10.4137/EBO.S23873

Koshlaf E, Shahsavari E, Aburto-Medina A, Taha M, Haleyur N, Makadia TH, Morrison PD, Ball AS (2016) Bioremediation potential of diesel-contaminated Libyan soil. Ecotoxicol Environ Saf 133:297–305. https://doi.org/10.1016/j.ecoenv.2016.07.027

Kumar SS, Shantkriti S, Muruganandham T, Murugesh E, Rane N, Govindwar SP (2016) Bioinformatics aided microbial approach for bioremediation of wastewater containing textile dyes. Ecol Inform 31:112–121. https://doi.org/10.1016/j.ecoinf.2015.12.001

Kumar S, Suyal DC, Yadav A, Shouche Y, Goel R (2019) Microbial diversity and soil physiochemical characteristic of higher altitude. PLoS One 14(3):e0213844

Kumari S, Das S (2019) Expression of metallothionein encoding gene bmtA in biofilm-forming marine bacterium *Pseudomonas aeruginosa* N6P6 and understanding its involvement in Pb (II) resistance and bioremediation. Environ Sci Pollut Res 26:28763–28774. https://doi.org/10.1007/s11356-019-05916-2

Kumavath RN, Deverapalli P (2013) Scientific swift in bioremediation: an overview, in: applied bioremediation – active and passive approaches. INTECH. https://doi.org/10.5772/56409

Li W (2009) Analysis and comparison of very large metagenomes with fast clustering and functional annotation. BMC Bioinformatics 10:359. https://doi.org/10.1186/1471-2105-10-359

Li J, Zhu H-J (2020) Liquid chromatography-tandem mass spectrometry (LC-MS/MS)-based proteomics of drug-metabolizing enzymes and transporters. Molecules 25:2718. https://doi.org/10.3390/molecules25112718

Liang P, Xu Y, Zhang X, Ding C, Huang R, Zhang Z, Lv J, Xie X, Chen Y, Li Y, Sun Y, Bai Y, Songyang Z, Ma W, Zhou C, Huang J (2015) CRISPR/Cas9-mediated gene editing in human tripronuclear zygotes. Protein Cell 6:363–372. https://doi.org/10.1007/s13238-015-0153-5

Liu S, Zhang F, Chen J, Sun G (2011) Arsenic removal from contaminated soil via biovolatilization by genetically engineered bacteria under laboratory conditions. J Environ Sci 23:1544–1550. https://doi.org/10.1016/S1001-0742(10)60570-0

Lovley DR (2003) Cleaning up with genomics: applying molecular biology to bioremediation. Nat Rev Microbiol 1:35–44. https://doi.org/10.1038/nrmicro731

Lueders T, Dumont MG, Bradford L, Manefield M (2016) RNA-stable isotope probing: from carbon flow within key microbiota to targeted transcriptomes. Curr Opin Biotechnol. https://doi.org/10.1016/j.copbio.2016.05.001

Mabonga L, Kappo AP (2019) Protein-protein interaction modulators: advances, successes and remaining challenges. Biophys Rev. https://doi.org/10.1007/s12551-019-00570-x

Mackenzie CO, Grigoryan G (2017) Protein structural motifs in prediction and design. Curr Opin Struct Biol 44:161–167. https://doi.org/10.1016/j.sbi.2017.03.012

Martín HG, Ivanova N, Kunin V, Warnecke F, Barry KW, McHardy AC, Yeates C, He S, Salamov AA, Szeto E, Dalin E, Putnam NH, Shapiro HJ, Pangilinan JL, Rigoutsos I, Kyrpides NC, Blackall LL, McMahon KD, Hugenholtz P (2006) Metagenomic analysis of two enhanced biological phosphorus removal (EBPR) sludge communities. Nat Biotechnol 24:1263–1269. https://doi.org/10.1038/nbt1247

Morgat A, Lombardot T, Coudert E, Axelsen K, Neto TB, Gehant S, Bansal P, Bolleman J, Gasteiger E, de Castro E, Baratin D, Pozzato M, Xenarios I, Poux S, Redaschi N, Bridge A

(2019) Enzyme annotation in UniProtKB using Rhea. Bioinformatics 36:1896–1901. https://doi.org/10.1093/bioinformatics/btz817

Muangchinda C, Rungsihiranrut A, Prombutara P, Soonglerdsongpha S, Pinyakong O (2018) 16S metagenomic analysis reveals adaptability of a mixed-PAH-degrading consortium isolated from crude oil-contaminated seawater to changing environmental conditions. J Hazard Mater 357:119–127. https://doi.org/10.1016/j.jhazmat.2018.05.062

Muth T, Kohrs F, Heyer R, Benndorf D, Rapp E, Reichl U, Martens L, Renard BY (2018) MPA portable: a stand-alone software package for analyzing Metaproteome samples on the go. Anal Chem 90:685–689. https://doi.org/10.1021/acs.analchem.7b03544

Naqvi AAT, Mohammad T, Hasan GM, Hassan MI (2019) Advancements in docking and molecular dynamics simulations towards ligand-receptor interactions and structure-function relationships. Curr Top Med Chem 18:1755–1768. https://doi.org/10.2174/1568026618666181025114157

Neuweger H, Albaum SP, Dondrup M, Persicke M, Watt T, Niehaus K, Stoye J, Goesmann A (2008) MeltDB: a software platform for the analysis and integration of metabolomics experiment data. Bioinformatics 24:2726–2732. https://doi.org/10.1093/bioinformatics/btn452

Nikel PI, Chavarría M (2015) Quantitative physiology approaches to understand and optimize reducing power availability in environmental bacteria. Springer, Berlin, pp 39–70. https://doi.org/10.1007/8623_2015_84

Oyewusi HA, Wahab RA, Huyop F (2021) Whole genome strategies and bioremediation insight into dehalogenase-producing bacteria. Mol Biol Rep. https://doi.org/10.1007/s11033-021-06239-7

Pande V, Pandey SC, Sati D, Pande V, Samant M (2020) Bioremediation: an emerging effective approach towards environment restoration. Environ Sustain 3:91–103. https://doi.org/10.1007/s42398-020-00099-w

Pandey A, Tripathi PH, Tripathi AH, Pandey SC, Gangola S (2019) Omics technology to study bioremediation and respective enzymes. In: Smart bioremediation technologies: microbial enzymes. Elsevier, pp 23–43. https://doi.org/10.1016/B978-0-12-818307-6.00002-0

Park H, Choi IG (2020) Genomic and transcriptomic perspectives on mycoremediation of polycyclic aromatic hydrocarbons. Appl Microbiol Biotechnol. https://doi.org/10.1007/s00253-020-10746-1

Parmar K, Dafale N, Pal R, Tikariha H, Purohit H (2018) An insight into phage diversity at environmental habitats using comparative metagenomics approach. Curr Microbiol 75:132–141. https://doi.org/10.1007/s00284-017-1357-0

Pazos F, Guijas D, Valencia A, De Lorenzo V (2005) MetaRouter: bioinformatics for bioremediation. Nucleic Acids Res. academic.oup.com

Qiao W, Zhang Y, Xie Z, Luo Y, Zhang X, Sang C, Xie S, Huang J (2019) Toxicity of perfluorooctane sulfonate on *Phanerochaete chrysosporium*: growth, pollutant degradation and transcriptomics. Ecotoxicol Environ Saf 174:66–74. https://doi.org/10.1016/j.ecoenv.2019.02.066

Rawat M, Rangarajan S (2019) Omics approaches for elucidating molecular mechanisms of microbial bioremediation. In: Smart bioremediation technologies: microbial enzymes. Elsevier, pp 191–203. https://doi.org/10.1016/B978-0-12-818307-6.00011-1

Rodríguez A, Castrejón-Godínez ML, Salazar-Bustamante E, Gama-Martínez Y, Sánchez-Salinas E, Mussali-Galante P, Tovar-Sánchez E, Ortiz-Hernández ML (2020) Omics approaches to pesticide biodegradation. Curr Microbiol. https://doi.org/10.1007/s00284-020-01916-5

Rodriguez-R LM, Gunturu S, Harvey WT, Rosselló-Mora R, Tiedje JM, Cole JR, Konstantinidis KT (2018) The microbial genomes atlas (MiGA) webserver: taxonomic and gene diversity analysis of archaea and bacteria at the whole genome level. Nucleic Acids Res 46:W282–W288. https://doi.org/10.1093/nar/gky467

Rosanti D, Wibowo YG, Safri M, Maryani AT, Ramadhan BS (2020) Bioremediations technologies on wastewater treatment: opportunities, challenges and economic perspective. Sainmatika J Ilm Mat Dan Ilmu Pengetah Alam 17:142. https://doi.org/10.31851/sainmatika.v17i2.5085

Rylott EL, Bruce NC (2020) How synthetic biology can help bioremediation. Curr Opin Chem Biol 58:86–95. https://doi.org/10.1016/j.cbpa.2020.07.004

Sarkar J, Kazy SK, Gupta A, Dutta A, Mohapatra B, Roy A, Bera P, Mitra A, Sar P (2016) Biostimulation of indigenous microbial community for bioremediation of petroleum refinery sludge. Front Microbiol 7:1407. https://doi.org/10.3389/fmicb.2016.01407

Saxena P, Hiwrale I, Das S, Shukla V, Tyagi L, Pal S, Dafale N, Dhodapkar R (2021) Profiling of emerging contaminants and antibiotic resistance in sewage treatment plants: an Indian perspective. J Hazard Mater 408:124877. https://doi.org/10.1016/j.jhazmat.2020.124877

Schomburg I, Jeske L, Ulbrich M, Placzek S, Chang A, Schomburg D (2017) The BRENDA enzyme information system–from a database to an expert system. J Biotechnol. https://doi.org/10.1016/j.jbiotec.2017.04.020

Seemann T (2014) Prokka: rapid prokaryotic genome annotation. Bioinformatics 30:2068–2069. https://doi.org/10.1093/bioinformatics/btu153

Sharma B, Shukla P (2020) Designing synthetic microbial communities for effectual bioremediation: a review. Biocatal Biotransformation. https://doi.org/10.1080/10242422.2020.1813727

Sharov AA, Dudekula DB, Ko MSH (2005) A web-based tool for principal component and significance analysis of microarray data. Bioinformatics 21:2548–2549. https://doi.org/10.1093/bioinformatics/bti343

Simon C, Daniel R (2011) Metagenomic analyses: past and future trends. Appl Environ Microbiol. https://doi.org/10.1128/AEM.02345-10

Singh OV (2006) Proteomics and metabolomics: the molecular make-up of toxic aromatic pollutant bioremediation. Proteomics. https://doi.org/10.1002/pmic.200600200

Singh R, Srivastava M, Shukla A (2016) Environmental sustainability of bioethanol production from rice straw in India: a review. Renew Sustain Energy Rev. https://doi.org/10.1016/j.rser.2015.10.005

Singh M, Singh D, Rai P, Suyal DC, Saurabh S, Soni R, Giri K, Yadav AN (2021) Fungi in remediation of hazardous wastes: current status and future. In: Yadav AN (ed) Recent trends in mycological research, fungal biology. Springer Nature, Cham

Srivastava S, Dafale NA, Jakhesara SJ, Joshi CG, Patil NV, Purohit HJ (2020a) Unraveling the camel rumen microbiome through metaculturomics approach for agriculture waste hydrolytic potential. Arch Microbiol 1:3. https://doi.org/10.1007/s00203-020-02010-x

Srivastava S, Dafale NA, Purohit HJ (2020b) Functional genomics assessment of lytic polysaccharide mono-oxygenase with glycoside hydrolases in *Paenibacillus dendritiformis* CRN18. Int J Biol Macromol 164:3729–3738. https://doi.org/10.1016/j.ijbiomac.2020.08.147

Srivastava S, Dafale NA, Tulsani N, Jakhesara SJ, Patil NV, Joshi CG, Purohit HJ (2021) Evaluation of designed consortium SNH-1 for efficient hydrolysis of agriculture waste to benefit bioethanol production. J Clean Prod 288:125601. https://doi.org/10.1016/j.jclepro.2020.125601

Stein HP, Navajas-Pérez R, Aranda E (2018) Potential for CRISPR genetic engineering to increase xenobiotic degradation capacities in model fungi. Springer, Cham, pp 61–78. https://doi.org/10.1007/978-3-030-02369-0_4

Sun S, Chen J, Li W, Altintas I, Lin A, Peltier S, Stocks K, Allen EE, Ellisman M, Grethe J, Wooley J (2011) Community cyberinfrastructure for advanced microbial ecology research and analysis: the CAMERA resource. Nucleic Acids Res 39:D546–D551. https://doi.org/10.1093/nar/gkq1102

Sun S, Xie S, Chen H, Cheng Y, Shi Y, Qin X, Dai SY, Zhang X, Yuan JS (2016) Genomic and molecular mechanisms for efficient biodegradation of aromatic dye. J Hazard Mater 302:286–295. https://doi.org/10.1016/j.jhazmat.2015.09.071

Suyal DC, Soni R, Singh DK, Goel R (2021) Microbiome change of agricultural soil under organic farming practices. Biologia 76:1315–1325

Tay PKR, Nguyen PQ, Joshi NS (2017) A synthetic circuit for mercury bioremediation using self-assembling functional amyloids. ACS Synth Biol 6:1841–1850. https://doi.org/10.1021/acssynbio.7b00137

Tcherepanov V, Ehlers A, Upton C (2006) Genome annotation transfer utility (GATU): rapid annotation of viral genomes using a closely related reference genome. BMC Genomics 7:1–10. https://doi.org/10.1186/1471-2164-7-150

Techtmann SM, Hazen TC (2016) Metagenomic applications in environmental monitoring and bioremediation. J Ind Microbiol Biotechnol 43:1345–1354. https://doi.org/10.1007/s10295-016-1809-8

Terashi G, Takeda-Shitaka M (2015) CAB-align: a flexible protein structure alignment method based on the residue-residue contact area. PLoS One 10:e0141440. https://doi.org/10.1371/journal.pone.0141440

Thiele I, Palsson B (2010) A protocol for generating a high-quality genome-scale metabolic reconstruction. Nat Protoc 5:93–121. https://doi.org/10.1038/nprot.2009.203

Turner D, Reynolds D, Seto D, Mahadevan P (2013) CoreGenes3.5: a webserver for the determination of core genes from sets of viral and small bacterial genomes. BMC Res Notes 6:140. https://doi.org/10.1186/1756-0500-6-140

Ufarté L, Laville É, Duquesne S, Potocki-Veronese G (2015) Metagenomics for the discovery of pollutant degrading enzymes. Biotechnol Adv. https://doi.org/10.1016/j.biotechadv.2015.10.009

Uhlik O, Leewis MC, Strejcek M, Musilova L, Mackova M, Leigh MB, Macek T (2013) Stable isotope probing in the metagenomics era: a bridge towards improved bioremediation. Biotechnol Adv. https://doi.org/10.1016/j.biotechadv.2012.09.003

Van Bel M, Proost S, Van Neste C, Deforce D, Van de Peer Y, Vandepoele K (2013) TRAPID: an efficient online tool for the functional and comparative analysis of de novo RNA-Seq transcriptomes. Genome Biol 14:R134. https://doi.org/10.1186/gb-2013-14-12-r134

Vaquerizas JM, Conde L, Yankilevich P, Cabezón A, Minguez P, Díaz-Uriarte R, Al-Shahrour F, Herrero J, Dopazo J (2005) GEPAS, an experiment-oriented pipeline for the analysis of microarray gene expression data. Nucleic Acids Res 33:W616–W620. https://doi.org/10.1093/nar/gki500

Vilchez-Vargas R, Geffers R, Suárez-Diez M, Conte I, Waliczek A, Kaser VS, Kralova M, Junca H, Pieper DH (2013) Analysis of the microbial gene landscape and transcriptome for aromatic pollutants and alkane degradation using a novel internally calibrated microarray system. Environ Microbiol 15:1016–1039. https://doi.org/10.1111/j.1462-2920.2012.02752.x

Wang Y, You FM, Lazo GR, Luo MC, Thilmony R, Gordon S, Kianian SF, Gu YQ (2013) PIECE: a database for plant gene structure comparison and evolution. Nucleic Acids Res 41:D1159–D1166. https://doi.org/10.1093/nar/gks1109

Wang X, Su X, Cui X, Ning K (2015) MetaBoot: a machine learning framework of taxonomical biomarker discovery for different microbial communities based on metagenomic data. PeerJ 2015:e993. https://doi.org/10.7717/peerj.993. peerj.com

Wichadakul D, Numnark S, Ingsriswang S (2009) D-Omix: a mixer of generic protein domain analysis tools. Nucleic Acids Res 37:417–421. https://doi.org/10.1093/nar/gkp329

Wicker J, Lorsbach T, Gütlein M, Schmid E, Latino D, Kramer S, Fenner K (2016) enviPath – the environmental contaminant biotransformation pathway resource. Nucleic Acids Res 44:D502–D508. https://doi.org/10.1093/nar/gkv1229

Wilkins MJ, Verberkmoes NC, Williams KH, Callister SJ, Mouser PJ, Elifantz H, Nguessan AL, Thomas BC, Nicora CD, Shah MB, Abraham P, Lipton MS, Lovley DR, Hettich RL, Long PE, Banfield JF (2009) Proteogenomic monitoring of Geobacter physiology during stimulated uranium bioremediation. Appl Environ Microbiol 75:6591–6599. https://doi.org/10.1128/AEM.01064-09

Williams W, Trindade M (2017) Metagenomics for the discovery of novel biosurfactants. In: Functional metagenomics: tools and applications. Springer International Publishing, pp 95–117. https://doi.org/10.1007/978-3-319-61510-3_6

Xavier JC, Costa PES, Hissa DC, Melo VMM, Falcão RM, Balbino VQ, Mendonça LAR, Lima MGS, Coutinho HDM, Verde LCL (2019) Evaluation of the microbial diversity and heavy metal resistance genes of a microbial community on contaminated environment. Appl Geochem 105:1–6. https://doi.org/10.1016/j.apgeochem.2019.04.012

Xue L, Ren H, Li S, Gao M, Shi S, Chang E, Wei Y, Yao X, Jiang Z, Liu J (2015) Comparative proteomic analysis in *Miscanthus sinensis* exposed to antimony stress. Environ Pollut 201:150–160. https://doi.org/10.1016/j.envpol.2015.03.004

Yadav M, Shukla AK, Srivastva N, Upadhyay SN, Dubey SK (2016) Utilization of microbial community potential for removal of chlorpyrifos: a review. Crit Rev Biotechnol 36:727–742. https://doi.org/10.3109/07388551.2015.1015958

Yang Y, Chen Y, Zhang G, Sun J, Guo L, Jiang M, Ou B, Zhang W, Si H (2020) Transcriptomic analysis of *Staphylococcus aureus* under the stress condition caused by *Litsea cubeba* L essential oil via RNA sequencing. Front Microbiol 11:1693. https://doi.org/10.3389/fmicb.2020.01693

Yun J, Malvankar NS, Ueki T, Lovley DR (2016) Functional environmental proteomics: elucidating the role of a c-type cytochrome abundant during uranium bioremediation. ISME J 10:310–320. https://doi.org/10.1038/ismej.2015.113

Ziganshina EE, Mohammed WS, Doijad SP, Shagimardanova EI, Gogoleva NE, Ziganshin AM (2018) Draft genome sequence of Brevibacterium epidermidis EZ-K02 isolated from nitrocellulose-contaminated wastewater environments. Data Br 17:119–123. https://doi.org/10.1016/j.dib.2017.12.053

Zwolinski MD (2007) DNA sequencing: strategies for soil microbiology. Soil Sci Soc Am J 71:592–600. https://doi.org/10.2136/sssaj2006.0125

Chapter 9
In Silico Approaches in Bioremediation Research and Advancements

Shabda Verma, Satinder Kour, and Rajesh Kumar Pathak

9.1 Introduction

Bioremediation is an approach that uses chiefly microbes, plants, or their enzymes, to remove toxic contaminants from environments. The concept comprises partial or total biodegradation, transformation, or detoxification of contaminants by microorganisms, which includes several approaches i.e. mineralization, a more restrictive term used for total conversion of an organic contaminant into inorganic components by one species or a group of microorganisms. Co-metabolism specifically refers to the transformation of a contaminant without carbon or energy production with the help of microorganisms. Bioremediation invariably uses living organisms for rapid degradation of hazardous pollutants from the environment to get healthy soil, deposits, and groundwater (Kumar et al. 2011; Giri et al. 2017a, b; Kumar et al. 2021). In other words, biodegradation refers to the recycling of waste or disintegration of organic matter to useable nutrients for other organisms (Alexander 1994). Bacteria, fungi, insects, worms, plants, among several other forms of life, carry out bioremediation by drawing nutrients from the contaminant, eventually transforming xenobiotics into environment-friendly compounds (Vidali 2001; Debbarma et al. 2017; Bhatt et al. 2020, 2021a, b). Bioremediation has a significant role in the cleaning of water reserves. Industrial effluents contaminate water bodies, thus reducing the availability of water for drinking and agricultural purpose. The bioremediation process quickens the rate of degradation of contaminants by augmenting the native microorganisms with nutrients, carbon sources, or electron donors

Shabda Verma and Satinder Kour have contributed equally.

S. Verma · S. Kour · R. K. Pathak (✉)
School of Agricultural Biotechnology, Punjab Agricultural University, Ludhiana, Punjab, India

© The Author(s), under exclusive license to Springer Nature Switzerland AG 2022

D. C. Suyal, R. Soni (eds.), *Bioremediation of Environmental Pollutants*,
https://doi.org/10.1007/978-3-030-86169-8_9

(biostimulation, bio restoration) or by introducing an enriched culture of microorganisms with peculiar characteristics that enable them to degrade the desired contaminant at a faster rate (bioaugmentation) (Goel et al. 2008; Dash et al. 2021; Singh et al. 2021). Bioremediation aims to minimize pollutant levels to undetectable or acceptable limits. Intensive agriculture and industrial waste have resulted in a discharge of a wide range of xenobiotic compounds to the environment. Uncontrolled piling of hazardous waste has led to a shortage of clean water and deterioration of soil, thus limiting crop production (Kamaludeen et al. 2003). This technology relies on promoting the growth of specific microflora that are native to the contaminated sites and can perform desired activities. For bioremediation to be effective, microbes must attack the contaminants with their enzymes and convert them to non-toxic products (Callaghan 2013; Kour et al. 2021). As bioremediation can be efficacious only where environmental conditions allow microbial activity, its application often requires the manipulation of environmental variables to allow microbial activity and degradation to proceed quickly. In silico approaches has huge potential to accelerate bioremediation process via identifying key components and engineering of microbial and plant system as per our need through genetic manipulation of key candidate gene(s).

9.2 Microbes and Plants: An Asset in Bioremediation

Microbes are playing a crucial role in protecting the environment by degrading contaminants and changing them to harmless forms via redox processes. The method is called microbial bioremediation, vibrant functioning, elastic nutritional needs and wide adaptability of microorganisms under extreme conditions make them the most qualified life forms for endurance. These qualities of the microbes make them useful to humanity, particularly when removal of contaminants or toxic materials from surroundings is concerned. A great number of microbes have been accounted for degradation of varied industrial wastes like coloring materials; hydrocarbons, particularly associated with petrochemical squanders (Kapley and Purohit 2009); tannery waste (Shrivastava et al. 2003); chlorinated aromatics (Banta and Kahlon 2007); distillery spent wash; pesticides, heavy metals, etc (Hivrale et al. 2016). Several contaminants in the environment have been eliminated by exploiting microorganisms. Methanotrophs have been found to reduce the emission of methane gas from soil sediments. The α-proteobacteria has shown up as vital microbes in the bioremediation of spilled oil and heavy metal that are grave concerns for the environment. Exogenous microorganisms are generally brought into the environment to speed up bioremediation. Likewise, chemotaxis and its application in bioremediation have also been reported in a model study (Paul et al. 2006). Some common genera of aerobic bacteria are capable of degrading pesticides and hydrocarbons belonging to alkanes and polyaromatic compounds are *Pseudomonas, Rhodococcus, Alcaligenes, Mycobacterium*, and *Sphingomonas*.

Plant-based bioremediation referred to as phytoremediation has shown promise for degrading a low level of long-lasting contaminants. In natural habitats, the plants act as sieves and break down substances produced by nature. Phytoremediation seems to be a very promising area that needs further research. Depending on the contaminant fate, it has been termed as phytoextraction, phytotransformation, phytostabilization, phytodegradation, and rhizofiltration, although a combination of these can be found in nature.

Phytoremediation is convenient for application on a field scale where other modes of remediation are cost-prohibitive or impracticable, especially the sites with trace amounts of contaminants. There are certain bottlenecks in the approach such as the requirement of a long duration of time for remediation, potential contamination of the vegetation and food chain, and difficulty with the growth of vegetation. Phytoremediation mainly includes three processes, namely phytostabilization, i.e. adsorption, reduction, and precipitation of the contaminants at the roots of the plants, phytovolatilization, i.e. converting contaminant to a gaseous state; and phytoextraction, i.e. recovery of dangerous elements through tolerant and accumulation.

Bioremediation is a sustainable solution to clean out the organic pollutants from contaminated areas. Both plants and microbes play significant roles in bioremediation processes. Out of phytoremediation and microbial bioremediation, the latter has received more attention and has wider application since microorganisms are found almost everywhere in nature and can survive under extreme weather conditions, where plants cannot thrive (Ma and Zhai 2012).

9.3 Environmental Factors Affecting Microbes and Plants

Microorganisms may not be present in sufficient numbers in contaminated soil for remediation of the site. Their growth and activity must be supported through the application of basic nutrients (nitrogen, phosphorus, carbon) which are the building blocks of life. Congenial conditions must also be provided to allow microbes to produce necessary enzymes for the breakdown of contaminants. The growth and development of microorganisms are readily affected by pH, temperature, and moisture conditions. Although microbes can live in harsh conditions, most of them grow ideally over a precise range, For instance, if the soil is too acidic, it may be reclaimed through the application of lime. Temperature influences the rate of various biological processes, which is doubled for each 10 °C rises in temperature. However beyond a certain range degradation of proteins occurs, and cells die. Plastic films can be employed to improve heating. An optimal moisture level also needs to be maintained. The right amount of oxygen is also required to be ensured for creating aerobic or anaerobic conditions. For instance, hydrocarbons break down readily under aerobic conditions, while, chlorurate compounds under an anaerobic state. Soil may be tilled to increase the amount of oxygen in the soil if desired. Hydrogen peroxide or magnesium peroxide can be added to the environment in some situations.

9.4 Omics Experimentation for Bioremediation

Developments in omics science and technology, such as genomics, proteomics, metabolomics, etc. have empowered scientists to understand physiology, ecology, and biochemistry of microbes and plants associated with the removal of contaminants from the natural habitats (Buermans and Den Dunnen 2014; Ghosal et al. 2016). It is now possible to study genes encoding bacterial inorganic transformations, which can facilitate the use of molecular genetics to boost metal tolerance (Mani and Kumar 2014). Using gene transfer technology, plants can be derived with increased bioremediation potential. This is attained by the addition or overexpression of the target gene(s) in the plant's DNA through molecular approaches. Genetically engineered endophytes and PGPR can effectively correct areas with metal contamination (Dixit et al. 2015, Divya and Kumar 2011). When genes governing antioxidant enzymes, or enzymes involved in the production of glutathione and other phytochelatins are over-expressed, this increases the extent of metal accumulation and toleration. Isolation, purification, and evaluation of microorganisms from polluted sites often yield insight into the metabolic potential of microbes in bioremediation. The methodology to study the microbial world is complex, as most of the microbes in the environment cannot be easily cultured in a research laboratory thus posing a limitation for basic research (Awasthi et al. 2020).

The breakthroughs in molecular science provided state-of-the-art tools to evaluate the organisms that cannot be cultured on routine media (Gilbert and Dupont 2011). The in silico methods and molecular sequencing techniques have changed the scenario in this way (Malla et al. 2018). The molecular approaches thus provided a moreclear insight into the microbial communities living in a specific ecological environment (Gupta et al. 2020; Suyal et al. 2021). Moreover, they have promoted the collection of huge biological data in an organized way and exploring them for the bioremediation purpose (Pandey et al. 2019).

Microbes-based bioremediation is an environmentally safe cost-effective method for restoration of an ecological system contaminated with chemical pollutants. Ideally, before the use of any microbial organism for degradation of a pollutant in the environment, there is a need for in silico analysis for predicting the possible degradative pathway and the reactions taking place inside the microorganism by employing different in silico tools. Several databases and computer programs are available to do computational analysis that will be helpful in bioremediation. They simplify the problem and provide a unique method to perform the task efficiently and successfully.

9.4.1 *In Silico Analysis of Omics Data and Their Integration for Bioremediation*

Omics technologies such as proteomics, transcriptomics, genomics metabolomics, and other omics are key biological approaches, which promote analyses of biological components like, RNA, DNA proteins, genes, lipids, and metabolites from organisms individually or in their communities (Chandran et al. 2020). Recent discoveries in molecular techniques have greatly expedited the study of microbial community structure (Gutleben et al. 2018) and have opened up new possibilities in protecting our environment (Rodríguez et al. 2020) by exploring new functional genes linked with destructive metabolism of contaminants and controlling their expression at broader level (Roume et al. 2015).

As we know that due to advances in several omics platform huge amount of omics data generated every day that hold enormous information concerning different biological application but it depends on in silico databases and tools that will enable us to decode these data as per objective. However, big amounts of data are also available in several databases that can also be analyzed for novel discovery.

Because of the above fact, we can utilize publically available omics data for master regulator identification in microbes and plant systems that hold bioremediation potential, which can further utilized to engineer microbes and crop plants as per

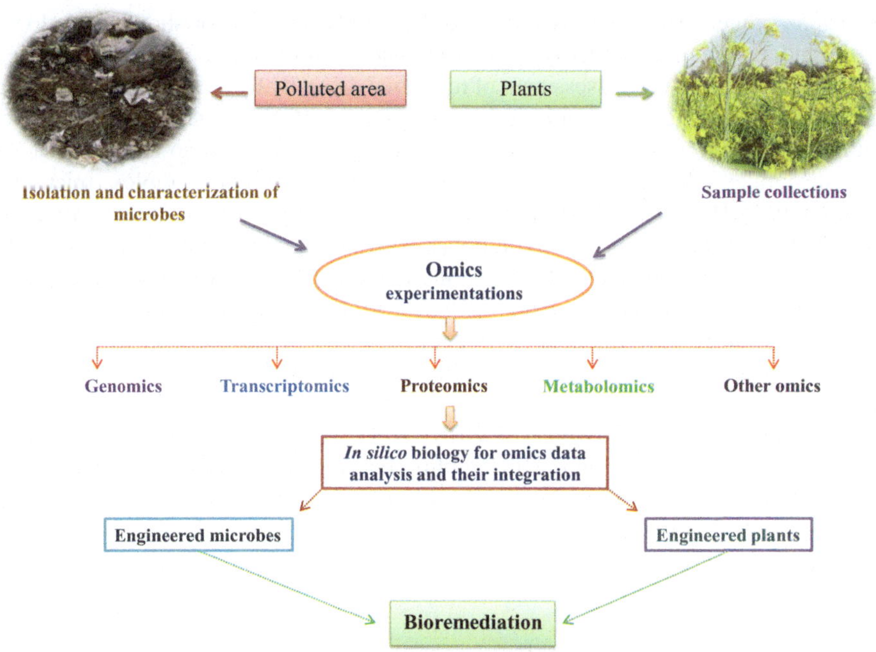

Fig. 9.1 Mechanism of bioremediation via integration and analysis of multi-omics data

the need for bioremediation (Fig. 9.1). The details of several omics data and in silico approaches used for their analysis are discussed in the following section.

9.4.1.1 Genome Assembly and Data Analysis

Genomics is the study of structure, function, evolution, sequence mapping, and editing of a complete set of DNA in an organism. Advanced techniques in genomics greatly helped in the identification of microorganisms that generated interest in the establishment of pure culture of the microorganisms for bioremediation of contaminants (Nierman and Nelson 2002). These days, the entire genomic sequence of a majority of the microorganisms is available and the sequencing technology is proving crucial in understanding the physiology of microorganisms (Bihari 2013), elucidation of the function of genes involved in pathways (Fulekar 2005), and development of genetically engineered organisms with an enhanced ability for bioremediation. Besides, it will also help in phytoremediation. The generated data from the genome sequencing project will be assembled and annotated through in silico approaches.

9.4.1.2 16S rRNA Based Approach in Bioremediation

The gene sequence, 16S rRNA is an excellent study material for showing phylogenetic and taxonomic similarity and differences among microorganisms (Lovley 2003). It is helpful in the assessment of microbial diversity by specifically replicating and mapping the highly variable segments of the 16SrRNA gene. It is an effective, and cost-efficient approach easily interpretable by various in silico tools and which has developed into an often-used technique for profiling complex microbial communities (Han et al. 2020). It can help in the identification of unique, difficult-to-culture microorganisms. An evolutionary tree of the microorganisms linked with biodegradation can be created by exploring the16S rRNA sequences obtained from polluted environments (Rogers and McClure 2003). 16S rRNA has also been used to understand the role of indigenous soil microbes on biodegradation of p*olycyclic aromatic hydrocarbons* in coal tar mixing plant (Kumar and Khanna 2010; Viant and Sommer 2013; Sakshi Haritash 2020) and areas contaminated with heavy metals like lead, zinc, and copper (Kou et al. 2018). 16S rRNA gene sequencing was used to screen bacterial species having the ability for degrading polyurethane polymer (Espinosa et al. 2020). Therefore, it is an innovative and most effective approach for bioremediation via identification of pollutant degrading microbes by sequencing and in silico analysis.

9.4.1.3 Transcriptome Data Analysis

The complete set of messenger or total RNA transcripts in an individual or a population of cells is known as transcriptome. Next-Generation sequencing technology is capable of producing transcriptome data at an affordable cost (Eiler et al. 2012). NGS technology has brought a transition in biotechnology of bioremediation (Ma and Zhai 2012). It helps in determining the relationship of various organisms using generated molecular data. The information will help to decode degradative pathways for further investigation.

It provides a strong connection between the genome, proteome, and cellular phenotype. It also provides an understanding of genes with increased or decreased expression under different environments in groups of organisms sharing a commonplace. Analysis of mRNAs given understanding of tissue- and cell-specific gene manifestations such as assessment of the transcript, protein isoforms, and influence of genotype on gene expression, through qualitative analysis. The extraction of transcriptome involve—RNA isolation and purification, synthesis of cDNA, the building of cDNA library, and generates data using the NGS platform. Therefore, transcriptome analysis in microorganisms and plants is an essential tool for the identification of key candidate genes linked with the bio-remediation process.

9.4.1.4 Proteomics Data Analysis

A complete set of proteins expressed at a given time are represented by proteome and its scientific study is known as proteomics. The proteomic analysis has empowered the analysis of the protein expression in the living cells associated with the polluted habitats (Kim et al. 2004). Further, the proteome analysis at a community level is known as metaproteomics (Hart et al. 2018). In biodegradation, proteomics has major applications such as analysis of bacterial response on different environmental pollutants, community structure analysis and identification of novel proteins/ enzymes involved in the metabolism of toxic compounds (Chauhan and Jain 2010). Advances in proteomics technology and in silico approaches enable us to dissect the complete proteome of an organism that will help bioremediation via engineering of microbe and plant systems.

9.4.1.5 Metabolomics Data Analysis

The total metabolites of a cell at a given time are represented by metabolome and its scientific study is known as metabolomics (Beale et al. 2017). Metabolites are low-molecular-weight organic compounds (<1000 Da) that are involved in general metabolic reactions or are required for the organism's growth, maintenance, and normal functioning. Under the influence of the external factors, any cell is able to change its metabolic behavior (Malla et al. 2018). Several models have been developed to predict the metabolic fluctuations in the cells under different environmental conditions. Metabolomics examines the metabolites released by cells in response to changing environmental factors, providing insight into how cells regulate themselves (Krumsiek et al. 2015). Therefore, such metabolites can be explored as a potential bioindicators monitoring the of pollutants. The progress in analytical, statistical, and in silico techniques has eased testing, extraction, and interpretation of various metabolites and in the elucidation of their pathway (Hill et al. 2015). The common metabolomic analysis approaches include metabolite profiling, metabolic fingerprinting and targeted analysis (Wang et al. 2010). Further, metabolite fingerprinting is a reliable, high throughput and fast method to identify and monitor the unknown metabolites in an ecosystem. The detailed workflow of such profiling involves sample collection, sample processing, identification and quantification of the metabolites. The quantification and identification of the analytes can be done by a combination of chromatography techniques, mass spectrometry, and nuclear magnetic resonance. The methods for analysis of metabolites should be simple, accurate, strong, and reproducible, and be capable of handling a large number of samples. The identified information through metabolome data analysis is helpful in initiating bioremediation program (Aldridge and Rhee 2014).

9.4.1.6 NMR Spectroscopy and Metabolite Profiling During Bioremediation

The majority of NMR studies in organic pollutant bioremediation have been performed in aquatic environments, with only a handful in soil or soil models. NMR has been used to demonstrate the biodegradation mechanism of morpholine, thiomorpholine, and piperidine by *Mycobacterium aurum* MO1 and *Mycobacterium* sp. RP1 (Combourieu et al. 2000, 2003). There are significant advantages to NMR-based analysis, such as little or no sample preparation, non-destructive sample analysis, detailed low-molecular-weight metabolite profiling, and inherent quantification, which allow studies of mass balance to be carried out on chemical reactions. It is less resilient than MS, however, and needs relatively large amounts of samples. The sensitivity problem has been resolved by developing magnets with increased field strength and improving the design of NMR detectors (Griffin 2004), but long acquisition times are still needed to detect low abundance metabolites with NMR.

Mass spectrometry creates spectra that consist of an arrangement of peaks that can be utilized for determination and quantification of metabolites. Spectral databases store these patterns, allowing for automated analysis and the production of metabolomic profiles. Metabolic data are generally analyzed using two approaches. The first approach uses statistics and algorithms for clustering, while the second method uses networks to envisage the properties of the data in space and time. MetScape Plugin, MAVEN, MetaMapp, Pathomx, and other network visualisation software tools were created to represent metabolic pathways graphically (Hill et al. 2015). Further, Mallick et al. (2019) have employed a computational tool—

MelonnPan, to predict microbial community metabolome. Metabolomics, in combination with in silico tools and databases, has allowed researchers to study the metabolic behaviors of the microorganisms. Therefore, it is a powerful tool and approach to discover novel metabolic pathways and characterize metabolic networks (McMahon et al. 2018). Metabolomics is thus rapidly evolving omics technology to provide intelligent insight for decoding various metabolites and their pathways for bioremediation.

9.4.2 Computational Systems Biology for Omics Data Integration and Novel Discovery

Computational systems biology is an interdisciplinary approach that combines several disciplines and methodology of natural sciences and engineering for solving the intricacy of biological systems (Pathak and Singh 2020). It is focused holistically rather than consider a single gene or protein. It can be utilized for integration of multi-omics data associated with bioremediation for identification of candidate genes for further research, besides, it will also enable us to analyze the behavior of biological systems i.e. plants or microbes concerning time through simulation analysis. The obtained information will accelerate the bioremediation research output for the benefit of society (Fig. 9.2).

9.4.3 Pathway Modeling

Pathway modeling and their analysis are useful methods for the integration of available information obtained from different sources i.e. *via* information obtained from omics experimentation or literature mining. Systems biology tools can integrate this available information by using the different symbols, known as systems biology graphical notation (SBGN) to build a model (Kumar et al. 2015). Further, the molecular data associated with each component in the model will be stored in systems biology markup language (SBML), and generates kinetic rate equation of model for simulation analysis to predict the behavior of a system in respect to time and investigate the key component linked with bioremediation processes.

9.4.4 Network Analysis

The network is a general term because it is usually used everywhere, in biology, it gains a lot of attention due to its importance so that network biology, a new discipline in biological science has come. Different types of in silico tools are

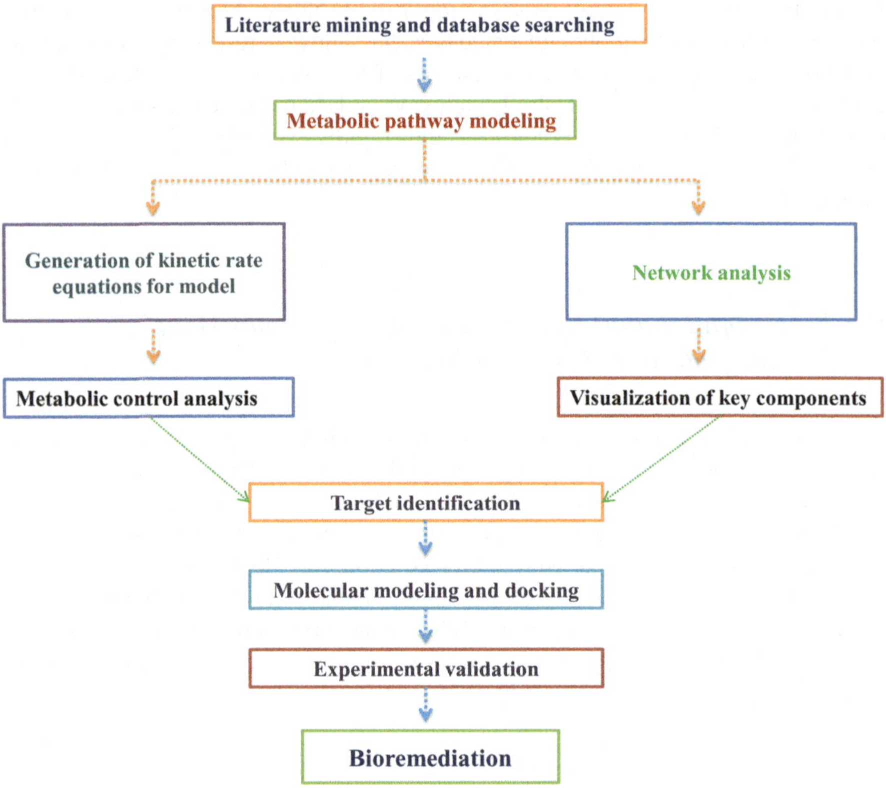

Fig. 9.2 In silico approaches used in bioremediation research for identification of key components

available for the analysis of the biological network. It will be represented by node and edge. Nodes are genes, proteins, metabolites, etc. whereas edges represent interaction in the form of bonding, regulatory effect, reaction, etc. (Pathak et al. 2017a). We can easily integrate multi-omics data for the identification of master regulators/hubs in biological systems. The resulted information will be utilized in a bioremediation research program.

9.4.5 Target Structure Modeling, Validation and Visualization

In silico approaches help predict the 2D and 3D structure of the molecular target, their stereochemical properties analysis, binding site prediction, and visualization. As we know that the experimental techniques used for structure determination i.e. X-ray crystallography and NMR are time-consuming and costly methods (Pathak et al. 2017b). Therefore, in silico approaches utilized sequence information in the determination of the 3D structure of macromolecules, its validation, and visualization. This will enable us to predict the interaction of small molecules having degradation potential.

9.4.6 Molecular Docking, Virtual Screening and Molecular Dynamics Simulation

Molecular docking, virtual screening, and molecular dynamics simulation are very powerful in silico approach that can be explored for various research activities. These approaches are used for visualization of macromolecular structure with small molecules (Receptor-ligand interaction) (Pathak et al. 2020). Calculation of binding energy and their stability in respect to time. In a molecular docking study, we can predict the receptor-ligand interaction and calculate its binding energy, besides virtual screening is used to screen small molecule databases against a particular target to find out agonist or antagonist molecule for further consideration. Molecular dynamics simulation is used to predict the behavior of macromolecular target structure before and after ligand binding concerning time computationally to provide a clue for experimental validation (Pathak et al. 2018). These techniques are very useful in the identification of mimicking molecules that will be used to accelerate bioremediation processes in plants and microbes.

9.5 Development of In Silico Platforms for Bioremediation Research

Pollution is a major problem for health and the environment. Therefore, it is a need of time to develop efficient, accurate, and fast in silico resources i.e. tools and databases that can be utilized in a bioremediation research program for data analysis and visualization. Some of the available resources are discussed in the following sections.

9.5.1 Database Development

Databases play a crucial role in the management of biological data generated for several experimental techniques. The omics data available in different databases will be further analyzed for novel discovery and it will also be utilized to compare newly generated data for their integration and further analysis to speed up the bioremediation program. Some important databases used in the bioremediation program are listed in the Table 9.1.

Table 9.1 List of some important database resources used in bioremediation

S. No.	Name of database	Utility	URL	References
1.	NCBI	It holds information/data about genes, proteins etc. that are generally utilized in bio-remediation program	https://www.ncbi.nlm.nih.gov/	NCBI Resource Coordinators (2012)
2.	BioRadBase	It provides information about bioremediation of radioactive waste	http://biorad.igib.res.in/	Reena et al. (2012)
3.	Bionemo	This database provides information about proteins and genes that are directly linked with biodegradation metabolism	http://bionemo.bioinfo.cnio.es	Carbajosa et al. (2009)
4.	BioSurfDB	It provides information about metabolic pathways involved in bioremediation	https://www.biosurfdb.org/#/	Oliveira et al. (2015)
5.	MetaCyc	This database provides information about metabolic pathway, related metabolites, enzymes, reactions, and genes from all domains of life	https://metacyc.org/	Caspi et al. (2020)

9.5.2 *Software/Tool Development*

Due to advances in several experimental techniques, big amounts of biological data are generated, which requires software for their analysis. Therefore, the development of accurate and exportable software is a need of time. The analysis of experimental data through in silico tools will boost research output cost-effectively. Some important tools currently used in omics data analysis are listed in Table 9.2.

9.6 Limitation of In Silico Approaches

Due to unavailability of specialized databases for bioremediation is a major concern that limit research program, besides, the accuracy of freely available software's are also one of the limitations for predicting and visualizing the correct behavior of complex biological systems. In recent years, improvement in several omics platforms and its accurate data generation capacity and availability of big data through these techniques will support the scientific community to utilized available resources for the development of efficient and accurate algorithms for software development, development of specialized databases for future bioremediation research and development.

Table 9.2 List of some In silico program utilized in bioremediation research program

S. N.	Name of tools/ Software's	Utility	URL	References
1.	FastQC	This tool is utilized to check the quality of sequencing data generated through various next-generation sequencing platform	https://www.bioinformatics.babraham.ac.uk/projects/fastqc/	Andrews (2010)
2.	Trimmomatic	It is an freely available program for trimming next generation sequencing data	http://www.usadellab.org/cms/?page=trimmomatic	Bolger et al. (2014)
3.	SPAdes	It is an open source program used for genome assembly	https://cab.spbu.ru/software/spades/	Prjibelski et al. (2020)
4.	Velvet	It is a freely available program used for genome assembly	https://www.ebi.ac.uk/~zerbino/velvet/	Zerbino and Birney (2008)
5.	Trinity	It is used for assembly of RNAseq data	https://github.com/trinityrnaseq/trinityrnaseq/wiki	Haas et al. (2013)
6.	FGENESH	It is a program used for predicting genes in genomic DNA sequences	http://www.softberry.com/	Conklin et al. (2005)
7.	BLAST2GO	This software is used for functional annotation, visualization and analysis of high-throughput sequencing data	https://www.blast2go.com/	Conesa et al. (2005)
8.	CellDesigner	It is one of the highly cited tools for modeling and simulation of bio-molecular pathway	http://www.celldesigner.org/	Funahashi et al. (2003)
9.	Cytoscape	It is well known tool used for analysis and visualization of biological networks	https://cytoscape.org/	Shannon et al. (2003)
10.	AutoDock	It is well known and widely used software for molecular docking and protein-ligand interaction study	http://autodock.scripps.edu/	Morris et al. (2009)

9.7 Future Prospects

Bioremediation holds great potential for detoxifying polluted sites. Nevertheless, the technique is full of obstacles that need to be removed. The key step is to understand the biological systems and their interactions with their environment so that their genes, enzymes and other metabolites can be explored for bioremediation purpose. Further, innovative and multidisciplinary approaches are needed to unravel the key concepts of bioremediation. Genomic approaches for microbes and plants mediated bioremediation have been illustrated in previous studies. Hence, the aim of this chapter is to provide a high-level overview of the most widely used multi-omics techniques and in silico approaches used for investigating and understanding the different molecular aspects of the living cells towards environmental cleaning. Given the importance of genetically modified microbes in significantly improving degradation and detoxification of heavy metal pollutants and xenobiotics. Since their survival is currently poor when released into the environment for bioremediation, therefore, further research should be done to improve their survival.

In addition, further investigation is needed to entirely comprehend the metabolic pathways of genetically modified plants and microorganisms exploited in bioremediation to assess their effectiveness and side effects. Hyperaccumulator plants with high biomass production should also be selected and produced through genetic engineering to efficiently remove heavy metals from the atmosphere through phytoextraction, which has proven to be an effective phytoremediation method.

9.8 Conclusion

Ecosystem imbalance has led to the development of novel techniques for the degradation of contaminants from the environment. The omics approaches, like genomics, proteomics, transcriptomics, metabolomics, and other omics, are highly versatile and have removed the bottlenecks in our understanding of the mechanisms involved in various bioremediation processes. They have proved highly fruitful in identifying novel genes in microbes and plants that can efficiently degrade pollutants. The approaches are also highly economical, their strengths are far greater than their weaknesses, which is evident from the large number of research projects currently running for this purpose using in silico support. Bioremediation is undoubtedly preferred over traditional methods nowadays as they are helping in protecting two of our greatest resources, i.e. soil and water. The bioremediation technology, therefore, offers great potential in restoring the health of our ecosystem and, will be used as an excellent management option for saving life and environment from the toxic pollutant.

References

Aldridge BB, Rhee KY (2014) Microbial metabolomics: innovation, application, insight. Curr Opin Microbiol 19:90–96. https://doi.org/10.1016/j.mib.2014.06.009

Alexander M (1994) Biodegradation and bioremediation. Academic, San Diego, CA

Andrews S (2010) FastQC: a quality control tool for high throughput sequence data. http://www.bioinformatics.babraham.ac.uk/projects/fastqc

Awasthi MK, Ravindran B, Sarsaiya S, Chen H, Wainaina S, Singh E (2020) Metagenomics for taxonomy profiling: tools and approaches. Bioengineered 11:356–374. https://doi.org/10.1080/21655979.2020.1736238

Banta G, Kahlon RS (2007) Dehalogenation of 4-chlorobenzoic acid by Pseudomonas isolates. Ind J Micorbiol 47:139–143. https://doi.org/10.1007/s12088-007-0027-5

Beale DJ, Karpe AV, Ahmed W, Cook S, Morrison PD, Staley C (2017) A community multi-omics approach towards the assessment of surface water quality in an urban river system. Int J Environ Res Public Health 14:E303. https://doi.org/10.3390/ijerph14030303
Bhatt P, Rene ER, Kumar AJ, Kumar AJ, Zhang W, Chen S (2020) Binding interaction of allethrin with esterase: bioremediation potential and mechanism. Bioresour Technol 315:13845
Bhatt P, Bhatt K, Sharma A, Zhang W, Mishra S, Chen S (2021a) Biotechnological basis of microbial consortia for the removal of pesticides from the environment. Crit Rev Biotechnol 41(3):317–338
Bhatt P, Joshi T, Bhatt K, Zhang W, Huang Y, Chen S (2021b) Binding interaction of glyphosate oxidoreductase and C-P lyase: molecular docking and molecular dynamics simulation studies. J Hazard Mater 5:409:124927
Bihari Z (2013) Current trends in bioremediation and biodegradation: next-generation sequencing. J Bioremed Biodegr 4(08):e138. https://doi.org/10.4172/2155-6199.1000e138
Bolger AM, Lohse M, Usadel B (2014) Trimmomatic: a flexible trimmer for Illumina sequence data. Bioinformatics 30(15):2114–2120
Buermans H, Den Dunnen J (2014) Next generation sequencing technology: advances and applications. Biochem Biophys Acta 1842:1932–1941
Callaghan AV (2013) Metabolomic investigations of anaerobic hydrocarbon-impacted environments. Curr Opin Biotechnol 24:506–515. https://doi.org/10.1016/j.copbio.2012.08.012
Carbajosa G, Trigo A, Valencia A, Cases I (2009) Bionemo: molecular information on biodegradation metabolism. Nucleic Acids Res 37(Suppl 1):D598–D602
Caspi R, Billington R, Keseler IM, Kothari A, Krummenacker M, Midford PE, Ong WK, Paley S, Subhraveti P, Karp PD (2020) The MetaCyc database of metabolic pathways and enzymes-a 2019 update. Nucleic Acids Res 48(D1):D445–D453
Chandran H, Meena M, Sharma K (2020) Microbial biodiversity and bioremediation assessment through omics approaches. Front Environ Chem 1:570326. https://doi.org/10.3389/fenvc.2020.570326
Chauhan A, Jain RK (2010) Biodegradation: gaining insight through proteomics. Biodegradation 21(6):861–879
Combourieu B, Besse P, Sancelme M, Godin JP et al (2000) Common degradative pathways of morpholine, thiomorpholine and piperidine by *Mycobacterium aurum* MO1: evidence from ^{1}H-nuclear magnetic resonance and ionspray mass spectrometry performed directly on the incubation medium. Appl Environ Microbiol 66:3187–3193
Combourieu B, Haroune N, Besse P, Sancelme M, Delort AM (2003) In: Mohan RM (ed) Research advances in microbiology, vol 3. Global research network, pp 1–22
Conesa A, Götz S, García-Gómez JM, Terol J, Talón M, Robles M (2005) Blast2GO: a universal tool for annotation, visualization and analysis in functional genomics research. Bioinformatics 21(18):3674–3676
Conklin D, Haldeman B, Gao Z (2005) Gene finding for the helical cytokines. Bioinformatics 21 (9):1776–1781
Dash B, Sahu N, Singh AK, Gupta SB, Soni R (2021) Arsenic efflux in *Enterobacter cloacae* RSN3 isolated from arsenic-rich soil. Folia Microbiol 66:189–196
Debbarma P, Raghuwanshi S, Singh J, Suyal DC, Zaidi MGH, Goel R (2017) Comparative in situ biodegradation studies of polyhydroxybutyrate film composites. 3Biotech 7(178):1–9. https://doi.org/10.1007/s13205-017-0789-3
Divya B, Kumar MD (2011) Plant-microbe interaction with enhanced bioremediation. Res J Biotechnol 6:72–79
Dixit R, Malaviya D, Pandiyan K, Singh UB, Sahu A, Shukla R, Singh BP, Rai JP, Sharma PK, Lade H (2015) Bioremediation of heavy metals from soil and aquatic environment: an overview of principles and criteria of fundamental processes. Sustainability 7:2189–2212
Eiler A, Heinric F, Bertilsson S (2012) Coherent dynamics and association networks among lake bacterioplankton taxa. ISME J 6:330–342. https://doi.org/10.1038/ismej.2011.113

Espinosa MJC, Blanco AC, Schmidgall T, Atanasoff-Kardjalieff AK, Kappelmeyer U, Tischler D et al (2020) Toward biorecycling: isolation of a soil bacterium that grows on a polyurethane oligomer and monomer. Front Microbiol 11:404. https://doi.org/10.3389/fmicb.2020.00404

Fulekar MH (2005) Bioremediation technologies for environment. Indian J Environ Prot 25:358–364

Funahashi A, Morohashi M, Kitano H, Tanimura N (2003) CellDesigner: a process diagram editor for gene-regulatory and biochemical networks. Biosilico 1(5):159–162

Ghosal D, Ghosh S, Dutta TK, Ahn Y (2016) Current state of knowledge in microbial degradation of polycyclic aromatic hydrocarbons (PAHS): a review. Front Microbiol 7:1369

Gilbert JA, Dupont CL (2011) Microbial metagenomics: beyond the genome. Annu Rev Mar Sci 3:347–371. https://doi.org/10.1146/annurev-marine-120709-142811

Giri K, Rai JPN, Pandey S, Mishra G, Kumar R, Suyal DC (2017a) Performance evaluation of isoproturon-degrading indigenous bacterial isolates in soil microcosm. Chem Ecol 33 (9):817–825. https://doi.org/10.1080/02757540.2017.1393535

Giri K, Suyal DC, Mishra G, Pandey S, Kumar R, Meena DK, Rai JPN (2017b) Biodegradation of isoproturon by *Bacillus pumilus* K1 isolated from foothill agroecosystem of north west Himalaya. Proc Natl Acad Sci India Sect B-Biol Sci 87(3):839–848. https://doi.org/10.1007/s40011-015-0667-x

Goel R, Zaidi MGH, Soni R, Kusumlata SYS (2008) Implication of *Arthrobacter* and *Enterobacter* species for polycarbonate degradation. Int Biodeterior Biodegrad 61(2):167–172

Griffin JL (2004) Metabolic profiles to define the genome: can we hear the phenotypes? Philos Trans R Soc Lond B Biol Sci 359:857–871

Gupta K, Biswas R, Sarkar A (2020) Advancement of omics: prospects for bioremediation of contaminated soils. In: Shah M (ed) Microbial bioremediation & biodegradation. Springer, Singapore, pp 113–142

Gutleben J, Chaib De Mares M, van Elsas JD, Smidt H, Overmann J, Sipkema D (2018) The multi-omics promise in context: from sequence to microbial isolate. Crit Rev Microbiol 44:212–229. https://doi.org/10.1080/1040841X.2017.1332003

Haas BJ, Papanicolaou A, Yassour M, Grabherr M, Blood PD, Bowden J, Couger MB, Eccles D, Li B, Lieber M, MacManes MD (2013) De novo transcript sequence reconstruction from RNA-seq using the trinity platform for reference generation and analysis. Nat Protoc 8 (8):1494–1512

Han D, Gao P, Li R, Tan P, Xie J, Zhang R et al (2020) Multicenter assessment of microbial community profiling using 16S rRNA gene sequencing and shotgun metagenomic sequencing. J Adv Res. https://doi.org/10.1016/j.jare.2020.07.010

Hart EH, Creevey CJ, Hitch T, Kingston-Smith AH (2018) Metaproteomics of rumen microbiota indicates niche compartmentalisation and functional dominance in a limited number of metabolic pathways between abundant bacteria. Sci Rep 8:10504. https://doi.org/10.1038/s41598-018-28827-

Hill CB, Czauderna T, Klapperstück M, Roessner U, Schreiber F (2015) Metabolomics, standards, and metabolic modeling for synthetic biology in plants. Front Bioeng Biotechnol 3:167. https://doi.org/10.3389/fbioe.2015.00167

Hivrale AU, Pawar PK, Rane NR, Govindwar SP Application of genomics and proteomics in bioremediation. In: Toxicity and waste management using bioremediation 2016. IGI Global, pp 97–112

Kamaludeen SPB, Arunkumar KR, Avudainayagam S, Ramasamy K (2003) Bioremediation of chromium contaminated environments. Ind J Exp Bio 41:972985

Kapley A, Purohit HJ (2009) Diagnosis of treatment efficiency in industrial wastewater treatment plants: a case study at a refinery ETP. Environ Sci Technol 43:3789–3795. https://doi.org/10.1021/es803296r

Kim HJ, Ishidou E, Kitagawa E, Momose Y, Iwahashi H (2004) A yeast DNA microarray for the evaluation of toxicity in environmental water containing burned ash. Environ Monit Assess 92:253–272. https://doi.org/10.1023/B:EMAS.0000014504.03500.41

Kou S, Vincent G, Gonzalez E, Pitre FE, Labrecque M, Brereton NJB (2018) The response of a 16S ribosomal RNA gene fragment amplified community to lead, zinc, and copper pollution in a Shanghai field trial. Front Microbiol 9:366. https://doi.org/10.3389/fmicb.2018.00366

Kour D, Kaur T, Devi R, Yadav A, Singh M et al (2021) Beneficial microbiomes for bioremediation of diverse contaminated environments for environmental sustainability: present status and future challenges. Environ Sci Pollut Res 28:24917–24939

Krumsiek J, Mittelstrass K, Do KT, Stückler F, Ried J, Adamski J et al (2015) Gender-specific pathway differences in the human serum metabolome. Metabolomics 11:1815–1833. https://doi.org/10.1007/s11306-015-0829-0

Kumar M, Khanna S (2010) Diversity of 16S rRNA and dioxygenase genes detected in coal-tar-contaminated site undergoing active bioremediation. J Appl Microbiol 108:1252–1262. https://doi.org/10.1111/j.1365-2672.2009.04523.x

Kumar A, Bisht BS, Joshi VD, Dhewa T (2011) Review on bioremediation of polluted environment: a management tool. Int J Environ Sci 1:1079–1093

Kumar A, Pathak RK, Gupta SM, Gaur VS, Pandey D (2015) Systems biology for smart crops and agricultural innovation: filling the gaps between genotype and phenotype for complex traits linked with robust agricultural productivity and sustainability. Omics 19(10):581–601

Kumar P, Dash B, Suyal D C, Gupta SB, Singh AK, Chowdhury T, Soni R (2021) Characterization of arsenic-resistant Klebsiella pneumoniae RnASA11 from contaminated soil and water samples and its bioremediation potential. Curr Microbiol. https://doi.org/10.1007/s00284-021-02602-w

Lovley DR (2003) Cleaning up with genomics: applying molecular biology to bioremediation. Nat Rev Microbiol 1:35–44. https://doi.org/10.1038/nrmicro731

Ma J, Zhai G (2012) Microbial bioremediation in omics era: opportunities and challenges. J Bioremed Biodegr 3:e120. https://doi.org/10.4172/2155-6199.1000e120

Malla MA, Dubey A, Yadav S, Kumar A, Hashem A, Abd Allah EF (2018) Understanding and designing the strategies for the microbe-mediated remediation of environmental contaminants using omics approaches. Front Microbiol 9:1132. https://doi.org/10.3389/fmicb.2018.01132

Mallick H, Franzosa EA, Mclver LJ, Banerjee S, Sirota-Madi A, Kostic AD et al (2019) Predictive metabolomic profiling of microbial communities using amplicon or metagenomic sequences. Nat Commun 10:3136. https://doi.org/10.1038/s41467-019-10927-1

Mani D, Kumar C (2014) Biotechnological advances in bioremediation of heavy metals contaminated ecosystems: an overview with special reference to phytoremediation. Int J Environ Sci Technol 11:843–872

McMahon MA, Prakash TP, Cleveland DW, Bennett CF, Rahdar M (2018) Chemically modified Cpf1-CRISPR RNAs mediate efficient genome editing in mammalian cells. Mol Ther 26:1228–1240. https://doi.org/10.1016/j.ymthe.2018.02.031

Morris GM, Huey R, Lindstrom W, Sanner MF, Belew RK, Goodsell DS, Olson AJ (2009) AutoDock4 and AutoDockTools4: automated docking with selective receptor flexibility. J Comput Chem 30(16):2785–2791

Nierman WC, Nelson KE (2002) Genomics for applied microbiology. Adv Appl Microbiol 51:201–248. https://doi.org/10.1016/S0065-2164(02)51007-8

Oliveira JS, Araújo W, Lopes Sales AI, Brito Guerra AD, Silva Araújo SCD, de Vasconcelos ATR, Agnez-Lima LF, Freitas AT (2015) BioSurfDB: knowledge and algorithms to support biosurfactants and biodegradation studies. Database 2015

Pandey A, Tripathi PH, Tripathi AH, Pandey SC, Gangola S (2019) Omics technology to study bioremediation and respective enzymes. In: Bhatt P (ed) Smart bioremediation technologies: microbial enzymes. Academic, New Delhi, pp 23–43. https://doi.org/10.1016/B978-0-12-818307-6.00002-0

Pathak RK, Singh DB (2020) Systems biology approaches for food and health. In: Advances in agri-food biotechnology. Springer, Singapore, pp 409–426

Pathak RK, Baunthiyal M, Pandey N, Pandey D, Kumar A (2017a) Modeling of the jasmonate signaling pathway in *Arabidopsis thaliana* with respect to pathophysiology of Alternaria blight in brassica. Sci Rep 7(1):1–12

Pathak RK, Baunthiyal M, Shukla R, Pandey D, Taj G, Kumar A (2017b) In silico identification of mimicking molecules as defense inducers triggering jasmonic acid mediated immunity against alternaria blight disease in brassica species. Front Plant Sci 8:609

Pathak RK, Gupta A, Shukla R, Baunthiyal M (2018) Identification of new drug-like compounds from millets as xanthine oxidoreductase inhibitors for treatment of hyperuricemia: a molecular docking and simulation study. Comput Biol Chem 76:32–41

Pathak RK, Singh DB, Sagar M, Baunthiyal M, Kumar A (2020) Computational approaches in drug discovery and design. In: Computer-aided drug design. Springer, Singapore, pp 1–21

Paul D, Singh R, Jain RK (2006) Chemotaxis of Ralstonia sp. SJ98 towards p-nitrophenol i soil. Environ Microbiol 8:1797–1804. https://doi.org/10.1111/j.1462-2920.2006.01064.x

Prjibelski A, Antipov D, Meleshko D, Lapidus A, Korobeynikov A (2020) Using SPAdes de novo assembler. Curr Protocols Bioinform 70(1):e102

Reena R, Majhi MC, Arya AK, Kumar R, Kumar A (2012) BioRadBase: a database for bioremediation of radioactive waste. Afr J Biotechnol 11(35):8718–8721

Resource Coordinators NCBI (2012) Database resources of the national center for biotechnology information. Nucleic Acids Res 41(D1):D8–D20

Rodríguez A, Castrejón-Godínez ML, Salazar-Bustamante E, Gama-Martínez Y, Sánchez-Salinas E, Mussali-Galante P et al (2020) Omics approaches to pesticide biodegradation. Curr Microbiol 77:545–563. https://doi.org/10.1007/s00284-020-01916-5

Rogers SL, McClure N (2003) The role of microbiological studies in bioremediation process optimization. In: Head IM, Singleton I, Milner M (eds) Bioremediation: a critical review. Horizon Scientific Press, Norfolk, pp 27–59

Roume H, Heintz-Buschart A, Muller EEL, May P, Satagopam VP, Laczny CC et al (2015) Comparative integrated omics: identification of key functionalities in microbial community-wide metabolic networks. NPJ Biofilms Microb 1:15007. https://doi.org/10.1038/npjbiofilms.2015.7

Sakshi Haritash AK (2020) A comprehensive review of metabolic and genomic aspects of PAH-degradation. Arch Microbiol 202:2033–2058. https://doi.org/10.1007/s00203-020-01929-5

Shannon P, Markiel A, Ozier O, Baliga NS, Wang JT, Ramage D, Amin N, Schwikowski B, Ideker T (2003) Cytoscape: a software environment for integrated models of biomolecular interaction networks. Genome Res 13(11):2498–2504

Shrivastava R, Upreti RK, Chaturvedi UC (2003) Various cells of the immune system and intestine differ in their capacity to reduce hexavalent chromium. FEMS Immunol Med Microbio 38:65–70. https://doi.org/10.1016/S0928-8244(03)00107-X

Singh M, Singh D, Rai P, Suyal DC, Saurabh S, Soni R, Giri K, Yadav AN (2021) Fungi in remediation of hazardous wastes: current status and future. In: Yadav AN (ed) Recent trends in mycological research, fungal biology. Springer Nature, Cham

Suyal DC, Soni R, Singh DK, Goel R (2021) Microbiome change of agricultural soil under organic farming practices. Biologia 76:1315–1325

Viant MR, Sommer U (2013) Mass spectrometry based environmental metabolomics: a primer and review. Metabolomics 9:144–158. https://doi.org/10.1007/s11306-012-0412-x

Vidali M (2001) Bioremediation. An overview. Pure Appl Chem 73:1163–1172. https://doi.org/10.1351/pac200173071163

Wang JH, Byun J, Pennathur S (2010) Analytical approaches to metabolomics and applications to systems biology. Semin Nephrol 30:500–511. https://doi.org/10.1016/j.semnephrol.2010.07.007

Zerbino DR, Birney E (2008) Velvet: algorithms for de novo short read assembly using de Bruijn graphs. Genome Res 18(5):821–829

Chapter 10
Modern Landfilling Approaches for Waste Disposal and Management

Pooja Sharma, Ravindra Soni, Sudhir Kumar Srivastava, and Surendra Pratap Singh

10.1 Introduction

The highly energetic and goods energy consumption coupled with rising population expansion and high living standards contribute to large municipal solid wastes (MSWs) production, which poses a serious environmental hazard if they are not efficiently disposed of or recycled. MSWs means a variety of solid waste mixture, disposed of every day as waste, garbage, and waste by the public and private communities (Sharma and Kumar 2021). There are currently an estimated 2 billion tonnes, of which over 33% are uncollected by municipalities worldwide MSWs (Waste Atlas 2018). The average per capita garbage produced per day is approximately 0.74 kg. The World Bank anticipates that the production of MSWs would climb "to 3.4 billion tonnes by 2050 (The World Bank 2020)". Of the municipal garbage collected, approximately 70% of the municipal solids are used in waste disposal, 19% are recycled, and 11% for fuel recycling. Almost 3.5 billion individuals "are deprived of basic waste management facilities" from the present global

P. Sharma
Department of Environmental Microbiology, School for Environmental Sciences, Babasaheb Bhimrao Ambedkar (A Central) University, Lucknow, Uttar Pradesh, India

R. Soni
Department of Agri, Microbiology, College of Agriculture, Indra Gandhi Krishi Vishwavidyalaya, Raipur, India

S. K. Srivastava
Chemical Research Laboratory, Department of Chemistry, Dayanand Anglo-Vedic (PG) College, Chhatrapati Shahu Ji Maharaj University, Kanpur, India

S. P. Singh (✉)
Department of Botany, Plant Molecular Biology Laboratory, Dayanand Anglo-Vedic (PG) College, Chhatrapati Shahu Ji Maharaj University, Kanpur, India
e-mail: surendrapratapsingh_kn03@csjmu.ac.in

© The Author(s), under exclusive license to Springer Nature Switzerland AG 2022

D. C. Suyal, R. Soni (eds.), *Bioremediation of Environmental Pollutants*,
https://doi.org/10.1007/978-3-030-86169-8_10

population of 7.6 billion (Waste Atlas 2018). Also, it is predictable that by 2050, 5.6 billion people might be reached without adequate access to basic waste management systems. The solid waste from offices, households, small-scale organizations, and trade firms is the main source of MSWs. The content and classification of MSWs vary greatly around the world across towns, while it comprises biodegradable as well as non-biodegradable fractions composed of inorganic compounds. MSWs are generally made from the waste yard, glass, cardboard, food waste, paper, rubber, plastic, inert materials, metal, electronic waste, and various scrap products. The waste from the kitchen and garden together constitutes the organic part of municipal waste. Miscellaneous waste, which comprises textiles, fabrics, biomedicinal products (shaves, glasses, for example), personal health products, cosmetic, pharmaceutical, pet weight, leather, rubber, and polymer residues, is the most heterogeneous of all components in MSWs (Sharma et al. 2021a, b). High concentration of heavy metals such as Cu, Zn etc are present in the MSWs (Long et al. 2011). Several authors reported that the removal of metals from different sources for the prevention of soil and aquatic pollution (Dixit et al. 2018; Sharma et al. 2021c, d; Tripathi et al. 2021).

Owing to its content, the MSWs management process involved among municipalities, towns, and nations around the globe. The main stages for MSWs management are (1) waste generation; (2) waste collection, processing, and transfer; (3) waste disposal, waste processing, and treatment. In all phases of the municipal waste management system, the sequential stages were taken in developing and developed nations. The well-organized managing of municipally-owned MSWs, i.e. disposal, disposal, depends significantly on a country's gross and population domestic. The waste-to-energy technology for the conversion of fuel, heat, and power of MSWs is well established in industrialized nations (Moya et al. 2017; Nanda and Berruti 2020a, b). In developing countries, when population density requires the capacity available for garbage, the transportation, collection, and disposal of MSWs is nonetheless successfully and hygienically managed. In developing countries, population density, cultural aspects, socio-economic issues, unplanned management, average livelihoods, and the absence of strong environmental legislation generally hinder the appropriate remediation of urban and solid waste. Under two sustainable development targets, for example, Goal 11, sustainable urban and community management, and Goal 12, responsible use and production, the United Nations has defined effective municipal solid waste management (United Nations 2020).

MSWs are a renewable and affordable source with great potential for recovery of energy and precious assets, through waste energy transformation and other recovery techniques. Various waste-to-energy methods are available to turn MSWs into strong, gaseous, and liquid fuels, such as biological conversion and thermochemical technology, to meet the growing need for energy. Including the burning of MSWs to generate power, vapor and heat, and combined energy. In addition, it considerably reduces trash' volume. The incineration of municipal solid waste can decrease the influence and capacity (RenoSam and Rambøll 2006). Pyrolysis liquefaction and gasification are some of possibly the best methods of thermochemical conversion to bio-oil products for MSWs (Munir et al. 2019; Nanda et al. 2016c; Saidi et al. 2020; Fang et al. 2018; Katakojwala et al. 2020). MSWs can their bio-conversion methods like anaerobic digestion and aerobic composting, be transformed into biogas or biomethane and fertile compost (Jain et al. 2015; Singh et al. 2020a, b; Shah et al. 2019).

Some of the neighbourhood of the MSWs stored globally is transferred to waste treatment options like composting, biomass gasification and decomposition. The efficient way in which municipal and global solid waste is disposed of or treated is therefore essential, given that the patterns and practices of recycling programs vary from each country. Landfilling is a usual method used worldwide to bury non-recyclables, although garbage is disposed of in piles or disposed of in pits instead of covering with soil in some developing countries. Deposits seem to be a widespread structured municipal solid residue disposal method in the majority of developed countries but in developing countries are significantly less prevalent because of high population density, which has restrictions in an open area. Some deposits in underdeveloped countries also function as temporary garbage storage and feature waste consolidation, transmission, and processing containment facilities, including sorption, recycling, and treatment (Kour et al. 2021). Deposits can also be used as particular municipal garbage dumping sites that can be examined before tipping for waste sorting and processing of biodegradable substances. Although landfills are the chosen form of trash disposal in the municipality, there is a shortage of space for waste-to-energy recycling for municipal solid waste in emerging nations and exceedingly metropolitan regions. In addition to the existing state of the waste disposal treatment, dumping in soil-clothed pits. To discover prospects for efficient sustainable and implementation management techniques in MSWs, the existing status of waste technology has been assessed.

This chapter also examines the sites as a fresh version of traditional sites. Recent advances in the use of waste disposal leachates and upgrades of waste disposal gas to electricity and fuels have been tried. The advantages, constraints, and uncertainties of reclamation as a modern idea for improved energy recovery and material recycling from closed waste dumps are reviewed. The article also discusses the national and worldwide statistics on the generation, composition, management, disposal, and diversion of municipal solid waste. In addition, we evaluate the strengths, limitations, possibilities, and risks associated with municipal solid waste disposal and management strategies.

10.2 Solid Waste Composition

A World Bank compilation in 2012 indicated that three billion people in worldwide cities produced around 1.3 Gt of solid trash and projected an increase of 5.6 billion by 2015 to 2.2 Gt (Hoornweg and Bhada-Tata 2012). Furthermore, for many emerging and developed countries, national statistics on landfill tonnage and trash composition are severely imprecise. Eurostat tracks more than a dozen solid waste streams in the EU realistically, for example, industry, agriculture, forestry, and mining, one of which is frequently urban waste. For the USA, the annual estimate for deposited solid waste uses a material flow model is a well-known disparity compared to about twice the amount based on annual detection reports published by various regions (Bogner et al. 2008; Van Haaren et al. 2010). Most subsequently, a larger US estimate was also backed by the total yearly waste mass in 2011–2015 reported to the "USEPA Greenhouse Gas Reporting Program" by landfill proprietors (GHGRP) (Powell et al. 2016). Significant volumes of organically buried waste, dominated by lignocellulosic matter, are also stored in sites. Historically, the measured or presumed bulk organic carbon content of US garbage disposal ranges between 15 and 25%. Recent work favors the lower end of the spectrum (De la Cruz et al. 2013). Cellulosic (food plant debris, paper products, garden trash), lipids, and proteins in food waste are among key biodegradable components (Asnani 2006; Krishnamurthi and Chakrabarti 2013). But caution must be given to study waste data from nations that do not supply municipal systemic waste managing procedures. Different forms of industrial waste not dangerous, construction, demolition, bio-solids and other constituents commonly divided on single entrance weighbridges may be also available in MSW sites in the United States.

10.3 Management of Solid Waste

Landfilling supporting the disposal in a designated earth burial site or landfill of biodegradable and non-biodegradable garbage from suburban areas of a municipality. In many countries, garbage disposal has been a customary, most lucrative route. Incineration requires large investments in vast infrastructure, and extremely high-temperature conditions but requires expensive labor as well as the costs of operation and maintenance equipment, such as liquefaction, pyrolysis, anaerobic digestion, gasification, and composting. Because of its cost-effectiveness and lower workload methods, waste disposal is preferred over incineration of MSWs. In addition, a consolidated waste disposal system might also create money by using its waste gas and leachate for energy production. Connectivity of a waste dump with the recycling of leachates and upgrading of the waste gas to biogas followed by burning or flare to produce combined heat and electricity.

The number of available and closed sites in the Union in which municipal solid trash is generated in Europe ranges between 150,000 and 500,000 (Jones et al. 2013). More than 150,000 deposits hold 30–50 billion cubic meters of municipal solid trash throughout Europe, in particular (Wagner and Raymond 2015). Over 33 million tonnes of municipal solid garbage in the United States are burnt, and over 136 million tonnes of urban solid waste are deposited annually (USEPA 2016). Consequently, MSWs landfills fell "from 89% in 1980 to less than 53% in 2014 in the USA" as a result of advances in recycling, composition, combustion, and energy recovery. Several estimates imply that there are over 2000 operational garbage disposal plants in both the US and Canada, whereas Canada has over 2000 active waste disposal plants (Giroux 2014; Peters 2016). Almost 97% of residual solid waste, i.e. recycling, composting, and energy recycling, is deposited each year in Canada and is about 24 million tonnes (Giroux 2014). Approximately 60% of MSWs created by the OECD or the OECD countries are also deposition-led (Hoornweg and Bhada-Tata 2012). Such landfills can also be changed, by implementing efficient integrated technologies that produce green power and secondary material, from trash storage to energy-energy powerhouses.

The proportion of energy input in municipal solid waste landfill and incineration. Deposit and waste incineration need substantially less human effort compared with other major energy supplies via diesel, power, and transport. Latest techno-economic analyses and lifecycle assessments in Tehran, Iran of 8500 tonnes, have shown that incineration has required a higher electricity percentage of 41% of total energy input in comparison to sites that demand 29% power (Nabavi-Pelesaraei et al. 2017). Additional exactly, 8500 tonnes of MSWs were incinerated and electrical input of 422.2 kWh, with an output of 3827.3 kWh, was needed. In other words, roughly 1 GJ of energy is used for every 1.99 tonnes of city MSWs burned for electricity, while more efficient methods and systems might boost energy output. Second, transport-based energy investment accounts for the majority of waste disposals, i.e., 21,600 t.km or 58% of total energy input, or 392 2 t.km, or 15% of the total energy supply. The use of shorter routes can lower the energy consumption generated by the transport sector in landfills by consuming fuel-effective rubbish vehicles collection with a greater waste assemblage or built-in garage compaction apparatus.

Deposits may be categorized as open dumping deposits, semi-controlled sites, and sanitary deposits (Narayana 2009). An open dump waste disposal area is terrain for the disposal in open-air environments of MSWs material. Landfill sites are frequent in all rising nations when MSWs is unilaterally thrown into low-lying open spaces. Subsequently these sites are not being achieved properly, they become a scavenging niche for worms, flies, vermin, mosquitoes, rodents, and pathogenic micro-organisms like vegetables, eagles, crowds, falcons, and other birds. Furthermore, due to oxygen-deficient circumstances, these operational problems are alleviated in anaerobic digestion. Open dumping is prohibited under regulatory law in many developed countries as illegal. Since 79% of municipal solid garbage is deposited in Canada, stringent waste disposal legislation and illegitimate removal can consequence in penalties, enforcement, and prosecution. Climate and Environmental Change Ministers apply the rules (Kelleher et al. 2005; Government of Ontario 2020). In addition, the U.S. Agency for Environmental Protection (AE) forbids open thrash dumping, burning, bans, and disposal of (bio) unsafe waste structure, devastation as well as refurbishment of garbage in open land or flowing water by law, with a penalty of US$1500–3000 (Illinois Environmental Protection Agency 2020).

Half-measured sites are functioned facilities in which MSWs residues are shredded, sorted on-site, and compacted previously discarding at authorized dumping sites. Piles of rubbish are dispersed and straightened using bulldozers or rackets and covered every day with topsoil layers to prevent pollution, like birds breeding, animals, pests, and micro-organisms. While the semi-controlled sites due to their high soil cover are generally less malodorous, they are not designed to handle emissions of waste pipeline and leachate (Narayana 2009). On either hand, advanced forms of semi-controlled deposits are hygienic deposits. To collect liquid leachate and waste gas emissions, the sanitary waste fills have also been designed to provide the onsite waste sorting, segregations, reduction in size, densification, and top-soil cover. Sanitary landfills are distinguished by the routine placing of covered soil atop refreshingly dumped garbage inside a regional border away from residential areas, hence minimizing odour, fires, disease vectors, and scattering of MSWs. In industrialized countries, such kinds of sites remain communal with leachate intervention handling systems. Planned for a landscape extension are also sanitary landfills by excavating additional sites after saturation of existing sites has been limited.

10.3.1 Decentralization of MSWs Management

With active citizen contribution initiatives, a decentralized approach to MSWs management can be reached. The population becomes less dependent upon on waste collection/segregation system in the municipality that enhances the collection of main waste at the site. In developing and executing a decentralized managing approach for MSWs, consensus and strong cooperation among the community and management are essential (Srivastava et al. 2005). Even under a decentralized approach, economically feasible composting is incorporated to generate jobs by distributing the manure (Narayana 2009).

10.3.2 Separation of MSWs at Sites

In MSWs stream, non-biodegradable items, such as plastics and glass, present barriers in their procedures of treatment. This also raises the labor and operating costs associated with their separation and sorting at solid waste treatment facilities before waste-to-energy technologies are deployed and recycled. Such substances have a secondary level of market value, ideal for alternate recycling, including pyrolysis, liquefaction, or gasification. The finest MSWs organization schemes will not be implemented without the separation of MSWs into various streams, i.e. organic, metals, plastic, glass, inert materials, and papers. Give the first responsibility for MSWs segregation to the source. The downstream processing of solid waste could be reduced drastically. The separation of waste at the site could also increase the effectiveness and sanitation of the collection of MSWs. In addition, this technique promotes the decentralized MSWs management system. The transfer of organic and recyclable materials into domestic waste must be minimized. This method could be influenced by the novelty in garbage separation technologies and the application of restraints at the source.

The notion of waste management should be understood by residents: glass, paper, composites, recycling, non-recyclability, pharmaceuticals, non-flammable, flammable, inert, plastics, food waste, electronic waste, as well as building material and destruction and refurbishment trash. However, this innovation is not always known to large municipal garbage collection lorries and their built-in collection mechanisms. It will require intelligible inscriptions and pictograms on accommodating waste in different waste-collecting bags. The level of waste management is dependent on the municipalities in Canada and the United States. For instance, major urban areas are generally supplied with a garbage-retrieving service, and drop-off service is offered to small rural villages. It should also strengthen downstream technologies, product standards, customer relations, and the expansion of alternative products and by-product markets by the municipal recycling industry. Most MSWs administration companies in the industrialized nations, for example, offer housing MSWs, marketable and engineering waste management, and environmental solutions. They operate collecting operations, transfer stations, organic treatment plants, waste energy plants, and garbage disposal plants. It provides consumers with up-to-date evidence on waste disposal sites, collection schedules, waste pick-up, and container delivery for illustration of a recent MSWs disposal capability (Onishi 2005).

10.3.3 Disinfected and Secure Management of MSWs

MSWs comprise a range of opportunistic as well as harmful microbial communities. The faecal bacteria that typically occur in MSWs are *Enterobacter, Klebsiella, Escherichia coli, Salmonella, Streptococci, Shigella,* and *Yersinia* (Hassen et al. 2001). The waste manufacturers and farmers who use solid waste composts are quite contagious. Furthermore, the management of municipal solid trash and compost can cause bodily injury to organic contaminants, flammable, volatile, needles, shards, and sharp things. Even in some developing nations, workers in all sectors of work, including the management of municipal solid waste, have workplace-specific training duties in the area of work safety and awareness of health risks. Albeit in developing, nations such training on safety occurs, not strictly, which leads to a lack of adequate understanding and preparation for health hazards from operators' unsanitary MSWs treatment of waste.

10.3.4 Combustible Gases from Landfills

Deposits are composed unevenly in pockets that can sprinkle finished the covering of the ground, liner, or accumulated garbage. In addition, in underdeveloped nations, compaction and leveling of solid waste at the dump and the soil area are rarely followed (Sharholy et al. 2008). Deposits and anaerobic digesters cause gas emissions that potentially endanger operators by excessive CO_2, CH_4, and CO levels. Such gases can cause solid waste operators or persons living close to the local disposal facilities to asphyxiate, suffocate, asthma, and other respiratory and pulmonary problems. Furthermore, deposit gases may increase pressure to make them highly inflammable through uncontrolled flames or explosions.

MSWs disposal sites should have fine-appointed waste gas, lixiviation, and urgent prevention systems in place to prevent any vulnerabilities. MSWs incineration also results in air pollution and fly ash including dioxins, heavy metals, volatile organic carbons, furans, and dangerously toxic chemicals, and particle matter. The capturing and recuperation of particles and fly ash can reduce their atmospheric propagation persistence. Additionally, appropriate defensive procedures to prevent health dangers for municipal garbage handlers should be established. Environmental monitoring should also be carried out to inspect the leachate, landfill soil, and gases for the instant protection of the operators of the landfill and routine and seasonal chemical analysis (Sharma 2021; Sharma et al. 2020; Sharma and Singh 2021; Sharma and Rath 2021).

10.3.5 Soil Salinity from Compost Application

Agricultural farmlands can apply MSWs compost to boost organic matter, fertility, and demand for nitrogen, carbon, and other microelements (Suyal et al. 2021). In addition, the use of fertilizer can improve metal substances dramatically in the soil that represents poor environments for plant development (Dash et al. 2021; Kumar et al. 2021). The manure can contaminate the soil water by permeability through the soil profile by refractory organic components and other heavy metals. High salt concentrations may also define municipal solid waste-derived compost which may cause salinization when used in soil (Hargreaves et al. 2008). Water stress can have a detrimental effect on the soil's qualities such as pH, fertility, texture, capacity for cation exchange, conductivity, plant development, water retention, and success (Nanda and Abraham 2013). Municipal solid waste compost has also been observed to raise salt and chlorine levels in plants for those with low sodium dietary constraints (Hargreaves et al. 2008). Besides agricultural measures like irrigation, lamination, and application of "plant-growth-promoting rhizobacteria", the synthesis of critical phytohormones and antioxidants in plants can minimize salt and osmotic stress. Pre-treatment and refinement of municipal solid waste can help to lower compost heavy-metal concentrations.

10.4 Sustainable Landfill Management

Passing open dumps typical to sanitary waste disposal sites in underdeveloped and third world countries is a direct way to sustainable waste disposal management. Furthermore, the rehabilitation of closed sites and waste disposal sites through the phytoremediation process might also help to minimize the erosion of covering topsoil from sites and heavy metal and plant adsorption. For remediation and phytoremediation of contaminated soils around the landfill areas, energies like hybrid poplar, olive seed, and grass species can be used. The plantation of energy crops on shut-off waste is highly preferable because of some of these characteristics such as (i) low-price growth; (ii) fast growth; (iii) short-term harvesting of high biomass yields; (iv) non-seasonal available crops; (v) marginally-degraded soils growth (Singh et al. 2020a, b; Nanda et al. 2016b).

Biochar is recognized as a potential approach for landfill site rehabilitation, pollutant immobile treatment, and carbon sequestration, and especially for surface landfill cover (Kumar et al. 2011; Gunarathne et al. 2019; Gopinath et al. 2020). Gasification, pyrolysis, torrefaction, carbonization of biomass, and other waste products, such as MSWs, sewage sludge, and wastewater (Azargohar et al. 2019). Biochar is very mesoporous and carbon-rich, which works as an adsorbent in soil pollution and nutrient immobility material and as a habitat for critical plant growth-focusing micro-organisms (Nanda et al. 2016a). This seeks to improve plant growth, soil fertility, soil nutrients, and humidity bioavailability. Modification of biochar to sites of deposition can greatly improve pathway engineering, especially soil weight, internal friction angles, and cohesiveness (Kumar et al. 2011). In addition, biochar is a cost-effective and eco-friendly adsorbent material to remove both organic pollutants, such as aromatic, fungal substances, colorants, polycyclic, pesticides, and inorganic pollutants, such as metal ions and anions (Han et al. 2014; Singh et al. 2021).

10.5 Optional Marketing of Products from MSWs

Specific fuel products like bio-oils, gas, syngas, methane, and biology are produced by waste-to-energy processes like liquefaction, pyrolysis, gasification, and anaerobic composting. Bio-oil production is regarded as profitable, pyrolysis and liquefaction charcoal yield are minimized. Syngas and methane are also more profitable than charcoal and bio sludge for gasification and anaerobic digestion, respectively. The average bio-oil and biochar expenses are calculated at US$740 per tea and USD500 per tonne (Nanda et al. 2016a). Catalytically, organic oils can be transformed into synthetic transport fuels, while biochar can be used as solid fuel, adsorbing substances, specialized materials production, and applications in soil. The recovery of valuable compounds also includes bio-oils and MSWs leachate.

On a commercial basis, approximately 96% of hydrogen is synthesized by steam reform of gas or methane due to its low cost of production of 1:5 to 3:7 dollars (Balat and Kırtay 2010; Nanda et al. 2017). Hydrogen is a potential source of energy and energy to replace fossil fuels through the methane reform (Singh et al. 2018). In a steam reform process for the production of hydrogen fuel, a new market for highly concentrated methane derived from municipal sites and anaerobic digestion is possible. Given its high energy content, deposit gases can also be used to fire flares or to generate electricity with motors or turbines. Disposable gases create 4–5 kWh/m^3 of energy and can provide around 0.5–2 MW of electricity for medium waste disposal.

10.6 Application of "Pay as You Throw" Scheme

"Pay as you throw" is a waste metering and costing concept for domestic, commercial, and industrial garbage disposal. The 'pay as you toss' model is founded upon the two environmental policy guiding concepts, in particular, the 'polluter pays' and the 'common responsibility notion' (Batllevell and Hanf 2008). Following this approach, the inhabitants are charged a price based on the number of waste generated by the town or local authority for collection. The costs, limitations, and regulations differ according to the municipality. The waste collected is quantified by tags, size, bags, weight, garbage, waste containers, or other advanced techniques like the identification of radio frequencies. Hughes ID Devices or HID Global, Austin (USA), is a manufacturer of safe identifiers and waste management identifying tags for radiofrequency identifying (HID Global 2020). The waste collection truck scans the RFI tag implanted in the waste bin of the customer during the pick-up process. Three different ways can be implemented: partial price unit, full-price, and variable price systems (Kelleher et al. 2005). Further bags or containers for waste are nonetheless subject to extra fees. The waste management system is financed through the fees paid to the citizen by buying waste bags, tags, and particular sizes and numbers of containers or bins, through the full-unit pricing system. Residents are nevertheless advised to hire the smallest basket to minimize their garbage and save the cost of "pay as you lance". In most of Canada's cities, the excess amount of rubbish, recyclables, and waste is also paid for individually. It can lead to illegal dumping of waste, which in many countries is outlawed by law.

10.7 Conclusion

Successful management of MSWs and selecting the appropriate material recycling, landfilling choices, is required to understand the features, contents, and diversity of the waste discharged. The disposal of MSWs is one of the most favoured procedures. In addition, the scarcity of space for new waste disposal sites in highly urban areas has led to waste-to-energy recycling solutions for solid wastes being implemented. Conversion of MSWs energy waste is significantly higher than waste disposal because of less environmental implications, like poor greenhouse gases, great potential for energy harvesting, and emission reduction. Despite being a long-term geological storage facility for trash, deposits also raise environmental concerns concerning air emission, contamination by groundwater, global warming, and health repercussions. Monitoring procedures like depositary liner installations, flaring installations, soil coverage, gas recovery, leachate collection systems, and the rehabilitation of closed deposits are therefore necessary. Landfills also generate landfill gases and volatiles, that cause health risks to the disposal workers. Deposits of gas are very often fuel for their high concentration of gas, but their energy recovery has not been addressed properly. The incorporation of site mining in new site designs is extremely important to ensure that the stabilized waste can easily be accessed in time for mining, recovery of resources, and reclaiming.

References

Asnani PU (2006) Solid waste management. In: Rastogi A (ed) India Infrastructure Report. Urban Infrastructure. Oxford University Press, New Delhi, p 186

Azargohar R, Nanda S, Dalai AK, Kozinski JA (2019) Physico-chemistry of biochars produced through steam gasification and hydro-thermal gasification of canola hull and canola meal pellets. Biomass Bioenergy 120:458–470

Balat H, Kırtay E (2010) Hydrogen from biomass–present scenario and future prospects. Int J Hydrog Energy 35(14):7416–7426

Batllevell M, Hanf K (2008) The fairness of PAYT systems: some guidelines for decision-makers. Waste Manag 28(12):2793–2800

Bogner J, Pipatti R, Hashimoto S, Diaz C, Mareckova K, Diaz L, Kjeldsen P, Monni S, Faaij A, Gao Q, Zhang T (2008) Mitigation of global greenhouse gas emissions from waste: conclusions and strategies from the intergovernmental panel on climate change (IPCC) fourth assessment report. Working group III (mitigation). Waste Manag Res 26(1):11–32

Dash B, Sahu N, Singh AK, Gupta SB, Soni R (2021) Arsenic efflux in Enterobacter cloacae RSN3 isolated from arsenic-rich soil. Folia Microbiol 66:189–196

De la Cruz F, Chanton J, Barlaz M (2013) Measurement of carbon storage in landfills from the biogenic carbon content of excavated waste samples. Waste Manag 33:2001–2005. https://doi.org/10.1016/j.wasman.2012.12.012

Dixit R, Agrawal L, Singh SP, Singh PC, Prasad V, Chauhan PS (2018) Paenibacillus lentimorbus induces autophagy for protecting tomato from Sclerotium rolfsii infection. Microbiol Res 215:164–174

Fang S, Yu Z, Ma X, Lin Y, Chen L, Liao Y (2018) Analysis of catalytic pyrolysis of municipal solid waste and paper sludge using TG-FTIR, Py-GC/MS and DAEM (distributed activation energy model). Energy 143:517–532

Giroux L (2014) State of waste management in Canada prepared for: Canadian Council of Ministers of Environment

Gopinath KP, Vo DVN, Prakash DG, Joseph AA, Viswanathan S, Arun J (2020) Environmental applications of carbon-based materials: a review. Environ Chem Lett. https://doi.org/10.1007/s10311-020-01084-9 (Return to ref 2020 in article)

Government of Ontario (2020) Illegal waste dumping. https://www.ontario.ca/page/illegal-waste-dumping. Accessed 14 Feb 2020

Gunarathne V, Ashiq A, Ramanayaka S, Wijekoon P, Vithanage M (2019) Biochar from municipal solid waste for resource recovery and pollution remediation. Environ Chem Lett 17 (3):1225–1235

Han D, Tong X, Currell MJ, Cao G, Jin M, Tong C (2014) Evaluation of the impact of an uncontrolled landfill on surrounding groundwater quality, Zhoukou, China. J Geochem Explor 136:24–39

Hargreaves JC, Adl MS, Warman PR (2008) A review of the use of composted municipal solid waste in agriculture. Agric Ecosyst Environ 123(1–3):1–14

Hassen A, Belguith K, Jedidi N, Cherif A, Cherif M, Boudabous A (2001) Microbial characterization during composting of municipal solid waste. Bioresour Technol 80(3):217–225

HID Global (2020) RFID waste management. https://www.hidglobal.com/solutions/waste-management. Accessed 15 Feb 2020

Hoornweg D, Bhada-Tata P (2012) What a waste: a global review of solid waste management

Illinois Environmental Protection Agency (2020) Open dumps. https://www2.illinois.gov/epa/topics/waste-management/illegal-dumping/Pages/open-dumping.aspx. Accessed 14 Feb 2020

Jain S, Jain S, Wolf IT, Lee J, Tong YW (2015) A comprehensive review on operating parameters and different pretreatment methodologies for anaerobic digestion of municipal solid waste. Renew Sust Energ Rev 52:142–154

Jones PT, Geysen D, Tielemans Y, Van Passel S, Pontikes Y, Blanpain B, Quaghebeur M, Hoekstra N (2013) Enhanced landfill mining in view of multiple resource recovery: a critical review. J Clean Prod 55:45–55

Katakojwala R, Kopperi H, Kumar S, Mohan SV (2020) Hydrothermal liquefaction of biogenic municipal solid waste under reduced H2 atmosphere in biorefinery format. Bioresour Technol 310:123369

Kelleher M, Dixie J, Robins J (2005) Taking out the trash: how to allocate the costs fairly. Commentary-CD Howe Institute, vol 213, p 1

Kour D, Kaur T, Devi R, Yadav A, Singh M et al (2021) Beneficial microbiomes for bioremediation of diverse contaminated environments for environmental sustainability: present status and future challenges. Environ Sci Pollut Res 28:24917–24939

Krishnamurthi S, Chakrabarti T (2013) Diversity of bacteria and archaea from a landfill in Chandigarh, India as revealed by culture-dependent and culture-independent molecular approaches. Syst Appl Microbiol 36(1):56–68

Kumar S, Chiemchaisri C, Mudhoo A (2011) Bioreactor landfill technology in municipal solid waste treatment: an overview. Crit Rev Biotechnol 31(1):77–97

Kumar P, Dash B, Suyal D C, Gupta SB, Singh AK, Chowdhury T, Soni R (2021) Characterization of arsenic-resistant Klebsiella pneumoniae RnASA11 from contaminated soil and water samples and its bioremediation potential. Curr Microbiol. https://doi.org/10.1007/s00284-021-02602-w

Long YY, Shen DS, Wang HT, Lu WJ, Zhao Y (2011) Heavy metal source analysis in municipal solid waste (MSW): case study on Cu and Zn. J Hazard Mater 186(2–3):1082–1087

Moya D, Aldás C, Jaramillo D, Játiva E, Kaparaju P (2017) Waste-to-energy technologies: an opportunity of energy recovery from municipal solid waste, using Quito-Ecuador as case study. Energy Procedia 134:327–336

Munir MT, Mardon I, Al-Zuhair S, Shawabkeh A, Saqib NU (2019) Plasma gasification of municipal solid waste for waste-to-value processing. Renew Sust Energ Rev 116:109461

Nabavi-Pelesaraei A, Bayat R, Hosseinzadeh-Bandbafha H, Afrasyabi H, Chau KW (2017) Modeling of energy consumption and environmental life cycle assessment for incineration and landfill systems of municipal solid waste management-a case study in Tehran Metropolis of Iran. J Clean Prod 148:427–440

Nanda S, Abraham J (2013) Remediation of heavy metal contaminated soil. Afr J Biotechnol 12:3099–3109

Nanda S, Berruti F (2020a) Municipal solid waste management and landfilling technologies: a review. Environ Chem Lett 1–24

Nanda S, Berruti F (2020b) Thermochemical conversion of plastic waste to fuels: a review. Environ Chem Lett 1–26
Nanda S, Dalai AK, Berruti F, Kozinski JA (2016a) Biochar as an exceptional bioresource for energy, agronomy, carbon sequestration, activated carbon and specialty materials. Waste Biomass Valorization 7(2):201–235
Nanda S, Dalai AK, Kozinski JA (2016b) Supercritical water gasification of timothy grass as an energy crop in the presence of alkali carbonate and hydroxide catalysts. Biomass Bioenergy 95:378–387
Nanda S, Isen J, Dalai AK, Kozinski JA (2016c) Gasification of fruit wastes and agro-food residues in supercritical water. Energy Convers Manag 110:296–306
Nanda S, Rana R, Zheng Y, Kozinski JA, Dalai AK (2017) Insights on pathways for hydrogen generation from ethanol. Sustain Energy Fuels 1(6):1232–1245
Narayana T (2009) Municipal solid waste management in India: from waste disposal to recovery of resources? Waste Manag 29(3):1163–1166
Onishi N (2005) How do the Japanese dump trash? Let us count the myriad ways. The New York Times 12
Peters A (2016) Fast company. These maps show how many landfills there are in the U.S. https://www.fastcompany.com/3062853/these-maps-show-how-much-of-the-us-is-covered-in-landfills. Accessed 12 Apr 2020
Powell JT, Townsend TG, Zimmerman JB (2016) Estimates of solid waste disposal rates and reduction targets for landfill gas emissions. Nat Clim Chang 6(2):162–165
RenoSam, Rambøll (2006) The most efficient waste management system in Europe: waste-to-energy in Denmark. https://stateofgreen.com/files/download/275. Accessed 8 Feb 2019
Saidi M, Gohari MH, Ramezani AT (2020) Hydrogen production from waste gasification followed by membrane filtration: a review. Environ Chem Lett 1–28
Shah GM, Tufail N, Bakhat HF, Ahmad I, Shahid M, Hammad HM, Nasim W, Waqar A, Rizwan M, Dong R (2019) Composting of municipal solid waste by different methods improved the growth of vegetables and reduced the health risks of cadmium and lead. Environ Sci Pollut Res 26(6):5463–5474
Sharholy M, Ahmad K, Mahmood G, Trivedi RC (2008) Municipal solid waste management in Indian cities–a review. Waste Manag 28(2):459–467
Sharma P (2021) Efficiency of bacteria and bacterial assisted phytoremediation of heavy metals: an update bioresource technology, 124835
Sharma P, Kumar S (2021) Characterization and phytotoxicity assessment of organic pollutants in old and fresh municipal solid wastes at open dump site: a case study. Environ Technol Innov 24:101938
Sharma P, Rath SK (2021) Potential applications of fungi in the remediation of toxic effluents from pulp and paper industries. In: fungi bio-prospects in sustainable agriculture, environment and nanotechnology. Academic, pp 193–211
Sharma P, Singh SP (2021) Role of the endogenous fungal metabolites in the plant growth improvement and stress tolerance. In: Fungi bio-prospects in sustainable agriculture, environment and nanotechnology. Academic, pp 381–401
Sharma P, Singh SP, Pandey S, Thanki A, Singh NK (2020) Role of potential native weeds and grasses for phytoremediation of endocrine-disrupting pollutants discharged from pulp paper industry waste. In: Bioremediation of pollutants, pp 17–37. Elsevier
Sharma P, Tripathi S, Purchase D, Chandra R (2021a) Integrating phytoremediation into treatment of pulp and paper industry wastewater: field observations of native plants for the detoxification of metals and their potential as part of a multidisciplinary strategy. J Environ Chem Eng 9(4):105547

Sharma P, Kumar S, Pandey A (2021b) Bioremediated techniques for remediation of metal pollutants using metagenomics approaches: a review. J Environ Chem Eng 9(4):105684

Sharma P, Ngo HH, Khanal S, Larroche C, Kim SH, Pandey A (2021c) Efficiency of transporter genes and proteins in hyperaccumulator plants for metals tolerance in wastewater treatment: sustainable technique for metal detoxification. Environ Technol Innov 23:101725

Sharma P, Pandey AK, Udayan A, Kumar S (2021d) Role of microbial community and metal-binding proteins in phytoremediation of heavy metals from industrial wastewater. Bioresour Technol 326:124750

Singh S, Kumar R, Setiabudi HD, Nanda S, Vo DVN (2018) Advanced synthesis strategies of mesoporous SBA-15 supported catalysts for catalytic reforming applications: a state-of-the-art review. Appl Catal A Gen 559:57–74

Singh A, Nanda S, Berruti F (2020a) A review of thermochemical and biochemical conversion of Miscanthus to biofuels. In: Biorefinery of alternative resources: targeting green fuels and platform chemicals, pp 195–220

Singh AD, Upadhyay A, Shrivastava S, Vivekanand V (2020b) Life-cycle assessment of sewage sludge-based large-scale biogas plant. Bioresour Technol 309:123373

Singh M, Singh D, Rai P, Suyal DC, Saurabh S, Soni R, Giri K, Yadav AN (2021) Fungi in remediation of hazardous wastes: current status and future. In: Yadav AN (ed) Recent trends in mycological research, fungal biology. Springer Nature, Cham

Srivastava PK, Kulshreshtha K, Mohanty CS, Pushpangadan P, Singh A (2005) Stakeholder-based SWOT analysis for successful municipal solid waste management in Lucknow, India. Waste Manag 25(5):531–537

Suyal DC, Soni R, Singh DK, Goel R (2021) Microbiome change of agricultural soil under organic farming practices. Biologia 76:1315–1325

The World Bank (2020) Trends in solid waste management. https://datatopics.worldbank.org/what-a-waste/trends_in_solid_waste_management.html. Accessed 21 Apr 2020

Tripathi S, Sharma P, Purchase D, Tiwari M, Chakrabarty D, Chandra R (2021) Biodegradation of organo-metallic pollutants in distillery wastewater employing a bioaugmentation process. Environ Technol Innov 23:101774

United Nations (2020) About the sustainable development goals. https://www.un.org/sustainabledevelopment/sustainable-development-goals/. Accessed 14 Apr 2020

USEPA, U.S. Environmental Protection Agency (2016) Advancing sustainable materials management: 2014 fact sheet. USEPA, Washington, DC (Return)

Van Haaren R, Themelis N, Goldstein N (2010) State of garbage in America. Biocycle 51:16

Wagner TP, Raymond T (2015) Landfill mining: case study of a successful metals recovery project. Waste Manag 45:448–457

Waste Atlas (2018) What a waste: an updated look into the future of solid waste management. https://www.worldbank.org/en/news/immersive-story/2018/09/20/what-a-waste-an-updated-look-into-the-future-of-solid-waste-management. Accessed 15 Apr 2020

Chapter 11
Aerobic Granular Technology: Current Perspective and Developments

Jyoti Rajwar, Divya Joshi, Shilippreet Kour, and Prasenjit Debbarma

11.1 Introduction

Today's ecosystem and its long-term water supplies are also under pressure. Because of our society's rapid urbanization and industrialization, cutting-edge water treatment technologies are needed. In response to the strong consumption and demands that our new generation puts on water, a technology capable of providing reliable, filtered water. Aerobic granulation is the process of converting activated seed sludge into a broad granule with self-immobilizing properties that adsorb and consume polluted and harmful microorganisms. The activated sludge (AS) method is utilized in wastewater treatment facilities (WWTPs) all over the globe to treat urban and industrial wastewater via biological treatment. In 2014, the discovery of AS was commemorated with a centennial year to commemorate its fruitful journey in providing sanitation and mitigating environmental harm. However, because of its huge land footprint, recirculation flows, and complicated process structures involving several process units to remove biomass, nitrogen, and phosphorus from wastewater, it is no longer deemed viable (Sheik et al. 2014). Because of bottlenecks and new advances in sustainable wastewater treatment, the AS method is now facing

J. Rajwar (✉)
Department of Microbiology, School of Science and Information Technology, Skyline University, Kano, Nigeria
e-mail: jyoti@sun.edu.ng

D. Joshi
Uttarakhand Pollution Control Board, Regional Office, Kashipur, Uttarakhand, India

S. Kour
Department of Microbiology, Akal College of Basic Sciences, Eternal University, Baru Sahib, Sirmaur, Himachal Pradesh, India

P. Debbarma
School of Agriculture, Graphic Era Hill University, Dehradun, Uttarakhand, India

© The Author(s), under exclusive license to Springer Nature Switzerland AG 2022

D. C. Suyal, R. Soni (eds.), *Bioremediation of Environmental Pollutants*,
https://doi.org/10.1007/978-3-030-86169-8_11

extinction. The key drivers of these sustainable wastewater management systems are improvements in the treatment mechanism, minimization of material inputs, and resource recovery from wastewater (Debbarma et al. 2017; Kumar et al. 2021; Kour et al. 2021). To address the difficulties associated with nutrient reduction and solid-fluid separation that have plagued the conventional aerobic sludge (AS) system, aerobic granular sludge (AGS) was created. AGS Technology addresses crucial bottlenecks of the AS mechanism including high-level land use, high-energy biomass, and wastewater recirculation, and biomass-water water separation issues. AGS is expected to surpass AS and provide more sustainable care over a 100 years. Because of its potential uses in aerobic wastewater treatment since the first study, AGS has been drawing the increasing interest of microbiologists and wastewater engineers. AGS process has been the advanced technical approach for huge WWTPs for the processing of household and commercial wastewaters during the last two decades (Nancharaiah and Sarvajith 2019). The main factors affecting the acceptance of AGS technology are the substantial reduction of land footprint (up to 75%), capital and maintenance costs (up to 50%), and environmental effects (up to 50%) as opposed to traditional AS processes and other technologies (Bengtsson et al. 2018). With these economic benefits in addition to increased treatment efficiency, the AGS method may quickly become the industry norm for biodegradable biological wastewater treatment. A significant proportion of the aerobic granules is made up of highly concentrated microbial aggregates with densities much higher than that of ordinary activated sludge. Apart from that, it is known that aerobic granules contain the following characteristics:

1. The form is regular, flat, and nearly circular.
2. Excellent settling properties
3. Microbial framework that is thick and solid
4. Exceptional biomass retention
5. Capacity to tolerate elevated levels of organic loading
6. Resistance to toxins

Due to the distinct characteristics of the granules, the technique has lately been developing for the management of high-strength wastewaters including organics, phosphorus, radioactive chemicals, nitrogen, and xenobiotics (Adav et al. 2008). This chapter addresses the crucial aspects of this evolving method, including technical implications, granulation processes, the extracellular polymeric material (EPS) matrix, microbiology, long-term survival, and current perspective & developments.

11.2 What Is Aerobic Granular Sludge?

Aerobic granules are dense, spherical, self-immobilized microbial masses with a dense construction & outstanding settling capacity. When crushed, they acquire a distinct appearance and settle as discrete entities with a diameter more than 0.1 mm in circumference when the material is pressed together. While no standard definition

Table 11.1 Comparison between activated sludge and aerobic granular sludge

Parameter	Activated sludge	Aerobic granular sludge	References
Size	0.5–500 μm (1–10 μm dominant)	0.3–4.0 mm (2–3 mm dominant)	Kuśnierz (2018) and Long et al. (2019)
Shape	Irregular and small	Spherical and large	Moragaspitiya et al. (2019) and Bassin et al. (2019)
Specific gravity	Nearly 1.03 kg dry weight m^{-3}	1.004–1.1 kg dry weight m^{-3}	Jafarinejad 2017 and Gao et al. (2011)
Settling velocity	1–3.29 m h^{-1}	18–40 m h^{-1})	Zou et al. (2021) and Bengtsson et al. (2018)
EPS content	50 ± 20 mg g^{-1} volatile content of the sludge	253 ± 14 mg g^{-1} volatile content of the sludge	Kim et al. (2020)
Sludge Volume Index SVI	250 mL g^{-1}	SVI5 (50–80 ml g^{-1}) SVI30 (30–40 ml g^{-1})	Li et al. (2006) and Yousuf (2013)
Microenvironment	Not possible to have distinct microenvironment	Aerobic, anoxic and anaerobic microenvironments within single granule	Nancharaiah and Sarvajith (2019)
Toxicity tolerance	Low tolerance	Very high tolerance	Ren et al. (2017) and Daraei et al. (2019)
Energy consumption	High consumption	23% lower than activated sludge	Bengtsson et al. (2018)

of aerobic granular waste existed before the first aerobic granular sludge workshop, which took place in Munich, Germany in 2004, there was an agreement on what constitutes aerobic granular sludge in scientific and technical literature. Aerobic granules, according to the findings of this workshop, are microbial aggregates that do not accumulate when exposed to lower hydrodynamic stress and settle considerably more quickly than activated sludge flocs. This came about as a consequence of a lengthy debate. The second aerobic granular workshop, which occurred in 2006, included discussions about how to interpret the definition broadly to include masses of microbes, zero coagulation in lower hydrodynamic shear, settling quicker than activated sludge, a minimal amount of scale, and a harvesting process. In the case of an aggregate that has any of the characteristics mentioned above, the aggregate is known as aerobic granular sludge. As a consequence, laboratory findings are easier to understand, and the distinctions between aerobic granular sludge, activated sludge, and biofilms are clearer than they were before (Ni 2012) (Table 11.1).

11.3 Granule Formation and Characterization

Various operational parameters, including seed sludge, the configuration of the substrate, rate of organic loading, feeding strategy, settling time, reactor design, aeration speed, and exchange ratio, had an impact on granulation. It seems that the successful development of aerobic granules had a rather short "window" of opportunity. When the proper circumstances are met, it is possible to get the much sought "super" grains.

11.3.1 Granule Formation

11.3.1.1 Seed Sludge

Seed sludge, feed configurations, sequencing batch reactors (SBR), SBR operating parameters (pH, cycling period, temperature, and others), all have an impact on the granulation technique. The bulk of the tests carried out allowed sludge seed to grow aerobic granules in a controlled environment (Liu and Tay 2004). Because hydrophilic bacteria, which constitute the bulk of available bacteria in full-scale wastewater management plants are less likely to bind to sludge flocs than hydrophobic bacteria, which make up the majority of free bacteria in full-scale treatment plant effluent, the bacterial population found in active sludge is critical for the AG method. The higher the concentration of hydrophobic bacteria in the seed sludge, the quicker the aerobic granulation with good settleability happens (Wilen et al. 2007).

11.3.1.2 Feed Composition

Numerous substrates, including glucose, acetate, fructose, phenol, and industrial wastewater, are used to cultivate aerobic granules for a long time. Aerobic granules were also produced by adding an inorganic source of carbon and nitrifying bacteria in addition to the standard conditions. When it came to nitrification, these nitrifying granules worked extremely well. Aerobic granules were also successfully produced in research lab sequencing batch reactors (SBR) for the management of wastewater including large amounts of particle organic debris. The filamentous framework of the glucose-fed aerobic granules was seen, while the non-filamentous and very condensed bacterial structure of the acetate-fed aerobic granules was observed. It has been suggested that the shape of the carbon source affects the microstructure of granules and the variety of species. Positive, divalent, and trivalent ions of Ca^{2+}, Fe^{2+}, and Mg^{2+}, have been shown to bind to the nucleus of microorganisms. The addition of Ca^{2+} ions enhanced the granulation phase by reducing the forming time in half, resulting in a shorter granulation phase (Jang et al. 2003). Yang et al. (2008) discovered that aerobic granulation produced granulates with a size of about 7 mm in the presence of a fungus at pH 4.0, while the granulate size was only 4.8 mm when bacteria were used to monitor the process at pH 8.0. The implications of pH on the selection of inoculum and the production of aerobic granulation are yet to be explored.

11.3.1.3 SBR Operation

SBR cyclic activity involves a considerable amount of loading and aeration, as well as the settling and drainage of effluents, among other things (Fig. 11.1). The time settled and the liquid volume exchange ratio after each stage are the most important tests for the removal of nongranular biomass and are performed after each stage. The time settled and the liquid volume exchange ratio is the most important tests for the removal of non-granular biomass from the reactor. As cycle length increases, the hydraulic retention time (HRT) decreases, resulting in a reduction in selection strain and an increase in selection strain. As a result of the limited time available, bacterial growth will not be able to compensate for sludge losses as a result of hydraulic washouts in a reasonable amount of time (Pan et al. 2004). Aeration is at the core of the SBR process, and it happens in two stages: the degradation phase, where the amount of substrate consumed is decreased to the bare minimum, and then an aerobic starvation phase, where the amount of substrate ingested is no longer beneficial. As a consequence of the extended starvation period, the granule stability has been compromised.

11.3.2 Granule Characterization

To better understand the properties of aerobic granules, a wide variety of factors were studied. Physical, chemical, and biological factors are among the variables to consider. Research studies have also documented general sludge granule features

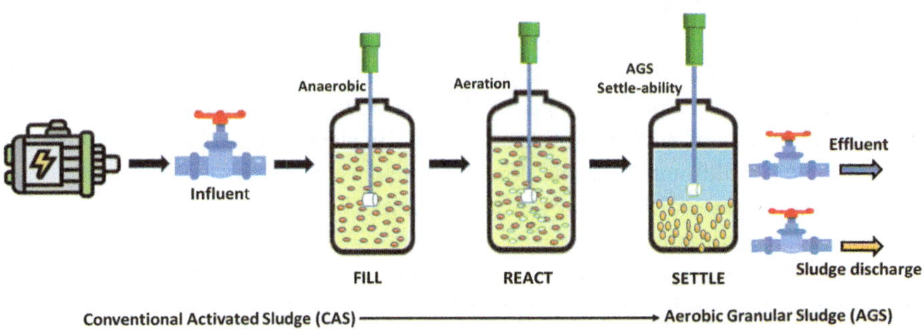

Fig. 11.1 The operation of aerobic granular sludge operation in a sequencing batch reactor

like dimensions and structure, settling capacity, stability against shear pressures, EPS material content, metabolism rates, and reactor efficiency.

11.3.2.1 Physical Parameters

Shape and Size are usually aerobic granules with a smooth outer face that are spherical or ellipsoidal (Chen et al. 2008). Two different phenomena have resulted in the average size of granulates in aerobic processes: (1) biomass growth and (2) hydrodynamic shear forces detaching cells (Liu and Tay 2002). When subjected to severe shear stress, the generation of aerobic granules and the stability of the granules were both improved. One of the key factors affecting the construction of aerobic granules was the low shear stress placed on the granules throughout the process. Higher local shear forces were generated in the SBR, resulting in denser granules with smaller diameters. Aerobic granules that were more uniform, spherical, and compressed were formed when the hydrodynamic shear force was increased. The total diameter of granules has been found to range from 0.2 to 5.0 mm and exceed the average sludge flock diameter. Instead of a rise in organic matter content overall, greater diameters could be attributed to higher inorganic content in granules (i.e., biomass 10 and EPS). There was also proof that a rise in the amount of granule above a certain threshold of the biomass (i.e. 4.0 mm in diameter) (Toh et al. 2003). The studies suggest that AGS will settle at a rate ranging from 25 to 70 m h^{-1}, which is much quicker than the settling rate of activated sludge flocs (7–10 m h^{-1}). The increased biomass concentrations caused by greater settling velocities improve removal capability by increasing active biomass and keeping slow-growing bacteria alive, which in turn increases removal capacity. There have been reports of values as low as 20 ml m^{-3} for granular sludge's sludge volume index (SVI). Because of the high density and compact structure of the granules, their SVI values are quite small. Grains have specific gravities ranging from 1.04 to 1.100 g cm^{-3}, while flocculent activated sludge has specific gravity ranges from 1.002 to 1.06 g cm^{-3} of water (Lashkarizadeh 2015).

11.3.2.2 Chemical Parameters

Liu et al. (2003) found a relationship between the hydrophobicity of the cell surface and the growth of heterotrophic and nitrifying granules. Granular sludge has double the hydrophobicity of normal bioflocs. Increases in shear force enhanced cell surface hydrophobicity, although organic loading rates were unaffected by cell surface hydrophobicity (Vu 2017). However, the exact mechanisms that cause the observed changes are undistinguishable.

Bacterial extracellular metabolites (EPS) are metabolic byproducts that accumulate on the surface of the cells and may alter the physicochemical features of the cell surface, such as its charge, hydrophobicity (Nishanth et al. 2021). Experiments have shown that extracellular polysaccharide production is significantly correlated with shear stress and that the stability of AG is correlated with extracellular polysaccharide synthesis. A significant role in the formation of all types of biofilms is believed to be played by extracellular polymeric substances (EPS), which are found in abundance in cell flocs and biofilms. These substances are thought to be involved in the formation of all types of biofilms, including flocculation and granulation. Aerobic granules showed more hydrophobicity on their cell surfaces than seed sludge, indicating that they were more resistant to water. The hydrophobicity of the cell surface was shown to be inversely proportional to an upsurge in the extracellular protein content of the sludge. Increased extracellular polysaccharide production in the presence of high shear may subsidize the dense and robust structure of AG, according to some researchers.

Extracellular proteins, according to Zhang et al. (2007), may perform a significant role in monitoring the growth and stability of aerobic granules throughout their formation and maturation. Also important is the fact that protein includes a higher percentage of negatively charged amino acids than carbs, which means that protein is highly involved in electrostatic interactions with multivalent cations than carbohydrates, which is critical for aggregate structure stability.

11.3.2.3 Biological Parameters

With the usage of scanning electron microscopy (SEM), optical and confocal scanning laser microscopy (CLSM), as well as fluorescent in situ hybridization (FISH), we were able to gain new horizons regarding the assembly of the aerobic granules (Tsitouras 2021). It has been found in aerobic granules that have been produced under several different conditions that bacteria that are heterotrophic, denitrifying, nitrifying, phosphorous, and glycogen-accumulating are identified (Liu et al. 2018). The type of aerobic granules, as well as the composition of the culture medium in which they were generated, were both significant variables in influencing the microbial diversity of aerobic granules (Rollemberg et al. 2019).

There is a significant connection between the microbial range of aerobic granules and the type of culture medium provided for their growth, according to the research. In contrast to aerobic granules grown on glucose, rod-shaped microbes predominated in granules developed on acetate, with filaments and some cocci bacteria making up the majority of the bacteria. The use of ribosomal-based molecular techniques as well as PCR cloning discovered variations in microbial variety across juvenile, mature, and elderly aerobic grains fed on sugar cane sugar, according to their findings. According to the results, aerobic granule formation seemed to be a dynamic process involving a collection of microorganisms that came together. According to the researchers, changes in the microbial range were accountable for physiological adaptation by numerous microorganisms throughout the aerobic granulation. Biological deviations, such as shifts in bacterial population and species richness, would be endorsed to interactions between different groups of bacteria and the microniches in which they dwell, according to this theory. This research has found that microorganisms linked with five operational taxonomic units were present in all of the granules examined at unlike phases of growth, suggesting that these bacteria may perform a key role in the creation of aerobic granules in the environment. It was also found that distinct operational taxonomic units predominated at various development stages and different growth stages. For example, the presence of many different types of bacteria that were more abundant in the older granules and missing from the younger granules are examples of this. Furthermore, fluctuations in relative abundance can be used as markers of granule formation or even as a prediction of the onset of granule degeneration and breakdown, depending on the circumstances (Tay et al. 2009).

11.4 Mechanism of Aerobic Granulation

Several circumstances must be met for bacteria to form granules, and the involvement of physiochemical, biochemical, and environmental variables in the granulation process should all be included in the same study. The following is a generalized model for the granulation process developed by Liu and Tay (2004), which includes the following features:

Step 1 Physical mobility is required for bacterium-to-bacterium contact to commence. The following are the forces involved:

- Hydrodynamic force
- Gravity force.
- Diffusion force
- Thermodynamic interactions like Brownian motion
- Cellular mobility is important. Cells may be directed via the use of cilia, flagella, and pseudopods, or they can be guided through the use of a signal transduction system.

Step 2 Preliminary attractive forces to keep steady multicellular interactions.

Physical Forces

- Van der Waals contacts
- Opposite charge attraction
- Thermodynamic forces comprised of free energy of surface and surface tension
- Hydrophobicity
- Filamentous bacteria serve as a connecting link, allowing distinct cells to remain structured.

Chemical Forces

- Hydrogen liaison
- Ionic pair formation
- Ionic triplet formation
- Inter-particulate bridge

Biochemical Forces

- Dehydration of cell surface
- Fusion of cell membrane
- Communication and cooperative behavior in microbial communities

Step 3 Microbial forces for making mature cell accumulation:

- Synthesis of the extracellular polymer by microbes, such as EPS, etc.
- Growth of cellular mass
- Metabolic adjustment and genetic competency prompted by environments, which enable the cellular interaction, and affects the formation of an extremely systematized microbial organization

Step 4 Stable 3-D configuration of bacterial aggregate designed by hydrodynamic shear forces.

A particular structured community of microbes would ultimately form as a result of the hydrodynamic shear stress that was applied. The external size and morphology of microbial aggregates are influenced by a variety of variables, including their interaction strength/pattern, microbial diversity, hydrodynamic shear force, and the pace at which the aggregates are loaded with the substrate (Wu et al. 2020). Notably, the hydrophobicity of the bacterial surface may have a substantial impact on the onset of aerobic granulation at the beginning of the process. It is believed that elevation in the hydrophobicity of cell surfaces leads to a decrement of the surface's excess Gibbs energy, which in turn increases cell-to-cell communication and also functions as a driving force for bacteria to self-aggregate out of the liquid phase, following thermodynamic theory. For cell attachment to be successful, it has been shown that hydrophobic affinity is required. Due to the increased hydrophobicity of the cellular membrane, there is a greater cell-cell contact, which results in a denser cell structure. Cell surface hydrophobicity has long been believed to be necessary for the development of bacterial biofilms and anaerobic granules, but this has not been proven. In contrast to sludge flocs, it has been found that the cell membrane hydrophobic nature of aerobic granules is significantly higher than that of sludge. This suggests that hydrophobicity of the cell surface may be important for aerobic grain granulation (Ali et al. 2018).

11.5 Applications of Aerobic Granulation Technology

11.5.1 Nutrient Removal by Aerobic Granules

It has been demonstrated that the efficiency of aerobic granules in removing nutrients (carbon, nitrogen, and phosphorus) is analogous to those of conventional methods used to decompose biological waste. In a single reactor-based process, aerobic granules are capable of conducting synchronized nitrification, denitrification, and phosphorus removal, according to some researchers. Aerobic granulation has a major technological advantage in this regard. In the case of concurrent elimination of nitrogen and phosphorous, the removal may take place in the absence of complete granule growth; nevertheless, full granule development is needed for efficient denitrification. A denitrification zone is present in the center of mature granules, which is most presumably responsible for the process of denitrification (Lemaire et al. 2008). A study conducted by Bao et al. in 2009 examined the effect of a cold temp of 10 °C on the production of aerobic granules and the removal of nutrients. It was observed that at such temperature, the formation of granules having 3.4 mm diameter takes place, and these granules have much more effective denitrification ability. When the influent chemical oxygen demand (COD) concentration was lowered, it was found that the denitrification capacity was reduced. These results suggest heterotrophic denitrification occurred primarily in the anoxic core of the granules, rather than elsewhere. Anaerobic ammonium oxidation (anammox) is now possible with aerobic granules, in addition to nitrification and eventual heterotrophic denitrification. Specifically, the anoxic staple area of the granules encourages the growth of anammox microbes. In this instance, aerobic granules containing an anammox microbial population must potentially be efficient for the growth of mild wastewater, such as that generated by municipal sewerage systems (Kagawa et al. 2015).

Kagawa et al. (2015) devised a prototype to describe the nutrient elimination capacity of aerobic granules. It has been discovered by scientists that the efficiency with which aerobic granules remove nutrients is dependent on the amount of DO present in the medium. Researchers Lashkarizadeh et al in 2016 studied the influence of pH on the competency with which aerobic granules collect nutrients from the environment. The use of an alkaline pH of 9 (high alkaline) in wastewater treatment facilities results in a permanent reduction in nitrogen removal effectiveness from 88 to 66% and a reduction in phosphorus removal efficiency from 98 to 50%, according to the study. A moderate acidic pH 6 had no long-term impact on the aerobic granules capacity to take nutrients from the surrounding environment. When dealing with low-strength wastewater, such as municipal wastewater, the removal of nutrients is challenging for aerobic granules. This is mostly owing to the period needed for granule production, which is relatively short. For nitrogen removal, Kang and Yuan (2017) investigated the potential of developing the granules in high-strength wastewater followed by applying them to low-strength wastewater to get the best possible outcomes. It has been determined that for the method to be effective, it is necessary to maintain an adequate food-to-microorganism ratio throughout.

Coma et al. (2012) examined the application of aerobic granules to remove phosphorus from wastewater from a residential setting. Furthermore, it was found that nitrite concentrations greater than 5 mg nitrite-nitrogen per liter significantly reduced the phosphorus removal ability. Water phosphorus removal is often done via the employment of phosphorus-accumulating microorganisms, which are used to extract the phosphorus from the wastewater (Erdal et al. 2006). In addition to this conventional technique of phosphorus removal, aerobic granules have the potential to precipitate phosphorus. The use of EDX to examine the SEM showed that phosphorus precipitated mostly as $Ca_5(PO_4)_3OH$ in the core region. Earlier it was discovered that precipitation was responsible for up to 45% of the overall phosphorus eliminated by granules and that this figure was higher in certain cases. It will need further research to completely grasp the process of phosphorus precipitation in aerobic granules. To get a better knowledge of phosphorus deposition and solubilization patterns, it is necessary to identify the factors that affect these processes.

11.5.2 Degradation of Pollutants

The utilization of the aerobic granulation technique is proved to be an effective technique for the biodegradation of a wide variety of organic impurities, including pesticides, in a controlled environment ever since it was first introduced in the late 1990s. Phenol (organic poisons) is also decomposed effectively by these granules. When Adav et al. (2007a) verified that aerobic granules were able to decompose phenol concentrations as high as 1000 mg L^{-1} at a frequency of 49 mg $gVSS^{-1}$ h^{-1}, it was regarded to be a significant achievement at the time. *Candida tropicalis* was discovered by using phenol-decomposing aerobic granules as a source of phenol and identifying a strain that was capable of degrading phenol at high starting concentrations. The surface layer of the granules has this species indicates the presence of pollutants. According to Yi et al. (2006), aerobic granules decompose 40.1 mg L^{-1} p-nitrophenol at the rate of 19.3 mg $gVSS^{-1}$ h^{-1}. In the laboratory, aerobic granules formed in 500 mg L^{-1} phenol degraded pyridine concentrations ranging from 250 to 2500 mg L^{-1} when exposed to high concentrations of the compound. This discovery was made when it was discovered that the maximum pyridine that was degradation rate was nearby 73 mg $gVSS^{-1}$ h^{-1} (Adav et al. 2007b). Biodegradation of p-cresol was studied by Basheer and Farooqi (2012) in their study.

Zhu et al. (2011) published a paper on the utilization of aerobic granules to degrade 4-chloroaniline. 4-Chloroaniline concentrations higher than or equal to 400 mg L^{-1} were found to have the highest specific degradation rates, with the greatest degradation rate being 270 mg $gVSS^{-1}$ h^{-1}. Several studies have shown that the granules can remove 8 g L^{-1} of 4-chloroaniline. Similarly, the use of aerobic granules for Acid Red 18 degradation in a controlled setting was examined. In-situ granule production and Acid Red 18 elimination were carried out in a batch reactor with oxygenic and an-oxygenic phases. It was discovered that the aerobic granules were capable of eliminating about 50 mg L^{-1} of dye without any difficulty. Doubling the dye concentration, on the other hand, had a negative influence on the longevity of the granules and their capacity to effectively remove dye from the water, as found by the researchers (Sadri Moghaddam and Alavi Moghaddam 2016).

Based on the above logic, we may conclude that aerobic granules can digest a vast variety of environmental pollutants. Following the findings of much research on pollutant biodegradation, it has been proposed that the aerobic granulation technique might be useful for the treatment of both industrial and municipal wastewater. It is conceivable that process parameters will need to be adjusted according to the pollutant being removed to be successful. Chemical pollutants such as dyes and moderately hydrophobic complexes seem to be removed via a process of concurrent decomposition and adsorption.

11.5.3 Heavy Metal Elimination Using Aerobic Granules

According to the researchers, aerobic granules can eradicate heavy metals from both municipal and industrial wastewater via the process of bio-sorption. It was discovered via a study utilizing an energy dispersive X-ray analysis and scanning electron microscope (SEM) that heavy metals may be adsorbents not only on their surfaces but also throughout the whole granule, along with the core region. When particles are broken down into granules, they include pores, which serve an essential function in transporting pollutants from the outside zones of the granules to the inner zones of the granules. The findings of the Fourier-transform infrared spectroscopy (FTIR) and X-ray Photoelectron Spectroscopy (XPS) studies showed that functional groups present in the aerobic granules, such as carboxylate, ether, and alcoholic groups, help in the binding heavy metals in the environment by providing attachment site (Ahn and Hong 2015). A connection between the early-stage heavy metal amount and its association with the initial aerobic granule concentrations was found in this research, which indicates how initial concentration affects bio-sorption efficacy. Researchers were able to show that aerobic granules were capable of adsorbing up to 270 mg of zinc (II) for every gram of aerobic granules used in the experiment. Therefore, it has the potential to be a helpful tool for industrial wastewater treatment due to its characteristics (Liu et al. 2002). Specifically, Copper ion bio-sorption was also investigated by Luo et al. (2016). Copper ion biosorption was shown to be dependent on two processes: complex formation and ion exchange, which were both determined to be significant after thorough study. A pattern similar to that seen for other metal ions often found on the surface of granules, such as Ca^{2+} and Mg^{2+}, was observed for the dispersion of copper ions on the granule surface throughout the investigation. In this research, it was found that there is a positive connection between pH, ionic strength, and Copper ion bio-sorption.

It takes many steps to produce aerobic granules, and bacteria and fungi both take part in the process. Depending on the process circumstances, it is conceivable that these microbes will be the dominant population in the system. In a laboratory environment, the heavy metal removal capability of bacteria-dominated granules is much more as compared to fungi-dominated granules (Singh et al. 2021). Several researchers have asserted that both of these granules have the potential to be used as bio-sorbents for the removal of Zn^{2+}, Cu^{2+}, and Ni^{2+} from the environment. However, when applied to antimony, they were not particularly effective. Using Fe (III) to alter the surfaces of both granule types was found to have the potential to improve their antimony removal efficiency, which was later verified. Nancharaiah et al. (2006), who carried out the study, discovered that aerobic granules can also carry uranium bio-sorption. It was found in this research that uranium results in the release of calcium and magnesium ions, which was unexpected. It raises the possibility that an ion-exchange mechanism is also governing the biosorption of uranium by these granules, which is consistent with previous findings. The above discussion indicates that aerobic granules have the potential to be a widely used bio-sorption method for heavy metal removal from wastewater.

11.5.4 Wastewater/Sewage Treatment

Pronk et al. (2015) showed the management of residential wastewater in a full-sized wastewater system that was built based on aerobic granules utilizing aerobic granules. Researchers carried out their investigation in The Netherlands, at the Garmerwolde wastewater treatment facility, where it took approximately half-year to construct an activated sludge bed containing mature granules. The bed's biomass concentration reached as high as 8 g L^{-1}, suggesting that it was very productive. During both the summer and winter, the granular sludge bed was shown to be efficient in removing nitrogen and phosphorus from the effluent. It was effective in keeping the element levels in the effluent below the required limits of 7 mg L^{-1} for nitrogen and 1 mg L^{-1} for phosphorus, respectively. According to the findings, the energy consumption of the full-size system was approximately 58–63% lower than that of the traditional method. According to Li et al. in 2014 a full-scale SBR with a working capacity of 50,000 m^3 per day was utilized to demonstrate wastewater treatment utilizing aerobic granules. Established aerobic granules were seen after about 337 days of process in the SBR. When equated to conventional wastewater treatment methods, the AGS produced in the reactor was dense and had a higher settling ability than traditional granular sludge. The thickness of these granules was estimated to be about 0.5 mm on average.

The mechanism of phosphorus precipitation within aerobic granules produced in a full-scale granular sludge manufacturing process was studied. The existence of SiO4 crystals (quartz) within the granules was confirmed using an XRD (X-ray diffraction) experiment. The mature granules, on the other hand, had no substantial amounts of precipitated phosphorus, which was unexpected. According to Niermans et al. (2014), a total of 20 full-scale aerobic granular sludge systems, marketed under the brand name Nereda®, are now being constructed in several countries at different phases of the development process. According to the stage of the SBR operating cycle at when the process is started, the quantity of precipitated phosphorus detected in granules changes significantly. Additionally, it has been shown that the Nereda technique may decrease energy use by about 40%. The outcome has been the successful implementation of a few full-scale aerobic granule-based wastewater treatment systems in real-world wastewater treatment applications. More information on the design, pricing, operating conditions, and effectiveness of these treatment facilities will be beneficial to the rapid proliferation of this technology.

11.6 Challenges Ahead

A method in which 100% of the sludge is in granular form cannot be developed due to technical constraints. As an example, a technique may generate sludge that contains about 50% acceptable granules and the remaining 50% dense microbial flocs, with the latter being the most often seen kind of sludge produced. A procedure involving 100% granular waste is anticipated to enhance settling ability, which is particularly significant when the sludge volume index is extremely low. This is especially true when the sludge volume index (SVI) is very low. Increasing the overall fraction of appropriate granules existing in the finished product, which may be difficult to accomplish, is a constant challenge for this technique. The SBR reactor is the best option for producing aerobic granules because of its high efficiency. In nearly all of the investigations on aerobic granules that have been published in the scientific literature so far, SBR has been utilized as the primary control. Continuous-flow reactors, on the other hand, are the most often employed for actual wastewater treatment all over the world, and they are also the most costly of the available options (Kishida et al. 2012). As a result, the generation of aerobic granules in a full-scale reactor system may be investigated in the future for further investigation. Grains are microspheres that contain a diverse variety of microorganisms in a concentrated form, and they are often used in food processing. It may take anything from a few weeks to many months for the granules to completely develop. In contrast, the disintegration of the granules as a consequence of unexplained process circumstances is a problem that requires more investigation. The granules, on the other hand, are not always steady enough to survive variations in process conditions (Sharma and Tay 2018).

However, even though they are already in use, reactors used for aerobic granulation technology are not appropriate for use in activated sludge treatment. If you find yourself in this position, the conversion of a prevailing activated sludge procedure to an aerobic granulation method would be a mammoth task (Mosquera-Corral et al. 2015). When dissolved oxygen concentrations are low, aerobic granules may be able to absorb nitrogen from the surrounding environment via a process known as simultaneous nitrification and denitrification. A greater amount of research and development is required in this field. Water granules cannot be formed in the presence of low oxygen concentrations, which is incompatible with their formation. So it may be difficult to produce finely divided granules when the dissolved oxygen content is low (Liu et al. 2007). Aerobic granules formation may take anywhere from a few weeks to many months depending on the circumstances. When the granule formation process is underway, nutrient removal efficiency is low, which is typical of the time period (Wilén et al. 2018). Due to the requirement of more granulation time, real-world use of aerobic granulation technology will encounter an extra time constraint in terms of processing time in the real world. It would be necessary to conduct further engineering technique optimizations as well as fundamental molecular microbiological research in order to overcome the challenges associated with this technology.

11.7 Future Prospects

The ability to maintain adequate structural integrity in granules is a major problem that limits the functional use of aerobic granulation in the actual world at this time. Given the present roadblock in the creation of aerobic granulations, more investigation into granule stability is required before full-scale functioning can be achieved. It is necessary to explore techniques or processes for generating granules with long-term structural integrity to improve their long-term performance. When it comes to the shape and functionality of the granules, it seems that the dehydration procedure had minimal impact. Drying granules as a new method of preserving them, according to this interesting finding, maybe a possibility. If wet granules can be kept and handled with ease, dry granules storage and handling are more difficult. Other applications for aerobic granules include use as an inoculant to speed up the formation and supplements to increase the treatment process. In order to evaluate the impact of drying properties, such as settle ability, hydrophobicity, density, strength surface, specific oxygen absorption, and EPS content, further research is needed (Show et al. 2012).

In order to conduct more study into the potential of these methods, the following focus fields are recommended for future research:

1. Increased augmentation and dispersion of specific EPS mechanisms during the granulation process, as well as granule solidity management throughout the reactor process and storage. Confocal laser scanning microscope (CLSM) in conjunction with various specific fluorochromes can be used to examine the 3-D distribution of proteins, polysaccharides, and lipids inside the granule, permitting for a more in-depth understanding of the granule internal structure and stability to be gained.
2. The bulk of research in this field has been performed using SBR systems, therefore it is essential to establish the operating and loading characteristics of a reactor that can handle a range of substrates in a continuous operating mode. It is better to operate the reactor continuously rather than in batches or sequences when it comes to a well-organized full-scale setup.
3. It seems that dehydration may be used as a granule preparation technique for storing purposes. Further research will be conducted to evaluate the steadiness granules and the influence of the process on other variables. Additional methods like encapsulation treatment, as well as additional research, need to be explored. Granules handled in a new way would allow for easier storage and handling, which would be beneficial for applications such as inoculants and supplements to enhance the treatment of prevailing methods, among others.
4. Combining granulation technology with other treatment techniques, such as membrane biological reactors (MBR), in order to optimize the benefits of both processes 4. An investigation (Juang et al. 2011) found that aerobic granulation has the capacity to minimize irreversible fouling on MBR membranes by reducing the growth of internal biofilm.

5. A single transformed bacterium is being used for cultivation containing genetically modified microbial species that have several targeted genes for the removal of various toxicants. This method is being explored. By incorporating genetically changed granules into the bioaugmentation process, it is possible to easily adapt the treatment capabilities of an aerobic granulation system to accommodate changes in loading rates, treatment goals, and wastewater composition.

11.8 Conclusion

When compared to traditional activated sludge-based wastewater treatment, aerobic granulation is a novel sewage treatment technique that offers a number of major benefits over traditional activated sludge-based wastewater treatment. It has been constructed many complete aerobic granule-based sewage/wastewater treatment set-up in order to demonstrate the viability of this technology in a commercial setting. These institutions may be found throughout the United States of America. Scientists are still baffled as to the exact process by which aerobic granules are generated in bacteria. To explain the process, a variety of theories have been advanced; however, no convincing experimental evidence has yet been shown to support any of these possibilities. Note that both fungus and bacteria are involved in the production of aerobic granules, with bacteria typically being the dominating species in this process. Because of its simplicity, the sequencing batch reactor is a popular kind of reactor for this sort of reaction to be carried out. Nevertheless, continuous-flow reactors will be better suited for broad use of this technology in the industrial setting. The use of traditional granulation technology for the management of high-strength industrial wastewater is not advised. The aerobic granulation technique is preferred over other methods. Increase the COD of the wastewater by adding external carbon sources such as volatile fatty acids in order to effectively treat low-strength household wastewater. Especially when it comes to removing nutrients from the soil, aerobic granules are very effective. The ability of aerobic granules, as previously mentioned, to remove phosphorus from a solution by causing precipitation inside the granules themselves is a significant advantage. A study at the most basic level of science is now being conducted to determine the precise process by which the aerobic granules precipitate the phosphorus. Aerobic granules are not only excellent at removing nutrients, but they are also effective in biodegrading organic pollutants such as phenol and at removing heavy metals from the environment. A significant problem with this technique is that the granules are disintegrated unexpectedly during the manufacturing process, which is a time-consuming operation. It would be necessary to do fundamental microbiological research in order to cope with this problem. Several key molecular microbiological processes, including quorum sensing QS and quorum quenching QQ, have the potential to directly influence granule formation and stability. QS & QQ are significant molecular microbiological processes that might have a direct impact on granule formation and stability. QS & QQ are two major molecular microbiological processes that may have a direct impact on granule construction and constancy. Quantitative symbiotic and quantitative symbiotic processes are significant molecular microbiological activities that may have an upfront influence on the formation and steadiness of granules in a range of settings.

A growing body of data suggests that aerobic granulation is a very successful method for treating high-concentration and/or possibly hazardous wastewater streams, maintaining the stability of aerobic granules in order to use them in practical applications is a significant problem that has yet to be addressed. In order to enhance the long-term performance of granules, it is essential to investigate methods or procedures for producing granules with long-term structural integrity. The capacity of filamentous bacteria to multiply, as well as the role played by extra polymeric chemicals, have both been called into question in the context of granule stability, with varying degrees of success. These strategies, which comprise of the collection of slow-growing bacteria, limiting the action of anaerobic bacteria, and reinforcing the granule's core, are explored in more depth in this article. There is reason to believe that the outcomes of granule culture with long-term stability in a continuous-flow reactor, as well as the findings of granule storage using a drying method, are promising results. Among other things, it is anticipated that the newly developed drying method for aerobic granules would make it simpler to store and manage granules for future use to improve the treatment of bioreactor systems, among other things. Additional investigation is required to determine whether or if there are additional effects of the drying process, as well as whether or not the granules are stable during long-term operation and storage in pilot and full-scale systems, among other things. Because of this, future investigation into these problems is expected to provide important information regarding the processes involved in granule formation.

References

Adav SS, Chen MY, Lee DJ, Ren NQ (2007a) Degradation of phenol by aerobic granules and isolated yeast Candida tropicalis. Biotechnol Bioeng 96(5):844–852. https://doi.org/10.1002/bit.21148

Adav SS, Lee DJ, Ren NQ (2007b) Biodegradation of pyridine using aerobic granules in the presence of phenol. Water Res 41(13):2903–2910. https://doi.org/10.1016/j.watres.2007.03.038

Adav SS, Lee DJ, Show KY, Tay JH (2008) Aerobic granular sludge: recent advances. Biotechnol Adv 26(5):411–423. https://doi.org/10.1016/j.biotechadv.2008.05.002

Ahn KH, Hong SW (2015) Characteristics of the adsorbed heavy metals onto aerobic granules: isotherms and distributions. Desalin Water Treat 53(9):2388–2402. https://doi.org/10.1080/19443994.2014.927125

Ali M, Shaw DR, Zhang L, Haroon MF, Narita Y, Emwas AH, Okabe S (2018) Aggregation ability of three phylogenetically distant anammox bacterial species. Water Res 143:10–18. https://doi.org/10.1016/j.watres.2018.06.007

Bao R, Yu S, Shi W, Zhang X, Wang Y (2009) Aerobic granules formation and nutrients removal characteristics in sequencing batch airlift reactor (SBAR) at low temperature. J Hazard Mater 168(2–3):1334–1340. https://doi.org/10.1016/j.jhazmat.2009.03.020

Basheer F, Farooqi IH (2012) Biodegradation of p-cresol by aerobic granules in sequencing batch reactor. J Environ Sci 24(11). https://doi.org/10.1016/S1001-0742(11)60988-1

Bassin JP, Tavares DC, Borges RC, Dezotti M (2019) Development of aerobic granular sludge under tropical climate conditions: the key role of inoculum adaptation under reduced sludge washout for stable granulation. J Environ Manag 230:168–182. https://doi.org/10.1016/j.jenvman.2018.09.072

Bengtsson S, de Blois M, Wilén BM, Gustavsson D (2018) A comparison of aerobic granular sludge with conventional and compact biological treatment technologies. Environ Technol 13:1479–1487. https://doi.org/10.1080/09593330.2018.1452985

Chen Y, Jiang W, Liang DT, Tay JH (2008) Aerobic granulation under the combined hydraulic and loading selection pressures. Bioresour Technol 99:7444–7449. https://doi.org/10.1016/j.biortech.2008.02.028

Coma M, Verawaty M, Pijuan M, Yuan Z, Bond PL (2012) Enhancing aerobic granulation for biological nutrient removal from domestic wastewater. Bioresour Technol 103(1):101–108. https://doi.org/10.1016/j.biortech.2011.10.014

Daraei H, Rafiee M, Yazdanbakhsh AR, Amoozegar MA, Guanglei Q (2019) A comparative study on the toxicity of nano zero valent iron (nZVI) on aerobic granular sludge and flocculent activated sludge: reactor performance, microbial behavior, and mechanism of toxicity. Process Safety Environ Protect 129:238–248. https://doi.org/10.1016/j.psep.2019.07.011

de Sousa Rollemberg SL, de Barros AN, Lira VNSA, Firmino PIM, Dos Santos AB (2019) Comparison of the dynamics, biokinetics and microbial diversity between activated sludge flocs and aerobic granular sludge. Bioresour Technol 294:122106. https://doi.org/10.1016/j.biortech.2019.122106

Debbarma P, Raghuwanshi S, Singh J, Suyal DC, Zaidi MGH, Goel R (2017) Comparative in situ biodegradation studies of polyhydroxybutyrate film composites. 3Biotech 7(178):1–9. https://doi.org/10.1007/s13205-017-0789-3

Erdal UG, Erdal ZK, Randall CW (2006) The mechanism of enhanced biological phosphorus removal washout and temperature relationships. Water Environ Res 78(7):710–715. https://doi.org/10.2175/106143006x101737

Gao D, Liu L, Liang H, Wu WM (2011) Aerobic granular sludge: characterization, mechanism of granulation and application to wastewater treatment. Crit Rev Biotechnol 31(2):137–152. https://doi.org/10.3109/07388551.2010.497961

Jafarinejad S (2017) Cost estimation and economical evaluation of three configurations of activated sludge process for a wastewater treatment plant (WWTP) using simulation. Appl Water Sci 7 (5):2513–2521. https://doi.org/10.1515/eces-2019-0024

Jang A, Yoon YH, Kim IS, Kim KS, Bishop PL (2003) Characterization and evaluation of aerobic granules in sequencing batch reactor. J Biotechnol 105:71–82. https://doi.org/10.1016/s0168-1656(03)00142-1

Juang YJ, Su A, Fang LH, Lee DJ, Lai JY (2011) Fouling with aerobic granule membrane bioreactor. Water Sci Technol 64(9):1870–1875. https://doi.org/10.2166/wst.2011.139

Kagawa Y, Tahata J, Kishida N, Matsumoto S, Picioreanu C, Van Loosdrecht MCM, Tsuneda S (2015) Modeling the nutrient removal process in aerobic granular sludge system by coupling the reactor-and granule-scale models. Biotechnol Bioeng 112(1):53–64. https://doi.org/10.1002/bit.25331

Kang AJ, Yuan Q (2017) Long-term stability and nutrient removal efficiency of aerobic granules at low organic loads. Bioresour Technol 234:336–342. https://doi.org/10.1016/j.biortech.2017.03.057

Kim NK, Mao N, Lin R, Bhattacharyya D, van Loosdrecht MC, Lin Y (2020) Flame retardant property of flax fabrics coated by extracellular polymeric substances recovered from both activated sludge and aerobic granular sludge. Water Res 170:115344. https://doi.org/10.1016/j.watres.2019.115344

Kishida N, Totsuka R, Tsuneda S (2012) Challenge for formation of aerobic granular sludge in a continuous-flow reactor. J Water Environ Technol 10(2):79–86. https://doi.org/10.2965/jwet.2012.79

Kour D, Kaur T, Devi R, Yadav A, Singh M et al (2021) Beneficial microbiomes for bioremediation of diverse contaminated environments for environmental sustainability: present status and future challenges. Environ Sci Pollut Res 28:24917–24939

Kumar P, Dash B, Suyal DC, Gupta SB, Singh AK, Chowdhury T, Soni R (2021) Characterization of arsenic-resistant Klebsiella pneumoniae RnASA11 from contaminated soil and water samples and its bioremediation potential. Curr Microbiol. https://doi.org/10.1007/s00284-021-02602-w

Kuśnierz M (2018) Scale of small particle population in activated sludge flocs. Water Air Soil Pollut 229:327. https://doi.org/10.1007/s11270-018-3979-7

Lashkarizadeh M (2015) Operating pH and feed composition as factors affecting stability of aerobic granular sludge. URI: http://hdl.handle.net/1993/30304

Lashkarizadeh M, Munz G, Oleszkiewicz JA (2016) Impacts of variable pH on stability and nutrient removal efficiency of aerobic granular sludge. Water Sci Technol 73(1):60–68. https://doi.org/10.2166/wst.2015.460

Lemaire R, Yuan Z, Blackall LL, Crocetti GR (2008) Microbial distribution of Accumulibacter spp. and Competibacter spp. in aerobic granules from a lab-scale biological nutrient removal system. Environ Microbiol 10(2):354–363. https://doi.org/10.1111/j.1462-2920.2007.01456.x

Li ZH, Kuba T, Kusuda T (2006) Aerobic granular sludge: a promising technology for decentralised wastewater treatment. Water Sci Technol 53(9):79–85: https://doi.org/10.2166/wst.2006.278

Li J, Ding LB, Cai A, Huang GX, Horn H (2014) Aerobic sludge granulation in a full-scale sequencing batch reactor. BioMed Res Int. https://doi.org/10.1155/2014/268789

Liu Y, Tay JH (2002) The essential role of hydrodynamic shear force in the formation of biofilm and granular sludge. Water Res 36:1653–1665. https://doi.org/10.1016/s0043-1354(01)00379-7

Liu Y, Tay JH (2004) State of the art of biogranulation technology for wastewater treatment. Biotechnol Adv 22(7):533–563. https://doi.org/10.1016/j.biotechadv.2004.05.001

Liu Y, Yang SF, Tan SF, Lin YM, Tay JH (2002) Aerobic granules: a novel zinc biosorbent. Lett Appl Microbiol 35(6):548–551. https://doi.org/10.1046/j.1472-765x.2002.01227.x

Liu Y, Yang SF, Liu QS, Tay JH (2003) The role of cell hydrophobicity in the formation of aerobic granules. Curr Microbiol 46:270–274. https://doi.org/10.1007/s00284-002-3878-3

Liu Y, Qin L, Yang SF (2007) Microbial granulation technology for nutrient removal from wastewater. Nova Publishers, New York

Liu X, Wu S, Zhang D, Shen J, Han W, Sun X, Wang L (2018) Simultaneous pyridine biodegradation and nitrogen removal in an aerobic granular system. J Environ Sci 67:318–329. https://doi.org/10.1016/j.jes.2017.09.016

Long B, Xuan X, Yang C, Zhang L, Cheng Y, Wang J (2019) Stability of aerobic granular sludge in a pilot scale sequencing batch reactor enhanced by granular particle size control. Chemosphere 225:460–469. https://doi.org/10.1016/j.chemosphere.2019.03.048

Luo H, Wang L, Tong Z, Yu H, Sheng G (2016) Approaching the binding between Cu (II) and aerobic granules by a modified titration and µ-XRF. Front Environ Sci Eng 10(2):362–367. https://doi.org/10.1007/s11783-015-0803-0

Moragaspitiya C, Rajapakse J, Millar GJ (2019) Effect of Ca: Mg ratio and high ammoniacal nitrogen on characteristics of struvite precipitated from waste activated sludge digester effluent. J Environ Sci 86:65–77. https://doi.org/10.1016/j.jes.2019.04.023

Mosquera-Corral A, Val del Río Á, Moralejo-Gárate H, Sánchez A, Méndez R, Campos JL (2015) The aerobic granulation as an alternative to the conventional activated sludge process. Sewage Treat Plant:95

Nancharaiah YV, Sarvajith M (2019) Aerobic granular sludge process: a fast growing biological treatment for sustainable wastewater treatment. Curr Opin Environ Sci Health 12:57–65. https://doi.org/10.1016/j.coesh.2019.09.011

Nancharaiah YV, Joshi HM, Mohan TVK, Venugopalan VP, Narasimhan SV (2006) Aerobic granular biomass: a novel biomaterial for efficient uranium removal. Curr Sci:503–509

Ni BJ (2012) Formation, characterization and mathematical modeling of the aerobic granular sludge, vol 131. Springer Science & Business Media, Cham

Niermans R, Giesen A, Loosdrecht MV, Buin BD (2014) Proceedings of the Water Environment Federation, pp 2347–2357

Nishanth S, Bharti A, Gupta H, Gupta K, Gulia U, Prasanna R (2021) Cyanobacterial extracellular polymeric substances (EPS): biosynthesis and their potential applications. In: Microbial and natural macromolecules. Academic Press, London, pp 349–369

Pan S, Tay JH, He YX, Tay SL (2004) The effect of hydraulic retention time on the stability of aerobically grown microbial granules. Lett Appl Microbiol 38(2):158–163

Pronk M, De Kreuk MK, De Bruin B, Kamminga P, Kleerebezem RV, Van Loosdrecht MCM (2015) Full scale performance of the aerobic granular sludge process for sewage treatment. Water Res 84:207–217. https://doi.org/10.1016/j.watres.2015.07.011

Ren Y, Ferraz F, Lashkarizadeh M, Yuan Q (2017) Comparing young landfill leachate treatment efficiency and process stability using aerobic granular sludge and suspended growth activated sludge. J Water Process Eng 17:161–167. https://doi.org/10.1016/j.jwpe.2017.04.006

Sadri Moghaddam S, Alavi Moghaddam MR (2016) Aerobic granular sludge for dye biodegradation in a sequencing batch reactor with anaerobic/aerobic cycles. Clean Soil Air Water 44 (4):438–443

Sarma SJ, Tay JH (2018) Aerobic granulation for future wastewater treatment technology: challenges ahead. Environ Sci Water Res Technol 4(1):9–15. https://doi.org/10.1039/C7EW00148G

Sheik AR, Muller EEL, Wilmes P (2014) A hundred years of activated sludge: time to rethink. Front Microbiol 5:–47. https://doi.org/10.3389/fmicb.2014.00047

Show KY, Lee DJ, Tay JH (2012) Aerobic granulation: advances and challenges. Appl Biochem Biotechnol 167:1622–1640. https://doi.org/10.1007/s12010-012-9609-8

Singh M, Singh D, Rai P, Suyal DC, Saurabh S, Soni R, Giri K, Yadav AN (2021) Fungi in remediation of hazardous wastes: current status and future. In: Yadav AN (ed) Recent trends in mycological research, fungal biology. Springer Nature, Switzerland

Tay JH, Liu Y, Tay STL, Hung YT (2009) Aerobic granulation technology. In: Wang LK, Shammas NK, Hung YT (eds) Advanced biological treatment processes. Humana Press, New York, pp 109–128

Toh SK, Tay JH, Moy BYP, Ivanov V, Tay STL (2003) Size-effect on the physical characteristics of the aerobic granule in a SBR. Appl Microbiol Biotechnol 60:687–695

Tsitouras A (2021) Sequencing batch moving bed biofilm reactors for treatment of cheese production wastewater. Doctoral dissertation, Université d'Ottawa/University of Ottawa. https://doi.org/10.20381/ruor-26348

Vu CT (2017) Comparison between granular and conventional activated sludge for trace metal elements sorption/desorption. Case of copper for landspreading application in France and in Vietnam. Doctoral dissertation, Université de Limoges

Wilen BM, Onuki M, Hermansson M, Lumley D, Mino T (2007) Microbial community structure in activated sludge floc analysed by fluorescence in situ hybridization and its relation to floc stability. Water Res 42:2300–2308. https://doi.org/10.1016/j.watres.2007.12.013

Wilén BM, Liébana R, Persson F, Modin O, Hermansson M (2018) The mechanisms of granulation of activated sludge in wastewater treatment, its optimization, and impact on effluent quality. Appl Microbiol Biotechnol 102(12):5005–5020. https://doi.org/10.1007/s00253-018-8990-9

Wu J, Francis L III, Ducoste JJ (2020) Modeling cell aggregate morphology during aerobic granulation in activated sludge processes reveals the combined effect of substrate and shear. Water Res 170:115384. https://doi.org/10.1016/j.watres.2019.115384

Yang SF, Li XY, Yu HQ (2008) Formation and characterisation of fungal and bacterial granules under different feeding alkalinity and pH conditions. Process Biochem 43:8–14

Yi S, Zhuang WQ, Wu B, Tay STL, Tay JH (2006) Biodegradation of p-nitrophenol by aerobic granules in a sequencing batch reactor. Environ Sci Technol 40(7):2396–2401. https://doi.org/10.1021/es0517771

Yousuf I (2013) Methods for estimation and comparison of activated sludge settleability. In 38th annual WIOA old water industry operations conference, Parklands, Gold Coast
Zhang L, Feng X, Zhuc N, Chena J (2007) Role of extracellular protein in the formation and stability of aerobic granules. Enzym Microb Technol 41:551. https://doi.org/10.1016/J.ENZMICTEC.2007.05.001
Zhu L, Yu Y, Xu X, Tian Z, Luo W (2011) High-rate biodegradation and metabolic pathways of 4-chloroaniline by aerobic granules. Process Biochem 46(4):894–899. https://doi.org/10.1016/j.procbio.2010.12.013
Zou J, Yu F, Pan J, Pan B, Wu S, Qian M, Li J (2021) Rapid start-up of an aerobic granular sludge system for nitrogen and phosphorus removal through seeding chitosan-based sludge aggregates. Sci Total Environ 762:144171. https://doi.org/10.1016/j.scitotenv.2020.144171

Chapter 12
Recent Perspectives of Immobilized Enzyme Reactors Used for Wastewater Treatment

Dinesh Chandola and Vasudha Agnihotri

12.1 Introduction

The water pollution is a huge concern throughout the world as it effects the aquatic life, and human health, gas emissions (methane, nitrous oxide gases etc.) occurs in the areas where such water is accumulated. The main cause of concern is the direct disposal of industrial and sewage effluents in the water bodies without proper treatment. The disposal of hotel, hospital, domestic waste effluents directly in the water bodies are also a major concern (Agnihotri and Thathola 2019). The runoff effluents from agricultural areas, where consumption of pesticides and other harmful chemicals is high, make the situation worst. The pollutant present in wastewater can cause harmful effects on human being through biomagnification of these pollutants through the food chain in aquatic life. So wherever possible, treatment of wastewater before disposal is recommended for controlling water pollution. Various processes are used for the treatment of wastewater. The majority of wastewater treatment procedures are classified as either physicochemical or biological procedures. Coagulation, lime softening, activated carbon adsorption, biofiltration, chlorination, membrane bioreactor, photocatalytic treatment, ultraviolet photodegradation, ozonization, and advanced oxidation (AOP) are some of the methods used for the treatment of wastewater. The biological means such as the use of microbes, plants, enzymes becoming more popular for wastewater treatment (Agnihotri 2020). Both the plants as well as microbes can perform remediation work with the help of enzymes secreted in the presence of the pollutants. So, the emphasis is given on direct utilization of enzymes either in free form or in immobilized form for wastewater treatment purposes. These large molecules are the protein which is used for the

D. Chandola · V. Agnihotri (✉)
G. B. Pant National Institute of Himalayan Environment (NIHE), Almora, Uttarakhand, India
e-mail: vasudha@gbpihed.nic.in

© The Author(s), under exclusive license to Springer Nature Switzerland AG 2022
D. C. Suyal, R. Soni (eds.), *Bioremediation of Environmental Pollutants*,
https://doi.org/10.1007/978-3-030-86169-8_12

increasing the rate of reactions, without being consumed or structurally changed (Kumar and Bharadvaja 2019; Bhatt et al. 2021).

Based on the mode of action, the enzymes may be classified into oxidoreductase, transferase, hydrolase, lyase, isomerase, ligase, translocase. Some famously reported families are the oxygenase, peroxidases, and phenol oxidases (tyrosinase and laccase), which initiate the disintegration of phenolic, aromatic, or inorganic pollutants via oxidation (Unuofin et al. 2019; Kour et al. 2021). The oxygenase enzymes have the potential to scavenge the oxygen while working against environmental pollutants. The peroxidase enzymes such as lignin peroxidase, manganese peroxidase, horseradish peroxidase, soybean peroxidase, bitter guard peroxidase, versatile peroxidase, etc. are successfully used for the removal of wastewater pollutants at a wider range of pH and temperature (Table 12.1).

With time, enzyme-based technologies have improved drastically. These are used in healthcare, pharma, biofuel, and food industries, in biosensors, treatment of solid waste and wastewater along with other industries (Apetrei et al. 2013; Atadashi et al. 2010; Das et al. 2012; Tong et al. 1997; Tonini and Astrup 2012). The enzymes such as lipase, laccase, peroxidase, amylase, etc., have shown the capability to degrade various pollutants present in wastewater such as dyes, oil and fat, emerging pollutants, heavy metal ions etc. The enzymes degrade organic waste immediately after their contact with the substrate and use them as carbon and nitrogen sources. Thus, as compared to the conventional wastewater treatment method, advanced treatments like enzyme-based wastewater treatment have better prospects in terms of high specificity and selectivity, environmental friendly, and biodegradability (Mugdha and Usha 2012).

Nowadays, the free enzyme has been substituted by the immobilized enzyme. The immobilization improves the stability of biomolecules, after which these molecules become reusable. This process improves the reusability of enzyme for an extended period and enables easier separation of the enzymes from the product. Additionally, immobilization improves many properties of enzymes such as performance in organic solvents, pH tolerance, heat stability, or functional stability. The increasing structural rigidity of the protein and stabilization of multimeric enzymes prevents dissociation-related inactivation. The first scientific evidence of immobilized enzymes was observed in 1916 (Gomes-Ruffi et al. 2012) where the invertase enzyme showed a similar activity when absorbed on charcoal or aluminum hydroxide as when uniformly dispersed through the solution. The main advantages and disadvantages of using immobilized enzymes are shown in Table 12.2.

In general, inert polymers and inorganic materials such as zeolites, ceramics, celite, silica, glass, activated carbon, charcoal are used support materials. Inorganic support materials are beneficial due to their thermal and mechanical resistance, microbial resistance, rigidity, and porosity while organic support, including both the natural (alginate, chitosan, collagen, carrageenan, gelatin, cellulose, starch, pectin, Sepharose etc) and synthetic support (polystyrene, polyvinyl chloride (PVC), polyacrylate, polyamide, polypropylene, diethyl amino-ethyl cellulose (DEAE cellulose), UV-activated polyethylene glycerol, etc), are beneficial due to their inert behavior, can be activated easily, works both in reversible as well as in an

Table 12.1 Removal of wastewater contaminant using immobilized enzyme system

S. No.	Enzyme source	Support material	Immobilized enzyme	Target pollutant	
1	Horseradish (HRP)	Calcium-alginate support using glutaraldehyde (GA) as a cross-linking reagent	Peroxidase	Reactive Red 120 (RR 120), Reactive Blue 4 (RB 4) and Reactive Orange 16 (RO 16)	Bilal et al. (2016)
2	Bitter gourd	Calcium alginate gel	Peroxidase	Disperse Brown 1 (DB 1) and Disperse Red 17 (DR 17)	Satar and Husain (2011)
3	Horseradish (HRP)	Polyaniline (PANI) grafted polyacrylonitrile (PAN) films	Peroxidase	Direct Black 38 Direct Blue 53	Bayramoglu et al. (2012)
4	Horseradish (HRP)	Kaolin	Peroxidase	Acid Violet 109	Šekuljica et al. (2016)
5	Tomato	Concanavalin A-cellulose	Peroxidase	Direct Red 23 Direct Blue 80	Mahreen Matto and Husain (2008)
6	Turnip	Concanavalin A-wood shavin	Peroxidase	Direct Red 23, Direct Red 239 and Direct Blue 80	Mahreen Matto and Husain (2009)
7	*Acinetobacter haemolyticus*	Eggshell membrane (ESM)	Lipase	Oil having high concentration of Oleic acid	Işık et al. (2021)
8	*Trametes versicolor*	TiO2-ZrO2 and TiO2-ZrO2-SiO2	Laccase	Alizarin Red S (ARS), Remazol Brilliant Blue R (RBBR), and Reactive Black 5 (RB5	Antecka et al. (2018)
9	Genetically modified *Aspergillus oryzae*	Granular activated carbon	Laccase	Bisphenol A (BPA), diclofenac (DCF), carbamazepine (CBZ) and sulfamethoxazole (SMX)	Nguyen et al. (2016)
10	*Coriolopsis polyzona*	Celite R-633	Laccase	Nonylphenol (NP), bisphenol A (BPA) and triclosan (TCS	Cabana et al. (2009a)
11	*Trametes versicolour*	TiO2sol–gel coated polyvinylidene fluoride (PVDF) membrane	Laccase	Bisphenol A (BPA)	Hou et al. (2014)

(continued)

Table 12.1 (continued)

S. No.	Enzyme source	Support material	Immobilized enzyme	Target pollutant	
12	*Bacillus licheniformis*	*Luffa operculata* fibers	α-amylase	Waste starch	Morais et al. (2013)
13	Pineapple (*Ananas comosus*)	Calcium alginate	Bromelain	Lead	Chatterjee et al. (2019)
14	Latex of papaya (*Carica papaya*)	Calcium alginate	Papain	Lead and cadmium	Chatterjee et al. (2019)

Table 12.2 Advantage and disadvantage of the use of an immobilized enzyme (Zhang and Xing 2011)

Advantages	Disadvantages
Increased enzyme efficiency	Enzymatic activity may be lost
Increase reusability	Requirement of support material is extra which needs expertise and separate efforts
Able to tolerate variable environmental conditions such as pH, temperature	Mass transfer is limited
Less labor intensive and more economic	Immobilization is time-consuming
Less product contamination	

irreversible manner, inexpensive, have high thermal and mechanical resistance. The immobilization of an enzyme to a solid support material can enhance its resistance to many environmental changes such as pH or temperature. The immobilized enzymes are used in different sectors including food application, biodiesel application, wastewater treatment, textile and detergent biomedical application industry, etc. (Homaei et al. 2013).

12.2 Modes of Immobilization

The enzymes are immobilized on the solid surface through both the reversible as well as irreversible processes. The reversible process includes adsorption/carrier-binding method, and *affinity binding/*encapsulation while irreversible process includes covalent bonding, entrapment, and cross-linking (Sirisha et al. 2016; Tran and Balkus 2011) as shown in Fig. 12.1. Immobilization through *adsorption* involves carrier-bound immobilization. The mineral, organic or ion exchange resin/modified Sepharose may be used as carrier or support for immobilization through adsorption. The weaker forces such as hydrogen bonds, hydrophobic bonds, Vander-Waals forces, salt linkages, etc. responsible for the process. In the *affinity binding/encapsulation* method, the enzyme is enclosed in the membrane, composed on nitro- cellulose or nylon. In the *covalent bonding* mode, functional group (such as α-carboxyl and amino group, hydroxyl, imidazole, thiol group etc.) present on the surface of enzyme as well as that on the support, interact with each other. The commonly used support for this mode are carbohydrate (agarose, cellulose etc.), Protein (collagen, gelatin etc.), inorganic carriers (porous glass, silica) etc. In the *entrapment* method, the enzymes are caged in the water-soluble polymeric network (such as poly acryl amide gels, agar, cellulose triacetate) through covalent or non-covalent bonds to prevent the enzyme loss. The cross-linking mode (also called copolymerization) involves the immobilization of enzymes on the polyfunctional reagent (glutaraldehyde, diazonium salts etc) through covalent linkages. The immobilized enzymes are beneficial over enzymes in solution, including economic convenience, higher stability, and the possibility to be easily separated from the reaction mixture leading to pure product isolation (Hakala et al. 2013).

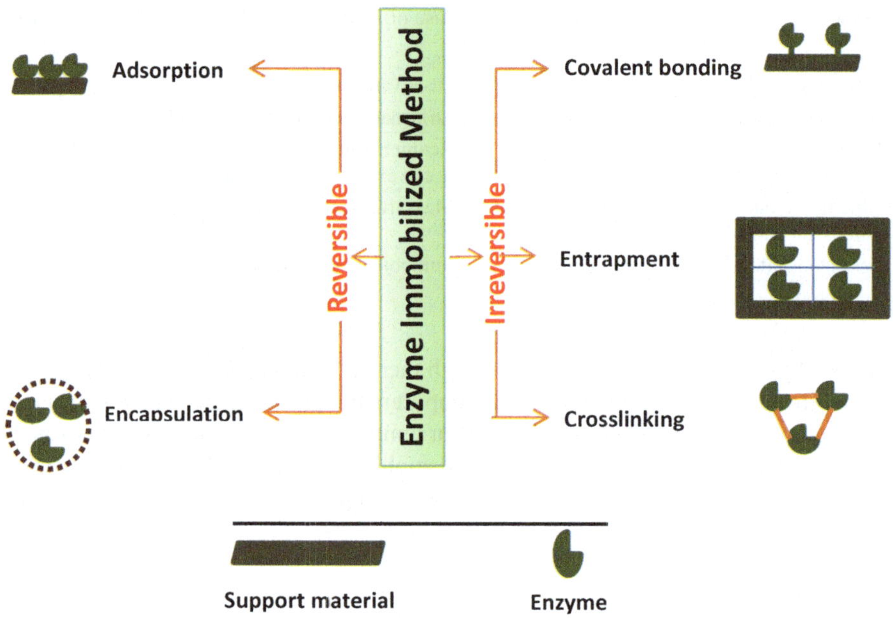

Fig. 12.1 Methods for enzyme immobilization

12.3 Enzyme Reactors

An enzyme reactor is a device inside which biochemical conversions or reactions are achieved with the help of enzymes for the production of expected end products. Mostly, all enzyme reactors deal with heterogeneous systems which involve two or more phases, for example solid, liquid, and gas. Hence, ideal conditions for the enzyme-catalyzed reactions require the effective transfer of heat, mass, and momentum from one phase to the other. Therefore, an enzyme reactor is supposed to provide optimum conditions of pH, agitation (baffles), temperature, etc. (Zhang and Xing 2011). Enzyme reactors are different from conventional chemical reactors in terms of support and control for biological activities, time-dependent properties of enzymes, and complex kinetic patterns. The first key difference is selectivity for producing the preferred products is much higher in enzyme bioreactors. Second, inactivation of the enzymes may often pose more severe concerns, than those from a chemical reactor (Zhang and Xing 2011).

There are different types of enzyme reactors, in which both the immobilized as well as free enzymes can be used for wastewater treatment process. The performance of enzyme reactors is affected by various factors such as type of enzyme used, substrate, mediator, surfactants, organic solvents, temperature, and pH conditions, agitation, immobilization, etc. which are optimized for enhancing the degradation process in the reactor. Selection of suitable enzyme is important as their catalytic potential effect the degradation process. The same enzyme originated from different sources works differently against the pollutants based on the varying redox potential. The laccase enzyme, originated from *Trametes versicolor* (0.78 V), works effectively against bisphenol A, nonsteroidal anti-inflammatory drugs, etc., while the same enzyme originated from *Mycliophthora thermophila* (0.47 V) with medium redox potential can degrade/bio-transform estrogen from wastewater (Lloret et al. 2013a, b). The substrate is required by the enzyme for the completion of the catalytic cycle (Adriana Arca-Ramos et al. 2018). The peroxidase needs a controlled amount of hydrogen peroxide which helps the enzyme for getting oxygen (Taboada-Puig et al. 2015), manganese peroxidase (MnP) require a controlled supply of Mn^{+2} ions (Mielgo et al. 2003), laccase and tyrosinase need direct oxygen supply, in most of the cases, for better degradation of pollutants such as polyaromatic hydrocarbons, dyes etc. (Arca-Ramos et al. 2014; da Silva et al. 2013). In the immobilized form, the enzymes can tolerate higher content of substrates in comparison to free forms. The additive is also added in the reactors which are helpful for the prevention of enzymes from inactivation during the wastewater treatment process (Modaressi et al. 2005).

Mediators help the enzymatic degradation process by modifying the redox potential of enzymes so that they can work on a wider range of pollutants (Husain and Husain 2008). The natural compounds such as syringaldehyde, vanillin, p-coumaric acid, etc. and synthetic compounds such as violuric acid and hydroxy benzotriazole (HBT), etc. are being used as a mediator for laccase, although the selection is based on the targeted contaminants (Lloret et al. 2010; Nguyen et al. 2014). Optimum temperature and pH are essential for getting the required enzymatic activity, although immobilized enzymes are less affected by pH and thermal variations as observed in the case of oxidoreductase enzymes (Fernández-fernández et al. 2013; Guzik et al. 2014). The laccase enzyme, isolated from *T. versicolor* and *M. thermophila* show the maximum enzymatic activity at 50 °C and 60 °C respectively (Han et al. 2005; Lloret et al. 2012), but the enzyme best perform at a comparatively lower temperature near 30 °C (Kurniawati and Nicell 2008). Horseradish peroxidase can tolerate the temperature up to 45 °C but perform effectively at lower temperatures (Jun et al. 2019). In the case of pH, many microbes can tolerate the pH at a wider range (Pandey et al. 2019), still, there is a specific pH range at which the activity of the enzymes is accelerated. Tyrosinase enzyme acts properly at 6–7 pH (Aytar and Bakir 2008), HRP in the range of 7–8 (Temoçin and Yigitoglu 2009), MnP at 4.5 pH, laccase acts perfectly at neutral pH (Eibes et al. 2007). This pH tolerance varies when used in the immobilized state in the enzyme reactors and also based on the type of targeted pollutant (Adriana Arca-Ramos et al. 2018). In the immobilized enzyme reactor, the agitation process is very important as it can affect the activity of the immobilized enzyme (Liu et al. 2021) as well as can harm the support. So, the agitation speed and type of enzyme should be compatible with the support which is used for enzyme immobilization. The presence of organic solvents and surfactants may also affect the enzymatic degradation process in the case of hydrophobic pollutants (Adriana Arca-Ramos et al. 2018).

Enzyme reactors can be run both in batch as well as in continuous mode. The batch mode is suitable for understanding the optimum conditions and kinetics of the

Table 12.3 Advantage and disadvantage of batch mode operation and continuous mode operation (Zhang and Xing 2011)

Batch mode operation		Continuous mode operation	
Advantage	Disadvantage	Advantage	Disadvantage
Simplicity and flexibility both in use and in bioprocess development	Higher operating costs	Automating the bioprocess and thus reducing labor costs	Higher investment costs for control and automation equipment
Closely controllable environment useful for slow reactions	Labor and service intensive	More productive	High operating cost
Lower capital investment	Variation in batchwise product qualtiy	Final product quality is consistent	Less flexible operating conditions

degradation process, where definite concentration of enzyme and the substrate are mixed in the same vessel for similar time duration under controlled conditions. Low volume of wastewater may be treated using this mode. This is also helpful for the evaluation of the feasibility of developed technology (Gasser et al. 2014). For the high flow of wastewater, the continuous mode is preferred where for maintaining the volume constant, continuous addition of substrate and simultaneous removal of reaction outputs are required (Schückel et al. 2011). A list of advantages and disadvantages of both the batch mode operation and continuous mode operation is shown in Table 12.3.

There are some existing immobilized enzyme technologies, which are being tested for the treatment of wastewater (Table 12.4). Most of these technologies are still at lab scale and few are at pilot scale. Some are described in the following paragraphs.

12.3.1 Batch Mode Reactor

Stirred-Tank Reactor (STR) works in the batch mode and its operational conditions are comparatively simple (Fig. 12.2). The reactor is fitted with baffles, which increases its stirring efficiency when the enzymes/immobilized enzymes and substrates are mixed in a tank under optimum conditions of pH and temperature (Narayanan and Narayanan 2019). The support used for immobilized enzymes has to be stable in the reactor as the stirring is involved. The immobilized enzymes may be recovered by the use of membranes at the initial level, through centrifugation, or through magnetic separation (if the magnetic particle has been used as support material) (Arca-Ramos et al. 2016). The centrifugation process has a limitation as it may reduce the catalytic efficiency of immobilized enzymes due to clump formation (Lopez et al. 2014).

Table 12.4 Examples of the immobilized enzyme-based reactors used for the removal of wastewater pollutants

S. No.	Type of reactor	Base material	Immobilized enzyme with source	Target contaminant/s	References
1	Enzymatic membrane reactor	Membrane	Laccase from *T. versicolor*	Tetracycline	De Cazes et al. (2014)
2	Continuous flow packed-bed reactor	Granular activated carbon (GAC-1300)	Laccase from *Aspergillus oryzae* (commercial)	Bisphenol A, diclofenac, carbamazepine and sulfamethoxazole	Nguyen et al. (2016)
3	Packed bed reactor	Epoxy-activated acrylic polymers (Eupergit C and Eupergit C 250 L)	Laccase from *Myceliophthora thermophila* (commercial)	Estrone (E1), 17b-estradiol (E2) and 17a-ethinylestradiol (EE2)	Lloret et al. (2012)
4	Fixed-bed bioreactor	Polymethacrylate-based polymers (Sepabeads EC-EP3 and Dilbeads NK)	Laccase from *Myceliophthora thermophila* (commercial)	Reactive Black 5 (RB−5), Acid Blue 25, Methyl Orange, Methyl Green, Acid Green 27, and Remazol Brilliant Blue B (RBBR)	Kunamneni et al. (2008)
5	Enzymatic membrane reactor	Poly ether sulfone membrane	Laccase from *Myceliophthora thermophila* (commercial)	Estrone (E1),17β-estradiol (E2), and 17α-ethinyl estradiol (EE2	Lloret et al. (2013a)
6	Spiral-wound module	Polyether-sulphone	Laccasse from *Pyricularia oryzae*	Synthetic industrial wastewater (phenols)	Lante et al. (2000)
7	Stirred batch process and continuous packed bed reactor	Calcium alginate Starch gel	Bitter gourd peroxidase (BGP)	Textile industrial effluents	Matto et al. (2009)
8	Stirred batch process and continuous packed bed reactor	Con-A wood shaving	Turnip peroxidase (TP)	Direct Red 23 and 80	Matto and Hussain (2009)
9	Packed bed bioreactor	Semi-permeable alginate membrane	Soybean peroxidase (SBP) isolated from soybean seed coat powder	Phenol	Rezvani et al. (2015)

(continued)

Table 12.4 (continued)

S. No.	Type of reactor	Base material	Immobilized enzyme with source	Target contaminant/s	References
10	Magnetically stabilized fluidized bed reactor	Magnetic beads of poly (glycidylmethacrylate-co-methylmethacrylate)	Horseradish peroxidase (HRP)	Phenol and chlorophenol	Banramo and Arica (2008)
11	Magnetically stabilized fluidized bed reactor	Magnetic mesoporous silica nanoparticles	Laccase	Phenol from coking wastewater	Wang et al. (2012)
12	Packed bed reactor	Calcium-alginate	Horseradish peroxidase (HRP)	Reactive Red 120 (RR120), Reactive Blue 4 (RB4) and Reactive Orange 16 (RO 16)	Bilal et al. (2016)
13	Enzymatic membrane reactor (EMR)	Gelatin layer previously deposited onto α-alumina tubular membranes	Laccase	2,6-dimethoxyphenol (DMP)	Chea et al. (2014)

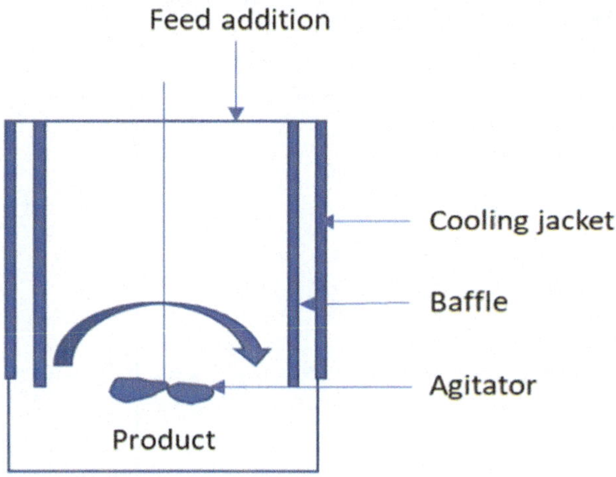

Fig. 12.2 Stirred tank reactor

12.3.2 Continuous Mode Reactor

The continuous mode enzyme reactors work both as a *continuous stirred tank reactor* (CSTR) as well as in plug flow mode such as a *packed bed reactor* (PBR). The *continuous fluidized bed reactor* (FBR) may also be considered under this category. *CSTR* has a stirring tank where the enzyme is mixed with the substrate and other contents and the process occurs under the controlled conditions of temperature and pH. In the case of the immobilized enzyme, a separate device is required for retaining the enzyme in the reactor. The membranes are generally used for the separation of immobilized enzymes without affecting their efficiency. Gasser et al. had used the ultrafiltration module for retaining the laccase immobilized on fumed silica nanoparticles while treating the bisphenol-A present in the wastewater (Gasser et al. 2014). The metallic filtration membrane-like structure is also used for the retention of immobilized enzyme in the reactor which is known as *perfusion basket reactor* (Cabana et al. 2009b). Kumar et al. had used this continuous reactor for the removal of dyes such as malachite green, brilliant blue R, and reactive black 5 in the presence of cross-linked enzyme aggregates (CLEA) laccase (Kumar et al. 2014).

In the PBR, the immobilized enzymes are packed inside the reactor. In this reactor, the reaction solution flows like a plug flow through the reactor; therefore it is also known as a plug-flow reactor (PFR) (Lilly et al. 1974) (Fig. 12.3). Although, long tube and lack of a stirring device inhibit proper mixing of the fluid in the reactor which causes the problem of pressure drop and bed compaction. The substrate stream flows with equal velocity, along the axis of the reactor axis with no back-mixing (Illanes and Altamirano 2008). It is also known as the tubular-flow enzyme reactor. Continuous PBR (CPBR) is a broadly used for immobilized enzymes, due to reduced labor costs, ease of automatic control and operation, ease of quality control of product, and stabilization of operating conditions (Tonini and Astrup 2012), but

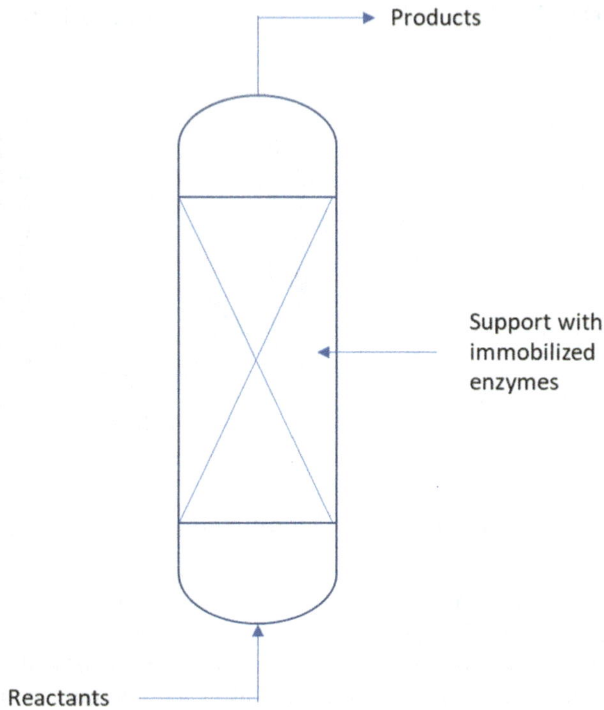

Fig. 12.3 Packed bed reactor

the rigid immobilized enzyme is preferred for avoiding particle deformations (Chaplin and Bucke 1990). The pH and temperature control are also a tough task in this reactor. The enzymes are immobilized in the bed of the reactor and the substrate solution is then passed over the enzyme bed for conversion into the product. A high concentration of immobilized enzymes may be used in PBR.

Lloret et al. (2012) had used laccase immobilized on epoxy-activated acrylic polymers (Eupergit and Eupergit 250 L) in PBR for the removal of estriol. Nguyen et al. (2016) had used laccase immobilized on granular activated carbon for the removal of bisphenol A, diclofenac, carbamazepine, and Sulfamethoxazole in PBR. The soybean peroxidase (SBP) isolated from soybean seed coat powder, immobilized on semi-permeable alginate membrane, when used in packed bed reactor, were able to remove 96.7% phenol from wastewater (Rezvani et al. 2015).

In FBR, the immobilized enzymes are used as a bed through which the substrate stream is fluidized in the up flow direction (Sarkar and Leonowicz 1989; Cai et al. 2021) (Fig. 12.4). The substrate stream should have sufficient velocity for carrying the particles having immobilized enzyme, although the porosity, size, shape, density of the particles along with viscosity and density of the solution also play the roles in the process. FBR is a combination of PBR and CSTR that behaves in an intermediary manner between those two types of reactors.

The laccase or tyrosinase enzyme immobilized on polyacrylonitrile beads have the capacity to remove different types of bisphenol (Bisphenol A (BPA), Bisphenol

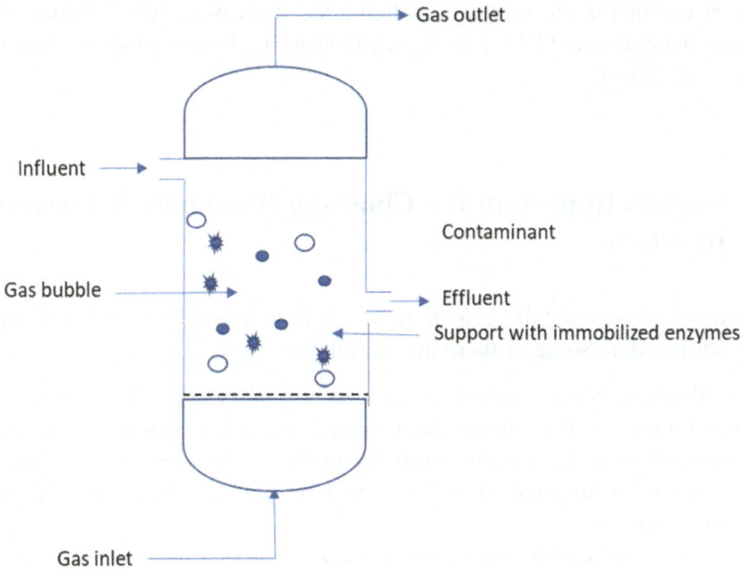

Fig. 12.4 Fluidized bed reactor

B (BPB), Bisphenol F (BPF), and Tetra chloro bisphenol A (TCBPA)) in FBR and found ≈100% and 90% removal of bisphenols using immobilized laccase and tyrosinase respectively. Soybean peroxidase, immobilized on a glass support, in FBR was able to remove 80% of phenol (Gomez et al. 2007).

12.3.3 Membrane Reactor (MR)

MR involves both the processes of membrane separation as well as enzyme catalysis. In this reactor, such semipermeable membrane is used which restrict the entry of enzymes and give passage to product molecules. It works in both the continuous as well as batch modes and allows easy recovery of the enzyme molecule from the mixture solution (Advances 2018). MR is generally a hollow reactor that contains hundreds of thin tubular fibers which have 200–300 μm inner diameter and 300–900 μm outer diameter. The substrate comes into contact with the enzyme through the tubular wall and converts it into products. Further diffusion through the semipermeable membrane makes the separation easy and retrieval. The product flows inside the tube due to the influence of the differential pressure along the tubular wall (Dwevedi 2016).

De Cazes et al. had used immobilized laccase, isolated from *Trametes versicolor,* in an enzymatic membrane reactor for the removal of tetracycline and obtained 56% degradation of the selected antibiotic, as a component of wastewater (De Cazes et al. 2014). hydrophilic PVDF microfiltration membrane immobilized with laccase enzyme is useful for the removal of herbicide derivative, N0, N0-(dimethyl)-N-(2-hydroxyphenyl)urea (2-HF) from wastewater in Frame plate reactor module (Jolivalt et al. 2000).

12.4 Factors Important for Choosing Immobilized Enzyme Reactors

Many factors determine the selection of an immobilized enzyme reactor while treating wastewater. Some of them are as follows:

- *Physical properties of substrates such as solubility can affect the wastewater treatment process.* The batch mode reactor is suitable for the pollutant present in low concentration due to low solubility in the wastewater while pollutant with higher solubility in water may be treated by both the batch as well as using continuous mode.
- *Operating conditions in immobilized enzyme reactors may affect the treatment process.* The controlled temperature and pH, proper gas supply etc. are required for getting proper results. For example, when maintenance of pH is required in the catalytic process, then STR is preferred while proper gas (eg. oxygen) supply is essential, FBR is a good choice. The stability of enzyme, substrate and products also needs to be considered prior selection f suitable process.
- *The enzyme stability in the immobilized form should be ensured prior selection of the reactor.* So the selection of appropriate carrier is required for immobilization of enzyme for inhibiting its inactivation.
- *The economic concern.* The total cost for running the reactor decides its applicability. The total cost includes the costs associated with downstream processing, substrate(s), labor, overheads, depreciation, and process development, in addition, the costs associated with building and running the enzyme reactor.

As discussed above, there is not a single standard rule for the choice of enzyme reactors; one should balance all the different concerns regarding the performance of the reactor and of the whole process, and then have to make a decision.

12.5 Conclusions

The immobilized enzyme-based reactors are better options for treating wastewater as in the immobilized form the catalytic efficiency of enzymes and their operational stabilities is high. The reusability and recovery of enzymes are also increased in immobilized form. It is observed that the catalytic efficiency of enzymes is also related to the source of origin. So, more resources need o explored for getting cheaper alternatives for enzyme production. The stability of support materials, which is used for enzyme immobilization, is also very important while treating wastewater in reactors. It indicates that more such materials are required according to the targeted reactors as the selection of an immobilized enzyme reactor is important concerning the economy involved in treating wastewater. Most of the wastewater treatment processes are being conducted in the lab-scale setup using immobilized enzyme reactors, so it is essential to test the existing technologies at the pilot scale as well as on the real site.

References

Advances R (2018) Industrial applications of enzymes: recent advances, techniques, and outlooks:20–29. https://doi.org/10.3390/catal8060238

Agnihotri V (2020) Advancement in bioremediation of pharmaceutical and personal care products. In: Waste management: concepts, methodologies, tools, and applications. IGI Global, Hershey, pp 1150–1168

Agnihotri V, Thathola P (2019) Pharmaceutical and personal care products (PPCPs) in wastewater/freshwater sources: distribution and related health concerns. Himal Ecol 27:63

Antecka K, Zdarta J, Siwińska-Stefańska K, Sztuk G, Jankowska E, Oleskowicz-Popiel P, Jesionowski T (2018) Synergistic degradation of dye wastewaters using binary or ternary oxide systems with immobilized laccase. Catalysts 8(9). https://doi.org/10.3390/catal8090402

Apetrei IM, Rodriguez-Mendez ML, Apetrei C, De Saja JA (2013) Enzyme sensor based on carbon nanotubes/cobalt(II) phthalocyanine and tyrosinase used in pharmaceutical analysis. Sensors Actuators B Chem 177:138–144. https://doi.org/10.1016/j.snb.2012.10.131

Arca-Ramos A, Eibes G, Moreira MT, Feijoo G, Lema JM (2014) Vegetable oils as NAPLs in two phase partitioning bioreactors for the degradation of anthracene by laccase. Chem Eng J 240:281–289. https://doi.org/10.1016/j.cej.2013.11.076

Arca-Ramos A, Kumar VV, Eibes G, Moreira MT, Cabana H (2016) Recyclable cross-linked laccase aggregates coupled to magnetic silica microbeads for elimination of pharmaceuticals from municipal wastewater. Environ Sci Pollut Res 23(9):8929–8939. https://doi.org/10.1007/s11356-016-6139-x

Arca-Ramos A, Eibes G, Feijoo G, Lema JM, Moreira MT (2018) Enzymatic reactors for the removal of recalcitrant compounds in wastewater. Biocatal Biotransformation 36(3):195–215. https://doi.org/10.1080/10242422.2017.1315411

Atadashi IM, Aroua MK, Aziz AA (2010) High quality biodiesel and its diesel engine application: a review. Renew Sust Energ Rev 14(7):1999–2008. https://doi.org/10.1016/j.rser.2010.03.020

Aytar BS, Bakir U (2008) Preparation of cross-linked tyrosinase aggregates. Process Biochem 43(2):125–131. https://doi.org/10.1016/j.procbio.2007.11.001

Banramo GG, Arica MY (2008) Enzymatic removal of phenol and p-chlorophenol in enzyme reactor: horseradish peroxidase immobilized on magnetic beads. J Hazard Mater 156:148–155

Bayramoglu G, Altintas B, Arica MY (2012) Cross-linking of horseradish peroxidase adsorbed on polycationic films: utilization for direct dye degradation. Bioprocess Biosyst Eng 35(8):1355–1365. https://doi.org/10.1007/s00449-012-0724-2

Bhatt P, Joshi T, Bhatt K, Zhang W, Huang Y, Chen S (2021) Binding interaction of glyphosate oxidoreductase and C-P lyase: molecular docking and molecular dynamics simulation studies. J Hazard Mater 5(409):124927

Bilal M, Iqbal HMN, Hussain Shah SZ, Hu H, Wang W, Zhang X (2016) Horseradish peroxidase-assisted approach to decolorize and detoxify dye pollutants in a packed bed bioreactor. J Environ Manag 183:836–842. https://doi.org/10.1016/j.jenvman.2016.09.040

Cabana H, Alexandre C, Agathos SN, Jones JP (2009a) Bioresource technology immobilization of laccase from the white rot fungus Coriolopsis polyzona and use of the immobilized biocatalyst for the continuous elimination of endocrine disrupting chemicals. Bioresour Technol 100 (14):3447–3458. https://doi.org/10.1016/j.biortech.2009.02.052
Cabana H, Jones JP, Agathos SN (2009b) Utilization of cross-linked laccase aggregates in a perfusion basket reactor for the continuous elimination of endocrine-disrupting chemicals. Biotechnol Bioeng 102(6):1582–1592. https://doi.org/10.1002/bit.22198
Cai QQ, Lee BCY, Ong SL, Hu JY (2021) Fluidized-bed Fenton technologies for recalcitrant industrial wastewater treatment—recent advances, challenges and perspective. Water Res 190:1–4. https://doi.org/10.1016/j.watres.2020.116692
Chaplin MF, Bucke C (1990) Immobilized enzymes and their use. In: Enzyme technology. Cambridge University Press, Cambridge, pp 167–195
Chatterjee S, Pandit S, Fosso-kankeu E (2019) Removal of heavy metal pollutants from wastewater using immobilized enzyme techniques: a review. In: Fosso-Kankeu E (ed) Nano and bio-based Technologies for Wastewater Treatment: prediction and control tools for the dispersion of pollutants in the environment. Scrivener Publishing LLC, Beverly, pp 459–480
Chea V, Paolucci-Jeanjean D, Sanchez J, Belleville MP (2014) Potentialities of a membrane reactor with laccase grafted membranes for the enzymatic degradation of phenolic compounds in water. Membranes 4(4):678–691. https://doi.org/10.3390/membranes4040678
da Silva LV, Tavares APM, Kameda E, Macedo EA, Coelho MAZ, Amaral PFF (2013) Factors affecting water colour removal by tyrosinase. Int J Environ Stud 70(2):316–326. https://doi.org/10.1080/00207233.2013.783267
Das R, Ghosh S, Bhattacharjee C (2012) Enzyme membrane reactor in isolation of antioxidative peptides from oil industry waste: a comparison with non-peptidic antioxidants. LWT Food Sci Technol 47(2):238–245. https://doi.org/10.1016/j.lwt.2012.01.011
De Cazes M, Belleville MP, Petit E, Llorca M, Rodríguez-Mozaz S, De Gunzburg J, Barceló D, Sanchez-Marcano J (2014) Design and optimization of an enzymatic membrane reactor for tetracycline degradation. Catal Today 236(PART A):146–152. https://doi.org/10.1016/j.cattod.2014.02.051
Dwevedi A (2016) Basics of enzyme immobilization, pp 21–44. https://doi.org/10.1007/978-3-319-41418-8
Eibes G, Lopez C, Moreira MT, Feijoo G, Lema JM (2007) Strategies for the design and operation of enzymatic reactors for the degradation of highly and poorly soluble recalcitrant compounds. Biocatal Biotransformation 25(2–4):260–268. https://doi.org/10.1080/10242420701444371
Fernández-fernández M, Sanromán MÁ, Moldes D (2013) Recent developments and applications of immobilized laccase. Biotechnol Adv 31(8):1808–1825. https://doi.org/10.1016/j.biotechadv.2012.02.013
Gasser CA, Ammann EM, Shahgaldian P, Corvini PFX (2014) Laccases to take on the challenge of emerging organic contaminants in wastewater. Appl Microbiol Biotechnol 98(24):9931–9952. https://doi.org/10.1007/s00253-014-6177-6
Gomes-Ruffi CR, da Cunha RH, Almeida EL, Chang YK, Steel CJ (2012) Effect of the emulsifier sodium stearoyl lactylate and of the enzyme maltogenic amylase on the quality of pan bread during storage. LWT Food Sci Technol 49(1):96–101. https://doi.org/10.1016/j.lwt.2012.04.014
Gómez JL, Bódalo A, Gómez E, Hidalgo AM, Gómez M, Murcia MD (2007) Experimental behaviour and design model of a fluidized bed reactor with immobilized peroxidase for phenol removal. Chem Eng J 127(1–3):47–57. https://doi.org/10.1016/j.cej.2006.09.021
Guzik U, Hupert-Kocurek K, Wojcieszynska D (2014) Immobilization as a strategy for improving enzyme properties—application to oxidoreductases. Molecules 19(7):8995–9018. https://doi.org/10.3390/molecules19078995
Hakala TK, Liitiä T, Suurnäkki A (2013) Enzyme-aided alkaline extraction of oligosaccharides and polymeric xylan from hardwood Kraft pulp. Carbohydr Polym 93(1):102–108. https://doi.org/10.1016/j.carbpol.2012.05.013

Han M-J, Choi H-T, Song H-G (2005) Purification and characterization of laccase from the white rot fungus Trametes versicolor. J Microbiol (Seoul, Korea) 43(6):555–560

Homaei AA, Sariri R, Vianello F, Stevanato R (2013) Enzyme immobilization: an update. J Chem Biol 6(4):185–205. https://doi.org/10.1007/s12154-013-0102-9

Hou J, Dong G, Ye Y, Chen V (2014) Enzymatic degradation of bisphenol-A with immobilized laccase on TiO2 sol-gel coated PVDF membrane. J Membr Sci 469:19–30. https://doi.org/10.1016/j.memsci.2014.06.027

Husain M, Husain Q (2008) Applications of redox mediators in the treatment of organic pollutants by using oxidoreductive enzymes: a review. Crit Rev Environ Sci Technol 38(1):1–42. https://doi.org/10.1080/10643380701501213

Illanes A, Altamirano C (2008) Enzyme reactors. In: Illanes A (ed) Enzyme biocatalysis. Springer Science + Business Media B.V, pp 205–251

Işık C, Saraç N, Teke M, Uğur A (2021) A new bioremediation method for removal of wastewater containing oils with high oleic acid composition: acinetobacter haemolyticus lipase immobilized on eggshell membrane with improved stabilities. New J Chem 45(4):1984–1992. https://doi.org/10.1039/D0NJ05175F

Jolivalt C, Brenon S, Caminade E, Mougin C, Pontié M (2000) Immobilization of laccase from Trametes versicolor on a modified PVDF microfiltration membrane: characterization of the grafted support and application in removing a phenylurea pesticide in wastewater. J Membr Sci 180(1):103–113. https://doi.org/10.1016/S0376-7388(00)00522-6

Jun LY, Yon LS, Mubarak NM, Bing CH, Pan S, Danquah MK, Abdullah EC, Khalid M (2019) An overview of immobilized enzyme technologies for dye and phenolic removal from wastewater. J Environ Chem Eng 7(2):102961. https://doi.org/10.1016/j.jece.2019.102961

Kour D, Kaur T, Devi R, Yadav A, Singh M et al (2021) Beneficial microbiomes for bioremediation of diverse contaminated environments for environmental sustainability: present status and future challenges. Environ Sci Pollut Res 28:24917–24939

Kumar L, Bharadvaja N (2019) Enzymatic bioremediation: a smart tool to fight environmental pollutants. In: Smart bioremediation technologies: microbial enzymes. Elsevier, San Diego. https://doi.org/10.1016/B978-0-12-818307-6.00006-8

Kumar VV, Sivanesan S, Cabana H (2014) Magnetic cross-linked laccase aggregates—bioremediation tool for decolorization of distinct classes of recalcitrant dyes. Sci Total Environ 487:830–839. https://doi.org/10.1016/j.scitotenv.2014.04.009

Kunamneni A, Ghazi I, Camarero S, Ballesteros A, Plou FJ, Alcalde M (2008) Decolorization of synthetic dyes by laccase immobilized on epoxy-activated carriers. Process Biochem 43(2):169–178. https://doi.org/10.1016/j.procbio.2007.11.009

Kurniawati S, Nicell JA (2008) Characterization of Trametes versicolor laccase for the transformation of aqueous phenol. Bioresour Technol 99(16):7825–7834. https://doi.org/10.1016/j.biortech.2008.01.084

Lante A, Crapisi A, Krastanov A, Spettoli P (2000) Biodegradation of phenols by laccase immobilised in a membrane reactor. Process Biochem 36:51–58. https://doi.org/10.1016/S0032-9592(00)00180-1

Lilly MD, Dunnii P, Dunnill P, Suckling KE (1974) [49] Immobilized-enzyme reactors. Methods Enzymol 6(8):717–738

Liu J, Tong S, Sun H, Chang Y, Luo H, Yu H, Shen Z (2021) Effect of shaking speed on immobilization of cephalosporin C acylase: correlation between protein distribution and properties of the immobilized enzymes. Biotechnol Prog 37(1):1–11. https://doi.org/10.1002/btpr.3063

Lloret L, Eibes G, Lú-Chau TA, Moreira MT, Feijoo G, Lema JM (2010) Laccase-catalyzed degradation of anti-inflammatories and estrogens. Biochem Eng J 51(3):124–131. https://doi.org/10.1016/j.bej.2010.06.005

Lloret L, Hollmann F, Eibes G, Feijoo G, Moreira MT, Lema JM (2012) Immobilisation of laccase on Eupergit supports and its application for the removal of endocrine disrupting chemicals in a packed-bed reactor. Biodegradation 23(3):373–386. https://doi.org/10.1007/s10532-011-9516-7

Lloret L, Eibes G, Moreira MT, Feijoo G, Lema JM (2013a) Removal of estrogenic compounds from filtered secondary wastewater effluent in a continuous enzymatic membrane reactor. Identification of biotransformation products. Environ Sci Technol 47(9):4536–4543. https://doi.org/10.1021/es304783k

Lloret L, Eibes G, Moreira MT, Feijoo G, Lema JM, Miyazaki M (2013b) Improving the catalytic performance of laccase using a novel continuous-flow microreactor. Chem Eng J 223:497–506. https://doi.org/10.1016/j.cej.2013.03.018

López C, Cruz-Izquierdo Á, Picó EA, García-Bárcena T, Villarroel N, Llama MJ, Serra JL (2014) Magnetic biocatalysts and their uses to obtain biodiesel and biosurfactants. Front Chem 2:72. https://doi.org/10.3389/fchem.2014.00072

Matto M, Husain Q (2008) Redox-mediated decolorization of Direct Red 23 and Direct Blue 80 catalyzed by bioaffinity-based immobilized tomato (Lycopersicon esculentum) peroxidase. Biotechnol J 3(9–10):1224–1231. https://doi.org/10.1002/biot.200800049

Matto M, Husain Q (2009) Decolorization of direct dyes by immobilized turnip peroxidase in batch and continuous processes. Ecotoxicol Environ Saf 72(3):965–971. https://doi.org/10.1016/j.ecoenv.2008.02.019

Matto M, Hussain Q (2009) Decolorization of direct dyes by immobilized turnip peroxidase in batch and continuous processes. Ecotoxicol Environ Saf 72:965–971

Matto M, Satar R, Hussain Q (2009) Application of calcium alginate–starch entrapped bitter gourd (Momordica charantia) peroxidase for the removal of colored compounds from a textile effluent in batch as well as in continuous reactor. Appl Biochem Biotechnol 158:512–523

Mielgo I, López C, Moreira MT, Feijoo G, Lema JM (2003) Oxidative degradation of azo dyes by manganese peroxidase under optimized conditions. Biotechnol Prog 19(2):325–331. https://doi.org/10.1021/bp020136w

Modaressi K, Taylor KE, Bewtra JK, Biswas N (2005) Laccase-catalyzed removal of bisphenol-A from water: protective effect of PEG on enzyme activity. Water Res 39(18):4309–4316. https://doi.org/10.1016/j.watres.2005.08.005

Morais RR, Pascoal AM, Caramori SS, Lopes FM, Fernandes KF (2013) Immobilization of α-amylase onto luffa operculata fibers. Enzyme Research 2013. https://doi.org/10.1155/2013/803415

Mugdha A, Usha M (2012) Enzymatic treatment of wastewater containing dyestuffs using different delivery systems. Sci Revs Chem Commun 2(1):31–40

Narayanan CM, Narayan V (2019) Biological wastewater treatment and bioreactor design: a review. Sustain Environ Res 1(1):1–17. https://doi.org/10.1186/s42834-019-0036-1

Nguyen LN, Hai FI, Kang J, Leusch FDL, Roddick F, Magram SF, Price WE, Nghiem LD (2014) Enhancement of trace organic contaminant degradation by crude enzyme extract from Trametes versicolor culture: effect of mediator type and concentration. J Taiwan Inst Chem Eng 45(4):1855–1862. https://doi.org/10.1016/j.jtice.2014.03.021

Nguyen LN, Hai FI, Dosseto A, Richardson C, Price WE, Nghiem LD (2016) Continuous adsorption and biotransformation of micropollutants by granular activated carbon-bound laccase in a packed-bed enzyme reactor. Bioresour Technol 210:108–116. https://doi.org/10.1016/j.biortech.2016.01.014

Pandey A, Jain R, Sharma A, Dhakar K, Kaira GS, Rahi P, Dhyani A, Pandey N, Adhikari P, Shouche YS (2019) 16S rRNA gene sequencing and MALDI-TOF mass spectrometry based comparative assessment and bioprospection of psychrotolerant bacteria isolated from high altitudes under mountain ecosystem. SN Appl Sci 1(3):278. https://doi.org/10.1007/s42452-019-0273-2

Rezvani F, Azargoshasb H, Jamialahmadi O, Hashemi-Najafabadi S, Mousavi SM, Shojaosadati SA (2015) Experimental study and CFD simulation of phenol removal by immobilization of soybean seed coat in a packed-bed bioreactor. Biochem Eng J 101:32–43. https://doi.org/10.1016/j.bej.2015.04.019

Sarkar JM, Leonowicz A (1989) Immobilization of enzymes and soils. Soil Biol Biochem 21:223–230

Satar R, Husain Q (2011) Catalyzed degradation of disperse dyes by calcium alginate-pectin entrapped bitter gourd (Momordica charantia) peroxidase. J Environ Sci 23(7):1135–1142. https://doi.org/10.1016/S1001-0742(10)60525-6

Schückel J, Matura A, van Pée KH (2011) One-copper laccase-related enzyme from Marasmius sp.: purification, characterization and bleaching of textile dyes. Enzym Microb Technol 48 (3):278–284. https://doi.org/10.1016/j.enzmictec.2010.12.002

Šekuljica N, Prlainović N, Jovanović JR, Stefanović AB, Djokić VR, Mijin D, Knezevic-Jugovic ZD (2016) Immobilization of horseradish peroxidase onto kaolin. Bioprocess Biosyst Eng 39 (3):461–472. https://doi.org/10.1007/s00449-015-1529-x

Sirisha VL, Jain A, Jain A (2016) Enzyme immobilization: an overview on methods, support material, and applications of immobilized enzymes. In: Advances in food and nutrition research, vol 79, 1st edn. Elsevier, San Diego, pp 179–211. https://doi.org/10.1016/bs.afnr.2016.07.004

Taboada-Puig R, Lu-Chau TA, Eibes G, Feijoo G, Moreira MT, Lema JM (2015) Continuous removal of endocrine disruptors by versatile peroxidase using a two-stage system. Biotechnol Prog 31(4):908–916. https://doi.org/10.1002/btpr.2116

Temoçin Z, Yiğitoğlu M (2009) Studies on the activity and stability of immobilized horseradish peroxidase on poly(ethylene terephthalate) grafted acrylamide fiber. Bioprocess Biosyst Eng 32 (4):467–474. https://doi.org/10.1007/s00449-008-0266-9

Tong Z, Qingxiang Z, Hui H, Qin L, Yi Z (1997) Removal of toxic phenol and 4-chlorophenol from waste water by horseradish peroxidase. Chemosphere 34(4):893–903. https://doi.org/10.1016/S0045-6535(97)00015-5

Tonini D, Astrup T (2012) Life-cycle assessment of a waste refinery process for enzymatic treatment of municipal solid waste. Waste Manag 32(1):165–176. https://doi.org/10.1016/j.wasman.2011.07.027

Tran DN, Balkus KJ (2011) Perspective of recent progress in immobilization of enzymes. ACS Catal 1:956–968

Unuofin JO, Okoh AI, Nwodo UU (2019) Aptitude of oxidative enzymes for treatment of wastewater pollutants: a laccase perspective. Molecules 24(11):1–36. https://doi.org/10.3390/molecules24112064

Wang F, Hu Y, Guo C, Huang W, Liu CZ (2012) Enhanced phenol degradation in coking wastewater by immobilized laccase on magnetic mesoporous silica nanoparticles in a magnetically stabilized fluidized bed. Bioresour Technol 110:120–124

Zhang C, Xing XH (2011) Enzyme bioreactors. In: Comprehensive biotechnology, vol 2, 2nd edn. Elsevier B.V, San Diego. https://doi.org/10.1016/B978-0-08-088504-9.00099-4

Chapter 13
Role of Chelating Compounds in Biodegradation and Bioremediation

Geeta Bhandari and Om Prakash

13.1 Introduction

Global advancements after the inception of industrial revolution and intensive farming practices have resulted into generation of may toxic contaminants thus creating many problems in protecting and preserving the environment (Bhatt et al. 2020; Bhandari et al. 2021). Heavy metal (HM) pollution is one such issue imposing adverse impact on environment, food security and human health. The heavy metals are present in form of soluble ions or colloids in the soil and are divided into essential and non-essential. The essential heavy metals are vital micronutrients for various life processes and consist of Co, Cr, Cu, Fe, Mn, Ni and Zn however these may become toxic on consuming higher quantities. Non-essential heavy metals comprise of Pb, Cd and Hg and are extremely hazardous to all the living beings (Sandeep et al. 2019; Kumar et al. 2021). Fifty-three elements have been recorded as heavy metals and are global contaminants with densities higher than 5 g/cm^3 (Prieto et al. 2018). However, these heavy metals are indigenous constituent of soil, but various anthropogenic factors cause enhancement in their concentration. Heavy metals can arise naturally in the soil due to activities for e.g., volcanic matter and weathering of rocks or may arise due to several human activities for e.g., mining, smelting, drilling, agrochemicals, military activities, fuel combustion, sewage sludge and industrial wastewater (Bhandari and Bhatt 2020; Kour et al. 2021). Heavy metals are nonbiodegradable, can't be mineralized or decomposed and can remain in soils/

G. Bhandari (✉)
Sardar Bhagwan Singh University, Dehradun, Uttarakhand, India

O. Prakash
Department of Chemistry, CBSH, G.B. Pant University of Agriculture and Technology, Pantnagar, Uttarakhand, India

Regional Ayurveda Research Institute Gwalior, Ministry of Ayush, Government of India, Gwalior, Madhya Pradesh, India

© The Author(s), under exclusive license to Springer Nature Switzerland AG 2022
D. C. Suyal, R. Soni (eds.), *Bioremediation of Environmental Pollutants*,
https://doi.org/10.1007/978-3-030-86169-8_13

sediment for long durations, having a documented half-life of greater than 20 years thus imposing an immense threat to various elements of the ecosystem and eventually adverse impact on various living organisms (Iheanacho et al. 2017; Singh et al. 2021). As reported by US EPA, heavy metal pollution in soil/sediment has resulted in health problems to more than ten million human beings worldwide (Ashraf et al. 2019).

Extensive research in the area of heavy metal remediation is the most important need of this hour for remediation of metal contaminated sites. Therefore, cost-effective and eco-friendly strategies are vital elements for the heavy metal removal in contaminated sites (Mani and Kumar 2014). The heavy metal elimination processes are divided into: (a) methods involving immobilization of these toxic metals on the soil, thus restricting their movement, and (b) methods that eliminate these metals from soil and recycling them for further usage (Nurchi et al. 2020). Till date, the remediation technologies developed for heavy metals contaminated sites consist of: (1) ex-situ or in-situ soil washing/leaching/flushing by use of chemicals, (2) immobilization/stabilization for reducing their solubility by addition of non-toxic compounds into the soils, (3) electrokinetic methods, (4) adsorption (5) dilution and (6) phytoremediation (Mani and Kumar 2014). The traditional physico-chemical processes can be employed in extensively polluted heavy metal sites, however high energy input and damaging effect on soil characteristics limits these methods (Kuppusamy et al. 2017). Microbial and plant based methods have been found to be more beneficial for reclamation of heavy metal contaminated sites (Ayangbenro and Babalola 2017). The plant-based strategy further confers several advantages such as greater extraction of toxic metals from soil without damaging the soil productivity (Lajayer et al. 2019).

13.2 Chelator-Assisted Phytoextraction

Phytoextraction is the process of utilizing specific, greatly adapted fast growing plants for absorbing, transporting and accumulating toxic heavy metals in harvestable plant parts from the environment. Several plants termed as metallophytes are capable of survival in high heavy metal pollution (Sheoran et al. 2010). Metallophytes generally comprise of plants belonging to Brassicaceae and may be classified into; excluders, indicators and hyperaccumulators (Ashraf et al. 2019). Metal excluders extract toxic metal ions from the soil and translocate them to roots, however further translocation to shoot region is limited (Malik and Biswas 2012). Metal indicators accumulate toxic metal ions in various parts of their aerial tissues (Sheoran et al. 2010). Metal hyper accumulator must accumulate at least 100 mg/kg of As and Cd, 1000 mg/kg Pb, Co, Cr, Cu, Mn and Ni (Ashraf et al. 2019). More than 400 plant species are documented as hyperaccumulators showing slow growth rate and lesser biomass generation (Kramer 2018) (Table 13.1).

Natural phytoextraction is nevertheless a highly limited process insufficient metal bioavailability in the rhizospheric zone which is dependent on soil pH, nutrient level,

Table 13.1 Metal hyperaccumulator plant species

Heavy metals	Plant species	Maximum Concentration in plants (mg/kg)	References
As, Cu, Pb, Zn	*Spartina maritime*	5000	Mesa et al. (2015)
Fe, Zn, Cd, Cu, B, and Cr	*Eichhorina crassipes*	40,000	Elias et al. (2014)
Cd	*Azolla pinnata*	1127	Rai (2008)
Cd	*Arabis paniculata*	1127	Zeng et al. (2009)
Cu	*Eleocharis acicularis*	20,200	Sakakibara et al. (2011)
Cu	*Aeolanthus biformifolius*	13,700	Chaney et al. (2010)
Cu	*Pteris vittata*	91,975	Wang et al. (2012b)
Zn	*Eleocharis acicularis*	11,200	Sakakibara et al. (2011)
Cr	*Pteris vittaia*	20,675	Kalve et al. (2011)
As	*Pteris vittaia*	8331	Kalve et al. (2011)
As	*Eleocharis acicularis*	1470	Sakakibara et al. (2011)
Hg	*Achillea millefolium*	18,275	Wang et al. (2012b)
Hg	*Erato polymnioides*	–	Chamba et al. (2017)
Pb	*Medicago sativa*	43,300	Koptsik (2014)
Pb	*Brassica juncea*	10,300	Koptsik (2014)
Pb	*Brassica nigra*	9400	Koptsik (2014)
Pb	*Helianthus annuus*	5600	Koptsik (2014)
Ni	*Alyssum heldreichii*	11,800	Bani et al. (2010)
Ni	*Alyssum markgrafii*	19,100	Bani et al. (2010)
Ni	*Ricinus communis*	–	Çelik and Akdaş (2019)

clay concentration, tolerance to metal ions and metal ion selectivity (Chibuike and Obiora 2014). For addressing these issues, chemical based/chelate enhanced phytoextraction has been exploited by several authors for improving metal solubility and bioavailability (Wang et al. 2017). Therefore, chelating compounds are employed for enhancing the mobility of heavy metal ions in soil and subsequent extraction in high amount by large biomass and rapidly-growing plants (Patra et al. 2018). Heavy metal accumulation ability of non-hyperaccumulating plants can also be improved by employing chelating compounds.

Chelating compounds are those compounds which can form coordinate bonds metal ions, therefore resulting into stable water soluble complexes (Flora and Pachauri 2010). The solubilized form of metal ions can be desorbed by chelating agents into the rhizospheric zone and thus making it available for uptake by the plant roots (Sun et al. 2016). Thereafter the solubilized metal can be extracted from the aerial tissues of plants by the process of phytoextraction (Lone et al. 2008). Chelating agents may be both inorganic or organic compounds; however all biochemical compounds are able to form coordinate bonds with metals; thus, all proteins and polysaccharides are excellent chelating agents for various heavy metal ions (Gupta and Diwan 2017). Heavy metal stress in plants results in synthesis of amino acids such as; glutamic acid, proline glycine, histidine (Irini et al. 2017; Jain and Chen 2018); glutathione, a peptide having metal-binding ability (Farooq et al. 2016; Sharma et al. 2017a, b); and amines such as; spermine, nicotianamine, spermidine, putrescine (Singh et al. 2016). All of these compounds are capable of solubilizing metal ions and thus are natural chelating agents (Wen et al. 2010). Some other low-molecular-weight organic acids (LMWOA), for e.g., citric acid, tartaric acid, malonic acid, lactic acid, succinic acid, malic acid and oxalic acid are also synthesized as root exudates and microbial residues, which further assist in metal ion chelation (Nworie et al. 2017; Chen et al. 2018). Therefore these LMWOAs and other synthetic chelators such as; ethylenediaminetetraacetic acid (EDTA), ethylenediaminedisuccinic acid (EDDS), nitrilotriacetic acid (NTA) and diethylenetriaminepentaacetic acid (DTPA) can be added in the rhizospheric zone for enhanced heavy metal removal (Ebrahimi 2014; Nanthavong and Sampanpanish 2015; Yang et al. 2013; Mujahid et al. 2013). The increased bioavailability and metal extraction from contaminated sites through plants by employing chelating compounds may be attributed to (1) increased amount of metal ions in soil solution (2) improved mobility of metal-chelators complexes in rhizosphere (3) poor interaction of metal-EDTA complex with cell wall components having negative charge (4) damage to physiological barriers present in plant roots owing to higher concentration of metal-EDTA complexes (5) enhanced movement of complexes in comparison to free ions thus causing higher metal extraction to aerial parts of plants (Evangelou et al. 2007).

The chelating compounds employed in chelant-assisted phytoextraction studies usually are grouped into: (1) synthetic aminopolycarboxylic acids (APCAs), such as; EDTA, hydroxy-EDTA, 1,2- cyclohexylenedinitrilotetraacetic acid (CDTA) and DTPA; (2) natural aminopolycarboxylates, such as; EDDS, NTA; and (3) LMWOAs, such as; citric acid, acetic acid, oxalic acid and gallic acid (Mir et al. 2017). Chelating agents are specifically efficient in improving the mobility of heavy metals at higher soil pH since the stability of metal-organic complex enhances by increasing the pH (Scheffer and Schachtschabel 1998).

13.2.1 Natural Chelating Agents

13.2.1.1 Aminopolycarboxylates as Chelating Agents

Nitrilotriacetic Acid (NTA)

Nitrilotriacetic acid is a tertiary amino-polycarboxylic acid capable of forming coordination complexes with metal ions so as to produce soluble complexes; and thus can be employed as a chelating agent for enhancing the efficiency of phytoextraction by improving the availability of metal ions globally (Reinoso-Maset et al. 2013). NTA has been found to enhance the metal ion concentration in aerial parts of Indian mustard with very limited leaching into the soil (Quartacci et al. 2006). Shilev et al. (2007) investigated the efficacy of NTA and EDDS for enhancing the extraction of Cd, Pb, and Zn by maize and sunflower. The study reported that Cd accumulates in the aerial parts of sunflower and maize on amending soil with NTA (5 mmol/L) EDDS respectively. No detrimental impact on soil characteristics and soil bacteria and fungi were reported. NTA is a tetradentate ligand since it comprise of four donor atoms and has been documented as a biodegradable chelating agents (Ruley et al. 2006). However it is less biodegradable in comparison to LMWOAs, thus it may result into detrimental effects on the ecosystem on leaching in the soil (Freitas and Nascimento 2016; Song et al. 2016).

Ethylenediaminedisuccinic Acid (EDDS)

EDDS is a biodegradable, low toxicity chelating agents employed in heavy metal phytoextarction studies as a substitute to EDTA (Ullmann et al. 2013; Saifullah et al. 2015; Wang et al. 2012a). EDDs possess excellent affinity to various metal ions and thus it can be employed for restoration of heavy metal polluted sites. It has been reported to enhance solubility and translocation of several metal ions for e.g., Cd, Ni, Cu, Zn, Hg and Pb to aerial tissue of several plant species (Xu and Thomson 2007; Yan et al. 2010; Yang et al. 2013). Studies conducted by several researchers on various plant varieties have reported that EDDS has better efficiency in accumulating Cu and Zn in comparison to EDTA but low efficiency in case of Pb has been reported (Luo et al. 2005; Evangelou et al. 2007; Epelde et al. 2008). Nonetheless, these reports suggested negative impact on plant development was reported in presence of EDDS. Some other researchers however did not evidence toxic impact of EDDS on plant development, thus suggesting that toxic effects are based on type of plant varieties, pH, soil characteristics and occurrence of other metal ions (Komarek et al. 2010; Meers et al. 2005). Xu and Thomson (2007) reported higher efficient of EDDS in enhancing phytoextraction and ex situ soil washing of heavy metal ion in comparison to EDTA and NTA. Bioavailability of Cu, Zn and Ni was higher in presence of EDDS in comparison to EDTA, whereas reverse has been found in case of Cd and Pb (Song et al. 2016). Despite high biodegradability and capability to enhance phytoextraction; EDDS has been found to leach to ground water level by various authors, thus imposing adverse impact on the environment (Fedje et al. 2013; Wang et al. 2012a; Yan et al. 2010).

13.2.1.2 Low Molecular Weight Organic Acids as Chelating Agents

Citric Acid

Citric acid is the first stable intermediate of TCA cycle being and commonly employed as an anticoagulant due to its calcium-binding ability and ability to synthesize coordinate compounds with divalent metal ions. This chelating characteristic of citric acid can be exploited for heavy metal extraction and accumulation by plants (Wuana et al. 2010). Sun et al. (2006) reported citric acid and acetic acid to be chelate Cd and accumulate it in *Solanum nigrum.* Enhanced heavy metal phytoextraction capability of citric acid has also been described in plant species such as *Brassica napus* and *Crotalaria juncea* (Ehsan et al. 2014; Alidoust et al. 2009) and reported that amending soil with citric acid assisted plant in accumulating higher amount of Cr along with improved antioxidant properties. Higher heavy metal phytoextraction by *Helianthus annuus* on amending soil with critic acid was also described by Turgut et al. (2004). Citric acid in comparison to EDTA was more effective in mobilizing Cr (Jean et al. 2008). Sinhal et al. (2010) conducted a similar phytoextraction studies for *Tagetes erecta* by employing citric acid and EDTA as chelating agents. Citric acid can also be employed for removal of heavy metal contaminants from sludge and wastewater (Dacera and Babel 2006). However the efficiency of citric acid as a biodegradable and nontoxic chelator for decontamination of heavy metal pollution is not very high in comparison to the synthetic chelating agents (Qu et al. 2011; Markovska et al. 2018).

Tartaric Acid

Tartaric acid is a low-molecular-weight organic acid which assists in fortification of heavy metal extraction by plants. It has also been documented to chelate various heavy metal ions including: Zn, Cu, Cr, Cd and Pb (Ke et al. 2006; Lin et al. 2009; Ding et al. 2014). Ling et al. (2011) used tartaric acid in soil as amendment in addition to malic and succinic acid for removal of Cr by maize plants from metal polluted sites. An enhanced phytextraction of Cr(VI) by *Spirodela polyrhiza* was reported on applying tartaric acid, glycerol and citric acid as chelating agents (Bala and Thukral 2011). However efficiency in enhancing phytoextraction of heavy metal is documented to be poor as compared to citric acid (Qu et al. 2011).

Oxalic Acid

Oxalic acid is a reducing agent with conjugate base $C_2H_2O_2^{4-}$. It can be employed as chelating agents (Ding et al. 2014). It is a widely employed agent for the removal of metal ions such as Pb, Cd, Cu and Zn (Prasad and Shivay 2017). Enhanced accumulation of several heavy metal ions has been observed on various plants, such as; *Beauveria caledonica*, *Zea mays*, *Leersia hexandra*, *Sedum alfredii* and *Senecio coronatus* (Fomina et al. 2005; Wang et al. 2012a; Nezami et al. 2016; Tao et al. 2016). Reduced Cr toxicity and improved oxidative stress mitigation was reported *Hibiscus sabdariffa* L. seedlings on application of oxalic acid (Ogunleye et al. 2016). Reported efficacy of oxalic acid as heavy metal chelators is comparatively low in comparison to citric acid and other synthetic chelating agents (Ogunleye et al. 2016; Tao et al. 2016).

13.2.2 Synthetic Chelating Agents

Besides natural chelating agents, various kinds of synthetic chelating compounds have been well documented in heavy metal phytoextraction studies (Liu et al. 2008; Saifullah et al. 2008). In last few decades, several researchers have exploited the role of various synthetic chelators for fortifying heavy metal uptake from metal polluted sites (Arabi et al. 2017). The next section describes various synthetic chelators which find significance in amending phytoextraction by various plant varieties (Sheoran et al. 2010).

13.2.2.1 Aminopolycarboxylic Acids as Chelating Agents

Ethylenediaminetetraacetic Acid (EDTA)

EDTA is a widely employed chelating agent for enhancing the potential of phytoextraction process (Ashraf et al. 2019). EDTA is an amino-polycarboxylic acid capable of forming six coordinate bonds with various heavy metals and thus works as a hexadentate ligand (Beck 2009). It has high binding affinity towards many metal ions and thus subsequently enhances the availability of these heavy metal ions in the soil for phytoextraction by plants (Lawal and Sauban 2014; Li et al. 2018). It has been documented for higher efficiency for phytoremediation in comparison to citric acid (Markovska et al. 2018; Zhang et al. 2018) and can increase heavy metal accumulation up to 100-fold (Ali and Chaudhury 2016). EDTA has been documented as a highly efficient metal ion chelator, but its large scale employment confers several disadvantages as following (Zia-ur-Rehman et al. 2015):

1. EDTA causes adverse affects on the physic-chemical properties of soil and rhizophersic microbes thus ultimately retarding plant growth and development (Vassilev et al. 2004).
2. EDTA is not highly biodegradable; therefore it can leach into the groundwater and retain in soil/sediments for long periods (Lu et al. 2017).
3. EDTA application in soil at higher concentrations may result into eutrophication due to extensive nitrogen release.
4. EDTA can also cause detrimental impact on soil nutrients owing to non-specific co-mobilization of macro- and micronutrients.

Diethylenetriaminepentaacetic Acid (DTPA)

Diethylenetriaminepentaacetic acid (pentetic acid), is another synthetic chelating agent employed in phytoextraction studies of various heavy metals for e.g., Cu, Zn, Ni, Cd, Hg, and Pb (Pastor et al. 2007; Liu et al. 2018). DTPA is however an expensive and less efficient chelator as compared to EDTA (Ghasemi et al. 2017). DTPA and EDTA are generally employed in total chlorine-free washing and bleaching of pulps by utilizing H_2O_2 for chelating metal ions such as Mn(II) and Fe(III) (Matzinger et al. 2007).

For prevention of potential leaching of metal-chelants into groundwater and affect of chelant on soil microflora, the choice and concentration of chelants and method and time of application and irrigation process is crucial (Evangelou et al. 2007; Luo et al. 2007). Efficacy of metal extraction by chelators depends upon the stability constants of the metal-chelate complexes and is of following order EDTA > NTA > citric acid > oxalic acid > acetic acid (Wenger et al. 1998). According to Elliott and Brown (1989), Ks ranks different chelants on the basis of efficiency, but it can't be employed for ranking of specific chelant toward several metals since it is also dependent on metal speciation in the soil matrix. The most accepted hypothesis for metal uptake by plant is the split-uptake method, according to which only free metal ions are allowed to be translocated to plant roots (Marschner et al. 1986). Another significant hypothesis describes that few intact metal-chelator complexes can also be up-taken by the plants (Nowack et al. 2006).

For any specific chelating agent, variation in application process may result in difference in phytoextraction efficiency and thus exploring the different methods of application of chelants is significant in optimization studies. It has been found that applying the chelating agent deep in soil in proximity of plant roots rather than on to the entire soil area results into significantly better heavy metal accumulation by the plants (Kayser et al. 1999). Application of chelators in different small doses also enhances the phytoextraction of Pb (Shen et al. 2002). The application of chelating agents in combination has also been documented to significantly increase the heavy metal phytoextraction efficacy. Blaylock et al. (1997) reported two-fold improve in Pb accumulation by aerial parts of Indian mustard on treating with EDTA and acetic acid in comparison to EDTA only. Another kind of combination involves the reaction between metals and several chelating agents, where the bio availability of heavy metal by chelant is enhanced by interaction with other chelating agents by inhibition of competition from other metals in soil. Luo et al. (2006) reported a synergistic effect of integrated application of EDTA and [S,S]-EDDS leading to a improved phytoextraction efficiency of Cu, Pb, Zn and Cd in comparison to single chelant only. Another kind of application involves employment of any agent for damaging the plant root structure so as to allow the direct extraction and translocation of metal-chelant complex to the aerial parts of the plants. In various studies, it has been reported that employment of glyphosate caused improved Pb accumulation (Mathis and Kayser 2001) due to obstruction in plant metabolic activity ultimately causing increased heavy metal translocation to plant shoots (Lestan et al. 2008).

13.3 Soil Washing Using Chelating Agents

Amongst the various methods of restoration of heavy metal polluted soil, the soil washing process has emerged as a potential alternative owing to several benefits such as: permanent and rapid elimination of metal ions, shorter time period for treatment, cost effectiveness, recovery, recyclability and energy generation (Cheng et al. 2020). Soil washing method was first described by Peters (1999), where the soil is initially excavated and then allowed to undergo physical separation, thereafter washed to eliminate the pollutants by employing chemical extraction. Finally the treated soil is moved back to the original site. Soil washing however can only be employed if significant metal extraction can be done form soil using the chemical extraction process. For achieving this, several kinds of extractant agents (surfactants, chelating agents, cyclodextrins and organic acids) are employed for solubilizing the target metals since heavy metal are usually present in an absorbed form with strong bonding to soil particles. Surfactants and cyclodextrins cause detrimental impact on soil characters and therefore only organic acids and chelating compounds are used (Nurchi et al. 2020). LMWOAs such as; oxalic, fumaric, succinic, citric, aconitic, malic, lactic, formic, maleic, acetic and malonic are the natural components of plant root exudates, bacterial secretions and plant and animal residues (Nurchi et al. 2020). Therefore metal solubilization by these LMWOAs is the reesentation of a mobile metal fraction that is bioavailable to biota. These chelating compounds are also able to dislodge the exchangeable, carbonate and reducible fractions of heavy metals by washing procedures (Peters 1999). Owing to limitations of LMWOAs, several biodegradable chelating agents such as N,N-bis (carboxymethyl)-L-glutamate (GLDA), iminodisuccinate (IDS), 3-hydroxy-2,20-iminodisuccinic acid (HIDS), S, S ethylenediamine-disuccinate (EDDS), methylglycine diacetate (MGDA), and nitrilotriacetate (NTA) have also emerged as potential alternative to LMWOAs in soil washing studies (Guo et al. 2018; Wang et al. 2018).

Acids carry out dissolution of carbonates and metal-containing soil material and interchange metal ions from soil surfaces to which H^+ ions are better attracted in comparison to cations of toxic metals. Chelators desorb metal ions from soil by synthesizing water-soluble stable metal-chelant complex. The metal-chelant complexes are highly stable, and inhibit the precipitation and sorption of metals, prevent discharge of metal ion from the complex until a significant reduction in soil pH occurs. The important factors to be considered while selecting chelants are (Gluhar et al. 2020):

- The chelant must be capable of forming stable complex with metal ions in a broad pH range. The chelator must also be less toxic and impose less damage to the environment.
- The metal-chelant complex must have poor adsorption affinity to solid soil particles.
- Chelating agents are expensive, thus is significant in soil-washing technology for recovering it in a cost-efficient manner. However if the recovered chelator is to be used multiple times then it must posses low biodegradability in soil.
- The process is closed one, so no toxic emissions or generation of leachate should ocuur. Since the treatment of these by-products to a acceptable level may add up to the cost.
- Soil being a non-renewable natural resource must be restored in a sustainable manner for achieving social acceptance of remediation process.

Yang et al. (2012) reported higher extraction efficiency at pH 5.5 than at pH 8.0 of EDDS, for Zn and Cd in soil washing studies. For replicating an in situ soil flushing condition, Mancini et al. (2011) employed soil columns for comparison of EDDS and EDTA for extraction of Pb, Cd, Zn, Cu and Ni. EDDS was found to be more efficient in mobilizing Cu, Ni, and Zn, whereas EDTA was better in case of Cd and Pb. Begum et al. (2012) conducted comparative study to analyze the efficacy of five biodegradable chelators: EDDS, HIDS, IDS, GLDA and MGDA for removal of Cd, Zn, Ni, Cu and Pb from contaminated soil. GLDA was reported to possess maximum extraction efficiency. However majority of the soil washing studies are carried under laboratory conditions and thus fail to exactly mimic the conditions in operational soil washing process. Therefore, Begum et al. (2013) carried out the same comparative study by employing EDDS, EDTA, GLDA, and HIDS in artificially polluted soils (Begum et al. 2013). Chelator/metal molar ratio and solution pH were changed, and outcomes suggested that GLDA has maximum extraction efficacy at low chelator concentration (0.01 M). Higher extraction efficiency was observed for GLDA in comparison to EDTA at pH 4, except for Pb (Begum et al. 2013). In a similar investigation, Yoo et al. (2018) initially employed EDDS for Pb and Cu extraction from polluted soil, followed by stabilizing with biochar and phosphogypsum for reducing the transport of left over toxic metal ions after soil washing. Lestan (2017) developed a soil washing technology ReSoil which fulfills all the sustainability parameters. ReSoil employs substitution, precipitation and adsorption for treating the process waters at a pH gradient developed by coset effective lime and H_2SO_4, employing the wastepaper as the metal adsorbent. EDTA was recovered and the treatment water was recycled in closed loop system, so as to minimize wastewater generation.

13.4 Chelant Enhanced Electrokinetic Extraction

Various remediation strategies have been adopted for efficient soil/sediment remediation for e.g., extraction, bioremediation, phytoremediation, thermal treatment, electrokinetic remediation (EKR) and integrated remediation technologies (Gan et al. 2009). Amongst all these technologies, EKR has emerged as a cost-efficient remediation process (Song et al. 2016). EKR employs low current or electrical potential through a series of electrodes (anodes and cathodes) for elimination of heavy metals form contaminated sites (Song et al. 2016). The employment of electric current causes induction of various chemical interactions and translocation processes in the polluted soil, that ultimately result in the solubilization and translocation of the metal ions towards the anode or cathode with successive elimination from the polluted soil. Electromigration and electro-osmosis are the transport methods in EKR (Cameselle et al. 2013), where electromigration is the transport of ions under the influence of electric field in the direction of oppositely charged electrode. Electro-osmosis is the overall flow of water caused by the electric field through the pores of soil and electro-osmotic flux is the outcome of the integrating electric field and the electric charge on the surface of the soil particles. Generally, the soil particles posses negative charge, and during electro-osmotic flow, they move in direction of cathode (Cameselle and Reddy 2012). The electric field also causes several chemical reactions at the electrodes and within the soil such as; electrolysis of water, adsorption/desorption of pollutants on the solid particle surfaces, redox reactions, and acid/base reactions. Nonetheless, EKR is only capable of extracting mobile (dissolved species or sorbed species on colloidal particles suspended in the pore fluid) pollutants from soil matrix. However the transport of sorbed species onto the soil particle and solid surfaces in form of precipitates needs strategies for solubilizing and keeping them in a mobile form (Yeung and Gu 2011). For enhancing the bioavailability of heavy metal in EKR, various types of chelating compounds may be employed such as; EDTA, NTA, EDDS and LMWOAs in a process termed as chelate-enhanced EKR (Alcántara et al. 2012; Song et al. 2016; Iannelli et al. 2015).

Owing to specific molecular structure of various chelating compounds, they are capable of forming coordination bonds with metal ion even from sorbed species and solid species. In EKR, heavy metals (M) are present in anionic complex and can be eliminated in form of M-EDTA$^-$ and M-citrate$^-$ (Yoo et al. 2015). In an EKR study done on both a spiked model sediment (Ammami et al. 2014) and a dredged sediment (Ammami et al. 2015), citric acid when employed as an electrolyte, was reported to be an efficient chelator for the elimination of several metal ions and polycyclic aromatic hydrocarbons.

A set of EKR were performed by Suzuki et al. (2014a, b) for investigation of heavy metal extraction efficacy of various chelating agents. EDDS was reported to be a highly efficient alternative of EDTA and worked as an electrolyte for enhancing the remediation process due to its biodegradable and strong chelating nature. The synthesis of Pb-EDDS and Cd-EDDS complexes resulted in increase in water solubility of these heavy metals and thus enhanced electromigration (Suzuki et al. 2014a, b). Tang et al. (2017) also investigated the heavy metal elimination efficiency of an EKR by employing a combination of biodegradable chelating agent tetrasodium of N, *N*-bis (carboxymethyl) glutamic acid (GLDA) and biodegradable biosurfactant (rhamnolipid). The results suggested that nature of sludge and interactions with different ammendments remarkably determined the EKR technology. A significant increase in elimination rate (60.0 ± 4.67%, 70.0 ± 3.51%, 70.6 ± 3.41%, 82.2 ± 5.21%,88.4 ± 4.43% 89.0 ± 3.34%, and for Pb, Mn, Cu, Zn, Ni and Cr respectively) were reported (Tang et al. 2017). Enhanced EKR was employed by Nasiri et al. (2020) for remediation of Cr-polluted soil utilizing a permeable reactive barrier (PRB) and chelating compounds. The study reported that EDTA (78%) was more efficient than citric acid (54%) in Cr elimination from the polluted soil and the employment of PRB in combination to EDTA resulted in better Cr elimination. In comparison to traditional EKR, the employment of chelating compounds caused significant (>90%) increase in Cr elimination in the following fractions: exchangeable phase, carbonate phase, and bond to Fe–Mn oxides. Along with electromigration, electroosmotic flow also plays vital role in Cr elimination in the EKR. Presently EKR is still at the developing stage and various demonstration and pilot-scale projects using EKR are available, however full-scale applications are still rare (Hansen et al. 2016).

The integration of phytoremediation to EKR has also been developed as an efficient strategy to reduce the limitations of phytoremediation (Mao et al. 2015). The coupled phytoremediation–EK technique involves implementation of low intensity electric field to the polluted soil in the proximity of plants. The electric field is supposed to improve the heavy metal removal by enhancing the availability through desorption and translocation of the metal ions even over smaller distances. Several factors such as; employment of either AC or DC current, voltage amount and mode of employment (continuous or periodic), soil pH, soil ammendments govern the EKR coupled phytoremediation technology (Figueroa et al. 2016). In the coupled phytoremediation–EKR method, the elimination of heavy metal is carried out by the plants however the electric field improves the plant metabolism by alleviating the availability of the metals (Fig. 13.1). Phytoremediation can also be employed at contaminated site post EKR to eliminate residual amounts of heavy metal (Wan et al. 2012). Additionally, the application of phytoremediation post EKR can effectively recover soil properties, soil structure damaged by the EK treatment. Kubiak et al. (2012) investigated the efficacy of the coupled electrophyto-remediation for the retention of arsenic in water. Bi et al. (2011) investigated the impact of coupled phytoremediation-EKR for heavy metal removal from polluted soil using rapeseed and tobacco. The effect of electrokinetic-coupled phytoremediation on soil physicochemical parameters, enzymatic and microbial activities of Cd, Zn, Cu and Pb contaminated soil was investigated by Cang et al. (2012).

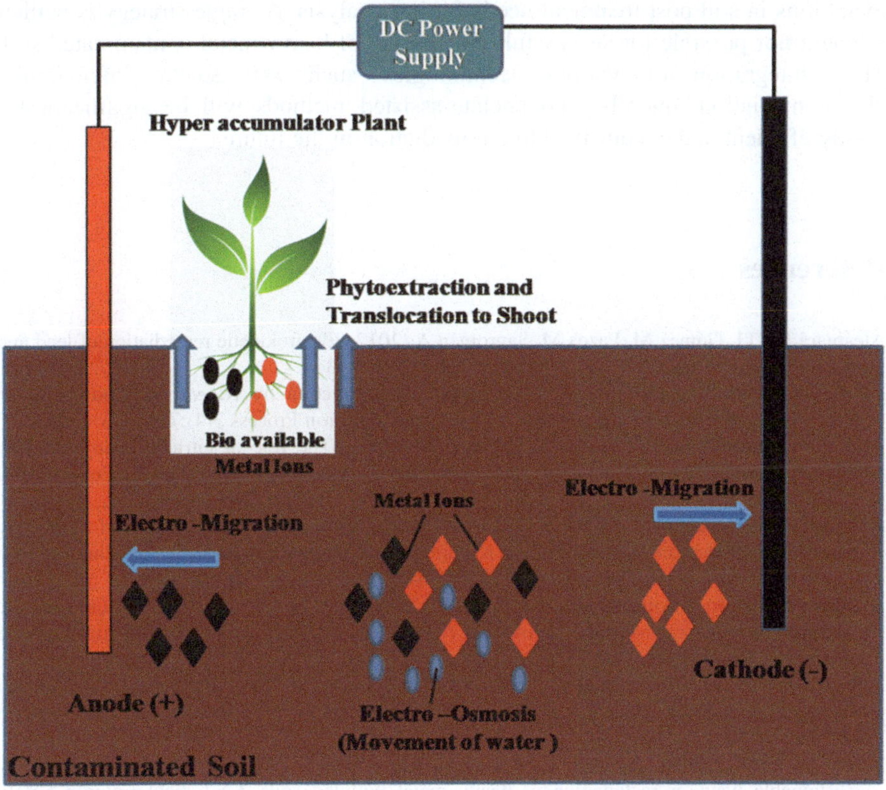

Fig. 13.1 The coupled phytoremediation–electrokinetic technology for heavy metal remediation

13.5 Conclusion

Heavy metals pollution is one of the most serious threats to the environment and human health. These metals are discharged into the soil and water from various sources. The enhancement of heavy metal contamination in various ecosystems has caused many researchers to emphasize on development of rapid, economical and more effective remediation methods. Traditional remediation technologies are expensive and non eco-friendly. The remediation of metal-pollutes sites using synthetic, natural and biodegradable chelating agents has emerged as one of most studied clean-up technologies in last decade. Chelating agents find use in soil washing, electrokinetics and phytoextraction studies for enhancing the heavy metal ion removal efficiency. However these techniques have some drawbacks like poor efficacy, leaching of metal-chelate complex into the soil, and accumulation of heavy metal in plant parts. Additionally, the strategies for recycling, recovery and reuse of the chelating agents need to be addressed in near future. The possible impact of abiotic and biotic soil parameter on the availability and mobility of residual toxic metal ions in soil post treatment needs further analysis. A single strategy is neither efficient nor possible for successful remediation of heavy metal-contaminated soil. Thus integration of various technologies such as; genetic engineering, phytoremediation, microbe and chelate-assisted methods will be significant for highly efficient and exhaustive phytoremediation in the future.

References

Alcántara MTTJ, Gómez M, Pazos M, Sanromán A (2012) Electrokinetic remediation of lead and phenanthrene polluted soils. Geoderma 173–174:128–133

Ali SY, Chaudhury S (2016) EDTA-enhanced phytoextraction by *Tagetes* sp. and effect on bioconcentration and translocation of heavy metals. Environ Process 3(4):735–746

Alidoust D, Suzuki S, Matsumura S, Yoshida M (2009) The role of citric acid in enhanced phytoextraction of heavy metals in an andosol by *Crotalaria juncea*. Fresenius Environ Bull 18:835–842

Ammami MT, Benamar A, Wang H (2014) Simultaneous electrokinetic removal of polycyclic aromatic hydrocarbons and metals from a sediment using mixed enhancing agents. Int J Environ Sci Technol 11:1801–1816. https://doi.org/10.1007/s13762-013-0395-9

Ammami MT, Portet-Koltalo F, Benamar A (2015) Application of biosurfactants and periodic voltage gradient for enhanced electrokinetic remediation of metals and PAHs in dredged marine sediments. Chemosphere 125:1–8

Arabi Z, Homaee M, Asadi ME, Kapourchal SA (2017) Cadmium removal from Cd-contaminated soils using some natural and synthetic chelates for enhancing phytoextraction. Chem Ecol 33 (9):389–402

Ashraf S, Alia Q, Zahira ZA, Ashraf S, Asghara HN (2019) Phytoremediation: environmentally sustainable way for reclamation of heavy metal polluted soils. Ecotoxicol Environ Safety 174:714–727

Ayangbenro AS, Babalola OO (2017) A new strategy for heavy metal polluted environments: a review of microbial biosorbents. Int J Environ Res Public Health 14(1):94. https://doi.org/10.3390/ijerph14010094

Bala R, Thukral AK (2011) Phytoremediation of Cr(VI) by *Spirodela polyrrhiza* (l.) Schleiden employing reducing and chelating agents. Int J Phytoremediation 13(5):465–491. https://doi.org/10.1080/15226511003758861

Bani A, Pavlova D, Echevarria G, Mullaj A, Reeves RD, Morel JL, Sulce S (2010) Nickel hyperaccumulation by the species of Alyssum and Thlaspi (Brassicaceae) from the ultramafic soils of the Balkans. Bot Serb 34:3–14

Beck W (2009) Metal complexes of biologically important ligands, CLXXII [1]. Metal ions and metal complexes as protective groups of amino acids and peptides—reactions at coordinated amino acids. Z Naturforsch 64b:1221–1245

Begum ZA, Rahman IMM, Tate Y, Sawai H, Maki T, Hasegawa H (2012) Remediation of toxic metal contaminated soil by washing with biodegradable aminopolycarboxylate chelants. Chemosphere 87:1161–1170

Begum ZA, Rahman IMM, Sawai H, Mizutani S, Maki T, Hasegawa H (2013) Effect of extraction variables on the biodegradable chelant assisted removal of toxic metals from artificially contaminated European reference soils. Water Air Soil Pollut 224:1381

Bhandari G, Bhatt P (2020) Concepts and application of plant–microbe interaction in remediation of heavy metals. In: Sharma A (ed) Microbes and signaling biomolecules against plant stress, rhizosphere biology. https://doi.org/10.1007/978-981-15-7094-0_4

Bhandari G, Bagheri AR, Bhatt P, Bilal M (2021) Occurrence, potential ecological risks, and degradation of endocrine disrupter, nonylphenol from the aquatic environment. Chemosphere 275:130013. https://doi.org/10.1016/j.chemosphere.2021.130013

Bhatt P, Gangola S, Bhandari G, Zhang W, Maithani D, Mishra S, Chen S (2020) New insights into the degradation of synthetic pollutants in contaminated environments. Chemosphere. https://doi.org/10.1016/j.chemosphere.2020.128827

Bi R, Schlaak M, Siefert E, Lord R, Connolly H (2011) Influence of electrical fields (AC and DC) on phytoremediation of metal polluted soils with rapeseed (*Brassica napus*) and tobacco (*Nicotiana tabacum*). Chemosphere 83:318–326

Blaylock MJ, Salt DE, Dushenkov S, Zakharova O, Gussman C, Kapulnik Y, Ensley BD, Raskin I (1997) Enhanced accumulation of Pb in Indian mustard by soil-applied chelating agents. Environ Sci Technol 31:860–865

Cameselle C, Reddy KR (2012) Development and enhancement of electro-osmotic flow for the removal of contaminants from soils. Electrochim Acta 86:10–22

Cameselle C, Chirakkara RA, Reddy KR (2013) Electrokinetic-enhanced phytoremediation of soils: status and opportunities. Chemosphere 93:626–636

Cang L, Zhou D, Wang Q, Fan G (2012) Impact of electrokinetic-assisted phytoremediation of heavy metal contaminated soil on its physicochemical properties, enzymatic and microbial activities. Electrochim Acta 86:41–48

Çelik Ö, Akdaş EY (2019) Tissue-specific transcriptional regulation of seven heavy metal stress-responsive miRNAs and their putative targets in nickel indicator castor bean (*R. communis L.*) plants. Ecotoxicol Environ Saf 170:682–690

Chamba I, Rosado D, Kalinhoff C, Thangaswamy S, Sánchez-Rodríguez A, Gazquez MJ (2017) Eratopolymnioides—a novel Hg hyperaccumulator plant in ecuadorian rainforest acid soils with potential of microbe associated phytoremediation. Chemosphere 188:633–641

Chaney RL, Broadhurst CL, Centofanti T (2010) Phytoremediation of soil trace elements. In: Hooda PS (ed) Trace elements in soils. Wiley, Chichester, pp 311–352

Chen H, Dou J, Xu H (2018) The effect of low-molecular-weight organic-acids (LMWOAs) on treatment of chromium-contaminated soils by compost-phytoremediation: kinetics of the chromium release and fractionation. J Environ Sci 70:45–53

Cheng S, Lin Q, Wang Y, Luo H, Huang Z, Fu H (2020) The removal of Cu, Ni, and Zn in industrial soil by washing with EDTA-organic acids. Arab J Chem 13:5160–5170. https://doi.org/10.1016/j.arabjc.2020.02.015

Chibuike GU, Obiora SC (2014) Heavy metal polluted soils: effect on plants and bioremediation methods. Appl Environ Soil Sci:12. https://doi.org/10.1155/2014/752708

Dacera DDM, Babel S (2006) Use of citric acid for heavy metals extraction from contaminated sewage sludge for land application. Water Sci Technol 54(9):129–135

Ding YZ, Song ZG, Feng RW, Guo JK (2014) Interaction of organic acids and pH on multi heavy metal extraction from alkaline and acid mine soils. Int J Environ Sci Technol 11:33–42. https://doi.org/10.1007/s13762-013-0433-7

Ebrahimi M (2014) Effect of EDTA and DTPA on phytoremediation of Pb-Zn contaminated soils by eucalyptus camaldulensis dehnh and effect on treatment time. Desert 19(1):65–73

Ehsan S, Ali S, Noureen S, Mahmood K, Farid M, Ishaque W, Shakoor MB, Rizwan M (2014) Citric acid assisted phytoremediation of cadmium by Brassica napus L. Ecotoxicol Environ Saf 106:164–172

Elias SH, Mohamed M, Ankur AN, Muda K, Hassan MAHM, Othman MN, Chelliapan S (2014) Water hyacinth bioremediation for ceramic industry wastewater treatment-application of rhizofiltration system. Sains Malays 43(9):1397–1403

Elliott HA, Brown G (1989) Comparative evaluation of NTA and EDTA for extractive decontamination of Pb-polluted soils. Water Air Soil Pollut 45:361–369

Epelde L, Hernandez-Allica J, Becerril JM, Blanco F, Garbisu C (2008) Effects of chelates on plants and soil microbial community: comparison of EDTA and EDDS for lead phytoextraction. Sci Total Environ 401:21–28

Evangelou MWH, Ebel M, Schaeffer A (2007) Chelate assisted phytoextraction of heavy metals from soil. Effect, mechanism, toxicity, and fate of chelating agents. Chemosphere 68 (6):989–1003

Farooq MA, Gill RA, Ali B, Wang J, Islam F, Ali S, Zhou W (2016) Subcellular distribution, modulation of antioxidant and stress-related genes response to arsenic in Brassica napus L. Ecotoxicol 25(2):350–366

Fedje KK, Yillin L, Stromvall AM (2013) Remediation of metal polluted hotspot areas through enhanced soil washing—evaluation of leaching methods. J Environ Manag 128:489–496

Figueroa A, Cameselle C, Gouveia S, Hansen HK (2016) Electrokinetic treatment of an agricultural soil contaminated with heavy metals. J Environ Sci Health A Tox Hazard Subst Environ Eng 51:691–700

Flora SJS, Pachauri V (2010) Chelation in metal intoxication. Int J Environ Res Public Health 7 (7):2745–2788. https://doi.org/10.3390/ijerph7072745

Fomina M, Hillier S, Charnock JM, Melville K, Alexander IJ, Gadd GM (2005) Role of oxalic acid over excretion in transformations of toxic metal minerals by *Beauveria caledonica*. Appl Environ Microbiol 71:371–381

Freitas E, Nascimento C (2016) Degradability of natural and synthetic chelating agents applied to a lead-contaminated soil. J Soils Sediments 17(5):1272–1278. https://doi.org/10.1007/s11368-015-1350-9

Gan S, Lau EV, Ng HK (2009) Remediation of soils contaminated with polycyclic aromatic hydrocarbons (PAHs). J Hazard Mater 172:532–549. https://doi.org/10.1016/j.jhazmat.2009.07.118

Ghasemi F, Ebrahimi M, Shirazi MP (2017) Lead phytoremediation capacity of *Puccinellia distans* (Jacq.) Parl. using EDTA and DTPA and associated potential leaching risk. Global NEST J 19:359–366

Gluhar S, Kaurin A, Lestan D (2020) Soil washing with biodegradable chelating agents and EDTA: technological feasibility, remediation efficiency and environmental sustainability. Chemosphere 257:127226

Guo X, Zhao G, Zhang G, He Q, Wei Z, Zheng W, Qian T, Wu Q (2018) Effect of mixed chelators of EDTA, GLDA, and citric acid on bioavailability of residual heavy metals in soils and soil properties. Chemosphere 209:776–782

Gupta P, Diwan B (2017) Bacterial exopolysaccharide mediated heavy metal removal: a review on biosynthesis, mechanism and remediation strategies. Biotechnol Rep 13:58–71. https://doi.org/10.1016/j.btre.2016.12.006

Hansen HK, Ottosen LM, Ribeiro AB (2016) Electrokinetic soil remediation: an overview, 854. In: Ribeiro A, Mateus E, Couto N (eds) Electrokinetics across disciplines and 855 continents. Springer, Cham, pp 3–18

Iannelli R, Masi M, Ceccarini A, Ostuni MBB, Lageman R, Muntoni A, Spiga D, Polettini A, Marini A, Pomi R (2015) Electrokinetic remediation of metalpolluted marine sediments: experimental investigation for plant design. Electrochim Acta 181:146–159. https://doi.org/10.1016/j.electacta.2015.04.093

Iheanacho EU, Ndulaka JC, Onuh CF (2017) Environmental pollution and heavy metals. Eur J Biotechnol Biosci 5(5):73–78

Irini NO, Panagiotis M, Kissoudis C, Voulgari G, Chronopoulou E, Tsaftaris A, Labrou NE (2017) Plant glutathione transferase-mediated stress tolerance: functions and biotechnological applications. Plant Cell Rep 36(6):791–805

Jain A, Chen WN (2018) Involvement of organic acids and amino acids in ameliorating Ni (II) toxicity induced cell cycle dysregulation in Caulobacter crescentus: a metabolomics analysis. Appl Microbiol Biotechnol 102(10):4563–4575. https://doi.org/10.1007/s00253-018-8938-0

Jean L, Bordas F, Gautier-Moussard C, Vernay O, Hitmi A, Bollinger JC (2008) Effect of citric acid and EDTA on chromium and nickel uptake and translocation by *Datura innoxia*. Environ Pollut 153:555–563

Kalve S, Sarangi BK, Pandey RA, Chakrabarti T (2011) Arsenic and chromium hyperaccumulation by an ecotype of *Pteris vittata*-prospective for phytoextraction from contaminated water and soil. Curr Sci 100:888–894

Kayser A, Schulin R, Felix H (1999) Field trials for the phytoremediation of soils polluted with heavy metals. In: Umweltbundesamt (ed) Proc. Int. Workshop am Fraunhofer Institut für Umweltchemic und Ökotoxikologie, Schmallenberg, Germany. 1–2 Dec. 1997. Erich Schmidt Verlag, Berlin, pp 170–182

Ke X, Li PJ, Zhou Q, Zhang Y, Sun TH (2006) Removal of heavy metals from a contaminated soil using tartaric acid. J Environ Sci 18:727–733

Komarek M, Vanek A, Mrnka L, Sudova R, Szakova J, Tejnecky V, Chrastny V (2010) Potential and drawbacks of EDDS-enhanced phytoextraction of copper from contaminated soils. Environ Pollut 158:2428–2438

Koptsik G (2014) Problems and prospects concerning the phytoremediation of heavy metal polluted soils: a review. Eurasian Soil Sci 47:923–939. https://doi.org/10.1134/S1064229314090075

Kour D, Kaur T, Devi R, Yadav A, Singh M et al (2021) Beneficial microbiomes for bioremediation of diverse contaminated environments for environmental sustainability: present status and future challenges. Environ Sci Pollut Res 28:24917–24939

Kramer U (2018) The plants that suck up metal. Ger Res 40(3):18–23

Kubiak JJ, Khankhane PJ, Kleingeld PJ, Lima AT (2012) An attempt to electrically enhance phytoremediation of arsenic contaminated water. Chemosphere 87:259–264

Kumar P, Dash B, Suyal DC, Gupta SB, Singh AK, Chowdhury T, Soni R (2021) Characterization of arsenic-resistant Klebsiella pneumoniae RnASA11 from contaminated soil and water samples and its bioremediation potential. Curr Microbiol. https://doi.org/10.1007/s00284-021-02602-w

Kuppusamy S, Thavamani P, Venkateswarlu K, Lee YB, Naidu R, Megharaj M (2017) Remediation approaches for polycyclic aromatic hydrocarbons (PAHs) contaminated soils: technological constraints, emerging trends and future directions. Chemosphere 168:944–968

Lajayer BA, Moghadam NK, Maghsoodi MR, Ghorbanpour M, Kariman K (2019) Phytoextraction of heavy metals from contaminated soil, water and atmosphere using ornamental plants: mechanisms and efficiency improvement strategies. Environ Sci Pollut Res. https://doi.org/10.1007/s11356-019-04241-y

Lawal M, Sauban OO (2014) Ethylene diamine tetra-acetate (EDTA) chelate effects on pennisetum purpureum as heavy metal phytoextraction from contaminated soil in Mokwa/New Bussa Highway Road, Niger State. J Trop Forest Res 30:127–136

Lestan D (2017) Novel chelant-based washing method for soil contaminated with Pb and other metals: a pilot-scale study. Land Degrad Dev 28:2585–2595

Lestan D, Luo C, Li XD (2008) The use of chelating agents in the remediation of metal contaminated soils: a review. Environ Pollut 153:3–13. https://doi.org/10.1016/j.envpol.2007.11.015

Li Y, Luo J, Yu J, Xia L, Zhou C, Cai L, Ma X (2018) Improvement of the phytoremediation efficiency of *Neyraudia reynaudiana* for lead-zinc mine-contaminated soil under the interactive effect of earthworms and EDTA. Sci Rep 8(1):6417. https://doi.org/10.1038/s41598-018-24715-2

Lin L, Li GD, Jing P, Yuan YF, Zhou NN, Jiao W (2009) Removal effect of EDTA and organicacids on Pb in soil. Health Environ Res Online 9(4):51–53

Ling G, Cheng MZF, Wang QH (2011) Effects of natural organic acids on growth of maize and uptake of chromium by maize in chromium contaminated soil. Adv Mater Res 281:21–24

Liu D, Islam E, Li T, Yang X, Jin X, Mahmood Q (2008) Comparison of synthetic chelators and low molecular weight organic acids in enhancing phytoextraction of heavy metals by two ecotypes of Sedum alfredii Hance. J Hazard Mater 153:114–122

Liu Z, Wang L-A, Ding S, Xiao H (2018) Enhancer assisted-phytoremediation of mercury contaminated soils by *Oxalis corniculata* L., and rhizosphere microorganism distribution of *Oxaliscorniculata* L. Ecotoxicol Environ Saf 160:171–177. https://doi.org/10.1016/j.ecoenv.2018.05.041

Lone MI, He Z, Stoffella PJ, Yang X (2008) Phytoremediation of heavy metal polluted soils and water: progresses and perspectives. J Zhejiang Univ (Sci) 9(3):210–220. https://doi.org/10.1631/jzus.B0710633

Lu Y, Luo D, Lai A, Liu G, Liu L, Long J, Zhang H, Chen Y (2017) Leaching characteristics of EDTA-enhanced phytoextraction of Cd and Pb by Zea mays L. in different particle-size fractions of soil aggregates exposed to artificial rain. Environ Sci Pollut Res 24(2):1845–1853

Luo CL, Shen ZG, Li XD (2005) Enhanced phytoextraction of Cu, Pb, Zn and Cd with EDTA and EDDS. Chemosphere 59:1–11

Luo CL, Shen ZG, Li XD, Baker AJM (2006) Enhanced phytoextraction of Pb and other metals from contaminated soils through the combined application of EDTA and EDDS. Chemosphere 63:1773–1784

Luo CL, Shen ZG, Li XD (2007) Plant uptake and the leaching of metals during the hot 695 EDDS-enhanced phytoextraction process. Int J Phytoremediation 9:696181196

Malik N, Biswas AK (2012) Role of higher plants in remediation of metal contaminated sites. Sci Rev Chem Commun 2:141–146

Mancini G, Bruno M, Polettini A, Pomi R (2011) Chelant-assisted pulse flushing of a field Pb-contaminated soil. Chem Ecol 27(3):251–262

Mani D, Kumar C (2014) Biotechnological advances in bioremediation of heavy metals contaminated ecosystems: an overview with special reference to phytoremediation. Int J Environ Sci Technol 11:843–872. https://doi.org/10.1007/s13762-013-0299-8

Mao X, Han FX, Shao X, Guo K, McComb J, Njemanze S (2015) Electro-kinetic enhanced phytoremediation for the restoration of multimetal(LOID) contaminated soils. Adv Environ Res 42:25–52

Markovska Y, Stancheva I, Petrov P, Georgieva T, Geneva M, Georgieva KM, Ivanova K (2018) EDTA and citrate impact on heavy metals phytoremediation using paulownia hybrids. Int J Environ Pollut 63:31. https://doi.org/10.1504/IJEP.2018.10014111

Marschner H, Romheld V, Kissel M (1986) Different strategies in higher-plants in mobilization and uptake of iron. J Plant Nutr 9:695–713

Mathis P, Kayser A (2001) Plant uptake of heavy metals following glyphosate treatment. In: International Society of Trace Element Biogeochemistry (ed) Proceedings of the 6th International Conference on the Geochemistry of Trace Elements (ICOBTE), Guelph, ON, Canada. 29 July–2 Aug. 2001. International Society of Trace Element Biogeochemistry, Vienna, pp .484

Matzinger I, Brunthaler J, Fürhapper C, Pollak U, Aschacher G (2007) Biodegradation of new complexing agents in compliance with Austrian waste water emission legislation for the pulp & paper industry. In: Nowack B, Giger W (ed) Complexing agents between science, industry, authorities and users. Monte Verità, Ascona, Switzerland, March 11–16, 2007, pp 19

Meers E, Ruttens A, Hopgood MJ, Samson D, Tack FMG (2005) Comparison of EDTA and EDDS as potential soil amendments for enhanced phytoextraction of heavy metals. Chemosphere 58:1011–1022

Mesa J, Rodriguez-Llorente JD, Pajuelo E, Piedras JMB, Caviedes MA, Redondo-Gomez S, Mateos-Naranjo E (2015) Moving closer towards restoration of contaminated estuaries: bioaugmentation with autochthonous rhizobacteria improves metal rhizoaccumulation in native spartina maritima. J Hazard Mater 300:263–271. https://doi.org/10.1016/j.jhazmat.2015.07.006

Mir ZA, Bharose R, Lone AH, Malik RA (2017) Role of chelating agents in phytoremediation of heavy metals. Trends Biosci 10(7):1449–1454

Mujahid F, Ali S, Shakoor M, Bharwana S, Rizvi H, Ehsan S, Tauqeer HM, Iftikhar U, Hannan F (2013) EDTA assisted phytoremediation of cadmium, lead and zinc. Int J Agron Plant Product 4:2833–2846

Nanthavong K, Sampanpanish P (2015) Effect of NTA and EDTA on arsenic uptake from contaminated soil by mimosa pudica. Mod Appl Sci 9(9):280–291

Nasiri A, Jamshidi-Zanjani A, Darban AK (2020) Application of enhanced electrokinetic approach to remediate Cr-contaminated soil: effect of chelating agents and permeable reactive barrier. Environ Pollut 266:115197

Nezami S, Mohammad MJ, Malakouti J, Samani AB, Maragheh MG (2016) Effect of low molecular weight organic acids on the uptake of ^{226}Ra by corn (*Zea mays* L.) in a region of high natural radioactivity in Ramsar-Iran. J Environ Radioact 164:145–150

Nowack B, Schulin R, Robinson BH (2006) Critical assessment of chelant-enhanced metal phytoextraction. Environ Sci Technol 40:5225–5232

Nurchi VM, Cappai R, Crisponi G, Sanna G, Alberti G, Biesuz R, Gama S (2020) Chelating agents in soil remediation: a new method for a pragmatic choice of the right chelator. Front Chem 8:597400. https://doi.org/10.3389/fchem.2020.597400

Nworie OE, Qin J, Lin C (2017) Differential effects of low-molecular-weight organic acids on the mobilization of soil-borne arsenic and trace metals. Toxics 5(3):18. https://doi.org/10.3390/toxics5030018

Ogunleye JA, Dagari MS, Balogun OL, Maigari A (2016) Effect of oxalic acid on Cr^{3+} ion uptake, accumulation and oxidative stress by sorrel (*Hibiscus sabdariffa* L.) seedlings in hydroponic solution. Int J Chem Mater Environ Res 3(3):49–55

Pastor J, Aparicio AM, Gutierrez-Maroto A, Hernández AJ (2007) Effects of two chelating agents (EDTA and DTPA) on the autochthonous vegetation of a soil polluted with Cu, Zn and Cd. Sci Total Environ 378:114–118. https://doi.org/10.1016/j.scitotenv.2007.01.022

Patra DK, Pradhan C, Patra HK (2018) Chelate based phytoremediation study for attenuation of chromium toxicity stress using lemongrass: *Cymbopogon flexuosus* (nees ex steud.). W Watson Int J Phytoremediat 20(13):1324–1329

Peters RW (1999) Chelant extraction of heavy metals from contaminated soils. J Hazard Mater 66:151–210. https://doi.org/10.1016/S0304-3894(99)00010-2

Prasad R, Shivay YS (2017) Oxalic acid/oxalates in plants: from self-defence to phytoremediation. Curr Sci 112:1665–1667. https://doi.org/10.18520/cs/v112/i08/1665-1667

Prieto MJ, Acevedo SOA, Prieto GF (2018) Phytoremediation of soils contaminated with heavy metals. Biodivers Int J 2(4):362–376. https://doi.org/10.15406/bij.2018.02.00088

Qu J, Lou CQ, Yuan X, Wang XH, Cong HQ, Wang L (2011) The effect of sodium hydrogen phosphate/citric acid mixtures on phytoremediation by alfalfa & metals availability in soil. J Soil Sci Plant Nutr 11(2):85–95

Quartacci MF, Argilla A, Baker AJM, Navari-Izzo F (2006) Phytoextraction of metals from a multiply contaminated soil by Indian mustard. Chemosphere 63:918–925

Rai PK (2008) Phytoremediation of Hg and Cd from industrial effluents using an aquatic free floating macrophyte *Azolla pinnata*. Int J Phytoremediation 10:430–439

Reinoso-Maset E, Worsfold PJ, Keith-Roach MJ (2013) Effect of organic complexing agents on the interactions of Cs^+, Sr^{2+} and with silica and natural sand. Chemosphere 91:948–954

Ruley AT, Sharma NC, Sahi SV, Singh SR, Sajwan KS (2006) Effects of lead and chelators on growth, photosynthetic activity and Pb uptake in *Sesbania drummondii* grown in soil. Environ Pollut 144:11–18

Saifullah U, Abdul G, Manzoor Q (2008) Lead phytoextraction by wheat in response to the EDTA application method. Int J Phytoremediation 11:268–282. https://doi.org/10.1080/15226510802432702

Saifullah U, Shahid M, Zia-Ur-Rehman M, Sabir M, Ahmad HR (2015) Phytoremediation of Pb-contaminated soils using synthetic chelates soil remediation and plants: prospects and challenges. In: Hakeem K, Sabir M, Ozturk M, Murmet A (eds) Soil remediation and plants, 1st edn. Elsevier, Amsterdam, pp 397–414. https://doi.org/10.1016/B978-0-12-799937-1.00014-0

Sakakibara M, Ohmori Y, Ha NTH, Sano S, Sera K (2011) Phytoremediation of heavy metal contaminated water and sediment by Eleocharis acicularis. Clean Soil Air Water 39:735–741

Sandeep G, Vijayalatha KR, Anitha T (2019) Heavy metals and its impact in vegetable crops. Int J Chem Stud 7(1):1612–1621

Scheffer F, Schachtschabel P (1998) Lehrbuch der Bodenkunde. Enke, Stuttgart

Sharma R, Bhardwaj R, Thukral AK, Bali S, Kohli S, Kaur P, Ohri P, Kaur R, Vig A, Sharma M (2017a) Role of phytochelators in land reclamation: recent progress and perspectives. Bentham Sci Curr Environ Eng 4(1):42–52. https://doi.org/10.2174/2212717804666161220160825

Sharma J, Shamim K, Dubey SK, Meena RM (2017b) Metallothionein assisted periplasmic lead sequestration as lead sulfite by *Providencia vermicola* strain SJ2A. Sci Total Environ 579:359–365. https://doi.org/10.1016/j.scitotenv.2016.11.089

Shen ZG, Li XD, Wang CC, Chen HM, Chua H (2002) Lead phytoextraction from contaminated soil with high-biomass plant species. J Environ Qual 31:1893–1900

Sheoran V, Sheoran AS, Poonia P (2010) Role of hyperaccumulators in phytoextraction of metals from contaminated mining sites: a review. Crit Rev Environ Sci Technol 41(2):168–214. https://doi.org/10.1080/10643380902718418

Shilev S, Naydenov M, Tahsin N, Sancho ED, Bennlloch M, Vancheva V, Sapundjieva K, Kuzmanova J (2007) Effect of easily biodegradable amendments on heavy metal solubilization and accumulation in technical crops-a field trial. J Environ Eng Landsc 15:237–242

Singh S, Parihar P, Singh R, Singh VP, Prasad SM (2016) Heavy metal tolerance in plants: role of transcriptomics, proteomics, metabolomics, and ionomics. Front Plant Sci 6:1143. https://doi.org/10.3389/fpls.2015.01143

Singh M, Singh D, Rai P, Suyal DC, Saurabh S, Soni R, Giri K, Yadav AN (2021) Fungi in remediation of hazardous wastes: current status and future. In: Yadav AN (ed) Recent trends in mycological research, fungal biology. Springer Nature, Cham

Sinhal VK, Srivastava A, Singh VP (2010) EDTA and citric acid mediated phytoextraction of Zn, Cu, Pb and Cd through marigold (*Tagetes erecta*). J Environ Biol 31:255–259

Song Y, Ammami MT, Benamar A, Mezazigh S, Wang H (2016) Effect of EDTA, EDDS, NTA and citric acid on electrokinetic remediation of As, Cd, Cr, Cu, Ni, Pb and Zn contaminated dredged marine sediment. Environ Sci Pollut Res 23(11):10577–10586

Sun R, Zhou Q, Jin C (2006) Cadmium accumulation in relation to organic acids in leaves of Solanum nigrum L. as a newly found cadmium hyperaccumulator. Plant Soil 285:125–134

Sun Y, Sun G, Xu Y, Liu W, Liang X, Wang L (2016) Evaluation of the effectiveness of sepiolite, bentonite, and phosphate amendments on the stabilization remediation of cadmium-contaminated soils. J Environ Manag 166:204–210. https://doi.org/10.1016/j.jenvman.2015.10.017

Suzuki T, Kawai K, Moribe M, Niinae M (2014a) Recovery of Cr as Cr(III) from Cr(VI)-contaminated kaolinite clay by electrokinetics coupled with a permeable reactive barrier. J Hazard Mater 278:297–303

Suzuki T, Niinae M, Koga T, Akita T, Ohta M, Choso T (2014b) EDDS-enhanced electrokinetic remediation of heavy metal-contaminated clay soils under neutral pH conditions. Colloid Surface Physicochem Eng Aspect 440:145–150

Tang J, He J, Liu T, Xin X, Hu H (2017) Removal of heavy metal from sludge by the combined application of a biodegradable biosurfactant and complexing agent in enhanced electrokinetic treatment. Chemosphere 189:599–608

Tao Q, Hou D, Li T (2016) Oxalate secretion from the root apex of *Sedum alfredii* contributes to hyperaccumulation of cadmium. Plant Soil 398(1):139–152

Turgut C, Pepe MK, Cutright TJ (2004) The effect of EDTA and citric acid on phytoremediation of Cd, Cr, and Ni from soil using *Helianthus annuus*. Environ Pollut 131(1):147–154

Ullmann A, Brauner N, Vazana S, Katz Z, Goikhman R, Seemann B, Marom H, Gozin M (2013) New biodegradable organic-soluble chelating agents for simultaneous removal of heavy metals and organic pollutants from contaminated media. J Hazard Mater 260:676–688

Vassilev A, Schwitzguebel JP, Thewys T, Van Der Lelie D, Vangronsveld J (2004) The use of plants for remediation of metal-contaminated soils. Sci World J 4:9–34

Wan QF, Deng DC, Bai Y, Xia CQ (2012) Phytoremediation and electrokinetic remediation of uranium contaminated soils: a review. He-Huaxue yu Fangshe Huaxue/J Nucl Radiochem 34:148–156

Wang D, Zhang X, Liu J, Zhu Y, Zhang H, Zhang A, Jin X (2012a) Oxalic acid enhances Cr tolerance in the accumulating plant *Leersia hexandra* Swartz. Int J Phytoremediation 14 (10):966–977

Wang J, Feng X, Anderson CW, Xing Y, Shang L (2012b) Remediation of mercury contaminated sites—a review. J Hazard Mater 221:1–18. https://doi.org/10.1016/j.jhazmat.2012.04.035

Wang S, Wu W, Liu F, Liao R, Hu Y (2017) Accumulation of heavy metals in soil-crop systems: a review for wheat and corn. Environ Sci Pollut Res 24:15209–15225. https://doi.org/10.1007/s11356-017-8909-5

Wang G, Zhang S, Zhong Q, Xu X, Li T, Jia Y, Zhang Y, Peijnenburg WJGM, Vijver MG (2018) Effect of soil washing with biodegradable chelators on the toxicity of residual metals and soil biological properties. Sci Total Environ 625:1021–1029. https://doi.org/10.1016/j.scitotenv.2018.01.019

Wen XP, Ban Y, Inoue H, Matsuda N, Moriguch T (2010) Spermidine levels are implicated in heavy metal tolerance in a *spermidine synthase* overexpressing transgenic European pear by exerting antioxidant activities. Transgenic Res 19(1):91–103

Wenger K, Hari T, Gupta MD, Krebs R, Rammelt R, Leumann CD (1998) Possible approaches for in situ restoration of soils contaminated by zinc. Adv Geoecol 31:745–754

Wuana RA, Okieimen FE, Imborvungu JA (2010) Removal of heavy metals from a contaminated soil using organic chelating acids. Int J Environ Sci Technol 7(3):485–496

Xu X, Thomson NR (2007) An evaluation of the green chelant EDDS to enhance the stability of hydrogen peroxide in the presence of aquifer solids. Chemosphere 69:755–762

Yan DYS, Yui MMT, Yip TCM, Tsang DCW, Lo IMC (2010) Influence of EDDS-tometal molar ratio, solution pH, and soil-to-solution ratio on metal extraction under EDDS deficiency. J Hazard Mater 178:890–894

Yang RX, Luo CL, Zhang G, Li XD, Shen ZG (2012) Extraction of heavy metals from e-waste contaminated soils using EDDS. J Environ Sci (China) 24:1985–1994

Yang L, Luo CL, Liu Y, Quan LT, Chen YH, Shen ZG (2013) Residual effects of EDDS leachates during EDDS-assisted phytoremediation oKf copper contaminated soil. Sci Total Environ 444:263–270

Yeung AT, Gu Y (2011) A review on techniques to enhance electrochemical remediation of contaminated soils. J Hazard Mater 195:11–29. https://doi.org/10.1016/j.jhazmat.2011.08.047

Yoo JC, Yang JS, Jeon EK, Baek K (2015) Enhanced-electrokinetic extraction of heavy metals from dredged harbor sediment. Environ Sci Pollut Res 22:9912–9921. https://doi.org/10.1007/s11356-015-4155-x

Yoo JC, Beiyuan J, Wang L, Tsang DCW, Baek K, Bolan NS, Ok YS, Li XD (2018) A combination of ferric nitrate/EDDS-enhanced washing and sludgederived biochar stabilization of metal-contaminated soils. Sci Total Environ 616–617:572–582

Zeng X, Ma LQ, Qiu R, Tang Y (2009) Responses of non-protein thiols to cd exposure in Cd hyperaccumulator Arabis *paniculata* Franch. Environ Exp Bot 66:242–248

Zhang X, Zhong B, Shafi M, Guo J, Liu C, Guo H, Peng D, Wang Y, Liu D (2018) Effect of EDTA and citric acid on absorption of heavy metals and growth of Moso bamboo. Environ Sci Pollut Res 25(19):18846–18852

Zia-ur-Rehman M, Sabir M, Rizwan M, Saifullah Ahmad HR, Nadeem M (2015) Remediating cadmium-contaminated soils by growing grain crops using inorganic amendments. In: Hakeem KR, Sabir M, Ozturk M, Murmet A (eds) Soil remediation and plants: prospects and challenges. Elsevier Inc., Academic Press, pp 367–396

Chapter 14
Spectroscopy and Its Advancements for Environmental Sustainability

Om Prakash, Abhishek Pathak, Ajay Kumar, Vijay Kumar Juyal, Hem C. Joshi, Saurabh Gangola, Kiran Patni, Geeta Bhandari, Deep Chandra Suyal, and Viveka Nand

14.1 Introduction

The study of the interaction between matter and electromagnetic radiation as a function of the wavelength and frequency of the radiation is referred to as spectroscopy. Monitoring the fate of petroleum hydrocarbons in the environmental samples is a retrospective method of evaluating the biodegradation process progress (Fig. 14.1) (Debbarma et al. 2017; Kour et al. 2021). Common analytical tools like chromatography and spectroscopy have been used extensively in this field of study (Fig. 14.2). This chapter discusses the importance of analytical chemistry and

O. Prakash (✉)
Department of Chemistry, G. B. Pant Agricultural University, Pantnagar, Uttarakhand, India

Regional Ayurveda Research Institute Gwalior, Ministry of Ayush, Government of India, Gwalior, Madhya Pradesh, India

A. Pathak · A. Kumar · V. K. Juyal · V. Nand
Department of Chemistry, G. B. Pant Agricultural University, Pantnagar, Uttarakhand, India

H. C. Joshi
Department of Plant Physiology, G. B. Pant Agricultural University, Pantnagar, Uttarakhand, India

S. Gangola
School of Agriculture, Graphic Era Hill University, Bhimtal, Uttarakhand, India

K. Patni
School of Allied Sciences, Graphic Era Hill University, Bhimtal, Uttarakhand, India

G. Bhandari
Sardar Bhagwan Singh University, Balawala, Dehradun, Uttarakhand, India

D. C. Suyal
Department of Microbiology, Eternal University, Baru Sahib, Sirmour, Himachal Pradesh, India

© The Author(s), under exclusive license to Springer Nature Switzerland AG 2022
D. C. Suyal, R. Soni (eds.), *Bioremediation of Environmental Pollutants*,
https://doi.org/10.1007/978-3-030-86169-8_14

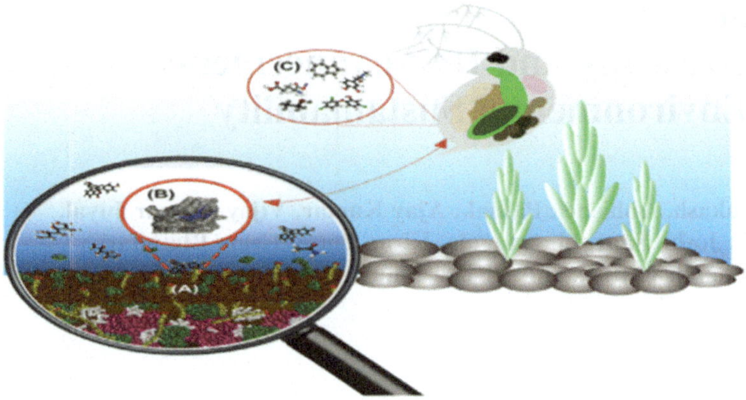

Fig. 14.1 (**a**) The structure of sedimentary organic matter which shows its critical functioning with nutrient dynamics and contaminant interactions. (**b**) The sediment interface shows the binding of xenobiotics and its interior sediment structure which shows its permanent sequestration. (**c**) Anthropogenic chemicals' dynamics and binding and their effect on living organisms

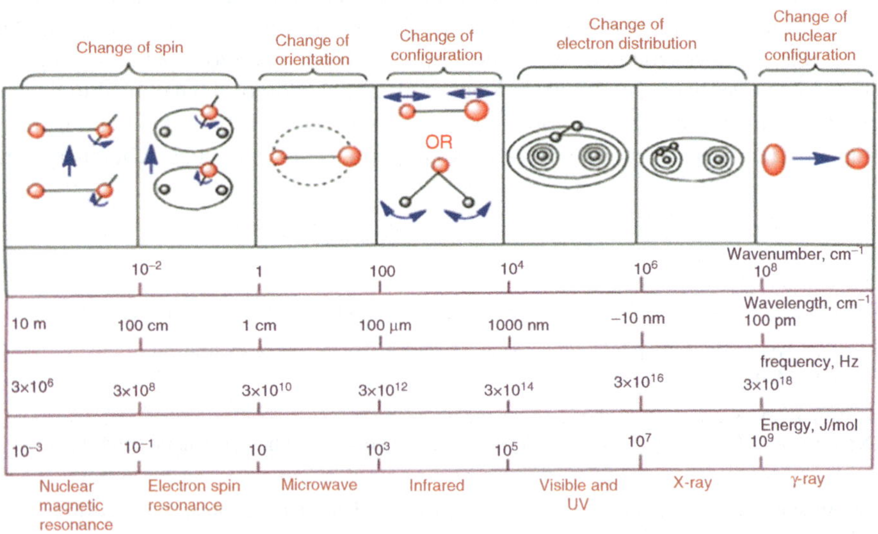

Fig. 14.2 Operation of different electromagnetic radiation

spectroscopy in the field of bioremediation. The use of various types of spectroscopies, as well as other hyphenated platforms, are also discussed to extend the research scenario in bioremediation studies.

14.2 UV-Visible (UV/VIS) Spectroscopy

The absorption of light by materials is analyzed by UV-Vis spectroscopy, which is commonly used in nearly every market segment as well as in scientific research studies (Fig. 14.3). It is mainly based on incident light and wavelength absorbed by the sample and provides valuable information. For more than 60 years, ultraviolet/

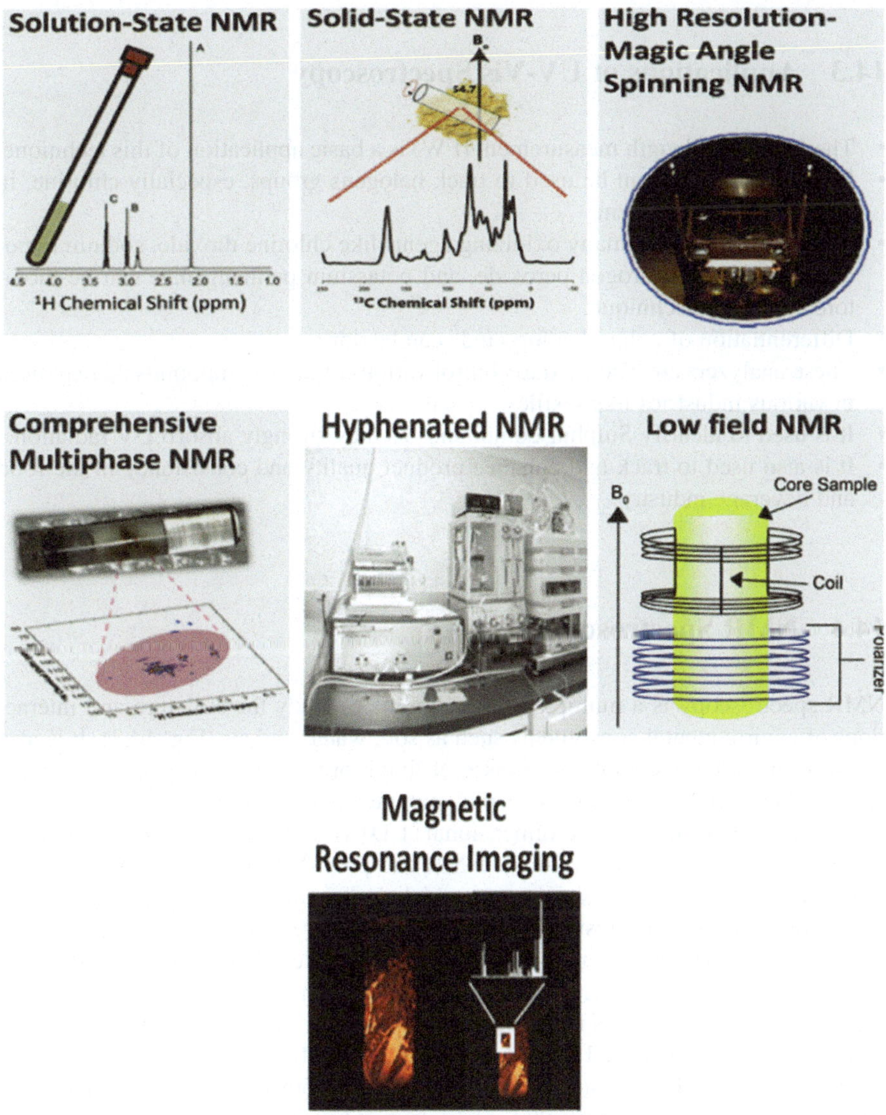

Fig. 14.3 Glimpses of spectroscopy techniques used in bioremediation studies

visible light has been used to track and regulate chemical processes. UV/Visible spectroscopy is based on the absorption of light by an unknown sample when illuminated with electromagnetic radiations either in the UV or visible range. Some part of the light is absorbed by the sample while the remaining is transmitted. This transmitted light is recorded by the suitable detectors and a spectrum is generated. Filter photometric analyzers and process spectrophotometers are two basic UV/Vis analyzers that are widely used (Krzewicka 2010).

14.3 Applications of UV-Vis Spectroscopy

- The fixed wavelength measurement (FW) is a basic application of this technique.
- UV-Vis analyzers can be used to track halogens groups, especially chlorine, in gaseous and liquid form.
- The concentration of many oxidizing agents like chlorine dioxide, sodium hypochlorite, ozone, hydrogen peroxide, and potassium permanganate can be monitored with this technique.
- Differentiation of colored compounds can be done.
- These analyzers are also used to control various organic compounds that are used in various industries like textiles.
- It is used to identify Sulphur compounds as they strongly absorb UV radiation.
- It is also used to track and enhance product quality and consistency in the food and beverage industry.

14.4 NMR Spectroscopy

NMR spectroscopy is a multipurpose technique to study the structure and interactions of environmental constituents such as soil, water, and air (Fig. 14.3). It is also useful to track the metabolic responses of living organisms in response to climate change (Levitt 2013). Furthermore, spin-spin interactions and bond connectivity can be defined with the help of one-dimensional (1-D) as well as three-dimensional (3-D) NMR experiments (Hertkorn 2007). NMR can provide specific information at the molecular level in many experiments which is unavailable with other analytical techniques. Further, the results obtained by NMR are highly reproducible under different magnetic field strengths in different laboratories (Akoka et al. 1999). NMR can detect any compound with NMR active nuclei such as ^{13}C or ^{1}H non-selectively which makes it an excellent technique to study non-targeted constituents. Unknown compounds can be detected with the help of 2-D NMR experiments without their prior knowledge. This is a significant advantage over other techniques such as mass spectrometry or chromatographic separation where some prior knowledge is required about the compounds to optimize the experiments.

The environmental system is dynamic and highly complex. Diversified ecosystems possess a variety of complex chemical, physical, and biological processes. Further, climate change, urbanization, agriculture, and industrial activity, are few factors that affect the functioning of an ecosystem. Environmental chemists are concerned with resolving these complex processes, and NMR spectroscopy is a key tool in this effort. NMR spectroscopy is also used to evaluate wastewater treatment effectiveness (Alves et al. 2007, 2015).

NMR has aided in the identification and detection of anthropogenic chemicals, and their transformations within the biosphere. In addition, to identify the structures of particles in the air and tracking reactions in the atmosphere, NMR has great potential in the study of mineral's bonding and their interactions with organic matter (Williams et al. 1981; Nielsen et al. 2005). NMR not only provides information about the structure but also, provides information about the non-covalent interactions which are crucial in environmental science.

Enormous experiments were carried out with NMR spectroscopy. The majority of these experiments can be extended with ^{2}H, ^{1}H, ^{7}Li, ^{14}N, ^{13}C, ^{15}N, ^{19}F, ^{29}Si, ^{27}Al, ^{31}P, ^{113}Cd, ^{111}Cd, ^{195}Pt, ^{207}Hg, and ^{199}Hg which are NMR active and important for analysis. The most common types of techniques used in the identification of constituents in the environment are discussed below in this chapter. Figure 14.1 shows the interaction of pollutants and living organisms.

14.4.1 Solution-State NMR

The solution-state NMR technique provides high resolution and quantitative measurement of soluble components (Akoka et al. 1999; Michel and Akoka 2004; Dumas et al. 2006; Ferreira et al. 2007; Hertkorn 2007) (Fig. 14.3). When the amount of sample is low then a cryogenically cooled probe with helium or nitrogen is used to increase the signal to noise ratio. For a large number of samples, room temperature probe (approx.1 mm diameter) is used to collect the NMR. The 5 mm cryoprobe is used for the detection of compounds in the range of nanograms but the environmental mixture contains components in the range of approximately 50–100 mg. Hence, 1-D and 2-D NMR are useful for the identification of environmental components (Alves et al. 2015; Simpson et al. 2011).

As a result, studies involving solution-state NMR are highly relevant to environmental science. Diffusion ordered spectroscopy (DOSY) NMR studies have shown that dissolved organic matter (DOM) is in an aggregated state even at low concentrations (Lam and Simpson 2009; Simpson 2002; Šmejkalová and Piccolo 2008). As a result, novel water suppression techniques have been established, allowing researchers to study DOM in altered samples at natural abundance (Lam and Simpson 2008). These methods have been used to study lakes, oceans, rivers, groundwater, and Antarctic glacial ice. Pautler et al. (2011) discovered that the accumulation of components in pond water varied as per the concentration of DOM.

14.4.2 Solid-State NMR

The solid-state NMR provides the environmental analysis of different types of carbon present in the solid sample (Conte et al. 2004; Preston 2007; Mao et al. 2017). The carbon line shape is improved due to ^{1}H decoupling at high power and narrow line shape when the sample is spun at a magic angle of 54.7° to the external magnetic field. This technique provides the quantitative analysis of solid environmental samples. The sample amount required for the commonly used 4 mm rotor is approximately 80–100 mg. A large amount of sample is required for the 7 mm rotor i.e., approximately 400 mg. However, a reduction in RF homogeneity is observed and the maximum spinning speed of the rotor is ~7 kHz. Hence, the spinning sidebands are observed in the solid-state NMR. The spinning sidebands in the solid-state NMR can be suppressed by the method Total Suppression of Sidebands (TOSS) (Dixon et al. 1982). Further, ^{1}H HR-MAS and ^{13}C solid-state techniques were used to identify the components that remained solids even in the presence of water. It is possible to study a multiphase sample with two probes if the phases do not change dynamically (Hughes et al. 2014; Lam et al. 2005; Simpson et al. 2006).

14.4.3 High-Resolution Magic Angle Spinning NMR

The high-resolution magic angle spinning NMR was invented in 1996 (Longstaffe et al. 2010). It is also called "gel phase NMR" as swollen compounds can be studied with similar resolution as analyzed in the solution state NMR (Farooq et al. 2013; Stark et al. 2007). The ^{1}H NMR can analyze swollen components and solution but unable to identify the crystallinity in the solid. This is due to the inaccessibility of decoupling of strong ^{1}H-^{1}H dipolar interaction in crystalline solids. This approach has wide utilization in the study of soil samples, plants samples, air particulates, and small organisms in their swollen state. Swollen samples such as soil + water or a piece of carrot + water provide information about components swollen by water. Here, solid-state NMR rotors are used with additional liquid for sealing to experiment.

Magnetic susceptibility distortions coupled with intact samples, as well as other undesirable line broadening interactions, can be reduced using magic angle spinning (MAS). As a result, in vivo organisms may produce high-resolution NMR data. HR-MAS NMR probes presently used in applications on bacterial cells (Li 2006; Gudlavalleti et al. 2006), fruit flies (Righi et al. 2010, 2014), earthworms (Bon et al. 2006), and *Daphnia magna* (Bunescu et al. 2010) in the last 10 years, demonstrating the broad applicability of this method.

14.4.4 Comprehensive Multiphase NMR

Comprehensive Multiphase NMR is the combination of solid-state and solution-state NMR was first introduced in 2012 (Courtier-Murias et al. 2012; Simpson et al. 2013). This approach is similar to High-resolution Mass spectrometry (HR-MAS) where solid-state samples are analyzed under high power. The probe can analyze all the components, liquid samples, gel samples, and solid samples in their swollen state. This technique represents the unique information about the conformation and structure of the components in the natural state. In this technique, the solid samples converted into liquid and vice versa. Few examples of this technique are flocculation, swelling, drying, growth, reactions, conformation change, or sequestration of liquid contaminants. Non-dynamic multiphase systems can be studied by this technique provide similar results as provided by HR-MAS and solid probes. However, for analysis of dynamic systems (component dynamically changing phase), the CMP probe is an ideal technique.

The comprehensive multiphase NMR spectroscopy (CMP NMR) is a relatively new technique suitable for the dynamic multiphase system. Solutions, gels, and solids can be distinguished in situ and unaltered natural samples using a series of editing experiments (Courtier-Murias et al. 2012). This method has been used to monitor plant germination (Simpson et al. 2018), seed structure (Lam et al. 2014), and oil-contaminated soils (Farooq et al. 2015). Masoom et al. (2016) showed that using a combination of solution-, gel-, and solid-state NMR methods, were able to gain a thorough understanding of how components are organized in soil and how their domains and associations change with pH and solvent.

14.4.5 Hyphenated NMR

Hyphenated NMR technique approaches to combine the NMR with other techniques. This technique became commercially available in 1990 (Godejohann 2007). The most common hyphenated NMR technique is the combination of HPLC with NMR. The advanced hyphenated NMR technique includes the integration of NMR with solid-phase extraction (SPE) and Mass spectrometry. Hyphenated NMR-Mass spectroscopy is useful to identify unknown new compounds. In practice, complex environmental samples cannot be analyzed with the help of a single technique such as MS or NMR. 2-D HPLC-NMR can be utilized for the analysis of dissolved organic matter (DOM) even after 10,000 fractions. The dissolved organic matter (DOM) is still too complex after 10,000 fractions to be analyzed by MS or NMR (Woods et al. 2012). Hyphenated NMR is utilized for the non-targeted screening of less complex matrices like wastewater as well as the detection of targeted unknowns in the mixtures.

The detection of contaminants and their degradation products is an essential aspect of environmental chemistry where NMR has played an important role. Taves et al. (1976) used ^{19}F NMR to identify perfluorinated organic fluorine compounds in human plasma in 1976, decades before MS application. The combination of liquid chromatography (LC) and nuclear magnetic resonance (NMR) has also made it possible to identify components of more complex pollutant mixtures (Godejohann 2007).

14.4.6 Low Field NMR

Low field NMR includes many areas such as mobile NMR (Danieli et al. 2007) and field cycling (Conte and Alonzo 2007). Field cycling NMR makes the graph between relaxation time versus field strength and provides beneficial information on the dynamics of the samples. Portable NMR is operated at low field and can reproduce the results of higher field NMR experiments. It is cost-effective due to its reduced size. Low field NMR spectroscopy provides relaxation information to oil well loggers which will help to identify the quality and existence of crude oil deposits. Low field NMR has substantial potential for environmental research. GC-MS and LC-MS analysis are extremely costly for monitoring the sites under remediation daily or weekly or monthly. Low Field NMR can provide complementary and real-time monitoring of the sites under remediation with low costs. The only drawback of Low Field NMR is the reduced resolution and sensitivity as compared to high field resolution.

14.4.7 Magnetic Resonance Imaging (MRI)

Magnetic Resonance Imaging (MRI) spectroscopy builds based on the chemical environment of nuclei within a sample. It produces visual images based on the distribution of water within the sample (Nestle et al. 2007). MRI is a useful technique as it provides a visual image and internal structure of the sample which is inaccessible in other techniques. The resolution of MRI is up to 1–5 μm. MRI is highly useful in the monitoring and quantification of significant processes for example transportation of contaminants, distribution of water in the sample, and can provide detailed information about pore size and dynamics of the sample. MRI is also useful as localized spectroscopy where NMR spectroscopy is used to collect sub-voxels within the sample.

14.5 Study of Non-covalent Interaction

Any change in the chemical atmosphere around nuclei can be detected using NMR spectroscopy. As a consequence, it can be used to classify non-covalent interactions as well as chemical arrangement (i.e. covalent bonds). Non-covalent interactions are important in environmental science because they play a key role in contaminant sequestration, transport, and fate (Mazzei and Piccolo 2015). For example, the interaction of herbicide with soil determines its effectiveness (object plant vs. in binds to the soil), bioavailability (does it attach irreversibly to soil and become non-bioavailable), transport (does it stick to soil or pass into the aquatic environment), and reactivity (bound chemicals can be shielded from photochemistry). Non-covalent pollutant interactions with soil have been studied using ^{13}C or ^{19}F solid-state NMR and ^{1}H HR-MAS NMR (Hatcher et al. 1993; Kohl et al. 2000; Nanny et al. 1997) and covalent binding using ^{15}N NMR (Thorn et al. 1996; Thorn and Kennedy 2002; Sachleben et al. 2004). Further, pyrene attaches strongly to mobile aliphatic domains in plant cuticles analyzed through ^{13}C NMR. Kohl et al. (2000) used solid-state, ^{19}F solid-state NMR to look into the molecular environment of hexafluorobenzene and explored that after sorption activity to the soil, hexafluorobenzene had a high degree of mobility, implying that the pollutant was bound to specific soil organic matter components.

14.6 Study of Metabolite

Metabolomics is the study of metabolite fluxes that can be related to changes in basic metabolic function (Lindon and Holmes 1999; Viant 2005, 2009). The number of studies focusing on environmental organism interactions such as pollutant, temperature, nutrient, salinity, and pH has increased dramatically in recent decades, focusing on ecology and ecotoxicity (Bundy et al. 2009; Lankadurai et al. 2013). Because of the large number of species that live in soil and water, environmental metabolomic studies are diverse (Charlton et al. 2006; Jones and Dias 2007; Nagato and Simpson 2007). Since the metabolites of interest in most studies were unknown, NMR analysis was chosen because of its ability to provide a comprehensive presentation of all metabolites. Furthermore, limited sample preparation for NMR allows for high-throughput analysis, resulting in large numbers of NMR spectra that can be analyzed using multivariate statistical approaches including principal component analysis (PCA). Bundy et al. (2002) for example, used ^{1}H NMR to identify a novel compound (2-hexyl-5-ethyl-3-furansulfonate) after earthworm exposure to 2-fluoro-4-methyl aniline, which was later verified by LCMS/MS.

14.7 X-Ray Fluorescence (XRF)

X-ray methods are widely used in the analysis in various areas like mineralogy, pedology, trace analysis of metals, medical diagnosis, forensic investigation, geology, archeological sampling, and various safety checks, etc. One of the commonly used methods is X-Ray Fluorescence (XRF) which can be used in trace analysis of metals as well as non-metals. The method is widely popular due to its non-destructive analysis, low-cost sample preparation, and relative ease of the process. XRF works on the principle of the interaction of matter with the electron beam and X-Rays. When matter interacts with the short wavelength and long frequency radiation, they become ionized. The high-energy X-Rays photons eject the inner core electron from the atom and create electron voids, the atom becomes unstable, and outer shell electrons cascade down to fill the inner shell electron void. Six key transitions commonly happen in XRF (Margui and Van Grieken 2013) (Fig. 14.4).

- LIII → K (Kα1)
- LII → K (Kα2)
- MIII → K (Kβ1)
- MV → LIII (Lα1)
- MIV → LIII (Lα2)
- MIV → LII (Lβ1)

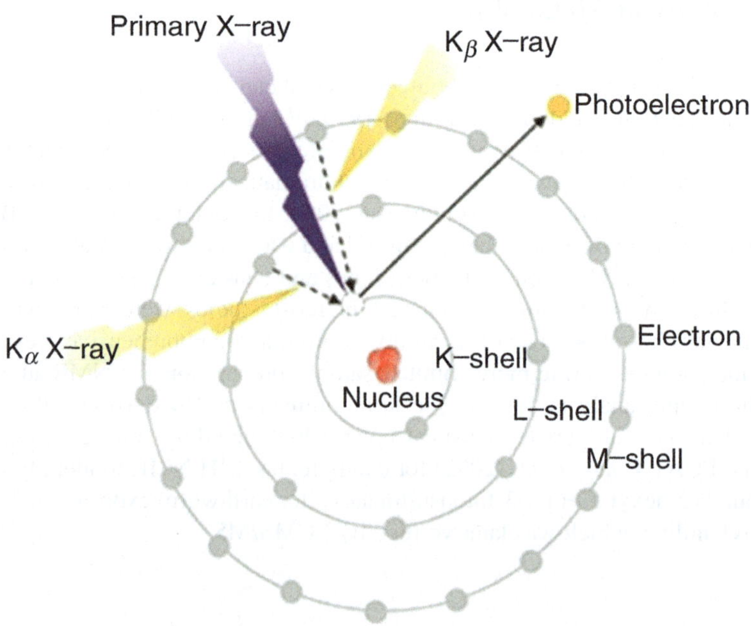

Fig. 14.4 Atomic model depicting the XRF principle

X-Ray fluorescence is the name assigned to this secondary energy emission (XRF). Every element has its characteristic energy value that is associated with its electronic configuration, so the energy emitted in fluorescence is element-specific. Qualitatively, the quantity of energy emitted is used to identify specific elements having characteristic peaks (e.g., Pb ~ 13 keV Ca ~ 4 keV; Zn ~ 9 keV). The amount of emitted fluorescent radiation energy (E) corresponds to a certain wavelength (λ) according to the formula $E = hc/\lambda$. Where h is Planck's constant and c is the speed of light. The wavelength of X-rays falls in the range of 0.01–10.0 nm of the electromagnetic spectrum, with corresponding energies of 0.125–125 keV (Margui and Van Grieken 2013).

The further classification of the XRF techniques into two categories, first one is an energy dispersive XRF (EDXRF) and the second one is wavelength-dispersive XRF (WDXRF). In EDXRF, electronic dispersion and a pulse height analyzer are used to separate the lines using photon energies. In WDXRF, diffraction, crystals, or geometric dispersion are used to separate lines based on their wavelengths. Detector plays a very important role in XRF instruments. There are many types of detectors in XRF. Gas-filled and scintillation detectors have conventionally been used in XRF instruments. But solid-state detectors, especially PIN diodes, are now commonly used to measure the energy of incoming photons through the ionization produced in the detector. WDXRF used gas-filled or scintillation detectors, however give poor resolution, but show high efficiency. EDXRF used improved Peltier-effect cooling systems like SDDs or PIN diodes to provide a better analytical response as compared to conventional liquid-nitrogen-cooled Si (Li) detectors.

14.7.1 Use of XRF in Environmental Application

Generally used spectroscopic techniques for elemental analysis are ICP-OES and AAS, but they are quite dangerous and time-consuming. XRF is the easy alternative for these techniques. XRF technique is used to detection of heavy metal contamination in soil, water, and in any sample. Some of the applications of XRF in environment cleaning are discussed below.

14.7.2 Detection of Heavy Metals in Water by XRF

Lelièvre et al. (2020) did work on the remediation of meal polluted water by using bacterial biofilms, where they used XRF as a promising technique analyzing the accurate concentration of metals in bacterial biofilms. They use EDXRF to measure the concentration of Cu, Cd, Mn, Fe, Zn, and Ni in the matrix of bacteria *Pseudomonas fluorescens* and its metabolites. Metal concentration was also analyzed by ICP-MS and ICP-AES and compared with XRF. Both techniques show a good correlation with an R2 of greater than 0.985. Zhou et al. (2018) used Portable

XRF (PXRF) in the detection of heavy metal (Cu and Pb) in water under indoor and fields studies. The PXRF was found to be accurate and results were compared with the actual concentrations of prepared metal solutions (R2 > 0.99 for both Cu and Pb). The field PXRF results possessed a promising linear correlation to the consequent laboratory results. Tighe et al. (2020) worked on the estimation of Pb in drinking water and compared the results with ICP-OEC and results show that XRF is also a good method as compared to the ICP-OEC.XRF. Ahmed et al. (2012) used XRF techniques to find the chemical composition of the sediment in a lake. The concentrations of Pb, As, Cu, N, TOC, Zn, and P were analyzed. The pollution level of sediments was evaluated by applying contamination factor, geo accumulation index, and pollution load index.

14.8 Inductively Coupled Plasma (ICP)

Inductively coupled plasma (ICP) is a widely used technique to estimates the actual concentration of all metals and some non-metals. It is the advanced version of AAS which have some limitation of using metal-specific hollow cathode tube. Plasma is the fourth state of matter that contains free electrons, positively charged ions, and neutral atoms and molecules. The plasma that is ionized by heating a gas inductively with an electromagnetic coil is known as inductively coupled plasma (ICP). Argon (Ar) gas is typically heated to a high-temperature plasma state.

The technique of ICP is of many types depending upon the analyzer used. If we use an atomic emission spectroscopy/optical emission spectroscopy then it is denoted as ICP-AES/ICP-OES. Further, if a mass spectrometer is attached then it is denoted as ICP-MS. When plasma energy attacks the sample and it is component atom gets excited. When an atom comes to its ground state emission rays are released corresponds to some specific wavelength. Which make the basis of ICP-AES and ICP-OES (Ghosh et al. 2013). The instrument of ICP-MS contains two parts ICP region and the Mass spectrometer region, which use mass spectrometry as one of the components. In a mass spectrometer, chemical species are ionized and analyze the ions based on their mass to charge ratio. The liquid sample is a requirement of that process. If the sample is in the solid form, then firstly it is converted to liquid form by multi-acid treatment in controlled temperature and pressure with the help of microwave-assisted digestion. The step-by-step process can be described as-

1. Liquid sample to form aerosol in the nebulizer.
2. Introduction of Ar gas to ICP torch, which supplies energy.
3. Radiofrequency (Rf) field causes a collision of Ar atom, generating a high-energy plasma.
4. In plasma (6000–10,000 K), sample aerosol decomposes into simultaneously ionized analyte atoms.
5. Ions from plasma are pumped into the analyzer.
6. The analyzer analyzes the ions.

Further advancement can be done by adding other instrumentation such as laser ablation, Gas chromatography, and liquid chromatography, etc. to the ICP system.

14.8.1 Application of ICP-MS in Environments

ICP-MS is widely used in the detection of heavy metals in environmental samples. Which enjoy wide acceptability due to wide elemental coverage, very low detection limits (up to ppb level), and fast analysis times (all elements at once). Some of the applications of ICP in environment cleaning are discussed below:

Yassin et al. (2021) used inductively coupled plasma optical emission spectroscopy (ICP-OES) to investigate Zn, Cu, Pb, and Cd in Egyptian olive oil obtained from different regions of Egypt. The process of digestion was done by using 10% HNO_3. Results were found that heavy metal content was under permissible limits as recommended by FAO/WHO. The method shows good accuracy, high selectivity, and good sensitivity. The method also investigates the effect of temperature and pressure. ICP-AES, ICP-MS, and CVAAS in ICP-AES and paddy in different soils of Cuba were used to estimate the contents of Cd, As, Cu, Cr, Ni, Hg, Zn, and Pb. To determine the safe intake level of these metals, an estimated weekly intake was determined. Merusomayajula et al. (2021) investigated the technique of inductively coupled plasma optical emission spectroscopic (ICP–OES) to estimate the elemental impurities in the voriconazole drug substances. The linearity results were obtained >0.9990 for all three impurities that indicate that these methods are potent candidates for regular sample analysis.

Kuznetsova et al. (2020) used the inductively coupled plasma–mass spectrometry (ICP-MS) technique as a quantification technique for several metals in marine sediments using a four-step microwave-assisted digestion protocol. It was found that the accuracy was improved on increasing time and temperature variables. The developed method was used to estimate the longitudinal metal distribution in the sediments obtained from the continental shelf of the East Siberian Sea 50–600 km range. The developed method was found to be a standard tool for accurate and precise sediment analysis. Bolea-Fernandez et al. (2020) used the inductively coupled plasma-mass spectrometry (ICP-MS) operated in single-particle mode (SP-ICPMS) as a method to detection and quantification of microplastics. Aydin et al. (2020) used inductively coupled plasma mass spectrometry (ICP-MS) for the estimation of uranium in the solid phase extraction in which multi-walled carbon nanotube/Cu2O-CuO ball-like hybrid material (MWCNTs/Cu2O-CuO) was used as a sorbent. The proposed procedure was very useful for the analysis of uranium (U) at very ultra-trace levels in geological rock samples and environmental water. Moreno-Andrade et al. (2020) used a tool that combines inductively coupled plasma-optical emission spectrometry (LC-ICP-OES) with liquid chromatography to analyze various compounds of antimony (Sb). This method was found to be reliable to quantify and separate Sb(V) and Sb (III) in aqueous samples.

Flores et al. (2021) studied the toxicological effect of nanoparticles in the environment and permeability in a biological system with the use of a Single-particle Inductively Coupled Plasma- Mass Spectrometer (SP-ICP-MS). Laser ablation inductively-coupled plasma mass spectrometry (LA-ICP-MS) techniques are widely used for the imaging of the surfaces of ice-core and soil-core (Zaeem et al. 2021). To establish the spatial distribution of selected elements in soil cores, researchers developed a hyperspectral imaging method using LA-ICP-MS. The surface of the cores was ablated using a 213 nm laser. 2D images of soil cores were made for Mg, Ca, K, P, Zn, Na, Fe, Mn, and Co by using the iolite software. This new imaging method was effective to demonstrate that different fertilizer applications and crop management systems change the spatial distribution and levels of metal concentration tested soil cores. Whereas, Bohleber et al. (2020) used laser ablation inductively-coupled plasma mass spectrometry (LA-ICP-MS) for imaging the impurity distribution in ice cores. That technique provides important information about the paleoenvironmental changes from very thinned ice layers. Jiménez-Lamana et al. (2020) established a system for detecting and quantifying nano-plastics (NPTs) at environmentally appropriate levels. They conjugate nano-plastics with functionalized gold (Au) containing nanoparticles, allowing them to be detected using single-particle inductively coupled plasma mass spectrometry (ICP-MS).

Leśniewska et al. (2020) studied the anthropogenic cycle of palladium (Pd), where they use inductively coupled plasma mass spectrometry (ICP-MS) for the detection and quantification of Pb in environmental water samples. For the separation of Pd (II) ions they used mesoporous silica materials (MCM41) which were functionalized with 3-mercaptopropyltrimethoxysilane (MPTMS). The method was found to be very useful for estimates the quantities of Pd at Pg ml^{-1} level. Ali et al. (2020) developed a green sorbent adsorbent using chitosan as a cation sorption material and thiocarbamate as an additive that increased active sites for cationic species. Herein they used ICP-OES instruments for quantification of cation during the sorption process. The chitosan with thiocarbamate was found to be terrific material for cations cleansing from water. They study the adsorption process via Langmuir, Temkin, and Freundlich isotherms methods. Linear regression results were found to be very significant for the adsorption process. The spectroscopic techniques can be used in various scientific fields such as pharmaceutical, organic synthesis, material science, biological science, and agricultural science Table 14.1. Such techniques are used for the analysis of available toxic metals and organic contaminants in the water body which coming from industry and household Table 14.2. Organic and inorganic contaminants of water, soil, and air can detect from available advance spectroscopic techniques.

Table 14.1 Applications of the Myriads Spectroscopic technique

Spectroscopic techniques	Application	References
UV-visible (UV-vis) spectroscopy	(a) Analysis of organic compounds—qualitative as well as quantitative (b) Revelation of compounds—structural study	Dyer (1965)
Fluorescence spectroscopy	(a) Microorganism tracing—via patho-physiological steps (b) Diagnostic tool—against bacteria at various level like group, genus and species	Shahzad et al. (2009) and Leblanc and Dufour (2002)
Atomic absorption spectroscopy (AAS)	(a) Determination—highly sensitive to detect certain metals	Moore et al. (1966)
Energy dispersive X-ray spectroscopy	(a) Characterization—compounds and elements analysis	Holmes et al. (1995)
Time-resolved spectroscopy	(a) Kinetic and mechanistic details—study of chemical processes (10^{12}–10^{15} s) (b) Tracking intermediate states—conformational changes, fluorescence processes, photo-chemical reaction and transfer process (c) Characterization—energy transfer processes and electronic states in nanoparticles	Dunn et al. (1993), Hou (2013) and Burda and El-Sayed (2000)
Femtosecond spectroscopy	(a) Characterization—population of DNA (b) Photoisomerization—study of the cis-trans rhodopsin (c) Charge transfer processes—study of photosynthetic reaction	Sundstrom (2008)
Attosecond spectroscopy	(a) Drive atomic or molecular dynamics—the quantitative motions of electrons	Anscombe (2008)
Mossbauer spectroscopy	(a) Study—enzymes, proteins (iron-containing) (b) Characterization—intermediates involves in oxygen activation by iron protein	Costas et al. (2004)
Multimodal spectroscopy	(a) Detecting or diagnosis tool—breast cancer, atherosclerosis and various diseases	Scepanovic et al. (2009)

14.9 Future Perspectives

Spectroscopic techniques under electromagnetic radiation assist in the study of synthesized as well as natural molecules along with their interactions. Although, spectroscopic techniques have revolutionized science and technology the instrument size, and complexity makes them unreachable for the larger research community. Therefore, portability, cost-effectiveness, easy to handle and higher sensitivity are

Table 14.2 Spectroscopic techniques and Microbes utilized in bioremediation of some toxic metals and organic compounds

Metals/organic compounds	Microorganisms	Techniques	References
Cr^{3+}, Cu^{2+}, Ni^{2+}, Cr^{6+}	*Spirulina* sp.	AAS, FTIR	Doshi et al. (2007)
Cr^{+3}, Ni^{+2}, Cu^{+2}	*Chlorella* sp.	AAS, FTIR	Doshi et al. (2008)
Cd^{+2}	*Pseudomonas aeruginosa*	UV-Vis, FTIR, AAS	Chakraborty and Das (2014)
Ur, Th	*Pseudomonas* sp.	FTIR, atomic force microscopy (AFM), transmission electron microscopy (TEM), X-ray diffractometry and energy dispersive X-ray (EDX)	Kazy et al. (2009)
Pb^{2+}	*Aeromonas caviae* strain KS-1	Energy dispersive X-ray spectroscopy, AAS	Shamim et al. (2013)
p-nitrophenol	*Pseudomonas putida*	UV-Vis, FTIR and GC-MS	Samuel et al. (2014)
Phenanthrene	*Sphingomonas* sp.	UV-Vis, LC-MS and GC-MS	Tao et al. (2007)
Benzo (a)pyrene	*Sphingomona syanoikuyae* JAR02	UV-Vis, HPLC/MS and HPLC/MS/MS	Rentz et al. (2008)
Anthracene	*Martelella* sp. AD-3	HPLC, NMR and GC-MS	Cui (2014)
Biphenyl	*Arthrobacter, Rhodococcus, Serratia, Rhizobium* and *Pseudomonas aeruginosa*	MALDI-TOF mass spectrometry, UV-Vis and GC-MS	Chakraborty and Das (2016) and Uhlik et al. (2011)
Polychlorinated	*Sinorhizobium meliloti* Environment sample	UV-Vis, Raman scattering spectroscopy, Surface-enhanced and GC-MS	Tu et al. (2011) and Zhou et al. (2011)

the target areas for the future. Such advanced techniques will encourage researchers and scientists to study and analysis of novel compounds, chemicals, hazardous substances, and environmental pollutants. Furthermore, the innovative spectroscopic techniques are fascinating to detect and analyze myriads fields of environmental parameters.

14.10 Conclusion

Spectroscopic techniques are sensitive and reliable methods for the detection and monitoring of pollutants. They can reveal a lot about chemical structure and transformations. Advanced techniques like NMR are capable of determining chemical structure from scratch, without the use of databases or libraries. Instrumentation costs, perceived awareness, cryogenic maintenance, and technological obstacles, as well as relatively low sensitivity, are all factors of NMR for its less use in environmental research. In outline, NMR spectroscopy has a wide range of applications, from determining structure (atmospheric particles, contaminants, and SOM) to molecular interactions (drug, agrochemical, and contaminant fate) to biological effects (detection and clarification of environmental stress). Similarly, the use of XRF and its advanced version PXRF in trace analysis of heavy metals from environmental samples is increasing due to its portability, simple sample preparation, and non-destructive nature of the analysis. XRF technique and its advancement PXRF is a good alternative of other sensitive spectroscopic techniques like AAS, ICP-AES which involves dangerous procedure and are times taking processes. However, XRF techniques are not able to provide exposure limits as low as other spectroscopic techniques such as ICP, but their portability and pace can assist in obtaining timely information at the field level. The acceptability of ICP techniques is increasing due to their high sensitivity, wide elemental coverage, very low detection limits (up to ppb level), and fast analysis times (all elements at once). Conclusively, spectroscopic techniques offer a large wealth of knowledge that can't be generated by any other technique but its advancement is necessary.

References

Ahmed F, Bibi MH, Asaeda T, Mitchell CPJ, Ishiga H, Fukushima T (2012) Elemental composition of sediments in Lake Jinzai, Japan: assessment of sources and pollution. Environ Monit Assess 184:4383–4396

Akoka S, Barantin L, Trierweiler M (1999) Concentration measurement by proton NMR using the ERETIC method. Anal Chem 71(13):2554–2557

Ali N, Khan A, Malik S, Badshah S, Bilal M, Iqbal HM (2020) Chitosan-based green sorbent material for cations removal from an aqueous environment. J Environ Chem Eng 8(5):104064

Alves FEG, Alexandre e Silva LM, Ferreira AG (2007) Screening organic compounds in urban wastewater using a hyphenated system and NMR pattern recognition. eMagRes:173–186

Alves FEG, Alexandre e Silva LM, Ferreira AG (2015) Advancements in waste water characterization through NMR spectroscopy. Magn Reson Chem 53(9):648–657

Anscombe N (2008) Attosecond analysis. Nat Photonics Tech Focus Spectrosc 2:548

Aydin F, Yilmaz E, Ölmez E, Soylak M (2020) Cu2O-CuO ball like/multiwalled carbon nanotube hybrid for fast and effective ultrasound-assisted solid phase extraction of uranium at ultra-trace level prior to ICP-MS detection. Talanta 207:120–295

Bohleber P, Roman M, Šala M, Barbante C (2020) Imaging the impurity distribution in glacier ice cores with LA-ICP-MS. J Anal At Spectrom 35(10):2204–2212

Bolea-Fernandez E, Rua-Ibarz A, Velimirovic M, Tirez K, Vanhaecke F (2020) Detection of microplastics using inductively coupled plasma-mass spectrometry (ICP-MS) operated in single-event mode. J Anal At Spectrom 35(3):455–460

Bon D, Gilard V, Massou S, Pérès G, Malet-Martino M, Martino R, Desmoulin F (2006) In vivo 31 P and 1 H HR-MAS NMR spectroscopy analysis of the unstarvedAporrectodeacaliginosa (Lumbricidae). Biol Fertil Soils 43(2):191–198

Bundy JG, Lenz EM, Bailey NJ, Gavaghan CL, Svendsen C, Spurgeon D, Hankard PK, Osborn D, Weeks JM, Trauger SA, Speir P, Sanders I, Lindon JC, Nicholson JK, Tang H (2002) Metabonomic assessment of toxicity of 4-fluoroaniline, 3, 5-difluoroaniline and 2-fluoro-4-methylaniline to the earthworm Eisenia veneta (Rosa): identification of new endogenous biomarkers. Environ Toxicol Chem 21(9):1966–1972

Bundy JG, Davey MP, Viant MR (2009) Environmental metabolomics: a critical review and future perspectives. Metabolomics 5(1):3–21

Bunescu A, Garric J, Vollat B, Canet-Soulas E, Graveron-Demilly D, Fauvelle F (2010) In vivo proton HR-MAS NMR metabolic profile of the freshwater cladoceran Daphnia magna. Mol BioSyst 6(1):121–125

Burda C, El-Sayed MA (2000) High-density femtosecond transient absorption spectros copy of semiconductor nanoparticles. A tool to investigate surface quality. Pure Appl Chem 72 (1–2):165–177

Chakraborty J, Das S (2014) Characterization and cadmium-resistant gene expression of biofilm-forming marine bacterium Pseudomonas aeruginosa JP-11. Environ Sci Pollut Res 21 (24):14188–14201

Chakraborty J, Das S (2016) Characterization of the metabolic pathway and catabolic gene expression in biphenyl degrading marine bacterium *Pseudomonas aeruginosa* JP-11. Chemosphere 144:1706–1714

Charlton AJ, Donarski JA, Jones SA, May BD, Thompson KC (2006) The development of cryoprobe nuclear magnetic resonance spectroscopy for the rapid detection of organic contaminants in potable water. J Environ Monit 8(11):1106–1110

Conte P, Alonzo G (2007) Environmental NMR: fast-field-cycling Relaxometry. EMagRes:389–398

Conte P, Spaccini R, Piccolo A (2004) State of the art of CPMAS 13C-NMR spectroscopy applied to natural organic matter. Prog Nucl Magn Reson Spectrosc 44(3–4):215–223

Costas M, Mehn MP, Jensen MP, Que L Jr (2004) Dioxygen activation at mononuclear nonheme iron active sites: enzymes, models, and intermediates. Chem Rev 104(2):939–986

Courtier-Murias D, Farooq H, Masoom H, Botana A, Soong R, Longstaffe JG, Simpson MJ, Maas WE, Fey M, Andrew B, Struppe J, Hutchins H, Krishnamurthy S, Kumar R, Monette M, Stronks HJ, Hume A, Simpson AJ (2012) Comprehensive multiphase NMR spectroscopy: basic experimental approaches to differentiate phases in heterogeneous samples. J Magn Reson 217:61–76

Cui C, Ma L, Shi J, Lin K, Luo Q, Liu Y (2014) Metabolic pathway for degradation of anthracene by halophilic Martelellasp. AD-3. Int Biodeterior Biodeg 89:67–73

Danieli E, Blümich B, Casanova F (2007) Mobile nuclear magnetic resonance. eMagRes 1:849–861

Debbarma P, Raghuwanshi S, Singh J, Suyal DC, Zaidi MGH, Goel R (2017) Comparative in situ biodegradation studies of polyhydroxybutyrate film composites. 3Biotech 7(178):1–9. https://doi.org/10.1007/s13205-017-0789-3

Dixon WT, Schaefer J, Sefcik MD, Stejskal EO, McKay R (1982) Total suppression of sidebands in CPMAS C-13 NMR. J Magn Reson 49(2):341–345

Doshi H, Ray A, Kothari IL (2007) Bioremediation potential of live and dead spirulina: spectroscopic, kinetics and SEM studies. Biotechnol Bioeng 96(6):1051–1063

Doshi H, Ray A, Kothari IL (2008) Bioremediation potential of chlorella: spectroscopic, kinetics, and SEM studies. Int J Phytoremediation 10(4):264–277

Dumas ME, Maibaum EC, Teague C, Ueshima H, Zhou B, Lindon JC, Nicholson JK, Stamler J, Elliott P, Chan Q, Holmes E (2006) Assessment of analytical reproducibility of 1H NMR spectroscopy based metabonomics for large-scale epidemiological research: the INTERMAP study. Anal Chem 78(7):2199–2208

Dunn RC, Xie X, Simon JD (1993) Real-time spectroscopic techniques for probing conformational dynamics of heme proteins. Methods Enzymol 226:77–198

Dyer JR (1965) Applications of absorption spectroscopy of organic compounds. Prentice Hall, Englewood Cliffs, NJ

Farooq H, Courtier-Murias D, Soong R, Bermel W, Kingery WM, Simpson AJ (2013) HR-MAS NMR spectroscopy: a practical guide for natural samples. Curr Org Chem 17(24):3013–3031

Farooq H, Courtier-Murias D, Simspon MJ, Maas WE, Fey M, Andrew B, Struppe J, Hutchins H, Krishnamurthy S, Kumar R, Monette M, Stronks HJ, Simpson AJ (2015) Characterisation of oil contaminated soils by comprehensive multiphase NMR spectroscopy. Environ Chem 12 (2):227–235

Ferreira AG, Lião LM, Monteiro MR (2007) Biofuels. eMagRes:529–540
Flores K, Turley RS, Valdes C, Ye Y, Cantu J, Hernandez-Viezcas JA, Gardea-Torresdey JL (2021) Environmental applications and recent innovations in single particle inductively coupled plasma mass spectrometry (SP-ICP-MS). Appl Spectrosc Rev 56(1):1–26
Ghosh S, Prasanna VL, Sowjanya B, Srivani P, Alagaraja M, Banji D (2013) Inductively coupled plasma–optical emission spectroscopy: a review. Asian J Pharm Anal 3(1):24–33
Godejohann M (2007) Environmental NMR: hyphenated methods. eMagRes 477–492
Gudlavalleti SK, Szymanski CM, Jarrell HC, Stephens DS (2006) In vivo determination of Neisseria meningitidis serogroup a capsular polysaccharide by whole cell high-resolution magic angle spinning NMR spectroscopy. Carbohydr Res 341(4):557–562
Hatcher PG, Bortiatynski JM, Minard RD, Dec J, Bollag JM (1993) Use of high-resolution carbon-13 NMR to examine the enzymatic covalent binding of carbon-13-labeled 2, 4-dichlorophenol to humic substances. Environ Sci Technol 27(10):2098–2103
Hertkorn N (2007) Environmental NMR: solution-state methods. eMagRes:55–74
Holmes JD, Smith PR, Evans-Gowing R, Richardson DJ, Russell DA, Sodeau JR (1995) Energy-dispersive X-ray analysis of the extracellular cadmium sulfifide crystallites of Klebsiella aerogenes. Arch Microbiol 163(2):143–147
Hou L (2013) Steady state and time resolved spectroscopy of photoswitchable systems. Doctoral dissertation, University of Groningen, Netherlands
Hughes CE, Williams PA, Harris KD (2014) "CLASSIC NMR": An in-situ NMR strategy for mapping the time-evolution of crystallization processes by combined liquid-state and solid-state measurements. Angew Chem 126(34):9085–9089
Jiménez-Lamana J, Marigliano L, Allouche J, Grassl B, Szpunar J, Reynaud S (2020) A novel strategy for the detection and quantification of nanoplastics by single particle inductively coupled plasma mass spectrometry (ICP-MS). Anal Chem 92(17):11664–11672
Jones OA, Dias DA (2007) Environmental metabolomics of soil organisms. eMagRes:1–12
Kazy SK, D'Souza SF, Sar P (2009) Uranium and thorium sequestration by a Pseudo monas sp.: mechanism and chemical characterization. J Hazard Mater 163(1):65–72
Kohl SD, Toscano PJ, Hou W, Rice JA (2000) Solid-state 19F NMR investigation of hexafluorobenzene sorption to soil organic matter. Environ Sci Technol 34(1):204–210
Kour D, Kaur T, Devi R, Yadav A, Singh M et al (2021) Beneficial microbiomes for bioremediation of diverse contaminated environments for environmental sustainability: present status and future challenges. Environ Sci Pollut Res 28:24917–24939
Krzewicka B (2010) *Umbilicaria rhizinata* comb. nov. (*Lichenized ascomycota*) Lichenologist 42:491–493
Kuznetsova OV, Bychkova YV, Timerbaev AR (2020) Development and validation of a sector-field inductively coupled plasma–mass spectrometry (ICP-MS) method for analyzing the diagenesis-designating metals in marine sediments. Anal Lett 53(4):563–573
Lam B, Simpson AJ (2008) Direct 1H NMR spectroscopy of dissolved organic matter in natural waters. Analyst 133(2):263–269
Lam B, Simpson AJ (2009) Investigating aggregation in Suwannee River, USA, dissolved organic matter using diffusion-ordered nuclear magnetic resonance spectroscopy. Environ Toxicol Chem 28(5):931–939
Lam B, Diamond ML, Simpson AJ, Makar PA, Truong J, Hernandez-Martinez NA (2005) Chemical composition of surface films on glass windows and implications for atmospheric chemistry. Atmos Environ 39(35):6578–6586
Lam L, Soong R, Sutrisno A, de Visser R, Simpson MJ, Wheeler HL, Campbell M, Maas WE, Fey M, Gorissen A, Hutchins H, Andrew B, Struppe J, Krishnamurthy S, Kumar R, Monette M, Stronks HJ, Hume A, Simpson AJ (2014) Comprehensive multiphase NMR spectroscopy of intact 13C-labeled seeds. J Agric Food Chem 62(1):107–115
Lankadurai BP, Nagato EG, Simpson MJ (2013) Environmental metabolomics: an emerging approach to study organism responses to environmental stressors. Environ Rev 21(3):180–205

Leblanc L, Dufour E (2002) Monitoring the identity of bacteria using their intrinsic fluorescence. FEMS Microbiol Lett 211(2):147–153

Lelièvre C, Rouwane A, Poirier I, Bertrand M, Gallon RK, Murat A (2020) ED-XRF: a promising method for accurate and rapid quantification of metals in a bacterial matrix. Environ Technol:1–9

Leśniewska B, Arciszewska Ż, Wawrzyńczak A, Jarmolińska S, Nowak I, Godlewska-Żyłkiewicz B (2020) Method development for determination of trace amounts of palladium in environmental water samples by ICP-MS/MS after pre-concentration on thiol-functionalized MCM-41 materials. Talanta 217:121004

Levitt MH (2013) Spin dynamics: basics of nuclear magnetic resonance. Wiley, Chichester

Li W (2006) Multidimensional HRMAS NMR: a platform for in vivo studies using intact bacterial cells. Analyst 131(7):777–781

Lindon JC, Holmes E (1999) Metabonomics: understanding the metabolic responses of living systems to pathophysiological stimuli via multivariate statistical analysis of biological NMR spectroscopic data. Xenobiotica 29(11):1181–1189

Longstaffe JG, Simpson MJ, Maas W, Simpson AJ (2010) Identifying components in dissolved humic acid that bind organofluorine contaminants using 1H {19F} reverse heteronuclear saturation transfer difference NMR spectroscopy. Environ Sci Technol 44(14):5476–5482

Mao J, Cao X, Olk DC, Chu W, Schmidt-Rohr K (2017) Advanced solid-state NMR spectroscopy of natural organic matter. Prog Nucl Magn Reson Spectrosc 100:17–51

Margui E, Van Grieken R (2013) X-ray fluorescence spectrometry and related techniques: an introduction. Momentum Press, New York

Masoom H, Courtier-Murias D, Farooq H, Soong R, Kelleher BP, Zhang C, Maas WE, Fey M, Kumar R, Monette M, Stronks HJ, Simpson MJ, Simpson AJ (2016) Soil organic matter in its native state: unravelling the most complex biomaterial on earth. Environ Sci Technol 50 (4):1670–1680

Mazzei P, Piccolo A (2015) Interactions between natural organic matter and organic pollutants as revealed by NMR spectroscopy. Magn Reson Chem 53(9):667–678

Merusomayajula KV, Tirukkovalluri SR, Kommula RS, Chakkirala SV, Vundavilli JK, Kottapalli PKS (2021) Development and validation of a simple and rapid ICP-OES method for quantification of elemental impurities in voriconazole drug substance. Futur J Pharm Sci 7(1):1–12

Michel N, Akoka S (2004) The application of the ERETIC method to 2D-NMR. J Magn Reson 168 (1):118–123

Moore EJ, Milner OI, Glass JR (1966) Application of atomic absorption spectroscopy to trace analyses of petroleum. Microchem J 10(1):148–157

Moreno-Andrade I, Regidor-Alfageme E, Durazo A, Field JA, Umlauf K, Sierra-Alvarez R (2020) LC-ICP-OES method for antimony speciation analysis in liquid samples. J Environ Sci Health A 55(4):457–463

Nagato EG, Simpson MJ (2007) NMR-based metabolomics of daphnia magna: insights into aquatic ecosystem health. eMagRes:315–324

Nanny MA, Bortiatynski JM, Hatcher PG (1997) Non-covalent interactions between acenaphthenone and dissolved fulvic acid as determined by 13C NMR T 1 relaxation measurements. Environ Sci Technol 31(2):530–534

Nestle N, Morris R, Baumann T (2007) Environmental NMR: magnetic resonance imaging. eMagRes:575–586

Nielsen UG, Paik Y, Julmis K, Schoonen MA, Reeder RJ, Grey CP (2005) Investigating sorption on iron−oxyhydroxide soil minerals by solid-state NMR spectroscopy: a 6Li MAS NMR study of adsorption and absorption on goethite. J Phys Chem B 109(39):18310–18315

Pautler BG, Simpson AJ, Simpson MJ, Tseng LH, Spraul M, Dubnick A, Sharp MJ, Fitzsimons SJ (2011) Detection and structural identification of dissolved organic matter in Antarctic glacial ice at natural abundance by SPR-W5-WATERGATE 1H NMR spectroscopy. Environ Sci Technol 45(11):4710–4717

Preston CM (2007) Environmental NMR: solid-state methods. eMagRes:29–42

Rentz JA, Alvarez PJ, Schnoor JL (2008) Benzo [a] pyrene degradation by Sphingomo nasyanoikuyaeJAR02. Environ Pollut 151(3):669–677

Righi V, Apidianakis Y, Mintzopoulos D, Astrakas L, Rahme LG, Tzika AA (2010) In vivo high-resolution magic angle spinning magnetic resonance spectroscopy of Drosophila melanogaster at 14.1 T shows trauma in aging and in innate immune-deficiency is linked to reduced insulin signaling. Int J Mol Med 26(2):175–184

Righi V, Apidianakis Y, Psychogios N, Rahme LG, Tompkins RG, Tzika AA (2014) In vivo high-resolution magic angle spinning proton NMR spectroscopy of Drosophila melanogaster flies as a model system to investigate mitochondrial dysfunction in Drosophila GST2 mutants. Int J Mol Med 34(1):327–333

Sachleben JR, Chefetz B, Deshmukh A, Hatcher PG (2004) Solid-state NMR characterization of pyrene cuticular matter interactions. Environ Sci Technol 38(16):4369–4376

Samuel MS, Sivaramakrishna A, Mehta A (2014) Bioremediation of p-Nitrophenol by Pseudomonas putida 1274 strain. J Environ Health Sci Eng 12(1):53

Scepanovic OR, Volynskaya Z, Kong CR, Galindo LH, Dasari RR, Feld MS (2009) A multimodal spectroscopy system for real-time disease diagnosis. Rev Sci Instrum 80(4):043103

Shahzad A, Kohler G, Knapp M, Gaubitzer E, Puchinger M, Edetsberger M (2009) Emerging applications of fluorescence spectroscopy in medical microbiology field. J Transl Med 7:99

Shamim K, Naik MM, Pandey A, Dubey SK (2013) Isolation and identifification of Aeromonas caviaestrain KS-1 as TBTC-and lead-resistant estuarine bacteria. Environ Monit Assess 185 (6):5243–5249

Simpson AJ (2002) Determining the molecular weight, aggregation, structures and interactions of natural organic matter using diffusion ordered spectroscopy. Magn Reson Chem 40(13):S72–S82

Simpson AJ, Lam B, Diamond ML, Donaldson DJ, Lefebvre BA, Moser AQ, Williams AJ, Larin NI, Kvasha MP (2006) Assessing the organic composition of urban surface films using nuclear magnetic resonance spectroscopy. Chemosphere 63(1):142–152

Simpson AJ, McNally DJ, Simpson MJ (2011) NMR spectroscopy in environmental research: from molecular interactions to global processes. Prog Nucl Magn Reson Spectrosc 58(3–4):97–175

Simpson AJ, Courtier-Murias D, Longstaffe JG, Masoom H, Soong R, Lam L, Sutrisno A, Farooq H, Simpson MJ, Maas WE, Fey M, Andrew B, Struppe J, Hutchins H, Krishnamurthy S, Kumar R, Monette M, Stronks HJ (2013) Environmental comprehensive multiphase NMR. eMagRes:399–414

Simpson AJ, Simpson MJ, Soong R (2018) Environmental nuclear magnetic resonance spectroscopy: an overview and a primer. Anal Chem 90(1):628–639

Šmejkalová D, Piccolo A (2008) Aggregation and disaggregation of humic supramolecular assemblies by NMR diffusion ordered spectroscopy (DOSY-NMR). Environ Sci Technol 42 (3):699–706

Stark RE, Yu B, Zhong J, Yan B, Wu G, Tian S (2007) Environmental NMR: high-resolution magic-angle spinning. eMagRes:377–388

Sundstrom V (2008) Femtobiology. Annu Rev Phys Chem 59:53–77

Tao XQ, Lu GN, Dang Z, Yang C, Yi XY (2007) A phenanthrene-degrading strain Sphingomonassp. GY2B isolated from contaminated soils. Process Biochem 42(3):401–408

Taves DR, Grey WS, Brey WS (1976) Organic fluoride in human plasma—its distribution and partial identification. Toxicol Appl Pharmacol 37:120

Thorn KA, Kennedy KR (2002) 15N NMR investigation of the covalent binding of reduced TNT amines to soil humic acid, model compounds, and lignocellulose. Environ Sci Technol 36 (17):3787–3796

Thorn KA, Goldenberg WS, Younger SJ, Weber EJ (1996) Covalent binding of aniline to humic substances: comparison of nucleophilic addition, enzyme-, and metal-catalyzed reactions by 15N NMR. In: Gaffney JS, Marley NA, Clark SB (eds) Humic and fulvic acids. Am Chem Soc, Washington, DC, pp 299–326

Tighe M, Bielski M, Wilson M, Ruscio-Atkinson G, Peaslee GF, Lieberman M (2020) A sensitive XRF screening method for lead in drinking water. Anal Chem 92(7):4949–4953

Tu C, Teng Y, Luo Y, Li X, Sun X, Li Z, Christie P (2011) Potential for biodegradation of polychlorinated biphenyls (PCBs) by Sinorhizobiummeliloti. J Hazard Mater 186 (2):1438–1444

Uhlik O, Strejcek M, Junkova P, Sanda M, Hroudova M, Vlcek C, Macek T (2011) Matrix-assisted laser desorption ionization (MALDI)-time of flight mass spectrometry-and MALDI biotyper-based identification of cultured biphenyl-metabolizing bacteria from contaminated horseradish rhizosphere soil. Appl Environ Microbiol 77(19):6858–6866

Viant MR (2005) Environmental metabolomics using 9^1H-NMR spectroscopy. Environ Genomics:137–150

Viant MR (2009) Applications of metabolomics to the environmental sciences. Metabolomics 5:1–2

Williams RJ, Giles RG, Posner AM (1981) Solid state phosphorus NMR spectroscopy of minerals and soils. J Chem Soc Chem Commun 20:1051–1052

Woods GC, Simpson MJ, Simpson AJ (2012) Oxidized sterols as a significant component of dissolved organic matter: evidence from 2D HPLC in combination with 2D and 3D NMR spectroscopy. Water Res 46(10):3398–3408

Yassin AA, Ghandour MA, Khalil MM (2021) Health risk assessment and micro determination of trace elements content in Egyptian olive oil using ICP-OES. IOP Conf Ser Mater Sci Eng 1046 (1):012028

Zaeem M, Nadeem M, Pham TH, Ashiq W, Ali W, Gillani SSM, Thomas R (2021) Development of a hyperspectral imaging technique using LA-ICP-MS to show the spatial distribution of elements in soil cores. Geoderma 385:114831

Zhou Q, Zhang X, Huang Y, Li Z, Zhang Z (2011) Rapid detection of polychlorinated biphenyls at trace levels in real environmental samples by surface-enhanced Raman scattering. Sensors 11 (11):10851–10858

Zhou S, Yuan Z, Cheng Q, Zhang Z, Yang J (2018) Rapid in situ determination of heavy metal concentrations in polluted water via portable XRF: using Cu and Pb as example. Environ Pollut 243:1325–1333

Chapter 15
Role of Biochar in Wastewater Treatment and Sustainability

Balram Sahu, Anisha Srivastava, Deep Chandra Suyal, Raj Kumar, and Ravindra Soni

15.1 Introduction

Biochar is a substitute obtained from the combustion of carbonaceous-rich biomass viz. algal biomass, agricultural and forest residues, manures, etc. (Feng and Zhu 2017; Abbas et al. 2018). Biochar has gained worldwide attention due to its role in soil fertility, bio-energy production, environmental remediation, and carbon sequestration processes (Suliman et al. 2017). It has been observed that biochar has the outstanding ability to immobilize the organic and inorganic contaminants from the soil as well as water environment (Wei et al. 2018). Moreover, it is an economical way to reduce antibiotics, aromatic dyes, and agrochemicals from the environment (Qiao et al. 2018; Dash et al. 2021; Kumar et al. 2021). It thus helps in reducing bioaccumulation, which otherwise leads entering of persistent, non-biodegradable inorganic and organic compounds into the food chain and causes various health problems. This characteristic makes the biochar more popular along with its diverse applications involving adsorption, microporosity, ion exchangeability, etc. (Rehman et al. 2016; Kour et al. 2021).

Anthropogenic activities are imposing harmful effects on nature and deteriorating it day by day. Agrochemicals, biomedical waste, and industrial effluents are among

Balram Sahu and Anisha Srivastava contributed equally with all other contributors.

B. Sahu · A. Srivastava · R. Soni (✉)
Department of Agricultural Microbiology, College of Agriculture, Indira Gandhi Krishi Vishwavidyalaya, Raipur, Chhattisgarh, India

D. C. Suyal
Department of Microbiology, Eternal University, Baru Sahib, Sirmaur, Himachal Pradesh, India

R. Kumar
Department of Agricultural Biotechnology and Molecular Biology, College of Basic Sciences and Humanity, Dr. Rajendra Prasad Central Agriculture University, Pusa, Samastipur, Bihar, India

© The Author(s), under exclusive license to Springer Nature Switzerland AG 2022

D. C. Suyal, R. Soni (eds.), *Bioremediation of Environmental Pollutants*,
https://doi.org/10.1007/978-3-030-86169-8_15

the top that needs to be monitored and regulated for making the environment healthy and safe. In this perspective, making use of biochar could be a safe and sustainable alternative.

15.2 Organic Contaminants

Biochar-assisted removal of organic contaminants like dyestuff, methylene blue, and phenol, has been extensively studied (Rafatullah et al. 2010). Nowadays, several emerging contaminants like pharmaceutical compounds, microplastics, novel agro-chemicals, and aromatic hydrocarbons are of great concern. Several factors are involved in the bioremediation of organic contaminants using biochar i.e. (a) adsorbent's nature, porosity, and ash content; (b) chemical structure of adsorbate, pH, molecular weight, polarity, and solubility and (c) nature of the solution, ionic strength, pH, polarity and concentration. In wastewater treatment, the biggest concern is that the contaminants are found in mixed forms. The presence of a variety of pollutants in the same environment makes adsorption tough by reducing the polarity and binding energy while increasing electrostatic repulsion. Naproxen (NPX), diclofenac (DCF), and ibuprofen (IBP) are some contaminants removed through activated biochar which is made up of loblolly pine chips (Jung and Ahn 2016). Moreover, Ranitidine hydrochloride (RH) that is used to treat gastric and intestinal problems was successfully adsorbed by using fixed-bed columns of the biochar (Mondal et al. 2016). This contaminant is excreted through patient feces and can pollute the disposal sites.

For the preparation of biochar, pyro-gasification operating diversity of feedstock and activation method plays an important role. Biochar synthesis is done at the temperature range between 350 and 700 °C. Besides having different compositions, lignocellulosic precursors and broiler litter possess low and high ash content, respectively (Table 15.1). High temperature and activation of desisopropyl atrazine increase the adsorption capacity, surface area, porosity, and aromaticity. These factors are important characteristic features for activated biochar (Park et al. 2013).

Table 15.1 Thermochemical processes, reactions, residence time and primary products

Thermochemical process	Temp. range (°C)	Heating rate	Pressure	Residence time	Desired product
Slow pyrolysis	350–800	Slow, (<10 °C/min)	Atmospheric	Hours–seconds	Biochar
Torrefaction	200–300	Slow, (<10 °C/min)	Atmospheric	Minutes–hours	Stabilized friable biomass
Fast pyrolysis	400–600	Very fast (~1000 °C/s)	Vacuum–atmospheric	Seconds	Biooil
Gasification	700–1500	Moderate-very fast	Atmospheric–elevated	Seconds–minutes	Syngas/producer gas
HTC	175–25	Slow	–	Hour	Hydrochar

For producing different properties in activated biochars the effective method is chemical activation and physicochemical activation. Further, different contaminants need different pore-sized biochar viz. absorption of iodine was successful with having smaller pores by activated biochars as compared to methylene blue (Oh and Park 2002).

15.3 Inorganic Contaminants

Among the inorganic contaminants, anions and cations of As, NH_4, Fe, Cu, Zn, Cd, Cr, and other heavy metals are important. They are widely generated from metallurgical processes, mining, industrial effluents, and electronic waste. Heavy metals are among the most dangerous pollutants of soil and water ecosystems.

Activated biochar can be used to remediate several inorganic ions including Cu, Zn, As, etc. The formation of metal complexes or chelates provides biochar greater efficiency for Cu^{2+} adsorption because activated biochar has more oxygenated functional groups. In both materials, through different mechanisms, adsorption of Cu^{2+} happened: inactivated biochar through π-cation interactions and intraparticle diffusion and in biochar through surface complexation with the metal. In a similar environment, various metals present that may adsorb the metal contaminants is quite a complex process, for each metal, it needs different pH ranges to remove it. (Banerjee et al. 2016) reported that from mining water normally adsorption of three types of pollutants found i.e., Cu^{2+}, Fe^{2+} and As (V) in between 2 and 7 pH range. Further, it has been observed that adsorption of Fe^{2+} by activated biochar was high at lower pH (equal to 3) whereas adsorption decreased at higher pH because of the formation of ferrous compounds $[Fe(OH)(H_2O)_5]^+$ and $[Fe(H2O)_6]^{2+}$ (Banerjee et al. 2016). Similarly, Cu^{2+} and As (V) contaminants are removed about 75 and 80% respectively, at 5 and 6 pH. The removal of inorganic contaminants and porosity of activated biochars is consequently, influenced by operating conditions and the type of pyro-gasification process. Regarding metal ion uptake and physicochemical properties, the same feedstock may show great differences if pyrolyzed with different speeds and duration (Lim et al. 2010).

In the porosity development of activated biochars, activation type becomes important for the sorption of inorganic contaminants. NH_3, CO_2, and KOH were used by Rambabu et al. (2015) for NH_4^+ removal from wastewater through activated biochar. In this case, chemical activation was found effective to develop the porosity in NH_4^+ sorption.

15.4 Application of Biochar for Wastewater Treatment

Biochar is frequently used to remove the different organic pollutants (such as PAHs, agrochemicals, VOCs, antibiotics/drugs, etc.) from water and soil and uptake of contaminants by microorganisms and plants and describes the mechanism of reduction of bioavailability of organic pollutants. It has been observed that the biochar produced from woody biomass, feedstock, and crop residues have shown higher surface area in comparison to that prepared from municipal wastes and animal manures. Further, its contaminant removal ability is also affected by biochar residence time, pH, biochar application rate, contaminant type. Further, a process called chemisorption plays an important role in the removal of organic pollutants through physisorption and electrophilic interaction between biochar and the functional group present on the contaminants.

The agrochemicals viz. fungicides, pesticides, herbicides, insecticides including pyrimethanil, carbofuran, atrazine, simazine, chlorpyrifos, etc. are of great concern in the aquatic ecosystem. Such agrochemicals from the water have been depleted and corrected by biochar amendment at a rate of 0.5–1.0%. Biochar production from hardwood at high temperature (>600 °C) was highly efficient in the reduction of Phyto-availability of pesticides that contaminates the soil as well as water. It was constant in numbers of nanopores; the surface area is high and capability to sequester these organic pollutants comparatively in low at temperature (<600 °C) of biochar production. In the whole process of removing the major absorption, mechanisms are partitioning and subsequent diffusion, half contribution in the removal of different non-ionic pollutants is played by electrostatic attraction (Kookana 2010).

The surface charge at a zero-point charge (ZPC) is an important parameter that plays an important role in the sorption of wastewater contaminants. It depends upon the pH. When derived at low pH, they comprised positive charge, hence methyl blue is sorbed less. More pH of biochars as compared to ZPC, containing negative surface charge on the biochar thus, higher electrostatic attractions lead to increase in sorption of methylene blue (−ve charged) (Yang et al. 2016). Similarly, pH-dependent interactions also work in the sorption of several polar antibiotics. Therefore, for biochar pH is an important factor to interact with the polar organic contaminants.

Several inorganic metals which remain persistent in the environment are of great concern for the living organisms and nature, when present in the higher concentration. Such inorganic metal ions viz. copper, lead, mercury, zinc, cadmium, etc. are part of municipal and industrial waste. Copper among all other metals gives more affinity towards functional groups of biochars which are derived from hardwood and crop straw. Further, it also depends upon pH and types of feedstocks. Biochar sorption on the surface was done mainly through cation exchange at lower pH of 6.0–7.0 whereas other mechanisms like biochar surface complexation electrostatic interaction and functional groups were involved in the removal of Cu^{2+} at higher 7.0–9.0 pH. Similarly, removing other contaminants such as Cd^{2+}, $Ni^{2+}$and Zn^{2+} along with Cu^{2+} biochars was effective which is derived from wheat straw, guayule shrub, sida-hermaphrodite, and soybean straw. Besides pH, different types of feedstocks play a crucial role in the removal efficiency of biochars in eliminating inorganic metal ions. They can also be produced from coconut shells, wood chips green waste, sewage sludge, cottonseed hull, and hardwood biochars by pyrolyzing them at ≤550 °C temperature. Wetland plants, such as *Pennisetum purpureum*, *Thalia dealbata* etc. are used for making the special biochar for removing the Cd^{2+}/NH_4^+ contamination (Cui et al. 2016). Further, biochars that are derived from rice straw are 1–5% (w/w) more effective than that derived from bamboo (Lu et al. 2017). Mercury is a highly toxic heavy metal and is of great concern in drinking water pollution. It can be removed by alkaline biochar is produced from different residues like switchgrass, soybean straw, cottonseed husk, corn cob, cocoa husk, corn stover, wheat shaft, etc. Further, biochars obtained from animal manure are known to contain high sulfur content which helps in precipitating mercury from the wastewater (Liu et al. 2016). A 35–92% reduction of Cu^{2+}, Cd^{2+}, Pb^{2+}, and Zn^{2+}has been reported by applying the biochar derived from agricultural waste such as dairy manure, hardwood, and straw at a rate of 1–15%. An increase in the application of biochar leads to increasing pH and surface area as a result metal is removed with high efficiency.

The biochar which is derived from hardwood is highly efficient in the removal of As^{3+}, Cu^{2+}, and Cd^{2+} via metal complexation, and biochar removed the As^{3+}, Cu^{2+}, and Cd^{2+} present in the soil through organic carbon dissolved however, other factors such as soil pH and high biochar dosage are also involved (Beesley et al. 2010). They involve the reduction of As^{5+} to $As^{3+}$and thus affect its immobilization.

Biochar having a higher porosity and larger surface area does not necessarily affect in sorption of NH_4^+ and Cd^{2+} Further, an increase in the dosage of biochar leads to an increase in the soil pH, which results in lowers mobility of Cd^{2+} and Zn^{2+}. It is seen that metal solubility is reduced at the high pH of the soil and restricts their mobility. Whereas, in the case of Sb^{3+} mobility electrostatic repulsion plays an important role. Studying the effect of feedstock in the removal of inorganic pollutants and different physicochemical characteristics of biochar derived from willow- and pinewood—(such as ash content, pH, C/N ratio, volatile matters, CEC, etc.) and found different at three similar pyrolysis temperatures i.e., 450, 550 and 650 °C. When biochar is applied in soil derived from willow shows high removal of B^+ Cu^{2+}, Sr^{2+}, Zn^{2+}, K^+, Na^+, Mg^{2+}, and Ca^{2+} and decreases in phyto-availability of NO. biochar derived from pine has higher phyto-availability for removal of Fe^{2+} and Ti^{3+}(Nelissen et al. 2014).

The indiscriminate use of antibiotics, analgesics, and anti-inflammatory drugs have been increased in recent years. These pharmaceutically active compounds have created dangerous situations for the environment as well as human health. Biochar has proven itself efficient to solve this problem. The malt rootlet-biochar and activated sodium persulfate (SPS) are known to degrade sulfamethoxazole (SMX) antibiotics (Zhu et al. 2018). In this study, the nitrogen content of the biochar was improved by using ammonium nitrate as a pretreating agent of the biomass at very temperatures. The biochar showed highly graphitic nanosheet structures and a large specific surface area. Thus, higher nitrogen content made electrostatic force with the contaminants and removed the contaminants efficiently.

15.5 Removal of the Pesticides by Biochar

Agrochemicals have made the life of farmers easy to get higher crop production and yield. However, on the other hand, their indiscriminate use has imposed several health and environmental threats (Giri et al. 2017a, b). In the world, atrazine and pentachlorophenol are the two most used pesticides in agriculture which work against various pests and try to manage them (Bhatt et al. 2021a, b). Some biochars which are produced by rice straw and phosphoric acid modified biochar have provided a high degree of adsorption of imidacloprid and atrazine (Mandal and Singh 2017). Similarly, soybean and some carbohydrate-rich biomass like corn straw are used to produce biochar that can be explored efficiently to absorb atrazine and sulfamethazine. Some agro-based industrial biochar viz. zero-valent iron magnetic paper mill sludge biochar (ZVI-MBC) are used to remove pentachlorophenol (PCP) compounds (Devi and Saroha 2014). Different kinds of agricultural contaminants in the form of pesticides i.e. diuron, carbaryl, and glyphosate from agricultural wastewater were reported to be controlled by the biochar and the performance has happened with functional materials, target contaminants, and biochar feedstock (Wei et al. 2018). The possible absorption mechanisms used for treating agricultural wastewater include surface complexation, electrostatic interactions, ion exchange-p interactions, and intermolecular interactions (Wei et al. 2018). Furthermore, the biochar shows different kinds of absorption behavior against various agricultural contaminates (Wei et al. 2018).

15.6 Biochar for Heavy Metal Contaminated Soils

The industrial sector contributes the highest in wastewater through various sources like smelting, chemical industry, mining, battery manufacturing, leather manufacturing, dyes, and others. Heavy metal and organic pollutants were the constituents of industrial wastewater. These wastewater industrial contaminants are possible to mitigate only by the use of biochar. Biochar consists of chitosan with cross-linking with beads, membranes, and solution and it can successfully use for absorption of heavy metals in industrial wastewater. The used ratio of biochar and chitosan mixture affected the absorption of mainly lead, cadmium, arsenic, copper, and other heavy metals in industrial wastewater (Hussain et al. 2017). Some different kind of biochar i.e. gliricidia biochar is effectively used for various metal with Crystal Violet (CV) in an aqueous environment in dye-based industries. CV sorption leads by the pH value, the volume of biochar, and the surface area (Wathukarage et al. 2017). The battery manufacturing industry is effectively treated by bagasse biochar. Among the maximum adsorption capacity can arrive at 12.7 mg/g and the adsorptive process is connected to contact time medium, pH value, and dosage (Poonam and Kumar 2018). Ghezzehei et al. (2014) was shown that the helpful for enriching nutrients with ammonium and phosphate-based wastewater in the dairy industry. Biochar absorbs ammonium 20–43% in around 24 h and phosphate in flushed dairy usually; any type of biochar application used to remove different types of pollutants is applied in-vitro condition (lab condition).

15.7 Mechanism of Removal of Contaminants by Modified Biochar

The mechanism of removing contaminants by different kinds of biochar is modified with a similar type of biochar-derived compounds. The basic working principle of modified biochar involves the following steps and the processes include: (a) Due to physical absorption the absorbate comes in contact with the absorbent's surface or absorbent area. (b) The precipitation and complexation steps are involved in the pore of the absorbent's surface. (c) Finally, the area of absorbent surface filling with absorbate into the pore (Fagbohungbe et al. 2017). This process mainly targets in three-zone: the first stage where is no absorption take place, it's called the clear zone. The second stage where the absorption progress is started is known as the mass transfer zone and the third stage is stability achieved known exhausted zone (de Ridder et al. 2012) when the saturated zone or exhausted zone increase rapidly the clear zone decreases frequently during the process. The term breakthrough is used for a saturation point and stops the process (Moreno-Castilla 2004).

The mechanisms involved in the removal of toxic metals include complexation, ion exchange, precipitation, and surface sorption.

- The complexation is the method in which a central atom makes a complex with several other metal atoms (called ligands) through the specific metal interaction. Biochars can make a multi atom structure with several heavy metals.
- In the ion-exchange method, the cations and anions are exchanged between biochar and the contaminants. By doing so, they make an electrostatic interaction.
- Precipitation is an important mechanism used to remove inorganic pollutants by using biochar. In this complete process, firstly biochar is produced by pyrolysis of cellulose and hemicelluloses material. Decomposition increases the temperature up to or more than 300 °C and makes the product alkaline. It results in the precipitation of the contaminants in the solution or on the surface of the biochar. Puga et al. (2016) have shown that some heavy metals Cd and Zn easily precipitate with the use of biochar prepared with the sugarcane and straw dust and they concluded that the precipitation degree of biochar depends on the efficiency of surface precipitation temperature of pyrolysis.
- Surface sorption is the physical phenomenon in which the toxic metals get adsorb on the pores of the biochar surface.

Similarly, the mechanisms involved in the removal of organic pollutants include hydrophobic interactions, partitioning, electrostatic interactions, electron donor and electron acceptor interactions, and pore filling.

- Hydrophobic interaction mechanism involves neutral and/or hydrophobic components for adsorption purposes. It requires less energy. Li et al. (2018) have observed the hydrophobic interactions involved in the decontamination of o-chlorobenzene acid, p-chlorobenzene acid, and benzoic acid. Similarly, perfluoro octane sulfonate sorption can be done by hydrophobic biochar produced from maize straw (Chen et al. 2011, 2017).
- In partitioning, contaminants are exposed on the biochar surface having carbonized and non-carbonized portions. The contaminants get partitioned in between these two based on crystalline structure, graphene fractions, and other characteristics of biochar. Sun et al. (2011) has prepared biochar from wood and grass that can significantly enhance absorption partitioning of norflurazon and fluridone.
- The electrostatic interaction mechanism involves the interaction between ionizable organic pollutants and the charged biochar. This method is mainly depending upon the pH and ionic capacity of the solution. Inyang et al. (2014) have prepared biochar from the bagasse and used it effectively against the removal of methylene blue.
- Interaction between electron donor and electron acceptor interactions represents the graph-like representation of the absorption of aromatic compounds of biochar at a temperature greater than 1100 °C. The electron density depends on the biochar temperature. If the biochar temperature remains $\geq$500 °C, π aromatic acts as electron acceptor; while, at a high temperature ($\leq$500 °C), the biochar act as a donor.
- In pore filling methods, the microporosity of the biochars is explored. Biochars containing mesopores (2–50 nm size) and micropores ($\geq$2 mm) on their surface are preferred in this method. The pore-filling mechanism is more efficient as compare to other absorption mechanisms.

15.8 Conclusion

Biochar-assisted bioremediation is an easy and economical way to remove various pollutants including organic waste, inorganic waste, heavy metals, pesticides, dyes, etc. In the present scenario, where anthropogenic activities have imposed serious threats to nature and humankind, adopting such eco-friendly techniques becomes very important.

References

Abbas T, Rizwan M, Ali S, Adrees M, Mahmood A, Rehman MZ, Ibrahim M, Arshad M, Qayyum M (2018) Biochar application increased the growth and yield and reduced cadmium in drought-stressed wheat grown in an aged contaminated soil. Ecotoxicol Environ Saf 148:825–833

Banerjee S, Mukherjee S, LaminKa-ot A, Joshi SR, Mandal T, Halder G (2016) Biosorptive uptake of Fe2+, Cu2+ and As5+ by activated biochar derived from Colocasia esculenta: isotherm, kinetics, thermodynamics, and cost estimation. J Adv Res 7(5):597–610. https://doi.org/10.1016/j.jare.2016.06.002

Beesley L, Jiménez EM, Eyles JLG (2010) Effects of biochar and green waste compost amendments on mobility, bioavailability and toxicity of inorganic and organic contaminants in a multi-element polluted soil. Environ Pollut 158:2282–2287

Bhatt P, Bhatt K, Sharma A, Zhang W, Mishra S, Chen S (2021a) Biotechnological basis of microbial consortia for the removal of pesticides from the environment. Crit Rev Biotechnol 41(3):317–338

Bhatt P, Joshi T, Bhatt K, Zhang W, Huang Y, Chen S (2021b) Binding interaction of glyphosate oxidoreductase and C-P lyase: molecular docking and molecular dynamics simulation studies. J Hazard Mater 5(409):124927

Chen X, Chen G, Chen L, Chen Y, Lehmann J, McBride MB, Hay AG (2011) Adsorption of copper and zinc by biochars produced from pyrolysis of hardwood and corn straw in aqueous solution. Bioresour Technol 102:8877–8884. https://doi.org/10.1016/j.biortech.2011.06.078

Chen J, Zhang D, Zhang H, Ghosh S, Pan B (2017) Fast and slow adsorption of carbamazepine on biochar as affected by carbon structure and mineral composition. Sci Total Environ 579:598–605

Cui X, Hao H, Zhang C, He Z, Yang X (2016) Capacity and mechanisms of ammonium and cadmium sorption on different wetland-plant derived biochars. Sci. Total Environ 539:566–575

Dash B, Sahu N, Singh AK, Gupta SB, Soni R (2021) Arsenic efflux in Enterobacter cloacae RSN3 isolated from arsenic-rich soil. Folia Microbiol 66:189–196

De Ridder M, De Jong S, Polchar J, Lingemann S (2012) Risks and opportunities in the global phosphate rock market: robust strategies in times of uncertainty. Hague Centre for Strategic Studies, Netherlands. https://library.wur.nl/WebQuery/titel/2016340

Devi P, Saroha AK (2014) Synthesis of the magnetic biochar composites for use as an adsorbent for the removal of pentachlorophenol from the effluent. Bioresour Technol 169:525e531

Fagbohungbe MO, Herbert BM, Hurst L, Ibeto CN, Li H, Usmani SQ, Semple KT (2017) The challenges of anaerobic digestion and the role of biochar in optimizing anaerobic digestion. Waste Manag 61:236–249. https://doi.org/10.1016/j.wasman.2016.11.028

Feng Z, Zhu L (2017) Sorption of phenanthrene to biochar modified by base. Front Environ Sci Eng 12:1–7

Ghezzehei TA, Sarkhot DV, Berhe AA (2014) Biochar can be used to capture essential nutrients from dairy wastewater and improve soil physico-chemical properties. Solid Earth 5:953e962

Giri K, Rai JPN, Pandey S, Mishra G, Kumar R, Suyal DC (2017a) Performance evaluation of isoproturon-degrading indigenous bacterial isolates in soil microcosm. Chem Ecol 33(9):817–825. https://doi.org/10.1080/02757540.2017.1393535

Giri K, Suyal DC, Mishra G, Pandey S, Kumar R, Meena DK, Rai JPN (2017b) Biodegradation of isoproturon by *Bacillus pumilus* K1 isolated from foothill agroecosystem of North West Himalaya. Proc Natl Acad Sci India Sect B Biol Sci 87(3):839–848. https://doi.org/10.1007/s40011-015-0667-x

Hussain A, Maitra J, Khan KA (2017) Development of biochar and chitosan blend for heavy metalsuptake from synthetic and industrial wastewater. Appl Water Sci 7(8):4525–4537

Inyang M, Gao B, Zimmerman A, Zhang M, Chen H (2014) Synthesis, characterization, and dye sorption ability of carbon nanotube–biochar nanocomposites. Chem Eng J 236:39–46. https://doi.org/10.1016/j.cej.2013.09.074

Jung KW, Ahn KH (2016) Fabrication of porosity-enhanced MgO/biochar for removal of phosphate from aqueous solution: application of a novel combined electrochemical modification method. Bioresour Technol 200:1029–1032

Kookana RS (2010) The role of biochar in modifying the environmental fate, bioavailability, and efficacy of pesticides in soils: a review. Soil Res 48(7):627–637

Kour D, Kaur T, Devi R, Yadav A, Singh M et al (2021) Beneficial microbiomes for bioremediation of diverse contaminated environments for environmental sustainability: present status and future challenges. Environ Sci Pollut Res 28:24917–24939

Kumar P, Dash B, Suyal DC, Gupta SB, Singh AK, Chowdhury T, Soni R (2021) Characterization of Arsenic-Resistant Klebsiella pneumoniae RnASA11 from contaminated soil and water samples and its bioremediation potential. Curr Microbiol. https://doi.org/10.1007/s00284-021-02602-w

Li H, Cao Y, Zhang D, Pan B (2018) pH-dependent KOW provides new insights into understanding the adsorption mechanism of ionizable organic chemicals on carbonaceous materials. Sci Total Environ 618:269–275. https://doi.org/10.1016/j.scitotenv.2017.11.065

Lim IM, Boateng AA, Klasson KT (2010) Physicochemical and adsorptive properties of fast-pyrolysis bio-chars and their steam activated counterparts. J Chem Technol Biotechnol 85:1515–1521. https://doi.org/10.1002/jctb.2461

Liu P, Ptacek CJ, Blowes DW, Landis RC (2016) Mechanisms of mercury removal by biochars produced from different feedstocks determined using X-ray absorption spectroscopy. J Hazard Mater 308:233–242

Lu K, Yang X, Gielen G, Bolan N, Ok YS, Niazi NK, Xu S, Yuan G, Chen X, Zhang X, Liu D, Song Z, Liu X, Wang H (2017) Effect of bamboo and rice straw biochars on the mobility and redistribution of heavy metals (Cd, Cu, Pb and Zn) in contaminated soil. J Environ Manag 186 (2):285–292

Mandal A, Singh N (2017) Optimization of atrazine and imidacloprid removal from water using biochars: designing single or multi-staged batch adsorption systems. Int J Hyg Environ Health 220:637e645

Mondal S, Aikat K, Halder G (2016) Ranitidine hydrochloride sorption onto superheated steam activated biochar derived from mung bean husk in fixed bed column. J Environ Chem Eng 4 (1):488–497. https://doi.org/10.1016/j.jece.2015.12.005

Moreno-Castilla C (2004) Adsorption of organic molecules from aqueous solutions on carbon materials. Carbon 42(1):83–94. https://doi.org/10.1016/j.carbon.2003.09.022

Nelissen V, Ruysschaert G, Müller-Stöver D, Bodé S, Cook J, Ronsse F, Shackley S, Boeckx P, Haugard-Nielsen H (2014) Short-term effect of feedstock and pyrolysis temperature on biochar characteristics, soil and crop response in temperate soils. Agronomy 4:52–73

Oh GH, Park CR (2002) Preparation and characteristics of rice-straw-based porous carbons with high adsorption capacity. Fuel 81(3):327–336. https://doi.org/10.1016/S0016-2361(01)00171-5

Park J, Hung I, Gan Z, Rojas OJ, Lim KH, Park S (2013) Activated carbon from biochar: influence of its physicochemical properties on the sorption characteristics of phenanthrene. Bioresour Technol 149:383–389

Poonam BSK, Kumar N (2018) Kinetic study of lead ($Pb2^+$) removal from battery manufacturing wastewater using bagasse biochar as biosorbent. Appl Water Sci 8:119

Puga AP, Abreu CA, Melo LCA, Beesley L (2016) Biochar application to a contaminated soil reduces the availability and plant uptake of zinc, lead and cadmium. J Environ Manag 159:86–93. https://doi.org/10.1016/j.jenvman.2015.05.036

Qiao K, Tiana W, Baia J, Dong J, Zhao J, Gong X, Liu S (2018) Preparation of biochar from Enteromorpha prolifera and its use for the removal of polycyclic aromatic hydrocarbons (PAHs) from aqueous solution. Ecotoxicol Environ Saf 149:80–87

Rafatullah M, Sulaiman O, Hashim R, Ahmad A (2010) Adsorption of methylene blue on low-cost adsorbents: a review. J Hazard Mater 177(1–3):70–80. https://doi.org/10.1016/j.jhazmat.2009.12.047

Rambabu N, Rao BVSK, Surisetty VR, Das U, Dalai AK (2015) Production, characterization, and evaluation of activated carbons from de-oiled canola meal for environmental applications. Ind Crop Prod 65:572–581. https://doi.org/10.1016/j.indcrop.2014.09.046

Rehman MZ, Rizwan M, Ali S, Fatima N, Yousaf B, Naeem A, Sabir M, Ahmad HR, Ok YS (2016) Contrasting effects of biochar, compost and farmmanure on alleviation of nickel toxicity in maize (Zea mays L.) in relation to plant growth, photosynthesis and metal uptake. Ecotoxicol Environ Saf 133:218–225

Suliman W, Harsh JB, Abu-Lail NI, Fortuna AM, Dallmeyer I, Garcia-Pérez M (2017) The role of biochar porosity and surface functionality in augmenting hydrologic properties of a sandy soil. Sci Total Environ 574:139–147

Sun K, Ro K, Guo M, Novak J, Mashayekhi H, Xing B (2011) Sorption of bisphenol A, 17-α-Ethinyl estradiol and phenanthrene on thermally and hydrothermally produced biochars. Bioresour Technol 102:5757–5763. https://doi.org/10.1016/j.biortech.2011.03.038

Wathukarage A, Herath I, Iqbal MCM, Vithanage M (2017) Mechanistic understanding of crystal violet dye sorption by woody biochar: implications for wastewater treatment. Environ Geochem Health 41(4):1647–1661

Wei D, Ngo HH, Guo W, Xu W, Du B, Khan MS, Wei Q (2018) Biosorption performance evaluation of heavy metal onto aerobic granular sludge-derived biochar in the presence of effluent organic matter via batch and fluorescence approaches. Bioresour Technol 249:410–416

Yang G, Wu L, Xian Q, Shen F, Wu J, Zhang Y (2016) Removal of congo red and methylene blue from aqueous solutions by vermicompost-derived biochars. PLoS One 11(5):e0154562

Zhu S, Huang X, Ma F, Wang L, Duan X, Wang S (2018) Catalytic removal of aqueous contaminants on N-doped graphitic biochars: inherent roles of adsorption and nonradical mechanisms. Environ Sci Technol 52:8649–8658

Chapter 16
Bio-inoculants for Biodegradation and Bioconversion of Agrowaste: Status and Prospects

Vinay Kumar, Debasis Mitra, Anju Rani, Deep Chandra Suyal, Bhanu Pratap Singh Gautam, Lata Jain, Manjul Gondwal, Kishan Kumar Raj, Anup Kumar Singh, and Ravindra Soni

16.1 Introduction

Agro-waste can be defined as waste/residues produced from several agriculture activities such as crop residues, plant stalk, leaves, manures, bedding, hulls, vegetable and fruit matter, leaves, animal wastes (Lim and Matu 2015). Agro waste is also called as Agricultural waste is comprised of crop waste, animal waste, food processing waste, agro industrial waste, hazardous and toxic agricultural waste (Obi et al. 2016). It includes crop wastes from farms, harvest, poultry farms, slaughterhouses; fertilizer and pesticides run- off from fields which enter into soils, air and water (Goel et al. 2020). However, some groups also have another classification especially for agro-industrial waste (Fig. 16.1).

V. Kumar · L. Jain
ICAR-National Institute for Biotic Stress Management, Baronda, Raipur, Chhattisgarh, India

D. Mitra
Department of Microbiology, Raiganj University, Raiganj, Uttar Dinajpur, West Bengal, India

A. Rani
Department of Life Sciences, Graphic Era Deemed to be University, Dehradun, Uttarakhand, India

D. C. Suyal
Department of Microbiology, Eternal University, Baru Saheb, Sirmour, Himachal Pradesh, India

B. P. S. Gautam · M. Gondwal
Laxman Singh Mehar Govt PG College, Pithoragarh, Uttarakhand, India

K. K. Raj · A. K. Singh · R. Soni (✉)
Department of Agricultural Microbiology, College of Agriculture, Indira Gandhi Krishi Vishva Vidyalaya, Raipur, Chhattisgarh, India

© The Author(s), under exclusive license to Springer Nature Switzerland AG 2022

D. C. Suyal, R. Soni (eds.), *Bioremediation of Environmental Pollutants*,
https://doi.org/10.1007/978-3-030-86169-8_16

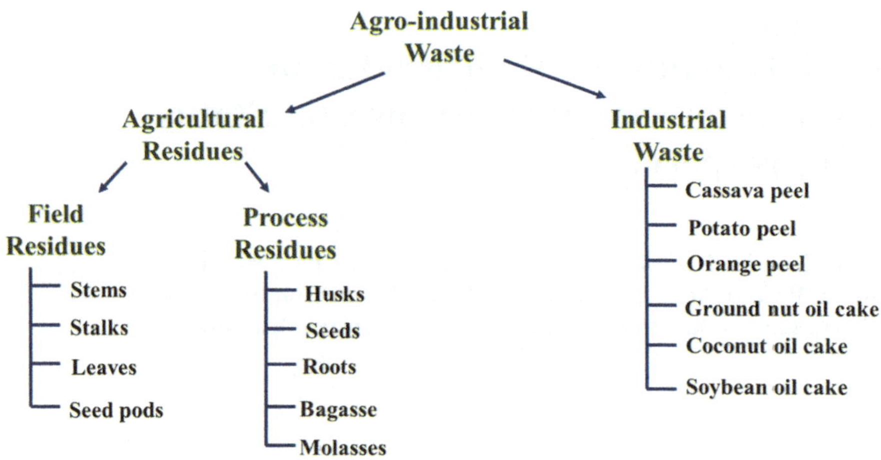

Fig. 16.1 Agro-industrial waste and their types

Now days biomass produced from agricultural waste are used to generate energy and this utilization of this energy refers to the conversion of agro-waste into clean energy is the important most effective method for disposing of agricultural waste (Xie et al. 2019). The major types include biogas, biomass, ethanol, bio-diesel, molding fuel, bio-kerosene, electricity and bio-gasoline (Wei et al. 2020). However, the major focus on research areas involved in conversion and utilization of energy is affected by several factors which include available technologies, policies and measures, laws and regulations and participation of NGOs (Wei et al. 2020).

Furthermore, it can be used in making organic fertilizer, in stabilizing of landfill soil and also have a potential to develop to a commercial level. Agricultural wastes mainly comprise of cellulosic fibres possessing high fixed carbon content and multifunctional groups. These wastes show considerable applicability due to its high strength, environmentally benign nature, low cost, and ease of availability and reusability (Rao and Rathod 2018).

Moreover, these agrowastes can also be utilized in various agro-based applications and other industrial processing in one or another forms (Bhuvaneshwari et al. 2019). Bioconversions into compost using potential beneficial bacterial and fungal microbes have ability of decomposition and fortification of organic matter (Singh et al. 2019). The microorganisms have unique properties for decomposing agricultural residues into valuable products are potential candidates (Kumar and Sai Gopal 2015), and some microbes possess enzyme activities directly linked to the decomposition of organic materials which under improved composting conditions yield better compost products.

16.2 Agro-Waste

16.2.1 Crop Waste

Agricultural wastes produced after harvesting of crops is named as crop waste or crop residues. Crop wastes includes paddy straw, rice husk, maize husk, cane trash, bagasse, peanut shells, cassava stem, coffee husk coconut shell and leaves, etc. It is estimated that the production of crop residue/waste about 3107 $\times$ 10^6 Mg/year for 17 cereals and legumes, 2802 $\times$ 10^6 Mg/year for cereal crops, and 3758 $\times$ 10^6 Mg/year for 27 food crops all over the world (Lal 2005). In India state wise crop waste data generated from 13 crops showed that the crop waste was mainly generated from cereal crops straw and the produced highest in North and western India. The harvest waste is also called as crop residue contains both the residues remain in the agricultural field or orchards once crop is harvested, these includes stalks and stubble leaves, and pods and seeds (Hoornweg and Bhada-Tata 2012). The crop residue reported from cereal crops i.e. rice and wheat system as 34% and 22%, respectively. More than 683 million tonnes (Mt) of crop residues of different crops are produced, of which a major part is used as fodder, fuel, and in various industrial processes. Despite this, about 178 Mt. of surplus crop residues are available around the country (Datta et al. 2020). Major part of this is burnt by the producers in the farms itself. However, during the year 2018–2019, 16.9% of total rice straw residue (7.93 Mt) in Haryana and nearly 50% of total rice straw residue (20.17 Mt) in Punjab state was burnt in situ (MoA&FW 2019).

A crop waste normally contains higher fiber and low level of protein, starch and fat. These wastes used for cattle feed, cooking purpose, animal-pen padding, mushroom production and as a source for organic contents. Crop/agro waste is also being used for the cultivation of mushrooms, *Pleurotus* spp. Subsequently, Bahia grass and banana stalks were used as starting materials more efficient for cultivation of *Pleurotus sajor-caju* mushroom without any additives with efficiencies of 74.4 and 74.12%, respectively. In addition, waste from coffee industries namely coffee spent wastes, coffee cherry husk, parchment husk and dried leaves alone or in combination with other agro wastes (wheat bran) were utilized for edible mushroom, *Pleurotus florida* cultivation.

Agro residues also used for production of bio-flavour production, nutrient-rich lignocellulosic waste generated from processing setup of agro industries was used for this purpose as well as substrate for the microbial production of flavor compounds.

16.2.2 Animal Waste

India is agriculture dominated country where livestock play a crucial role in farming system and Indian economy. Livestock are the key source of income of small and marginal farmers and contributes about 25.6 and 4.1% in Agriculture and National GDP (Gupta et al. 2017). Livestock waste produced from various activities such as manure, organic materials in the slaughter house; waste water from urine, cage wash water and waste water from the bathing of animals and sanitation in slaughterhouses, air pollutants such as H_2S and CH_4 and odours (Obi et al. 2016). Livestock wastes are also reported as an important source of pollution, greenhouse gas, disease causing microbes (pathogens) and odour. About 40% global methane is produced by agriculture and livestock by-products followed by 18% from waste disposal all over the world (EPA 1998). Animal waste are widely used as sources of biomass-based conversion processes for production of bio fertilizer, bioenergy and also served as source of plant nutrients (fertilizer) and animal nutrients (feedstuffs).

Milk production has been increasing and it is widely used for production of various milk products like paneer, cheese also produce certain residue/by products. It was reported that the 6–10 L of waste water per liter of the milk processed was generated from dairy industry. Furthermore, whey is produced after removing the milk clot in the production of cheese after the precipitation of the protein. It has been reported that about 47% of whey is discharged into the rivers, drainage and soils which causing serious contamination. With the use of biotechnological processes whey can be used for production of organic acids, fermented beverages and microbial protein using microorganism which uses buttermilk and whey nutrients to diminishing the pollution potential and production of useful products. Some microorganisms of genera *Bacillus*, *Lactococcus*, *Stenotrophomonas*, *Enterococcus*, *Serratia*, *Escherichia* and *Klebsiella* were tested as potential fat/protein-degrading bacteria. Additionally, microorganisms play an important role in rapid bioconversion of cellulosic and lignocellulosic wastes into organic materials (Kour et al. 2021). The huge amount of agricultural residual biomass generated from livestock production of manures and slurries which is applied to the land for fertility improvement. In India cattle manure is extensively used as manure in carp polyculture. Poultry wastes have been used in practical feeding for different classes of beef cattle, and potential exists for increased utilization of these by-products. Poultry industry generates a large amount of waste/by-products are feathers, soft meat, blood, heads, bones, legs, skin etc. These wastes represent a huge quantity of solid waste which needs to be properly managed to avoid loss of raw material for feed industry and environmental damage (Jayathilakan et al. 2012; Lasekan et al. 2013). These wastes are used for bioconversion in different valuable products like poultry feather used for preparation of preparation of nitrogen fertilizers or soil amendments (Kornillowicz-Kowalska and Bohacz 2011). Enzymatic processing may be useful to recycle protein-rich waste generated by poultry industry into valuable products and a high-quality meal from poultry by products is also a viable alternative for diets in aquaculture. Pig slurries and poultry manures have been used as common source of composting ingredient and composting of pig carcassesd processed in bins built from treated wood, over a concrete floor (Sorathiya et al. 2014). Manure produced from livestock waste seems to be the most perspective energy source as an alternative of non-renewable energy in coming days (Xue et al. 2020). Antibiotics are frequently applied for treatment of diseased animals due to these various types of antibiotic resistance genes (ARGs) are commonly apparent in livestock waste around the world. Conventional techniques used for treatment of waste were unable to completely remove ARGs, resulting in their release to soil and water environments. Hence, use of microorganism having potential to degrade antibiotic gene is essentially required to for bioconversion of livestock waste.

16.3 Biodegradation of Agrowaste

Huge amount of lignocellulosic wastes can be converted to the value-added products using biotechnological innovations. The microbial enzymes using fermentation processes (Himmel et al. 2010) can convert sugar rich lignocellulosic residues into chemicals, fuel, food additives, textile, paper, and agricultural inputs (Alonso Bocchini Martins et al. 2011; Joyce and Stewart 2012). The metabolites produced using microbes by solid state fermentation process may be categorized into two parts depending upon their application.

(a) Production of valuable products—enzymes, mushrooms, amino acids, bio-pesticides, bio-fuels, bio-surfactants, organic acids, SCP etc.
(b) Environment control—composting, production of animal feed, bioremediation/ biodegradation of toxic components and detoxification of agro-industrial wastes

Abundant and underutilized agro waste residues may serve as an excellent raw material for production of valuable products at industrial scale (Table 16.1). For bioconversion process strain selection is very important and selected microorganism must be non pathogenic, genetically stable, able to utilize various carbon substrates and be able to grow rapidly and vigorously. The most important economic characteristic in the selection of a microbe is its ability to produce high yields of the desired product (Dong et al. 2011).

Bioconversion of agricultural waste into compost is carried out by various microbial communities. The exceptional degradation capabilities of microorganisms have made them potential candidates for decomposing agricultural residues into compost (Kumar and Sai Gopal 2015; Suyal et al. 2021). Various microorganisms have been reported as fast decomposers, biodegraders, and bioconverters of non-useful products (Gautam et al. 2012; Singh et al. 2021).

Besides influencing decomposition of waste residue, microbial communities are also known to maintain nutrients pool in the soil (Rajwar et al. 2018; Joshi et al. 2019; Rawat et al. 2019). Microbes are recognized to immobilize mineralized nutrients into cell biomass and release it back to soil after microbe decomposition and thus influence bioavailability of nutrients to plants (Tomer et al. 2017; Kumar et al. 2018; Sahu et al. 2018; Negi et al. 2020). The compost/decomposed matter due to high carbon content makes native soil microflora more efficient when amended with soil (Rashid et al. 2016; Singh et al. 2019; Sahu et al. 2020). However, the decomposition ability of the microbial communities is affected by the type and size

Table 16.1 Microbial conversion of lignocellulosic waste into value added product through fermentation processes

Product	Type	Microorganisms
Biofuel	Bioethanol	*Sacchromyces cereviseae, Zymomonas mobilis, Pichia pastoriss*
	Butanol	*Clostridium acetobutylicum, Clostridium beijerinckii*
Enzymes	Cellulases	*Streptomyces* sp.
	Proteases	*Aspergillus oryzae*
	Xylanase	*Aspergillus terreus*; *Thermoascus aurantiacus*; *Bacillus* sp.
	Chitinases	*Penicillium aculeatum*
Organic acids	Lactic	*Lactobacillus pentosus, Lactobacillus delbrueckii, Lactobacillus brevis, Rhizopus* sp.
	Acetic	*Acetobacter, Aspergillus wentii, Aspergillus clavatus, Mucor piriformis, Citromyces*
	Citric	*Penicillium luteum, Penicilium citrinum, Aspergilus niger, Aspergillus wentii, Aspergillus clavatus, Mucor piriformis, Citromyces pfefferianus, Paecilomyces divaricatum, Trichoderma viride, Yarrowia lipolytica, Candida guilliermondii*
	Scuccinic	*Anaerobiospirillum succiniciproducens, Actinobacillus succinogenes, Mannheimia succiniciproducens*
SCP		*Chaetomium cellulolyticum, Pleurotus sajorcaju, Aspergillus* sp., *Penicillum* sp.

of waste residue, pH, moisture content, temperature and C/N ratio (Urbanová et al. 2015).

16.3.1 Enzymes Responsible for Biodegradation

Bioconversion of agricultural wastes into various by products (compost any other value added product) using potential microorganism (bacterial and fungal species) involved in decomposition and fortification organic matter, production of certain enzymes of immense importance (Singh et al. 2019; Giri et al. 2017a, b). Plant biomass is mainly consists of cellulose, hemicellulose, lignin and small amount of sugars, pectin, protein extractives, chlorophyll, nitrogenous material and inorganic waste (Chandra et al. 2012; Bhuvaneshwari et al. 2019; Sonwani et al. 2020). The biological lysis or degradation of cellulose (cellulolysis) is an enzymatically controlled synergistic process (Cui et al. 2019). Three types of glycoside hydrolases (GH): endo-β-1,4-glucanases (EG; EC. 3.2.1.4), β glucosidases (EC. 3.2.1.21) and exo-β-1,4-cellobiohydrolases (CBH; EC. 3.2.1.91) are involved in cellulose hydrolysis (Willis et al. 2010). On the other hand, xylan is a component of hemicelluloses from rice straw, which can be hydrolyzed by the enzyme-producing microbial xylanase (Lo et al. 2009). Further, lignin involved in strengthening of plant cell walls by adhesion of layers of cellulose microfibrils that allows flexibility to plants to enlarge considerably in body size, water transport of improve, resistance to pathogens and slowdown degradation of wood by microbes (Leisola et al. 2012; Labeeuw et al. 2015). In the prokaryotes, Laccase a ligninolytic enzyme is most frequently identified (Tian et al. 2014). Enzymes which are involved in degradation of lignin were classified in two main groups (1) lignin-degrading auxiliary enzymes and; (2) lignin-modifying enzymes (Janusz et al. 2017). In addition, enzymes produced the bacterial and fungal microbes namely peroxidases, lignin peroxidases, quinosine-reducing enzymes, laccases, polyphenol oxidases also used for degradation of lignin rich wastes. Thus microorganism directly or its enzymes play a crucial role in bioconversion of agricultural waste in to value added products. Several studies have reported the use of different agricultural residues namely maize bran, maize pericarp, wheat bran, soybean meal, sunflower meal and olive oil cake as growth substrates deployed for the production of enzymes of industrial use α-amylase, protease, cellulase and pectinase by using a new strain of *Bacillus* (Salim et al. 2017).

Pectinase comprises a heterogeneous group of enzymes that catalyze the breakdown of pectin-containing substrates. It is mainly used in food industry as several waste products from the agricultural and food industry containing pectin namely pulps of citrus sugar beet, lemon, apple pomace were used as carbon source for induction of pectinase by the microorganisms (Said et al. 1991). Bai et al. (2004) used Sugar beet pulp as carbon source and wastewater from monosodium glutamate production was used as nitrogen and water source for production of pectinase by solid state fermentation with *Aspergillus niger*.

Microbial enzymes for bio-conversion of poultry waste, such as proteases, lipases, combined enzyme preparations and, especially, keratinases for bioconversion of feathers. Some microbial enzymes can hydrolyze insoluble feather keratins, allowing their conversion into feedstuffs, fertilizers, and films.

16.4 Bioinoculants for Biodegradation and Bioconversion of Agro Waste

The enormous amount of agro-waste (over 300 million tonnes/year) is generated across the world which accounts for over 30% of renewable sources on earth. Most of the lignocellulose wastes remains unutilized and create environmental pollution if burnt upon (Singh and Nain 2014). In general the agro waste wastes of plant origin are chiefly made up of cellulose, hemicellulose and lignin. The amount of each component varies according to waste and plant species and age of the plant. The presence of sugars, proteins, minerals and water make the agro-industrial wastes suitable for the growth of microorganisms and make them attractive substrates for re-use in other processes. However presence of lignin makes it recalcitrant for microbial attack therefore effective conversion of the lignocellulose components into biomass have remained challenging (Loow et al. 2017). Microbes mainly used to the hydrolysis/breakdown of complex, lignin rich agro waste, and some rot causing fungi namely *Pleurotus ostreatus*, *Trametes versicolor*, *Ceriporiopsis subvermispora and Phanerochaete chrysosporium* mainly used for pretreatment of lignocellulosic biomass.

Pre-treatment is an important tool for breakdown of the structure of these recalcitrant residues i.e. cellulose, hemicellulose and lignin. A range of physical, chemical and biological pre-treatment methods are available for pretreatment. Major biological methods exploit the enzymatic potential of microbial strains of, e.g., *Aspergillus niger*, *A. awamori*, *Phanerohaete chrysosporium*, *P. sajor-caju*, *Bjerkendra adusta*, *Cyathus stercoreus*, *Pleurotus ostreatus*, *Trametes reesei* and *T. versicolor*. Bai et al. (2014) optimized solid-state fermentation process involving different agro-industrial wastes using producer microbe *Rhizopus microsporus var. oligosporus.*

Composting process occurs in three phases in which diverse microbial organisms (bacteria, actinomycetes and fungi) attack upon lignocellulosic components of the residue biomass. This converts waste into humus under mesophilic (*Streptomyces rectus*) and thermophilic (*Actinobifida chromogena*, *Thermomonospora fusca*, *Microbispora bispora*) conditions (Pan et al. 2012; Zeng et al. 2016). Co-inoculation of beneficial bacterial and fungal organisms like species of *Rhizobium*, *Azotobacter*, *Azospirillum*, *Pseudomonas*, *Bacillus*, *Burkholderia cepacia*, *sclerotiorum*, *Aspergillus niger*, *Fusarium oxysporum* (nonpathogenic), *Streptomyces griseoviridis* and *Trichoderma* spp., *Pythium oligandrum* etc. with compost can improve the soil health and crop productivity through various mechanisms (Singh et al. 2019). A large numbers of the bacteria are strict anaerobes *Bifidobacteria*, *Bacteriocides* and *Clostridia* and some of them are facultative anaerobes such as *Enterobacteriaceae* and *Streptococci* and are involved in the process. The bacterial community involved in solid state anaerobic digestion is comprised of *Anaerobacter* sp., *Bacillus* sp., *Clostridium* sp., *Ruminococcus* sp., *Oscillibacter* sp., *Sporobacter* sp., *Thiomargarita* sp., *Saccharofermentans* sp., *Sporobacterium* sp., etc.

16.4.1 Importance of Bio-inoculants in Agrowaste Degradation

Biodegradation or bioconversion is the way nature recycles waste, or breaks down biological matter into resources so certain species may use and reuse. In the microbiological context, "biodegradation" implies that a large variety of life forms consisting primarily microbes, yeast and fungi, and probably other species, are conducting the deterioration of all organic materials (Joutey et al. 2013).

The ways microorganisms were used to improve human and animal health, food manufacturing, food safety and sustainability, environmental conservation, crop development, and agricultural biotechnology have transformed them into substitutes for high-input farming activities (Singh et al. 2019). However, the metabolic capacity of microorganisms is a biotic consideration. The biotic factors that function on the microbial degradation of organic compounds involve direct inhibition of enzymatic activities and the degrading microorganism proliferation processes (Joutey et al. 2013; Kumar et al. 2019). Sometimes the extent of contaminant oxidation depends on the contaminant content and the volume of "catalyst" that is available. In this case, the sum of "catalyst" represents the number of cells capable of metabolizing the contaminant, as well as the quantity of enzymes each cell releases (Abatenh et al. 2017a, b; Joutey et al. 2013). However, the degree to which pollutants are metabolized is primarily a result of the individual enzymes involved and their "affinity" with the contaminant and the contaminant's availability. Additionally, there must be adequate quantities of nutrients and oxygen available in a accessible form and in suitable proportions for unregulated microbial production. Certain factors that affect the biodegradation process by regulating the enzyme catalyzed reaction levels are pH, temperature, and moisture (Siracusa 2019).

Microbes that are used as bioinoculants improve agro-, grain-, food-, manufacturing- and toxic waste degradation etc. (Santra et al. 2015). Microbial bacteria have the potential to destroy all organic compounds and inorganic waste compounds in a group process or in a separate type. Knowledge of efficient mechanisms for mineralizing and destroying toxic organic compounds can therefore play a key role in near future (Ahmad et al. 2018). For example, crop residue can be handled in situ by speeding microbial decomposition such that the zero-tilting machine can seed wheat without burning in the residues. Decomposition and release of N from crop residues rely on the autochthonous soil bacteria, the duration of the decomposition cycle and the conditions of soil and climate. Fungi are an essential part of soil micro-biota that makes up most of the plant biomass of water (Ainsworth and Bisby 1995) than bacteria, depending on depth and nutrient conditions of soil. Fungi that are filamentous in nature have a benefit in the decomposition of lignocellulosic waste because they have the capacity to generate abundant spores that can penetrate substrates rapidly and are supplied by a large range of enzymes with complementary catalytic activities (Lundell et al. 2010; Lynd et al. 2002). Fungi play a major role in crop waste degradation, such as wheat straw (Dagar et al. 2015; Dinis et al. 2009), maize stover (Wan and Li 2010), rice straw (Chang et al. 2012), and sugarcane residue (Maza et al. 2014). In fact, mixed cultures may have a greater effect on colonization of substrates owing to higher enzyme output and tolerance to contaminant microbes relative to pure crops. However, the most significant element in the usage of mixed cultures is the stability of species, which can influence the overall efficiency of bacteria, their organisation, spread, community scale and the ecosystems' ecological equilibrium. Hence, a compatible lignocellulolytic fungal consortium may play a significant role in the rapid degradation of rice straw (Kausar et al. 2010). Agriculture also results in accumulation of toxic substances like pesticides, heavy metals and other organic compounds. Microbial bioinoculants besides degrading agrowaste also help in bioremediation of tcontaminants present in soil. To date, microbes with the potential to degrade different organic compounds such as PCBs (polychlorinated biphenyls) have been isolated from various locations, and their pathways to encoding genes have also been examined. Microbes will efficiently eliminate pollutants such as TPHs (Total Petroleum Hydrocarbons), Polychlorinated Biphenyls,

Copper, Lead and Organophosphates, Organochlorines, and Carbamates by enzymatic activity. It has been documented that fungi used as bioinoculants such as *Agrocybe semiorbicularis*, *Hypholoma fasciculate*, *Cyathus bulleri*, *Phanerochaete chrysosporium*, *Auricularia auricular*, *Phanerochaete sordia*, *Coriolus versicolor*, and *Pleurotus ostreatus* degrade a broad variety of pesticide groups such as lindane, phenylurea, phenylamide, chlorinated, triazine and organophosphorus compounds (Ganash et al. 2016; Sharma and Pandit 2016; Singh and Walker 2006). Some microorganisms are capable of using different enzymes to biodegrade pesticides. Biodegradation of pesticides catalyzed by enzymes is a complex mechanism involving a sequence of biochemical reactions. The concentration and final fate of different pesticides in the environment, however, varies greatly, and depends on biotic and abiotic influences. Various species, including fungi, bacteria and actinomycetes, are capable of destroying and removing organic pesticides (Parte et al. 2017). Indigenous microflora or the enhancement of microbial communities may achieve bioremediation.

There have been many studies on the isolation and classification of microbial species that can function best in conjunction with existing rhizosphere microbes, beneficial fungi and biological control agents (Vishan et al. 2017; Kumar et al. 2021). Moreover, findings of long-term studies suggest that, while the introduction of agro-residues in soil greatly enhances soil quality, subsequent crop yields are decreased due to the development of microbial phytotoxins and allelochemicals and the immobilization of usable nitrogen (Singh and Nain 2014).

16.4.2 Renewable Energy Generation

16.4.2.1 Bioethanol Production

Ethanol obtained from various renewable resources as an alternative fuel to the present fossil fuel has been of relevance in current decades. There are usually four phases involved in the manufacturing of ethanol from any lignocellulosic biomass–pretreatment of feedstock, enzymatic saccharification, fermentation and regeneration of ethanol. Agricultural crops residue and wood is lignocellulose biomass e.g. straw and sugarbeet pulp. In lignocelluloses approximately 80% polysaccharides are present. Mostly yeast strains were reported for ethanol production from agrowaste or agroproduces. However, the fungal isolates *T. viride* and *A. terreus* were found to produce bioethanol by enzymatic saccharification of paddy straw. Similarly, ethanol production by fungal pre-treated wheat and paddy straw has also been studied. They used pretreatment with *Aspergillus awamori* and *Aspergillus niger* and subsequent fermentation to produce big quantity of ethanol from paddy and wheat straw. However, chemical treatment of straw was also used by several workers before sachharification by yeast. The scientists have reported successfully ethanol production from paddy straw (*Oryza sativa*) by simultaneous saccharification and co-fermentation using three different strain of *saccharomyces cerevisiae*. This group used diluted alkali solution (2%) for paddy straw pre-treatment and observed high degradation in cellulose, lignin and hemicellulose (90.6, 12.52 and 28.15%). However, alkali treatment was used by several worker (Takano and Hoshino 2018; Sonwani et al. 2020) but some also used acidic (sulfuric acid) treatment fast degradation of straw (Agrawal et al. 2015, 2018; Chiranjeevi et al. 2018).

Butanol contains four carbon with aliphatic saturated alcohol is considered as a superior biofuel and has ability mix together with gasoline in higher amount up to 80–85% without any alteration in conventional Otto-cycle engine. Bio-butanol can be produced from a variety of biomass including food materials with high starch contents and from agro industrial resides/wastes as feedstock for butanol production using *Clostridium beijerinckii.* Alkali-pretreated rice straw by co-culture of cellulolytic and butanol producing *Clostridium thermocellum* and *Clostridium saccharoperbutylacetonicum.*

16.4.2.2 Biogas Production

Biogas refers to the mixture of gases, and a type of biofuel that is naturally produced from the decomposition of organic waste, primarily consit of methane and carbon dioxide. Biogas has emerged as a promising renewable technology for conversion of agricultural, industrial, livestock and municipal wastes into energy (Mittal et al. 2018). Biogas is a multipurpose energy source that can be used as alternative to the fossil fuels in the heat and power generation (Weiland 2010). The energy production from renewable sources became essential to shrink the impact of greenhouse gases, mainly derived from combustion of fossil fuel (Olivier et al. 2017). Biomethane (methane-rich biogas) can be used to natural gas as a feedstock for producing material and chemicals (Weiland 2010). Biogas production through anaerobic digestion (AD) is one of the most prospective technologies in bio-energy portfolio as it can utilize all kinds of wastes from agriculture, livestock's, food industry and organic materials. These sources include crops and crop residues, fruit and vegetable wastes, manure can be used in small and large scale. Anaerobic digestion of agricultural resources such as agro-food biomasses produces biogas and can be used as gaseous fuel for motor vehicles. However the yield of specific methane is influenced by the chemical composition of crop/crop wastes that change as plants turning to maturing stage (Döhler et al. 2006). Microbes are being utilized for the efficient biogas production. These microbes can be grouped into three functional groups: methanogenic archaea, obligate hydrogen-producing acetogenic bacteria and hydrolysing and fermenting bacteria (Ahring 2003; Sárvári Horváth et al. 2016). A consortium of microorganism's involved in different steps of biogas production, they contributes in the hydrolysis and fermentation of agricultural waste/organic material (Angelidaki et al. 1993). Some of the microorganisms excrete hydrolytic enzymes like amylase, xylanase, cellulase, cellobiase, protease and lipase.

16.4.3 Compost Production

Composting is the process of biological degradation and stabilization of organic substrates under specific environment which allows the development of thermophilic temperature as a result of biologically produced heat. Composting process allows conversion of massive organic wastes into nutrient-enriched, stable products (Kharrazi et al. 2014). Composting has been widely used for the disposal of organic wastes such as paddy straw, rice husks, sugarcane trash, maize straw and other agro waste. Normally composting process takes more than 180 days to produce good quality and fully decomposed compost for agriculture waste/residues with high lignocellulose contents. Further, Sarkar and Chourasia (2017) done bioconversion of organic waste into biofortified compost through a microbial consortium within 30 days.

Composting of crop residues with the use of microorganisms have lignocellulolytic action is easier to recycles the lignocellulosic waste with high economic efficiency. The recycled material when applied to soil improves soil health, fertility and improves the carbon content in soil (Singh and Nain 2014). In the entire process of composting diverse microflora including fungi, bacteria and thermophilic *Actinomycetes*, mesophilic, *Streptomyces* and Microbispora ultimately converting organic waste to humus. Several microbes namely *Trichoderma harzianum*, *Phanerochaete chrysosporium*, *Polyporus ostriformis* and *Pleurotus ostreatus* have been applied for composting. It was also assumed that microbial activity indirectly may nutrient quality compost by several factors namely sulfur oxidizers, nitrogen fixers and increasing the availability of phosphate by hydrolys. The scientists have is of substrates by enzymes produced by microorganisms (Richardson and Simpson 2011). Studies on microbial mediated phosphate (P) availability to plants have been reported. The utilization of phytate P by legume and grass pasture was enhanced by soil microorganisms (Richardson et al. 2001), similarly genetically modified plants producing having gene for extracellular fungal phytase from *Aspergillus niger* in roots showed ability of plants to uptake P from phytate directly, and also synthesize some molecule, growth enhancers which act as plant growth hormones for improving the growth of plants. A number of microbial inoculants have been tested with the aim to improve composting processes, such as *Bacillus megaterium*, *Bacillus cereus*, *B. megaterium* + *B. cereus* and reported that the bacterial inocula altering the breakdown of cellulose and hemicelluloses and influenced the process of composting. While, Pan et al. (2012) found that the consortium of nitrogen fixing and cellulolytic bacteria enhanced the decomposition rate of wheat straw and compost became stable after 75 days with C: N ratio close to 25:1 and pH value of 7.0 ± 0.2. It was also indicated that the consortium of microbial inoculants is more effective than the individual isolate.

16.5 Conclusion

Increasing pollution due to crop residue burn becomes serious concern for several governments. However, several scientific groups continuously targeting this area but the development of many more eco-friendly bioinoculant formulations to reuse this unused biomass is the demand of hour. Not only reuse but reuse to generate renewable energy as an alternatives of current energy sources.

References

Abatenh E, Gizaw B, Tsegaye Z et al (2017a) Application of microorganisms in bioremediation—review. J Environ Microbiol 1(1):02–09

Abatenh E, Gizaw B, Tsegaye Z, Wassie M (2017b) The role of microorganisms in bioremediation—a review. Open J Environ Biol 2(1):038–046

Agrawal R, Satlewal A, Gaur R, Anshu M, Kumar R, Gupta RP et al (2015) Improved saccharification of pilot-scale acid pretreated wheat straw by exploiting the synergistic behavior of lignocellulose degrading enzymes. RSC Adv 5:71462–71471. https://doi.org/10.1039/C5RA13360B

Agrawal R, Bhadana B, Mathur AS, Kumar R, Gupta RP, Satlewal A (2018) Improved enzymatic hydrolysis of pilot scale pretreated Rice straw at high Total solids loading. Front Energy Res 6:115

Ahmad M, Pataczek L, Hilger TH, Zahir ZA, Hussain A, Rasche F, Schafleitner R, Solberg SØ (2018) Perspectives of microbial inoculation for sustainable development and environmental management. Front Microbiol 9:2992

Ahring BK (2003) Perspectives for anaerobic digestion. In: Ahring BK et al (eds) Biomethanation I. Advances in biochemical engineering/biotechnology, vol 81. Springer, Berlin

Ainsworth GC, Bisby GR (1995) Dictionary of the fungi. Commonwealth Mycological Institute, Kew, p 445

Alonso Bocchini Martins D, Ferreira Alves do Prado H, Simões Ribeiro Leite R, Ferreira H, de Souza MM (2011) Agroindustrial wastes as substrates for microbial enzymes production and source of sugar for bioethanol production. In: Integrated waste management, vol 2. IntechOpen, London

Angelidaki I, Ellegaard L, Ahring BK (1993) A mathematical model for dynamic simulation of anaerobic digestion of complex substrates: focusing on ammonia inhibition. Biotechnol Bioeng 42:159–166

Bai ZH, Zhang HX, Qi HY, Peng XW, Li BJ (2004) Pectinase production by Aspergillus niger using waste water in solid state fermentation for eliciting plant disease resistance. Bioresour Technol 95:49–52

Bai X, Shi P, Liu Y (2014) Society: realizing China's urban dream. Nature 509(7499):158–160

Bhuvaneshwari S, Hettiarachchi H, Meegoda JN (2019) Crop residue burning in India: policy challenges and potential solutions. Int J Environ Res Public Health 16:832

Chandra R, Takeuchi H, Hasegawa T (2012) Methane production from lignocellulosic agricultural crop wastes: a review in context to second generation of biofuel production. Renew Sust Energ Rev 16:1462–1476

Chang AJ, Fan J, Wen X (2012) Screening of fungi capable of highly selective degradation of lignin in rice straw. Int Biodeterior Biodegrad 72:26–30

Chiranjeevi T, Mattam AJ, Vishwakarma KK, Uma A, Peddy VR, Gandham S, Ravindra Velankar H (2018) Assisted single-step acid pretreatment process for enhanced delignification of rice straw for bioethanol production. ACS Sustain Chem Eng 6:8762–8774

Cui J, Mai G, Wang Z, Liu Q, Zhou Y, Ma Y, Liu C (2019) Metagenomic insights into a cellulose-rich niche reveal microbial cooperation in cellulose degradation. Front Microbiol 10:618

Dagar SS, Singh N, Goel N, Kumar S, Puniya AK (2015) Role of anaerobic fungi in wheat straw degradation and effects of plant feed additives on rumen fermentation parameters in vitro. Benefic Microbes 6(3):353–360

Datta A, Mrinal A. Emmanuel, N. K. Ram, Sunil Dhingra (2020) Crop residue management: solution to achieve better air quality. TERI, Discussion report (2020)

Dinis MJ, Bezerra RMF, Nunes F, Dias AA, Guedes CV, Ferreira LMM, Cone JW, Marques GSM, Barros ARN, Rodrigues MAM (2009) Modification of wheat straw lignin by solid state fermentation with white-rot fungi. Bioresour Technol 100:4829–4835

Döhler H, Eckel H, Frisch J (2006) Energiepflanzen. KTBL, Darmstadt

Dong Y, Butler EC, Philp RP, Krumholz LR (2011) Impacts of microbial community composition on isotope fractionation during reductive dechlorination of tetra chloroethylene. Biodegradation 22(2):431–444

EPA (1998) National air pollutant emission trends. 1990–1998

Ganash MA, Abdel Ghany TM, Reyad AM (2016) Pleurotus ostreatus as a biodegradator for organophosphorus insecticide malathion. J Environ Anal Toxicol 6(369):2161–0525

Gautam SP, Bundela PS, Pandey AK, Jamaluddin Awasthi MK, Sarsaiya S (2012) Diversity of cellulolytic microbes and the biodegradation of municipal solid waste by a potential strain. Int J Microbiol 1:325907. 12 p

Giri K, Rai JPN, Pandey S, Mishra G, Kumar R, Suyal DC (2017a) Performance evaluation of isoproturon-degrading indigenous bacterial isolates in soil microcosm. Chem Ecol 33 (9):817–825. https://doi.org/10.1080/02757540.2017.1393535

Giri K, Suyal DC, Mishra G, Pandey S, Kumar R, Meena DK, Rai JPN (2017b) Biodegradation of isoproturon by *Bacillus pumilus* K1 isolated from foothill agroecosystem of North West Himalaya. Proc Natl Acad Sci India Sect B Biol Sci 87(3):839–848. https://doi.org/10.1007/s40011-015-0667-x

Goel R, Debbarma P, Kumar P, Suyal DC, Kumar S, Mahapatra BS (2020) Assessment of soil chemical quality, soil microbial population and plant growth parameters under organic and conventional rice–wheat cropping system. Agric Res 10:193–204

Gupta SK, Gupta S, Panigrahy K, Panda S, Shinde KP, Dey D (2017) Importance of livestock in Indian scenario: a review. Bhartiya Krishi Anusandhan Patrika 32:245249

Himmel ME, Xu Q, Luo Y, Ding S-Y, Lamed R, Bayer EA (2010) Microbial enzyme systems for biomass conversion: emerging paradigms. Biofuels 1:323–341

Hoornweg D, Bhada-Tata P (2012) What a waste: a global review of solid waste management. World Bank, Washington, DC

Janusz G, Pawlik A, Sulej J, Swiderska-Burek U, Jarosz-Wilkolazka A, Paszczynski A (2017) Lignin degradation: microorganisms, enzymes involved, genomes analysis and evolution. FEMS Microbiol Rev 41(6):941–962. https://doi.org/10.1093/femsre/fux049

Jayathilakan K, Sultana K, Radhakrishna K, Bawa AS (2012) Utilization of byproducts and waste materials from meat, poultry and fish processing industries: A review. J Food Sci Technol 49:278–293

Joshi D, Chandra R, Suyal DC, Kumar S, Goel R (2019) Impact of bioinoculants *Pseudomonas jesenii* MP1 and *Rhodococcus qingshengii* S10107 on *Cicer arietinum* yield and soil nitrogen status. Pedosphere 29(3):388–399

Joutey NT, Bahafid W, Sayel H, El Ghachtouli N (2013) Biodegradation: involved microorganisms and genetically engineered microorganisms. Biodegr Life Sci 1:289–320

Joyce BL, Stewart CN Jr (2012) Designing the perfect plant feedstock for biofuel production: using the whole buffalo to diversify fuels and products. Biotechnol Adv 30:1011–1022

Kausar H, Sariah M, Saud HM, Alam MZ, Ismail MR (2010) Development of compatible lignocellulolytic fungal consortium for rapid composting of rice straw. Inter Biodeter Biodegrad 64:594–600
Kharrazi SM, Younesi H, Abedini-Torghabeh J (2014) Microbial biodegradation of waste materials for nutrients enrichment and heavy metals removal: an integrated composting-vermicomposting process. Int Biodeterior Biodegrad 92:41–48. https://doi.org/10.1016/j.ibiod.2014.04.011
Kornillowicz-Kowalska T, Bohacz J (2011) Biodegradation of keratin waste: theory and practical aspects. Waste Manag 31:1689–1701
Kour D, Kaur T, Devi R, Yadav A, Singh M et al (2021) Beneficial microbiomes for bioremediation of diverse contaminated environments for environmental sustainability: present status and future challenges. Environ Sci Pollut Res 28:24917–24939
Kumar BL, Sai Gopal DVR (2015) Effective role of indigenous microorganisms for sustainable environment. 3. Biotech 5:867–876
Kumar S, Suyal DC, Bhoriyal M, Goel R (2018) Plant growth promoting potential of psychrotolerant Dyadobacter sp. for pulses and finger millet and impact of inoculation on soil chemical properties and diazotrophic abundance. J Plant Nutr 41(8):1035–1046
Kumar P, Gupta SB, Anurag RS (2019) Bioremediation of Cadmium by mixed indigenous isolates Serratia liquefaciens BSWC3 and Klebsiella pneumoniae RpSWC3 isolated from Industrial and mining affected water samples. Pollution 5(2):351–360. https://doi.org/10.22059/poll.2018.268603.533
Kumar P, Dash B, Suyal DC, Gupta SB, Singh AK, Chowdhury T, Soni R (2021) Characterization of arsenic-resistant Klebsiella pneumoniae RnASA11 from contaminated soil and water samples and its bioremediation potential. Curr Microbiol. https://doi.org/10.1007/s00284-021-02602-w
Labeeuw L, Martone PT, Boucher Y et al (2015) Ancient origin of the biosynthesis of lignin precursors. Biol Direct 10:23
Lal R (2005) World crop residues production and implications of its use as a biofuel. Environ Int 31 (4):575–584
Lasekan A, Bakar A, Hashim D (2013) Potential of chicken by-products as sources of useful biological resources. Waste Manag 33:552–565
Leisola M, Pastinen O, Axe DD (2012) Lignin-designed randomness. Biocomplexity 2012:1–11
Lim S, Matu SU (2015) Utilization of agro-wastes to produce biofertilizer. Int J Energy Environ Eng 6:31–35. https://doi.org/10.1007/s40095-014-0147-8
Lo YC, Saratale GD, Chen WM, Bai MD, Chang JS (2009) Isolation of cellulose- hydrolytic bacteria and applications of the cellulolytic enzyme for cellulosic biohydrogen production. Enzym Microb Technol 44:417–425
Loow Y-L, Wu TY, Lim YS, Tan KA, Siow LF, Md Jahim J, Mohammad AW (2017) Improvement of xylose recovery from the stalks of oil palm fronds using inorganic salt and oxidative agent. Energy Convers Manag 138:248–260
Lundell TK, Makela MR, Hilden K (2010) Lignin-modifying enzymes in filamentous basidiomycetes—ecological, functional and phylogenetic review. J Basic Microbiol 50:5–20
Lynd LR, Weimer PJ, Van Zyl WH, Pretorius IS (2002) Microbial cellulose utilization: fundamentals and biotechnology. Microbiol Mol Biol Rev 66(3):506–577
Maza M, Pajot HF, Amoroso MJ (2014) Post-harvest sugarcane residue degradation by autochthonous fungi. Int Biodeterior Biodegrad 87:18–25
Mittal S, Ahlgrena EO, Shukla PR (2018) Barriers to biogas dissemination in India: a review. Energy Policy 112:361–370
MoA&FW (2019) Report of the committee on review of the scheme "promotion of agricultural mechanisation for in-situ management of crop residue in states of Punjab, Haryana, Uttar Pradesh and NCT of Delhi". Department of Agriculture, Cooperation & Farmers Welfare, Ministry of Agriculture and Farmers Welfare, Government of India
Negi H, Rani A, Joshi S, Sharma PK (2020) Biodiversity of microbial community: association with sustainable hill agroecosystems. In: Goel R, Soni R, Suyal DC (eds) Microbiological

advancements for higher altitude agro-ecosystems & sustainability. Springer Nature, Singapore, pp 163–182

Obi F, Ugwuishiwu B, Nwakaire J (2016) Agricultural waste concept, generation, utilization and management. Niger J Technol 35:957–964

Olivier JGJ, Schure KM, Peters JAHW (2017) Trends in global CO2 and total greenhouse gas emissions, report (2017). PBL publication number: 2674. PBL Netherlands Environmental Assessment Agency, The Hague

Pan I, Dam B, Sen SK (2012) Composting of common organic wastes using microbial inoculants. 3. Biotech 2:127–134

Parte SG, Mohekar AD, Kharat AS (2017) Microbial degradation of pesticide: a review. Afr J Microbiol Res 11:992–1012

Rajwar J, Chandra R, Suyal DC, Tomer S, Kumar S, Goel R (2018) Comparative phosphate solubilizing efficiency of psychrotolerant *Pseudomonas jesenii* MP1 and *Acinetobacter* sp. ST02 against chickpea for sustainable hill agriculture. Biologia 73(8):793–802

Rao P, Rathod V (2018) Valorization of food and agricultural waste: a step towards greener future. Chem Rec 19(9):1858–1871

Rashid MI, Mujawar LH, Shahzad T, Almeelbi T, Ismail IMI, Oves M (2016) Bacteria and fungi can contribute to nutrients bioavailability and aggregate formation in degraded soils. Microbiol Res 183:26–41

Rawat N, Sharma M, Suyal DC, Singh DK, Joshi D, Singh P, Goel R (2019) Psyhcrotolerant bio-inoculants and their co-inoculation to improve *Cicer arietinum* growth and soil nutrient status for sustainable mountain agriculture. J Soil Sci Plant Nutr 19(3):639–647

Richardson AE, Simpson RJ (2011) Soil microorganisms mediating phosphorus availability. Plant Physiol 156:989–996. https://doi.org/10.1104/pp.111.175448

Richardson AE, Hadobas PA, Hayes JE, O'Hara CP, Simpson RJ (2001) Utilization of phosphorus by pasture plants supplied with myo-inositol hexaphosphate is enhanced by the presence of soil microorganisms. Plant Soil 229:47–56

Sahu PK, Singh DP, Prabha R, Meena KK, Abhilash PC (2018) Connecting microbial capabilities with the soil and plant health: options for agricultural sustainability. Ecol Indic 105:601–612

Sahu B, Singh AK, Chaubey AK, Soni R (2020) Effect of rhizobium and phosphate solubilizing bacteria inoculation on growth and yield performance of Lathyrus (*Lathyrus sativus* L.) in Chhattisgarh Plains. Curr J Appl Sci Technol 39(48):237–247

Said S, Fonseca MJV, Siessere V (1991) Pectinase production byPenicillium frequentans. World J Microbiol Biotechnol 7:607–608

Santra SC, Mallick A, Samal AC (2015) Biofertilizer for bioremediation. Recent trends in biofertilizers. IK International Publishing House, New Delhi, pp 205–234

Sarkar P, Chourasia R (2017) Bioconversion of organic solid wastes into biofortified compost using a microbial consortium. Int J Recycl Org Waste Agric 6:321–334

Sárvári Horváth I, Tabatabaei M, Karimi K, Kumar R (2016) Recent updates on biogas production—a review. Biofuel Res J 10:394–402. https://doi.org/10.18331/BRJ2016.3.2.4

Sharma AK, Pandit J (2016) Biodegradation of chlorpyrifos by microbes—a review. Discov Biotechnol 7(18):1–10

Singh S, Nain L (2014) Microorganisms in the conversion of agricultural wastes to compost. Proc Indian Natn Sci Acad 80(2):473–481

Singh BK, Walker A (2006) Microbial degradation of organophosphorus compounds. FEMS Microbiol Rev 30(3):428–471

Singh DP et al (2019) Agrowaste bioconversion and microbial fortification have prospects for soil health, crop productivity, and eco-enterprising. Int J Recycl Org Waste Agric:1–16. https://doi.org/10.1007/s40093-019-0243-0. Corpus ID: 91983679

Singh M, Singh D, Rai P, Suyal DC, Saurabh S, Soni R, Giri K, Yadav AN (2021) Fungi in remediation of hazardous wastes: current status and future. In: Yadav AN (ed) Recent trends in mycological research, fungal biology. Springer Nature, Switzerland

Siracusa V (2019) Microbial degradation of synthetic biopolymers waste. Polymers 11(6):1066

Sonwani R, Gupta SB, Soni R (2020) Production of bioethanol from biodegraded alkali pretreated rice straw. Vegetos 33:128–134. https://doi.org/10.1007/s42535-019-00089-2
Suyal DC, Soni R, Singh DK, Goel R (2021) Microbiome change of agricultural soil under organic farming practices. Biologia 76:1315–1325
Tian JH, Pourcher AM, Bouchez T et al (2014) Occurrence of lignin degradation genotypes and phenotypes among prokaryotes. Appl Microbiol Biotechnol 98:9527–9544
Tomer S, Suyal DC, Rajwar J, Yadav A, Shouche Y, Goel R (2017) Isolation and characterization of phosphate solubilizing bacteria from Western Indian Himalayan soils. 3Biotech 7(2):1–8
Urbanová M, Šnajdr J, Baldrian P (2015) Composition of fungal and bacterial communities in forest litter and soil is largely determined by dominant trees. Soil Biol Biochem 84:53–64
Vishan I, Sivaprakasam S, Kalamdhad A (2017) Isolation and identification of bacteria from rotary drum compost of water hyacinth. Int J Recycl Org Waste Agric 6:245–253
Wan C, Li Y (2010) Microbial delignification of corn stover by Ceriporiopsis subvermispora for improving cellulose digestibility. Enzym Microb Technol 47:31–36
Wei J, Liang G, Alex J, Zhang T, Ma C (2020) Research progress of energy utilization of agricultural waste in China: bibliometric analysis by citespace. Sustain 12. https://doi.org/10.3390/su12030812
Weiland P (2010) Biogas production: current state and perspectives. Appl Microbiol Biotechnol 85:849–860. https://doi.org/10.1007/s00253-009-2246-7
Willis JD, Klingeman WE, Oppert C, Oppert B, Jurat-Fuentes JL (2010) Characterization of cellulolytic activity from digestive fluids of Dissosteira carolina (Orthoptera: Acrididae). Comp Biochem Physiol B: Biochem Mol Biol 157(3):267–272
Xie GH, Fang Y, Li S, Li M, Yang Y, Fu T, Bao W (2019) Review of the definition, classification, and resource assessment of biowaste. J China Agric Univ 24:1–9
Xue S, Song J, Wang X, Shang Z, Sheng C, Li C, Zhu Y, Liu J (2020) A systematic comparison of biogas development and related policies between China and Europe and corresponding insights. Renew Sust Energ Rev 117:109474
Zeng Y, De Guardia A, Dabert P (2016) Improving composting as a post-treatment of anaerobic digestate. Bioresour Technol 201:293–303

Chapter 17
Biochemical Parameters and Their Optimization Strategies for Microbial Bioremediation of Wastewater

Pooja Thathola and Vasudha Agnihotri

17.1 Introduction

Bioremediation involves the use of plants and microbes, to degrade, consume or uptake environmental pollutants. These biological entities have the adaptability and versatility to enabled bioremediation as a "biologically-friendly" strategy to degrade or detoxify many of the harmful contaminants (Kumar et al. 2021; Kour et al. 2021). The potential of bioremediation was recognized decades ago starting with efforts to treat environmental pollutants using microbial isolates. Researchers are expanding these studies for many purposes, particularly for environmental decontamination. Contaminants such as industrial waste products including polycyclic aromatic compounds, organic dyes, heavy metals, pharmaceuticals, and personal care products, and polyhalogenated compounds are commonly reported in the soil as well as in the aquatic environment which are the targets of various bioremediation strategies (Wu et al. 2017; Liu et al. 2018; Bhatt et al. 2021a, b). The process can degrade, mineralize, reduce, transform, or detoxify the contaminants present in the environment. Many types of pollutants have been reported including dyes, xenobiotic compounds, heavy metals, pesticides, agrochemicals, polyaromatic hydrocarbons, etc. The microorganisms that have been isolated from contaminated sources show the potential to consume or convert the target contaminant into nontoxic and simpler compounds (Gustavsson et al. 2016; Debbarma et al. 2017; Dash et al. 2021). The benefit of using the microbes for this purpose is that they can survive in extreme conditions of temperature, and pH in the presence of hazardous compounds or any waste stream (Singh et al. 2021). Different types of microbial strains such as *Bacillus*, *Pseudomonas*, *Mycobacterium*, *Arthrobacter*, *Corynebacterium*,

P. Thathola · V. Agnihotri (✉)
Centre for Land and Water Resource Management, G.B.Pant National Institute of Himalayan Environment, Almora, Uttarakhand, India
e-mail: vasudha@gbpihed.nic.in

© The Author(s), under exclusive license to Springer Nature Switzerland AG 2022
D. C. Suyal, R. Soni (eds.), *Bioremediation of Environmental Pollutants*,
https://doi.org/10.1007/978-3-030-86169-8_17

Nitrosomonas, etc. have shown their capacity to degrade different contaminants (Singh et al. 2014; Giri et al. 2017a, b). The microbes can work in both aerobic as well as anaerobic conditions.

The bioremediation process involves energy production which is generated due to various physiological activities occurring in the microbial cells. For the maintenance of lifecycle, the microbial cells require electron donors (energy sources), electron acceptors (such as nitrate, manganese, iron, oxygen, carbon dioxide, etc.), and nutrients. The many contaminants work as a source of electron donor and electron acceptor. During bioremediation, both biotic, as well as abiotic factors, play a major role. The biotic factor including microbial interaction (competition, succession, and predation), enzyme activity, horizontal gene transfer, mutation, population size, and its composition affect the bioremediation process (Boopathy 2000). Along with the biotic factors, environmental conditions such as temperature, pH, oxygen content, and redox potential, the bioavailability of pollutants, and their concentration, chemical structure, solubility, type, and toxicity of the pollutant, etc. affects the biodegradation process. The role of enzymes is very important in the bioremediation process which is produced during multiple metabolic pathways (Sharma et al. 2018; Dangi et al. 2019; Junghare et al. 2019). Some of the important biochemical parameters, which can affect the bioremediation of contaminants present in wastewater by affecting microbial growth, are described in the upcoming paragraphs.

17.2 Factors Affecting Microbial Growth and Bioremediation Process

The bioremediation process using microbes will only be successful if the microbes can grow in that atmosphere. There are some important parameters such as pH, temperature, oxygen level, and nutrients, which should be at optimum level so that microbes can perform better.

17.2.1 Nutrient/Substrate Content

Microbes need nutrients based on their cell composition (Baker and Herson 1994). Based on the microbial cell requirement, the nutrients/substrate may be classified into macro, micro, and trace nutrients. The macronutrients include carbon, nitrogen, and phosphorus, and micronutrients include sulfur, calcium, magnesium. Trace nutrients such as are iron, manganese, cobalt, copper, and zinc, etc. are required by some of the microbes. The optimum combination of these nutrients is required for the growth and further degradation activity of the microbes, for example, the combination of 150 mg of nitrogen and 30 mg phosphorus would be obligatory to degrade 1 g of a supposed hydrocarbon into cellular material. More carbon would be required to sustain microbes if the carbon source were simply and quickly converted into carbon dioxide (Cookson 1995). The detailed metabolic pathways of polyaromatic hydrocarbon (PAHs) indicated the evolution of carbon into carbon dioxide at the end of the bioremediation process. Therefore, additional supplements for carbon are not generally required until the completion of the bioremediation process. Nitrogen is the nutrient most added to bioremediation processes. This macronutrient is used for primarily cellular growth and as an alternative electron acceptor. It is frequently added as urea or as ammonium chloride, but it is also being supplied as any ammonia salt or ammonium nitrate. These forms are willingly assimilated in bacterial metabolism. But the main problem of addition of ammonium ion is that then the process requires increased oxygen supply. The second most added nutrient in bioremediation is phosphorus which is used by the microorganism for cellular growth. Potassium phosphate, sodium phosphate, or orthophosphoric and polyphosphate salts are added to the medium as the source of phosphorous.

The relationship between substrate content and microbial growth can be analyzed using the Monod equation (Bitton 1994)

$$\mu = \mu_{max} \frac{[S]}{K_s + [S]} \tag{17.1}$$

where S = substrate concentration (mg/L), K_s is half-saturation constant (mg/L) ($\approx \mu_{max}/2$), which shows the microbial affinity for the substrate, and μ_{max} is the maximum specific growth rate (h^{-1}). K_s and μ_{max} are affected by the temperature, carbon sources, and other parameters. The linear form of the Monod equation is shown as Eq. (17.2)

$$\frac{1}{\mu} = \frac{K_s}{\mu_{max}[S]} + \frac{1}{\mu_{max}} \tag{17.2}$$

Through the plot between $1/\mu$ and $1/S$, the value of K_s and μ_{max} may be calculated using the slope (K_s/μ_{max}), y-intercept ($1/\mu_{max}$) and x-intercept ($-1/K_s$).

17.2.2 Temperature

The microbial growth and the activities are affected by the temperature conditions. The temperature tolerance range may vary from ≤0 to 100 °C≥, and based on which the microbes may be classified into psychrophiles, mesophiles, thermophiles, and extreme thermophiles. The microbial growth rate may be represented using Eq. (17.3):

$$\mu = Ae^{-E/RT} \tag{17.3}$$

where E is activation energy (kcal/mole), A is constant, and t is the absolute temperature (K). The extreme temperature may alter the protein structure, mainly that of enzymes, and may also change the cell permeability. The bioremediation process is highly dependent on the temperature as micro-organisms, responsible for degradation, are active at a definite temperature range (Zekri and Chaalal 2005; Popp et al. 2006) and also affect properties of contaminants as well as the microbial enzymatic activity. Volatilization of short-chain alkanes retard at low temperature, which may increase the solubility of the pollutant in the aqueous phase and their microbial toxicity also increases, which may affect the biodegradation processes.

17.2.3 pH

The solution pH plays a vital role in the activity of any kind of biological molecule. pH is helpful for the ionization of chemicals through which the nutrients may be transported into the cell and it also affects the activity of enzymes, released by the microbes for their physiological activities. The optimum pH for the growth of most of the microbes is 7, although many of the microbes can tolerate a wider range of pH (Pandey et al. 2019). Some of the microbes can grow more effectively at lower or higher pH such as *Sulfolobus*, *Thiobacillus* are acidophilic which can grow optimally at pH lower than 2 pH, fungus preferably grow at pH lower than 5 pH, cyanobacteria prefer to grow at pH more than 7 (Bitton 1994). Many of the microbes, may increase or decrease the pH of the medium where they are growing. So, the optimization of pH must be beneficial in the rate of enzyme production for better degradation is necessary.

17.2.4 Oxygen Level

The microbes can survive both in the presence and absence of oxygen, based on that they may be classified into aerobes, facultative anaerobes, and strict anaerobes. In aerobic microbes, oxygen works as a terminal electron acceptor during respiration, while in anaerobic, sulfates, nitrates, or carbon dioxide may work as electron acceptors. Oxygen generates free radicals such as hydrogen peroxide (H_2O_2), superoxide (O_2^-), or hydroxyl radicals through the reduction process., which is destroyed by the cell with the help of enzymes e.g., O_2^- is deactivated by superoxide dismutase, while H_2O_2 is deactivated by peroxidase and catalase enzymes.

17.2.5 Enzymes

After going through all the factors which are important for microbial growth, it can be realized that the enzymes are a major parameter that is important for both the physiological and bioremediation activities shown by the microbes. There are many types of enzymes, which are secreted by the microbes for achieving different targets. The enzyme secretion is also dependent on environmental conditions. In the presence of contaminants, an adverse environment is created which may inhibit general microbial growth and function. Enzymes are substances that increase the rate of reactions and/or increase the reaction's activation energy without being present in the reaction products. They are naturally produced by nearly every known organism to support processes such as digestion, metabolism, and cell synthesis (Zhu et al. 2017). Most enzymes are made up of proteins, but some enzymes require a no-protein prosthetic group to perform their functions. Proteins are prepared from amino acid polymers associated by peptide bonds and may contain as many as 22 singular types of amino acids. The order of the amino acids in the protein and how the protein is folded determine the function of the enzyme. Each enzyme has defined active sites on the folded surface, and it is at these locations that the enzyme can bind to a substrate and catalyze a reaction. In the process of the reaction, a substrate relates to an active site and forms a temporary bond with the enzyme. This substrate–enzyme complex loosens mainly those bonds which hold the substrate together and allow the bonds to break down. When this reaction is over, substrate pieces are released and catalyze another reaction. Transformation is the rate-controlling step with the substrate enzyme complex to the separated enzyme and products. At the optimal conditions, the substrate saturates the enzyme, which minimizes the reaction time. The velocity of the reaction is always related to the concentration enzyme. The enzymatic activity tends to double for every 10 °C increase between 10 and 40 °C temperature and is highest when the process is running at optimal pH, specific to that enzyme.

Many microbes are being used in bioremediation, based on the targeted contaminants and this work is possible due to the release of different types of enzymes by these microbes. In the environment, the microorganism can break or cleave the carbon-carbon, carbon-hydrogen, carbon-nitrogen bonds, etc. The oxidoreductase and hydrolases are the major class for the enzyme involved in bioremediation Table 17.1. Oxidoreductase is the group of enzymes (oxygenases, laccases, and peroxidases) produced by microorganisms to degrade the contaminants by oxidation involving oxidative coupling of contaminants by electron transfer from oxidants from reluctant and release chloride ions, CO_2, and methanol as a final product. Degradation of pollutants by oxidoreductase generates heat or energy as a result, which is utilized for metabolic activities by microorganisms (Medina et al. 2016). Hydrolase's enzymes are commonly involved in the bioremediation of pesticides, insecticides for their toxicity reduction. Different hydrolytic enzymes disrupted major chemical bonds such as esters, peptide bonds, carbon-halide bonds, etc. hydrolyses are divided into five subcategories i.e., lipases, cellulases,

Table 17.1 The enzyme involved in the bioremediation process

S. No.	Enzyme	Producing microbial sps.	Mode of action	References
1	Oxygenases	*Arthrobacter, Burkholderia, Mycobacterium, Pseudomonas, Sphingomonas, Rhodococcus, Trametes versicolor, Pleurotus ostreatus, Aspergillus, Pseudomonas, Bacillus, Enterobacter, Deinococcus, Shewanella, Agrobacterium, Escherichia, Thermus*	Catalyze oxidation of toxic compounds such as chlorinated biphenyls, aliphatic olefins by adding one or two molecules of oxygen for further transformation and mineralization	Shah et al. (2017)
2	Laccases		Reduce one molecule of oxygen in the water to Cleave ring present in aromatic compounds and produces free radicals	Shraddha et al. (2011), Shraddha et al. (2011), and Bilal et al. (2017)
3	Peroxidases		In the presence of peroxides, catalyze reduction reactions such as hydrogen peroxide (H_2O_2) and after oxidation of organic compounds generate reactive free radicals	Bansal and Kanwar (2013) and Medina et al. (2016)
4	Lipases		Widely used for the treatment of wastewater, polyaromatic hydrocarbon degradation it breaks triglycerol into glycerol and fatty acid, etc.	Shome (2020)
5	Cellulases		It breaks complex cellulosic materials into simple sugars. Commonly used for the treatment of agricultural residues	Bhardwaj et al. (2017)
6	Carboxylesterases		With the addition of water, it catalyzes the hydrolysis of carboxyl ester bonds present in synthetic pesticides	Singh et al. (2012)
7	Phosphotriesterases		Catalyze hydrolysis of organophosphorus i.e., phosphodiester, the main components of organophosphorus compounds worldwide used in pesticides	Santillan et al. (2016)

(continued)

Table 17.1 (continued)

S. No.	Enzyme	Producing microbial sps.	Mode of action	References
8	Haloalkane dehalogenases		Mainly used for biodegradation of halogenated aliphatic compounds	Dvořák et al. (2017)

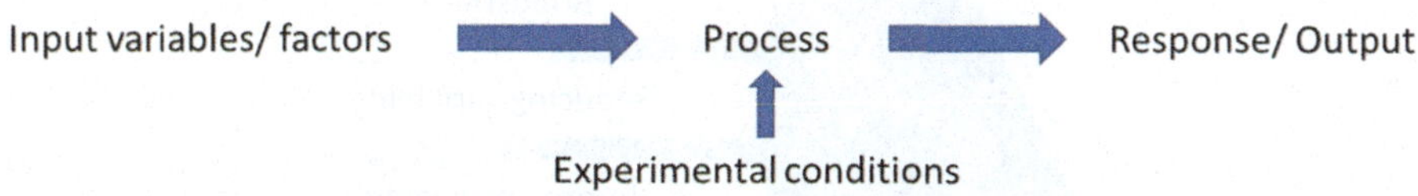

Fig. 17.1 Schematic presentation of DoE

carboxylesterases, phosphotriesterases. Along with the enzymes, the selection of suitable microbes and their inoculum size is also important.

17.3 Process Optimization

The optimization of any process is essential for analyzing the effect of various parameters on the efficiency and effectiveness of the process. In any process, there is an input variable, independent variable (experimental conditions), and output variables (Fig. 17.1).

Generally, the suitable combination of experimental conditions along with the input variable (dependent variable) may affect the output. So, the optimization of these dependent and independent variables (experimental conditions) can influence the whole process. This may be carried out using the statistical design of the experiment (DoE) tool. This may be helpful for the identification of uncontrollable or unknown factors, time and resource requirement may be reduced, so the benefit of the existing process may be explored more appropriately which can be helpful for the establishment of the effect of the selected variable through mathematical models. The number of experiments may also be reduced through DoE. Some of the benefits of process optimization is shown in Fig. 17.2.

There are different modes through which DoE for any process may be set up. The main types of DoE are factorial experiments, screening experiments, and response surface analysis (Fig. 17.3). Full factorial experiments involve the evaluation of the effect of individual parameters as well their interactions, so the number of experiments are quite high. With the increase in conditional parameters/factors along with the levels, the number of experiments increases. So, the statistical analysis becomes complex and time-consuming. The fractional factorial design evaluates the most significant effects and interactions but cannot analyse the effects of all the parameters. In the screening experiments using Plackett-Burman design and Taguchi's

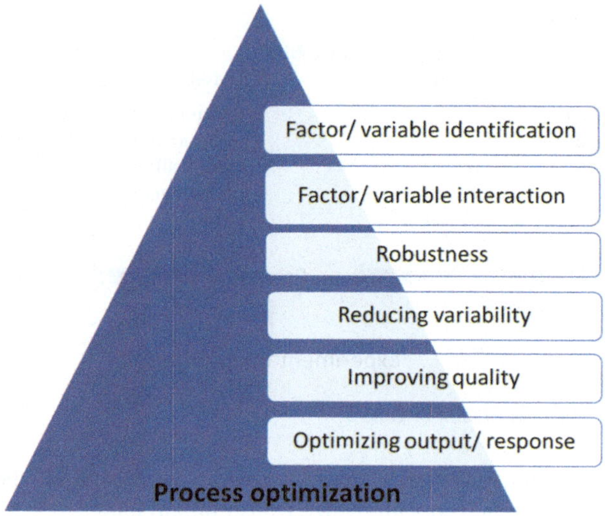

Fig. 17.2 Advantages of following design of experiments

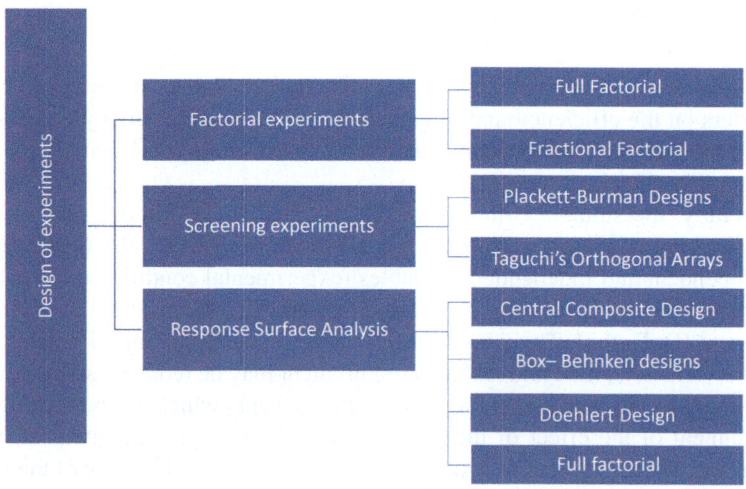

Fig. 17.3 Type of statistical design of experiments

orthogonal arrays method, the effect of critical variables is analysed. In the Response Surface Analysis method, the response of a series of full factorial experiments is noted and analysed. The response is generated in the form of a mathematical equation which is used for analysing the effect of the selected independent variable, which is affecting the final response (Weissman and Anderson 2015). The central composite, Box– Behnken designs, Doehlert Design (DD), and 3-level full factorial designs are frequently used under RSM.

17.4 Bioremediation Process Optimization

Wastewater treatment using microbes (bioremediation) is affected by different environmental conditions including temperature, pH, contaminant concentration, oxygen content in wastewater along with the type and amount of microbial biomass required for treating the contaminants present in wastewater. The optimization of these conditions may increase the potential of the bioremediation process and it may directly affect the cost of the process. Statistical designing has been widely applied in the bioremediation processes, mainly in developmental trials (Benyounis and Olabi 2008). In most lab-based studies, the major task is to optimize the experimental conditions for proper microbial growth as well as to increase their biodegradation potential for targeted contaminants or for treating real wastewater.

Different types of experimental designing methods are being used (Fig. 17.3; Table 17.2). Out of all the methods, the response surface methods (RSM) are used the most. It is a statistical and mathematical technique that is helpful for the prediction of response of bioremediation process at different environmental or experimental conditions with the help of polynomial Eq. (17.4) (Najib et al. 2017).

$$Y = \beta 0 + \Sigma\beta ixi + \Sigma\beta ii + \Sigma\beta ijxixj + \varepsilon \tag{17.4}$$

where, Y is the response variable; $\beta 0$ is the intercept; while βi, βij, and βii are the coefficients of the first-and second-order polynomial equation, binary interactions; xi and xj, are the factor variable (independent), and the model error is defined by ε. The optimum conditions for getting the maximum bioremediation are obtained by following three main steps including statistical designing of experiments, calculation of model coefficients, and then validation of the suitability of the applied models.

Full factorial design (FFD) under RSM can be effective if less than 5 factors are used which are affecting the process. 3 level FFD will have three values (high, middle, low) which can be used for showing the total number of designs for k factors as 3 k where k represents the number of variables (Sakkas et al. 2010). The main concern while using FFD is the difficulty of using second or higher-order polynomial models. Clofibric Acid (CLF), a xenobiotic compound that is persistently present in the aquatic environment and becomes a part of treated water, passing through the water treatment plants without removal. Ungureanu et al. used *Trametes pubescens* for its transformation or removal with the help of properly designing the experiments using Central Composite Design (CCD)-Response Surface Method (Table 17.2). After optimization, it was easier to found the suitable experimental condition doe getting better degradation (pH 5.5, agitation speed 135 rpm, 5 g/L glucose and mineral salts 15 g/L peptone, 3 g/L yeast extract, 2% inoculum, and 14 days duration) where 60% CLF biotransformation was observed (Ungureanu et al. 2020). RSM based on the Box–Behnken design is an experiential statistical technique used to investigate the influence of interactive effects of the selected parameters on wastewater contaminants. The data generated from the designed set of experiments are statistically analysed using the multiple regression method. Khatoon

Table 17.2 The design of experiment method used for microbial remediation of wastewater contaminants

S. No.	DoE type	Microorganism	Wastewater contaminants	Experimental variables	Total number of runs/experiments	References
1	Box Behnken design	*Bacillus badius ABP6*	Atrazine	pH, temperature, agitation speed, and atrazine concentration (3-level)	29	Khatoon and Rai (2020)
2	RSM/optimal design factorial method	*E. coli and Bacillus*	Reactive red 195 and reactive blue dyes	Temperature, incubation time, initial RR195 and RB49 dyes concentration, biomass loading, pH	31	M-Ridha et al. (2020)
4	Central composite design (CCD)	*Proteus mirabilis, Bacillus anthracis, Enterobacter hormaechei, Pseudomonas aeruginosa and Serratia rubidaea*	Aromatic amine 2-ABS	Temperature, pH, and initial 2-ABS concentration	20	Fatima et al. (2020)
5	Central composite design (CCD)	*Enterobacter hormaechei, Serratia rubidaea, Stenotrophomonas acidaminiphila*	Oil sludge	Time, pH, amount of oil sludge, and mass of microbial-assisted bio carrier matrix	30	Swathi et al. (2020)
6	Central composite design (CCD)	*Apergillus niger*	Diazinon	pH temperature (°C), reaction time, and concentration	31	Hamad (2020)
7	Box—Behnken design	*Aspergillus oryzae* M30011	Ochratoxin A	Temperature, pH, and inoculum volume	17	Xiong et al. (2021)
8	Doehlert design	*Serratia marcescens*	Tannin	Temperature, tannic acid concentration, and inoculum	15	de Sena et al. (2020)

10	Central Composite Design (CCD) (5-levels)	*Trametes pubescens*	Clofibric acid	Yeast extract concentration, peptone concentration, inoculum concentration, and the incubation time	30	Ungureanu et al. (2020)
11	Response Surface Methodology (RSM)	*Sphingomonas paucimobilis*, *Pseudomonas putida* and *Lactobacillus acidophilus*	CI Reactive Yellow 174 (CI RY 174)	*Sphingomonas paucimobilis*, *Pseudomonas putida* and *Lactobacillus acidophilus*	10	Ayed et al. (2020)
12	Plackett-Burman Design (PBD) (2-level) and Central Composite Design (5-level)	*Rhodococcus* sp. strain AQ5–14 t	Phenol	Salinity (g l-1), pH and temperature (°C)	12	Tengku-Mazuki et al. (2020)
13	Full factorial design (2-level)	Mixed culture	Phenol and m-cresol	Phenol and m-cresol	7	Saravanan et al. (2008)
14	Response surface methodology (RSM) (2 levels)	*E. coli* and *Bacillus* sp.	RR195 and RB49 by some bacteria like	Occupation time, solution pH, initial Dyes concentrations, biomass loading, and temperature	31	M-Ridha et al. (2020)
15	Response surface method	*Aspergillus niger* MK640786	Diazinon	Diazinon, reaction time, temperature, and pH	31	
16	Factorial design	Biological treatment	Fish canning wastewater	Hydraulic retention time (HRT) and organic matter content	9	Cristóvão et al. (2015)

and Rai had used the RSM-based Box–Behnken design (Table 17.2) for analysing the suitable conditions of pH, temperature, agitation speed and atrazine concentration (at three levels of their values) for degradation of atrazine using *Bacillus badius* ABP6 strain. Through this method, only 29 experiments were required for getting the suitable conditions for getting the best degradation of atrazine using the selected microbes (Khatoon and Rai 2020).

Many studies are being carried out for getting the suitable conditions for removing wastewater contaminants through bioremediation process. The proper optimization of the process can make this process easy to conduct and for analyzing the outputs.

17.5 Conclusion and Future Perspectives

The bioremediation is the process, which is also used by the nature for self-revival where a lot of environmental conditions are controlling the process. The man-made setups require maintaining the same types of conditions so that the biological sources such as microbes can work efficiently for the treatment of environmental pollutants. The optimization process helps screen out the best conditions for maintaining the growth of microbes so that the enzymes required for contaminant degradation can be released and the microbes can work efficiently. So wastewater treatment study should be conducted following the suitable experimental designs so that the optimum conditions of the wastewater treatment can be carried out in the limited resources as well as it may be replicable at pilot and industrial levels.

References

Ayed L, Ladhari N, Achour S, Chaieb K (2020) Decolorization of reactive yellow 174 dye in real textile wastewater by active consortium: experimental factorial design for bioremediation process optimization. J Text Inst:1–11. https://doi.org/10.1080/00405000.2020.1824416

Baker KH, Herson DS (1994) Bioremediation. McGraw-Hill, New York

Bansal N, Kanwar SS (2013) Peroxidase(s) in environment protection. Sci World J 2013. https://doi.org/10.1155/2013/714639

Benyounis KY, Olabi AG (2008) Optimization of different welding processes using statistical and numerical approaches—a reference guide. Adv Eng Softw 39:483–496. https://doi.org/10.1016/j.advengsoft.2007.03.012

Bhardwaj V, Degrassi G, Bhardwaj RK (2017) Bioconversion of cellulosic materials by the action of microbial cellulases. Int Res J Eng Technol 4:494–503

Bhatt P, Bhatt K, Sharma A, Zhang W, Mishra S, Chen S (2021a) Biotechnological basis of microbial consortia for the removal of pesticides from the environment. Crit Rev Biotechnol 41(3):317–338

Bhatt P, Joshi T, Bhatt K, Zhang W, Huang Y, Chen S (2021b) Binding interaction of glyphosate oxidoreductase and C-P lyase: molecular docking and molecular dynamics simulation studies. J Hazard Mater 5(409):124927

Bilal M, Iqbal HMN, Hu H, Wang W, Zhang X (2017) Development of horseradish peroxidase-based cross-linked enzyme aggregates and their environmental exploitation for bioremediation purposes. J Environ Manag 188:137–143. https://doi.org/10.1016/j.jenvman.2016.12.015

Bitton G (1994) Wastewater microbiology. Wiley-Liss, New York

Boopathy R (2000) Factors limiting bioremediation technologies. Bioresour Technol 74:63–67. https://doi.org/10.1016/S0960-8524(99)00144-3

Cookson JT Jr (1995) Bioremediation engineering: design and application. McGraw-Hill, New York

Cristóvão RO, Gonçalves C, Botelho CM, Martins RJE, Loureiro JM, Boaventura RAR (2015) Fish canning wastewater treatment by activated sludge: application of factorial design optimization. Biological treatment by activated sludge of fish canning wastewater. Water Resour Ind 10:29–38. https://doi.org/10.1016/j.wri.2015.03.001

Dangi AK, Sharma B, Hill RT, Shukla P (2019) Bioremediation through microbes: systems biology and metabolic engineering approach. Crit Rev Biotechnol 39:79–98. https://doi.org/10.1080/07388551.2018.1500997

Dash B, Sahu N, Singh AK, Gupta SB, Soni R (2021) Arsenic efflux in Enterobacter cloacae RSN3 isolated from arsenic-rich soil. Folia Microbiol 66:189–196

de Sena AR, de Melo Lopes LM, Gouveia MJ, Gouveia MJ, de Mello MRF, Leite TCC, de Souza Lima GM, Moreira KA, de Assis SA (2020) Tannin biodegradation by tannase from Serratia marcescens: optimization of production by response surface methodology and its partial characterization. Chem Eng Commun:1–16. https://doi.org/10.1080/00986445.2020.1780212

Debbarma P, Raghuwanshi S, Singh J, Suyal DC, Zaidi MGH, Goel R (2017) Comparative in situ biodegradation studies of polyhydroxybutyrate film composites. 3Biotech 7(178):1–9. https://doi.org/10.1007/s13205-017-0789-3

Dvořák P, Nikel PI, Damborský J, de Lorenzo V (2017) Bioremediation 3.0: engineering pollutant-removing bacteria in the times of systemic biology. Biotechnol Adv 35:845–866. https://doi.org/10.1016/j.biotechadv.2017.08.001

Fatima M, Saeed M, Aslam M, Lindström RW, Farooq R (2020) Application of novel bacterial consortium for biodegradation of aromatic amine 2-ABS using response surface methodology. J Microbiol Methods 174:105941. https://doi.org/10.1016/j.mimet.2020.105941

Giri K, Rai JPN, Pandey S, Mishra G, Kumar R, Suyal DC (2017a) Performance evaluation of isoproturon-degrading indigenous bacterial isolates in soil microcosm. Chem Ecol 33 (9):817–825. https://doi.org/10.1080/02757540.2017.1393535

Giri K, Suyal DC, Mishra G, Pandey S, Kumar R, Meena DK, Rai JPN (2017b) Biodegradation of isoproturon by *Bacillus pumilus* K1 isolated from foothill agroecosystem of North West Himalaya. Proc Natl Acad Sci India Sect B Biol Sci 87(3):839–848. https://doi.org/10.1007/s40011-015-0667-x

Gustavsson M, Hörnström D, Lundh S, Belotserkovsky J, Larsson G (2016) Biocatalysis on the surface of Escherichia coli: melanin pigmentation of the cell exterior. Sci Rep 6:1–9. https://doi.org/10.1038/srep36117

Hamad MTMH (2020) Biodegradation of diazinon by fungal strain Apergillus niger MK640786 using response surface methodology. Environ Technol Innov 18:100691. https://doi.org/10.1016/j.eti.2020.100691

Junghare M, Spiteller D, Schink B (2019) Anaerobic degradation of xenobiotic isophthalate by the fermenting bacterium Syntrophorhabdus aromaticivorans. ISME J 13:1252–1268. https://doi.org/10.1038/s41396-019-0348-5

Khatoon H, Rai JPN (2020) Optimization studies on biodegradation of atrazine by Bacillus badius ABP6 strain using response surface methodology. Biotechnol Rep 26:e00459. https://doi.org/10.1016/j.btre.2020.e00459

Kour D, Kaur T, Devi R, Yadav A, Singh M et al (2021) Beneficial microbiomes for bioremediation of diverse contaminated environments for environmental sustainability: present status and future challenges. Environ Sci Pollut Res 28:24917–24939

Kumar P, Dash B, Suyal DC, Gupta SB, Singh AK, Chowdhury T, Soni R (2021) Characterization of arsenic-resistant Klebsiella pneumoniae RnASA11 from contaminated soil and water samples and its bioremediation potential. Curr Microbiol. https://doi.org/10.1007/s00284-021-02602-w

Liu Q, Li Q, Wang N, Liu D, Zan L, Chang L, Gou X, Wang P (2018) Bioremediation of petroleum-contaminated soil using aged refuse from landfills. Waste Manag 77:576–585. https://doi.org/10.1016/j.wasman.2018.05.010

Medina JDC, Woiciechowski AL, Guimarães LRC, Karp SG, Soccol CR (2016) Peroxidases. Curr Dev Biotechnol Bioeng Prod Isol Purif Ind Prod:217–232. https://doi.org/10.1016/B978-0-444-63662-1.00010-5

M-Ridha MJ, Hussein SI, Alismaeel ZT, Atiya MA, Aziz GM (2020) Biodegradation of reactive dyes by some bacteria using response surface methodology as an optimization technique. Alexandria Eng J 59:3551–3563. https://doi.org/10.1016/j.aej.2020.06.001

Najib T, Solgi M, Farazmand A, Heydarian SM, Nasernejad B (2017) Optimization of sulfate removal by sulfate reducing bacteria using response surface methodology and heavy metal removal in a sulfidogenic UASB reactor. J Environ Chem Eng 5:3256–3265. https://doi.org/10.1016/j.jece.2017.06.016

Pandey A, Jain R, Sharma A, Dhakar K, Kaira GS, Rahi P, Dhyani A, Pandey N, Adhikari P, Shouche YS (2019) 16S rRNA gene sequencing and MALDI-TOF mass spectrometry based comparative assessment and bioprospection of psychrotolerant bacteria isolated from high altitudes under mountain ecosystem. SN Appl Sci 1:278. https://doi.org/10.1007/s42452-019-0273-2

Popp N, Schlömann M, Mau M (2006) Bacterial diversity in the active stage of a bioremediation system for mineral oil hydrocarbon-contaminated soils. Microbiology 152:3291–3304. https://doi.org/10.1099/mic.0.29054-0

Sakkas VA, Islam MA, Stalikas C, Albanis TA (2010) Photocatalytic degradation using design of experiments: a review and example of the Congo red degradation. J Hazard Mater 175:33–44. https://doi.org/10.1016/j.jhazmat.2009.10.050

Santillan JY, Dettorre LA, Lewkowicz ES, Iribarren AM (2016) New and highly active microbial phosphotriesterase sources. FEMS Microbiol Lett Adv 363:1–38. https://doi.org/10.1093/femsle/fnw276

Saravanan P, Pakshirajan K, Sana PK (2008) Kinetics of growth and multi substrate degradation by an indigenous mixed microbial culture isolated from a wastewater treatment plant in Guwahati, India. Water Sci Technol 58:1101–1106. https://doi.org/10.2166/wst.2008.463

Shah PC, Kumar VR, Dastager SG, Khire JM (2017) Phytase production by Aspergillus niger NCIM 563 for a novel application to degrade organophosphorus pesticides. AMB Express 7. https://doi.org/10.1186/s13568-017-0370-9

Sharma B, Dangi AK, Shukla P (2018) Contemporary enzyme based technologies for bioremediation: a review. J Environ Manag 210:10–22. https://doi.org/10.1016/j.jenvman.2017.12.075

Shome R (2020) Role of microbial enzymes in bioremediation. eLifePress 1:15–20

Shraddha SR, Sehgal S, Kamthania M, Kumar A (2011) Laccase: microbial sources, production, purification, and potential biotechnological applications. Enzyme Res 2011:1–11. https://doi.org/10.4061/2011/217861

Singh B, Kaur J, Singh K (2012) Biodegradation of malathion by Brevibacillus sp. strain KB2 and Bacillus cereus strain PU. World J Microbiol Biotechnol 28:1133–1141. https://doi.org/10.1007/s11274-011-0916-y

Singh R, Singh P, Sharma R (2014) Microorganism as a tool of bioremediation technology for cleaning environment: a review. Proc Int Acad Ecol Environ Sci 4:1–6

Singh M, Singh D, Rai P, Suyal DC, Saurabh S, Soni R, Giri K, Yadav AN (2021) Fungi in remediation of hazardous wastes: current status and future. In: Yadav AN (ed) Recent trends in mycological research, fungal biology. Springer Nature, Switzerland

Swathi KV, Muneeswari R, Ramani K, Sekaran G (2020) Biodegradation of petroleum refining industry oil sludge by microbial-assisted biocarrier matrix: process optimization using response surface methodology. Biodegradation 31:385–405. https://doi.org/10.1007/s10532-020-09916-9

Tengku-Mazuki TA, Subramaniam K, Zakaria NN, Convey P, Abdul Khalil K, Lee GLY, Zulkharnain A, Shaharuddin NA, Ahmad SA (2020) Optimization of phenol degradation by Antarctic bacterium Rhodococcus sp. Antarct Sci 32:486–495. https://doi.org/10.1017/S0954102020000358

Ungureanu CV, Favier L, Bahrim GE (2020) Improving biodegradation of clofibric acid by trametes pubescens through the design of experimental tools. Microorganisms 8:1–16. https://doi.org/10.3390/microorganisms8081243

Weissman SA, Anderson NG (2015) Design of Experiments (DoE) and process optimization. A review of recent publications. Org Process Res Dev 19:1605–1633. https://doi.org/10.1021/op500169m

Wu M, Li W, Dick WA, Ye X, Chen K, Kost D, Chen L (2017) Bioremediation of hydrocarbon degradation in a petroleum-contaminated soil and microbial population and activity determination. Chemosphere 169:124–130. https://doi.org/10.1016/j.chemosphere.2016.11.059

Xiong K, Wei ZH, Yun LJ, Yi WX, Yao ZZ, Gang PP, Deng L, Yue XS (2021) Detoxification of Ochratoxin A by a novel Aspergillus oryzae strain and optimization of its biodegradation. Rev Argent Microbiol 53:48–58. https://doi.org/10.1016/j.ram.2020.06.001

Zekri AY, Chaalal O (2005) Effect of temperature on biodegradation of crude oil. Energy Sources 27:233–244. https://doi.org/10.1080/00908310490448299

Zhu YG, Xue XM, Kappler A, Rosen BP, Meharg AA (2017) Linking genes to microbial biogeochemical cycling: lessons from arsenic. Environ Sci Technol 51:7326–7339. https://doi.org/10.1021/acs.est.7b00689

Chapter 18
Advanced Molecular Technologies for Environmental Restoration and Sustainability

Saurabh Gangola, Samiksha Joshi, Divya Joshi, Jyoti Rajwar, Shilippreet Kour, Jyoti Singh, and Saurabh Kumar

18.1 Introduction

Since the earliest identification of living beings or microbes, researchers have been developing systematic categorization methods in the field of evolution and phylogeny. This gets more problematic in the context of bacteria, the most common kind of microbe. Bacteria reproduce asexually, which explains the traditional concept of species as a collection of organisms capable of interbreeding and procreating fertile offspring is not universally applicable. Additionally, bacteria's tiny size contributes to their restricted variety of morphological features. Bacteria show a broad variety of biochemical variation in terms of cell structure and metabolism, and although this

S. Gangola
School of Agriculture, Graphic Era Hill University, Bhimtal, Uttarakhand, India

S. Joshi
Department of Biosciences, Shri Ram Group of Colleges, Muzaffarnagar, Uttar Pradesh, India

D. Joshi (✉)
Uttarakhand Pollution Control Board, Regional Office, Kashipur, Uttarakhand, India

J. Rajwar
Department of Microbiology, School of Science and Information Technology, Skyline University, Kano, Nigeria

S. Kour
Department of Microbiology, Akal College of Basic Sciences, Eternal University, Sirmaur, Himachal Pradesh, India

J. Singh
Department of Microbiology, G.B. Pant University of Agriculture and Technology, Pantnagar, Uttarakhand, India

S. Kumar
Division of Crop Research, Research Complex for Eastern Region, Patna, Bihar, India

© The Author(s), under exclusive license to Springer Nature Switzerland AG 2022

D. C. Suyal, R. Soni (eds.), *Bioremediation of Environmental Pollutants*,
https://doi.org/10.1007/978-3-030-86169-8_18

provides some background knowledge about their taxonomy, it is far from comprehensive. With the advent of molecular biology, a new revolution has occurred, and this new revolution has made major contributions to bacterial taxonomy and systematics, as well as to other fields of biological taxonomy.

In the 1970s, Carl Woese proposed a classification system based on the molecular comparisons of evolutionarily conserved ribosomal genes and segregated two domains, Bacteria and Archaea as they are different from the Eukaryotes (contains all the higher forms of the organism) (Woese 1987). Now the ribosomal based classification system is widely accepted by the microbiologist across the world. Still the bacterial systematic is still evolving and also the standardized concept of bacterial species constituents (Berg et al. 2020). However, molecular-based systematics has given a strong outline for designing classification schemes.

Despite the lack of a clear logical and precise definition of species, traditional techniques continue to be used in a wide variety of fields or sectors. However, modern molecular methods (genomics and proteomics) provide superior characterisation than traditional techniques. They rapidly produce multidirectional information on both microbial communities and their taxonomic relationships.

In most cases, bacteria can only be identified using restricted approaches that rely on genetic techniques that use microorganism genetic profiling and phenotypic techniques that use metabolic characteristics and chemical composition of the organism to identify the microbes in question. The benefits of genotypic methods above phenotypic techniques are not influenced by the physiological condition, medium composition, or growth phase of the organisms. The phenotypic techniques, on the other hand, revealed the functional features of organisms, such as metabolic processes, that are required for their survival, growth, and development.

18.2 Genomic Methods

These methods are based on the analysis of the genome that is represented by the haploid set of genes or chromosomes within an organism. It may be classified as structural genomics and functional genomics (Wang et al. 2020; Raghu et al. 2021; Soni et al. 2021). Structural genotyping involves the gene location, sequence, and physical characterization; while, functional genotyping involves gene regulation and protein expression (Soni et al. 2016; Suyal et al. 2019a). Moreover, the combinations of various "meta-" and "-omics" technologies have made it beneficial to humankind especially in environmental, medical, industrial, and agricultural fields (Suyal et al. 2015a, b, 2019c).

For the identification, genotypic methods are classified into two distinctive categories: (1) pattern- or fingerprint-based techniques and (2) sequence-based techniques. Pattern-based methods make use of a systematic process that generates a sequence of fragments from the chromosomal DNA of the organism being studied. The fragments are then segregated based on the size which generates a profile (or fingerprint) unique to that organism and its close relatives. Then, using the information gathered, researchers can create a database of the particular organism (fingerprint), which can then be used as a reference for the test organisms to compare against (Emerson et al. 2008). If two profiles of different organisms' match, they can be viewed as a close relative of each other, particularly at the level of strain or species. On the other hand, sequence-based techniques usually depend on the specific stretch of DNA or chromosome but do not always like the specific gene. In general, the approach is similar to the genotyping method: a specific sequence of DNA database created and then the test organism's sequences compared with it. The degree of homology or similarity or matched sequences between the compared organisms is an estimate of how closely linked the organisms are among the compared organisms. Several computer-based algorithms have been created to build the phylogenetic tree in which we can compare the multiple sequences of different organisms to one another at a time. Thus, by making use of sequence comparisons of ribosomal RNA (rRNA) gene, archaea and bacteria can be easily distinguished as separate branches or having different relationships among the microorganisms (Raina et al. 2019). Both the techniques discussed above have merits and demerits. Conventionally, for the establishment of phylogenetic relationship among the bacteria at phylum, order, family, genus level, 16S rRNA gene sequence was analysed whereas for the establishment of relatedness at the species level or genus level the fingerprinting-based methods are good but less dependable above those levels (Vandamme et al. 1996). Fingerprinting and sequence-based methods combined with phenotypic characters is called polyphasic technique, is the standard approach nowadays to describe a new species or genus (Carro and Nouioui 2017).

18.3 Specific Genotyping Methodologies

The current techniques for characterisation may make use of a variety of fingerprinting or sequence-based methods, which may be employed either individually or in combination. These methods are continuously evolving and improving in terms of accuracy. Some of the most frequently used methods are listed below.

18.3.1 Fingerprinting-Based Methodologies

Among the genotypic methods, fingerprinting techniques are the most widely used presently. Techniques like Amplified fragment length polymorphism (AFLP), repetitive element PCR (rep-PCR), and random amplification of polymorphic DNA, utilize PCR for amplification of desired short DNA fragments by using specific primer sets (Sharma et al. 2020). These methods use the advantages of polymorphism in the DNA of the concerned organism which might be formed from the evolutionary process. A unique set of primer is used for more than one organism in the multiplex PCR; based on the molecular weight of amplicon (size) these sets can be separated through electrophoresis. They enable the fast identification of many microorganisms from a single sample combination (Settanni and Corsetti 2007).

Riboprinting utilizes sensitive probes instead of PCR to detect the difference in gene sequence or pattern between species and strain (Bruce 1996). It is one of several molecular methods that generates comparative data which is independent of the complexity of the morphology of the organisms. Diversi Lab system for rep-PCR (http://biomerieux-usa.com/diversilab) (Dou et al. 2015) and DuPont's Ribo-Printer system (www2.dupont.com/Qualicon/en_US/) (Shintani 2013) have been exclusively developed commercial products bacterial identification. All the techniques discussed here are already mentioned in many kinds of literature as identification methods. These applications include source tracing, authentication of bacterial isolates for archiving reasons, taxonomy and systematics, as well as the identification of microbial population patterns, among other things.

18.3.2 Sequence-Based Methodologies

As the housekeeping genes are conserved and present universally in all the cells, the primer can be designed for the amplification of similar genes across the multiple genera. Multilocus sequencing (MLS) is a promising method developed to identify microbiological species. The principle is almost similar to 16S rRNA gene sequencing, but fragments of multiple "housekeeping" genes are sequenced. Later the combined sequences are put into one long sequence which is then compared with other sequences.

Since designing universal sets of primers is not possible, designing specific sets of primers for families or orders is a good concept. Two multilocus sequencing approaches used are multilocus sequence typing (MLST) and multilocus sequence analysis (MLSA). In MLST, a set of primers are used according to 6–10 genetic loci which allow the PCR amplification (amplicon size 400–600 bp). The concatenated sequences are then compared with the existing sequence database for the same organism. The result exhibits a very strong identification of a particular strain and showed a very close evolutionary relationship (Huebner et al. 2021). Among other things, this method may be used to monitor the spread of a disease and to demonstrate their usefulness in epidemiological research (Pérez-Losada et al. 2011). On the other hand, MLSA involves sequencing of multiple fragments of conserved protein-encoding genes, but with the more ad-hoc approach for gene selection for comparative analysis as it uses identification using a small subset of genes or loci (Glaeser and Kämpfer 2015). It identifies the organisms and finds relationships of species within genera of families in detail. One of the major limitations of this approach is the lack of standardization and central databases. Recently, several studies conducted with MLSA showed that instead of using a single common gene, different sets of genes were used for the identification of various bacterial phyla (Glaeser and

Kämpfer 2015; Palmer et al. 2018). Hence comparative analysis is impossible with this technique.

18.4 The Genomic Future

Whole-genome comparisons have proven to be more accurate and precise than DNA-DNA hybridization and has gained more popularity over the phenotypic traits concept for bacterial classification and identification. At the moment, the notion of "species" is defined by digital whole genome comparisons utilising average nucleotide identities (ANIs) or genome-to-genome distance computations (GGDCs). Since the advent of whole genome sequencing, phylogenomics has made significant contributions to the field of contemporary taxonomy (Lalucat et al. 2020). Complete genome comparisons identify a species at the genomic level based on 95% average nucleotide identity between two related strains (Olm et al. 2020). The advanced technology of next-generation sequencing has provided more rapid, economical, and easily available sequence-based methods for the identification and classification of bacteria at all levels. Another promising approach for the identification and characterization of microbes at the community level is Microarray. It works by probing several genes on a substrate (example: glass, silicon, nylon, etc.) and further hybridizing with DNA or RNA samples (Solieri et al. 2013). For rapid detection of hybridized samples with probes, fluorescent reporter molecules are used as markers on the microarray. In addition, use of microarray is of great importance for medical purposes such as disease diagnosis and pathogen identification (Herrera-Rodriguez et al. 2013). Some modifications such as phylochips are used to identify specific or various groups of bacteria directly from the environment samples and geochips for the identification of microbes responsible for biogeochemical processes (Liu et al. 2021).

18.5 Proteomics Technologies in Bacterial Identification and Characterization

Genotypic and phenotypic methods are not enough to understand the physiological and functional activities of an organism at the protein level. Proteomics a new approach is a rapid way to explore biomolecules and understand their activity. It is based on mass spectrometry and provides an integrative study of genotypic and proteomic data with all the vital information (Suyal et al. 2018, 2019b). Several of the most widely used technologies include electrospray ionisation mass spectrometry (ESI-MS), matrix-assisted laser desorption/ionization time-of-flight mass spectrometry (MALDI-TOF-MS), one- or two-dimensional sodium dodecyl sulfate-polyacrylamide gel electrophoresis (SDS-PAGE), and surface-enhanced laser desorption/ionization (SELDI) mass spectrometry, etc.

18.5.1 Mass Spectrometry-Based Bacterial Characterization and Identification

Thomson has invented mass spectrometry to determine the mass to charge ratio of electrons in the late nineteenth century. The method is used to identify, quantify, and deduce the structure of a wide range of molecules (Baghel et al. 2017). The twentieth century saw an expansion of technology and its applications to chemical characterization, physical measurement, and biological identification.

Some soft ionization methods in mass spectrometry such as ESI-MS and MALDI-TOF-MS have made it easier to analyze larger molecules. It allows direct use of samples in their native form for interrogation (Fenn et al. 1989). MS has shown better outcomes than traditional approaches: in resolving the time constraint and generating protein profiles. Applications of these abovementioned techniques for identification and characterization are described below:

18.5.1.1 Matrix-Assisted Laser Desorption/Ionization Time-of-Flight Mass Spectrometry

The mass spectral method gives: detailed overview of whole bacterial cells, spectrum patterns over a broad mass range, and identification and characterization are done by comparing with reference data. Initially, application of MALDI-TOF-MS in rapid identification of whole bacteria was shown by Holland et al. (1996) after which various strains such as *Mycobacteria* sp. (Pignone et al. 2006), *Staphylococcus* sp. (Edwards-Jones et al. 2000), and extremophilic bacteria and archaea (Krader and Emerson 2004) have been analyzed using the same.

One of the most famous examples was during the first outbreak of methicillin-resistant *S. aureus* (MRSA) in European Hospitals (1960). The threat of spreading resistant *S. aureus* urgently required some rapid identification method. Edward-Jones et al. (2000) developed matrix-assisted laser desorption/ionization time-of-flight mass spectrometry for the same purpose as well as for the differentiation of methicillin-sensitive Staphylococcus aureus (MSSA) from methicillin-resistant Staphylococcus aureus (MRSA). The procedure involves smear preparation of a single bacterial colony on slide followed by applying matrix which is then observed using MALDI-TOF-MS. The analysis shows distinct spectral peaks for MRSA and MSSA. Based on this technique, several other instruments such as Bruker Daltonics' MALDI BioTyper equipped with bioinformatics tools were developed. It also serves the same function by targeting some ribosomal proteins and proteins found in high amounts (Mellmann et al. 2008).

18.5.1.2 Electrospray Ionization Mass Spectrometry

ESI-MS is a potential approach for the characterization and analysis of various cellular components in microbes. It is considered more accurate in protein identification than MALDI-TOF-MS (as it is based on only molecular weight). ESI-MS uses peptide fragmentation fingerprints to search in the database and identifies the specific protein. The fingerprint is obtained by tandem mass spectrometry, in which target protein can be fragmented for second mass analysis. Modified approach, developed by integrating PCR with ESI-MS introduced Ibis Biosciences (the T5000 Biosensor System) for identification and characterization of bacteria (Sampath et al. 2007). Few major advantages associated with ESI-MS are rapid and fast process (Banerjee and Mazumdar 2012) provides specific identification of target bacteria in mixed culture; high resolution; and identification of virulence factors (Ho and Reddy 2011).

18.5.1.3 Surface-Enhanced Laser Desorption/Ionization

SELDI is a relatively new technique that separates proteins based on their binding affinity to a chip surface. Chemically and biologically modified chips are used for mass spectrometric analysis of complex protein mixtures. SELDI-MS generates a unique spectra pattern for proteins in the mixture based on their mass-to-charge ratio (You et al. 2013). Furthermore, different proteins can be identified from these profiles by comparing the respective peak intensities (Lu et al. 2010). Lundquist et al. (2005) demonstrated that SELDI-TOF-MS is one of the potential methods to produce distinct and reproducible protein profiles for the identification and discrimination of different species. For example, it made it possible to identify and distinguish the most infectious subspecies of *Francisella tularensis* out of four, the only subspecies found in North America causing tularemia in humans. This technique is used in the identification and characterization of bacteria, exploring bacterial proteomes, pathogen detection (Ho and Reddy 2011; Ardito et al. 2016), virulence factor identification, biomarker, and protein profiling in oncology (Langbein et al. 2006; Liu 2011), etc.

Although mass spectrometry plays a great role in the identification and characterization of bacteria by generating spectral patterns various factors cause difficulty in the reproducibility of protein profiles. Factors associated are physiological state of cell (García-Flores et al. 2012), growth medium of the cell (Wieme et al. 2014), sample preparation, the difference in instrument quality, and matrix selection (Wunschel et al. 2005; Vats et al. 2016). Scientists resolved this issue by introducing standard techniques for MALDI-TOF-MS of whole cells (Strejcek et al. 2018).

18.5.2 Gel-Based Method

SDS-PAGE is a widely used method for differentiating bacteria based on their protein contents. It separates the entire protein complement based on their charge and molecular weight. The difference in mobility of charged molecules leads to different migration patterns of proteins. This unique pattern helps to differentiate and characterize the variety of bacterial strains. It is considered promising fractionation technique and provides good resolution for proteins based on sizes, isoelectric points, and hydrophobic behaviour (Carruthers et al. 2015). The drawback of this approach is that it is time-consuming and tedious.

18.5.2.1 Two-Dimensional Gel Electrophoresis (2DE)

Combining SDS-PAGE with isoelectric focusing (IEF)—lead to the development of a new high-resolution technique named 2DE discovered by O'Farrell in 1975. It is capable of separating complicated protein mixtures in a single gel analysis. Two-dimensional gel electrophoresis begins with initial segregation on the basis of pH gradient associated with the isoelectric point of the proteins in the first dimension, followed by SDS-PAGE separation in the second dimension. Further staining of a gel with standard staining solutions for visualization of protein spots and analysis of protein gel patterns or 2DE maps (Soni et al. 2015; Kendrick et al. 2019). These patterns can be studied further and stored in reference databases for future use. It is often used for isolating and analyzing target protein from complex protein mixtures, and identification of unknown species by comparing differential expression 2DE maps, with a reference database. To obtain more efficient and complex proteome analysis 2DE is merged with mass spectrometry. Numerous reports demonstrated that this combined approach can be used to study the entire proteome or subproteome of a variety of species, including the exosporium of *Bacillus anthracis* spores (Redmond et al. 2004), *Bacillus subtilis*, *Helicobacter pylori*, *E. coli*, *Pseudomonas aeruginosa*, and *Staphylococcus aureus* (Hecker et al. 2003; Peng et al. 2005; Pieper et al. 2006). Databases with complete information of 2DE maps and mass spectra of known bacteria will provide rapid identification and efficient comparative study of unknown bacteria (Curreem et al. 2012). However, building such a database is generally a very tedious job.

18.6 Databases

To generate, archive, process and integrate large data sets of many samples with robust quality is a real challenge for both the (genomic and proteomic) approaches. There are a variety of databases and tools available which provide integrated data for the particular type of analysis. For genomic analysis databases based on 16SrRNA genes include green genes (DeSantis et al. 2006) and Ribosomal Database Project (Cole et al. 2009). On the other side, in-depth data analysis of proteomics has been carried out with tools like GlycoMod and databases such as Phospho Site (Gasteiger et al. 2003). Moreover, new algorithms have been developed that adapt to actual experimental phenomenon and parameters (Lees et al. 2016; San et al. 2020). Wilke et al. (2003) introduced the ProDB platform which provided enriched protein profile along with experimental set-up and parameters, like growth and culture conditions to check impact generation on mass-spectra profile.

18.7 Conclusion

The use of molecular technologies is at the core of the identification and characterisation of microorganisms. However, there are certain problems that need to be addressed, such as the functional knowledge of related instruments, their mobility, cost-effectiveness, and accessibility, among others. It will undoubtedly inspire students, researchers, and the scientific community to use a variety of technologies in order to achieve environmental sustainability.

References

Ardito F, Perrone D, Cocchi R, Lo Russo L, De Lillo A, Giannatempo G, Lo Muzio L (2016) Novel possibilities in the study of the salivary proteomic profile using SELDI-TOF/MS technology. Oncol Lett 11(3):1967–1972

Baghel US, Singh A, Singh D, Sinha M (2017) Application of mass spectroscopy in pharmaceutical and biomedical analysis. In: Spectroscopic analyses: developments and applications. IntechOpen, London, p 105

Banerjee S, Mazumdar S (2012) Electrospray ionization mass spectrometry: a technique to access the information beyond the molecular weight of the analyte. Int J Anal Chem 2012:282574

Berg G, Rybakova D, Fischer D, Cernava T, Vergès MCC, Charles T et al (2020) Microbiome definition re-visited: old concepts and new challenges. Microbiome 8(1):1–22

Bruce J (1996) Automated system rapidly identifies and characterizes microorganisms in food. Food Technol (Chicago) 50(1):77–81

Carro L, Nouioui I (2017) Taxonomy and systematics of plant probiotic bacteria in the genomic era. AIMS Microbiol 3(3):383

Carruthers NJ, Parker GC, Gratsch T, Caruso JA, Stemmer PM (2015) Protein mobility shifts contribute to gel electrophoresis liquid chromatography analysis. J Biomol Tech 26(3):103

Cole JR, Wang Q, Cardenas E, Fish J, Chai B, Farris RJ et al (2009) The Ribosomal Database Project: improved alignments and new tools for rRNA analysis. Nucleic Acids Res 37(suppl_1): D141–D145

Curreem SO, Watt RM, Lau SK, Woo PC (2012) Two-dimensional gel electrophoresis in bacterial proteomics. Protein Cell 3(5):346–363

DeSantis TZ, Hugenholtz P, Larsen N, Rojas M, Brodie EL, Keller K et al (2006) Greengenes, a chimera-checked 16S rRNA gene database and workbench compatible with ARB. Appl Environ Microbiol 72(7):5069–5072

Dou H, Xu Y, Li T (2015) Application of the DiversiLab system for tracing the source of the mixed infections caused by Cryptococcus neoformans var. grubii from a patient with systemic lupus erythematosus. Mycoses 58(3):149–159

Edwards-Jones V, Claydon MA, Evason DJ, Walker J, Fox AJ, Gordon DB (2000) Rapid discrimination between methicillin-sensitive and methicillin-resistant Staphylococcus aureus by intact cell mass spectrometry. J Med Microbiol 49(3):295–300

Emerson D, Agulto L, Liu H, Liu L (2008) Identifying and characterizing bacteria in an era of genomics and proteomics. Bioscience 58(10):925–936

Fenn JB, Mann M, Meng CK, Wong SF, Whitehouse CM (1989) Electrospray ionization for mass spectrometry of large biomolecules. Science 246(4926):64–71

García-Flores M, Juárez-Colunga S, Montero-Vargas JM, López-Arciniega JAI, Chagolla A, Tiessen A, Winkler R (2012) Evaluating the physiological state of maize (Zea mays L.) plants by direct-injection electrospray mass spectrometry (DIESI-MS). Mol BioSyst 8(6):1658–1660

Gasteiger E, Gattiker A, Hoogland C, Ivanyi I, Appel RD, Bairoch A (2003) ExPASy: the proteomics server for in-depth protein knowledge and analysis. Nucleic Acids Res 31 (13):3784–3788

Glaeser SP, Kämpfer P (2015) Multilocus sequence analysis (MLSA) in prokaryotic taxonomy. Syst Appl Microbiol 38(4):237–245

Hecker M, Engelmann S, Cordwell SJ (2003) Proteomics of Staphylococcus aureus—current state and future challenges. J Chromatogr B 787(1):179–195

Herrera-Rodriguez SE, Elizondo-Quiroga D, Alvarez-Maya I (2013) Infectious diseases detection by microarray: an overview of clinical relevant infections. J Biomed Sci Eng 2013

Ho YP, Reddy PM (2011) Advances in mass spectrometry for the identification of pathogens. Mass Spectrom Rev 30(6):1203–1224

Holland RD, Wilkes JG, Rafii F, Sutherland JB, Persons CC, Voorhees KJ, Lay JO Jr (1996) Rapid identification of intact whole bacteria based on spectral patterns using matrix-assisted laser desorption/ionization with time-of-flight mass spectrometry. Rapid Commun Mass Spectrom 10 (10):1227–1232

Huebner R, Mugabi R, Hetesy G, Fox L, De Vliegher S, De Visscher A et al (2021) Characterization of genetic diversity and population structure within Staphylococcus chromogenes by multilocus sequence typing. PLoS One 16(3):e0243688

Kendrick N, Darie CC, Hoelter M, Powers G, Johansen J (2019) 2D SDS PAGE in combination with Western blotting and mass spectrometry is a robust method for protein analysis with many applications. In: Advancements of mass spectrometry in biomedical research. Springer, Cham, pp 563–574

Krader P, Emerson D (2004) Identification of archaea and some extremophilic bacteria using matrix-assisted laser desorption/ionization time-of-flight (MALDI-TOF) mass spectrometry. Extremophiles 8(4):259–268

Lalucat J, Mulet M, Gomila M, García-Valdés E (2020) Genomics in bacterial taxonomy: impact on the genus Pseudomonas. Genes 11(2):139

Langbein S, Lehmann J, Harder A, Steidler A, Michel MS, Alken P, Badawi JK (2006) Protein profiling of bladder cancer using the 2D-PAGE and SELDI-TOF-MS technique. Technol Cancer Res Treat 5(1):67–71

Lees JA, Vehkala M, Välimäki N, Harris SR, Chewapreecha C, Croucher NJ et al (2016) Sequence element enrichment analysis to determine the genetic basis of bacterial phenotypes. Nat Commun 7(1):1–8

Liu C (2011) The application of SELDI-TOF-MS in clinical diagnosis of cancers. J Biomed Biotechnol 2011

Liu D, He X, Chater CC, Perez-Moreno J, Yu F (2021) Microbiome community structure and functional gene partitioning in different micro-niches within a sporocarp-forming fungus. Front Microbiol 12:629352

Lu HB, Zhou JH, Ma YY, Lu HL, Tang YL, Zhang QY, Zhao CH (2010) Five serum proteins identified using SELDI-TOF-MS as potential biomarkers of gastric cancer. Jpn J Clin Oncol 40 (4):336–342

Lundquist M, Caspersen MB, Wikström P, Forsman M (2005) Discrimination of Francisella tularensis subspecies using surface enhanced laser desorption ionization mass spectrometry and multivariate data analysis. FEMS Microbiol Lett 243(1):303–310

Mellmann A, Cloud J, Maier T, Keckevoet U, Ramminger I, Iwen P et al (2008) Evaluation of matrix-assisted laser desorption ionization-time-of-flight mass spectrometry in comparison to 16S rRNA gene sequencing for species identification of nonfermenting bacteria. J Clin Microbiol 46(6):1946–1954

O'Farrell PH (1975) High resolution two-dimensional electrophoresis of proteins. J Biol Chem 250 (10):4007–4021

Olm MR, Crits-Christoph A, Diamond S, Lavy A, Matheus Carnevali PB, Banfield JF (2020) Consistent metagenome-derived metrics verify and delineate bacterial species boundaries. Msystems 5(1):e00731–e00719

Palmer M, Steenkamp ET, Coetzee M, Blom J, Venter SN (2018) Genome-based characterization of biological processes that differentiate closely related bacteria. Front Microbiol 9:113

Peng X, Xu C, Ren H, Lin X, Wu L, Wang S (2005) Proteomic analysis of the sarcosine-insoluble outer membrane fraction of *Pseudomonas aeruginosa* responding to ampicilin, kanamycin, and tetracycline resistance. J Proteome Res 4(6):2257–2265

Pérez-Losada M, Porter ML, Viscidi RP, Crandall KA (2011) Multilocus sequence typing of pathogens. In: Genetics and evolution of infectious disease. Elsevier, London, pp 503–521

Pieper R, Gatlin-Bunai CL, Mongodin EF, Parmar PP, Huang ST, Clark DJ et al (2006) Comparative proteomic analysis of Staphylococcus aureus strains with differences in resistance to the cell wall-targeting antibiotic vancomycin. Proteomics 6(15):4246–4258

Pignone M, Greth KM, Cooper J, Emerson D, Tang J (2006) Identification of mycobacteria by matrix-assisted laser desorption ionization-time-of-flight mass spectrometry. J Clin Microbiol 44(6):1963–1970

Raghu S, Kumar S, Suyal DC, Sahu B, Kumar V, Soni R (2021) Molecular tools to explore rhizosphere microbiome. In: Nath M, Bhatt D, Bhargava P, Chaudhary DK (eds) Microbial metatranscriptomics belowground. Singapore, Springer Nature, pp 37–57. https://doi.org/10.1007/978-981-15-9758-9_2

Raina V, Nayak T, Ray L, Kumari K, Suar M (2019) A polyphasic taxonomic approach for designation and description of novel microbial species. In: Microbial diversity in the genomic era. Academic Press, London, pp 137–152

Redmond C, Baillie LW, Hibbs S, Moir AJ, Moir A (2004) Identification of proteins in the exosporium of Bacillus anthracis. Microbiology 150(2):355–363

Sampath R, Hall TA, Massire C, Li F, Blyn LB, Eshoo MW et al (2007) Rapid identification of emerging infectious agents using PCR and electrospray ionization mass spectrometry. Ann N Y Acad Sci 1102(1):109

San JE, Baichoo S, Kanzi A, Moosa Y, Lessells R, Fonseca V et al (2020) Current affairs of microbial genome-wide association studies: approaches, bottlenecks and analytical pitfalls. Front Microbiol 10:3119

Settanni L, Corsetti A (2007) The use of multiplex PCR to detect and differentiate food-and beverage-associated microorganisms: a review. J Microbiol Methods 69(1):1–22

Sharma A, Lee S, Park YS (2020) Molecular typing tools for identifying and characterizing lactic acid bacteria: a review. Food Sci Biotechnol 29(10):1301–1318

Shintani H (2013) Rapid assay of airborne microorganisms and bioburden using several procedures. Pharm Anal Acta 4(6):1–12

Solieri L, Dakal TC, Giudici P (2013) Next-generation sequencing and its potential impact on food microbial genomics. Ann Microbiol 63(1):21–37

Soni R, Suyal DC, Agrawal K, Yadav A, Souche Y, Goel R (2015) Differential proteomic analysis of Himalayan psychrotolerant diazotroph *Pseudomonas palleroniana* N26 Strain under low temperature diazotrophic conditions. CryoLetters 36(2):74–82

Soni R, Suyal DC, Sai S, Goel R (2016) Exploration of nifH gene through soil metagenomes of the western Indian Himalayas. 3 Biotech 6(1):25

Soni R, Suyal DC, Sahu B, Phulara SC (2021) Metagenomics: an approach to unravel the plant microbiome and its function. In: Verma A, Saini JK, Hesham AEL, Singh HB (eds) Phytomicrobiome interactions and sustainable agriculture. Wiley-Blackwell, Hoboken, pp 32–44

Strejcek M, Smrhova T, Junkova P, Uhlik O (2018) Whole-cell MALDI-TOF MS versus 16S rRNA gene analysis for identification and dereplication of recurrent bacterial isolates. Front Microbiol 9:1294

Suyal DC, Yadav A, Shouche Y, Goel R (2015a) Diversified diazotrophs associated with the rhizosphere of Western Indian Himalayan native red kidney beans (Phaseolus vulgaris L.). 3 Biotech 5(4):433–441

Suyal DC, Yadav A, Shouche Y, Goel R (2015b) Bacterial diversity and community structure of Western Indian Himalayan red kidney bean (Phaseolus vulgaris) rhizosphere as revealed by 16S rRNA gene sequences. Biologia 70(3):305–313

Suyal DC, Kumar S, Joshi D, Soni R, Goel R (2018) Quantitative proteomics of psychotrophic diazotroph in response to nitrogen deficiency and cold stress. J Proteome 187:235–242

Suyal DC, Joshi D, Debbarma P, Soni R, Das B, Goel R (2019a) Soil metagenomics: unculturable microbial diversity and its function. In: Mycorrhizosphere and Pedogenesis. Springer, Singapore, pp 355–362

Suyal DC, Joshi D, Kumar S, Soni R, Goel R (2019b) Differential protein profiling of soil diazotroph *Rhodococcus qingshengii* S10107 towards low-temperature and nitrogen deficiency. Sci Rep 9(1):1–9

Suyal DC, Kumar S, Joshi D, Yadav A, Shouche Y, Goel R (2019c) Comparative overview of red kidney bean (*Phaseolus vulgaris*) rhizospheric bacterial diversity in perspective of altitudinal variations. Biologia 74(10):1405–1413

Vandamme P, Pot B, Gillis M, De Vos P, Kersters K, Swings J (1996) Polyphasic taxonomy, a consensus approach to bacterial systematics. Microbiol Rev 60(2):407–438

Vats P, Verma SM, Monif T (2016) Mechanistic evaluation of matrix effect on three different model of mass spectrometer by using a model drug. J Anal Bioanal Tech 7:314

Wang D, Li F, Cao S, Zhang K (2020) Genomic and functional genomics analyses of gluten proteins and prospect for simultaneous improvement of end-use and health-related traits in wheat. Theor Appl Genet 133(5):1521–1539

Wieme AD, Spitaels F, Aerts M, De Bruyne K, Van Landschoot A, Vandamme P (2014) Effects of growth medium on matrix-assisted laser desorption–ionization time of flight mass spectra: a case study of acetic acid bacteria. Appl Environ Microbiol 80(4):1528–1538

Wilke A, Rückert C, Bartels D, Dondrup M, Goesmann A, Hüser AT et al (2003) Bioinformatics support for high-throughput proteomics. J Biotechnol 106(2–3):147–156

Woese CR (1987) Bacterial evolution. Microbiol Rev 51(2):221–271

Wunschel SC, Jarman KH, Petersen CE, Valentine NB, Wahl KL, Schauki D et al (2005) Bacterial analysis by MALDI-TOF mass spectrometry: an inter-laboratory comparison. J Am Soc Mass Spectrom 16(4):456–462

You J, Willcox MD, Madigan MC, Wasinger V, Schiller B, Walsh BJ et al (2013) Tear fluid protein biomarkers. Adv Clin Chem 62:151–196

The manufacturer's authorised representative in the EU is Springer
Nature Customer Service Centre GmbH, Europaplatz 3, 69115 Heidelberg,
Germany. If you have any concerns regarding our products, please
contact ProductSafety@springernature.com

Printed and bound by CPI Group (UK) Ltd, Croydon, CR0 4YY
28/09/2026
02224065-0005